NUMBER 155

J. Peter May

Classifying Spaces and Fibrations

MEMOIRS
OF THE AMERICAN
MATHEMATICAL SOCIETY

VOLUME 1 · ISSUE 1 · NUMBER 155 (last of 2 numbers) · JANUARY 1975 · CODEN: MAMCAU

MEMOIRS of the American Mathematical Society

This journal is designed particularly for long research papers (and groups of cognate papers) in pure and applied mathematics. It includes, in general, longer papers than those in the TRANSACTIONS.

Mathematical papers intended for publication in the Memoirs should be addressed to one of the editors. Subjects, and the editors associated with them, follow:

Real analysis (excluding harmonic analysis) **and applied mathematics** to FRANÇOIS TREVES, Department of Mathematics, Rutgers University, New Brunswick, NJ 08903.

Harmonic and complex analysis to HUGO ROSSI, Department of Mathematics, University of Utah, Salt Lake City, UT 84112.

Abstract analysis to ALEXANDRA IONESCU TULCEA, Department of Mathematics, Northwestern University, Evanston, IL 60201.

Algebra and number theory (excluding universal algebras) to DOCK S. RIM, Department of Mathematics, University of Pennsylvania, Philadelphia, PA 19104.

Logic, foundations, universal algebras and combinatorics to ALISTAIR H. LACHLAN, Department of Mathematics, Simon Fraser University, Burnaby, 2, B. C., Canada.

Topology to PHILIP T. CHURCH, Department of Mathematics, Syracuse University, Syracuse, NY 13210.

Global analysis and differential geometry to VICTOR W. GUILLEMIN, c/o Ms. M. McQuillin, Department of Mathematics, Harvard University, Cambridge, MA 02138.

Probability and statistics to HARRY KESTEN, Department of Mathematics, Cornell University, Ithaca, NY 14850.

All other communications to the editors should be addressed to the Managing Editor, HARRY KESTEN.

MEMOIRS are printed by photo-offset from camera-ready copy fully prepared by the authors. Prospective authors are encouraged to request booklet giving detailed instructions regarding reproduction copy. Write to Editorial Office, American Mathematical Society (address below). For general instructions see inside back cover.

Annual subscription is $34.50. Three volumes of 2 issues each are planned for 1975. Each issue will consist of one or more papers (or "Numbers") separately bound; each Number may be ordered separately. Prior to 1975 MEMOIRS was a book series; for back issues see the AMS Catalog of Book Publications. All orders should be directed to the American Mathematical Society; please specify by **NUMBER** when ordering.

TRANSACTIONS of the American Mathematical Society

This journal consists of shorter tracts which are of the same general character as the papers published in the MEMOIRS. The editorial committee is identical with that for the MEMOIRS so that papers intended for publication in this series should be addressed to one of the editors listed above.

Published bimonthly beginning in January, by the American Mathematical Society. Subscriptions for journals published by the American Mathematical Society should be addressed to American Mathematical Society, P. O. Box 1571, Annex Station, Providence, Rhode Island 02901.

Second-class postage permit pending at Providence, Rhode Island, and additional mailing offices.

Copyright © 1975 American Mathematical Society
All rights reserved
Printed in the United States of America

Memoirs of the American Mathematical Society

VOLUME 1 · ISSUE 1 · NUMBER 155 (last of 2 numbers)

JANUARY 1975

J. Peter May

Classifying Spaces and Fibrations

Published by the

AMERICAN MATHEMATICAL SOCIET

Providence, Rhode Island

Abstract

The basic theory of fibrations is generalized to a context in which fibres, and maps on fibres, are constrained to lie in any preassigned category of spaces $\mathcal{F}$. Then axioms are placed on $\mathcal{F}$ to allow the development of a theory of associated principal fibrations and, under several choices of additional hypotheses on $\mathcal{F}$, a classification theorem is proven for such fibrations. The same proof applies to the classification of bundles and generalizes to give a classification theorem for fibrations or bundles with additional structure, such as a reduction of the structural monoid, or a trivialization with respect to a coarser type of fibration, or an orientation with respect to an extraordinary cohomology theory. The proofs are constructive and are based on use of the two-sided geometric bar construction, the topological and homological properties of which are analyzed in detail. Related topics studied include the classification of fibrations by transports, the Eilenberg-Moore and Serre spectral sequences, and the group completion theorem.

AMS (MOS) subject classifications (1970). Primary 55F05, 55F10, 55F15, 55F20, 55F35, 55F40, 55F65, 55H10, 55H20, 57F30.

Key words and phrases. Fibration, bundle, principal fibration, lifting function, transport, quasifibration, classification theorem, classifying space, bar construction, Eilenberg-Moore spectral sequence, Serre spectral sequence, group completion

ISBN 0-8218-1855-4

CLASSIFYING SPACES AND FIBRATIONS

J. P. May

University of Chicago

Received by the editors June 11, 1974

The author was partially supported by NSF grant GP-29075

Introduction

"The theory of fibrations is thus fairly complete and well worked out on a conceptual level; the rest should be applications and computations." So ended Stasheff's 1970 survey article [35] on the classification of fibrations.

At the time, the conclusion seemed not unreasonable. The basic outlines of a complete theory were visible, and this theory did seem adequate for most applications. Even in Stasheff's very clear summary, the theory appeared technically to be extremely complicated, but this was felt to be intrinsic to the subject.

However, recent developments make this sanguine view of the adequacy of the theory untenable and, in the process of obtaining a theory which is adequate for the new applications, we shall also see how to avoid most of the previous technical complications.

A brief account of the existing classification theorems will be necessary in order to place our contribution in perspective.

The simplest and most conceptual method of classification is based on the observation that if $\mathcal{C}$ is a small topological category and if a space $\mathcal{B}\mathcal{C}$ is appropriately constructed from the associated simplicial space (technically, by use of face but not degeneracy operators in forming the geometric realization), then $\mathcal{B}\mathcal{C}$ classifies the functor defined on paracompact spaces X as the quotient obtained from the cohomology set $H^1(X; \mathcal{C})$ by identifying homotopic cohomology classes. This method is due to Segal [30], and an exposition has also been given by Stasheff [2, p. 86-94]. $\mathcal{B}\mathcal{C}$ is a generalization of Milnor's classifying space for topological groups [24], and this method of classification is a generalization of one found for bundles by tom Dieck [6]. It is particularly appropriate to the study of foliations via the classification of Haefliger structures (e.g. [2]). While this approach is very general, it is only useful when the structures one wishes to study are obtained by patching together local coordinates by means of cocycles with values in some category $\mathcal{C}$. In practice, this means that the morphisms of $\mathcal{C}$ must at least be homeomorphisms, so that $\mathcal{C}$ is a topological groupoid, and this approach is inapplicable to the classification of fibrations or of bundles with globally defined additional structure.

A second conceptual method of classification is based on appeal to Brown's representability theorem [4]. It has two defects. First, as applied to fibrations, there is no completely rigorous treatment in the literature. The point is that if one wishes to represent a set-valued functor, then one must first verify that one's proposed functor does indeed take well-defined sets as values. This is by no means obvious for the functors of interest in the theory of fibrations, and this set-theoretical question has been totally ignored in the literature. Second, and probably more fundamental, the basic purpose of a classification theorem is to enable one to calculate the represented functor, or at least to calculate invariants of the structures under study. A space constructed by appeal to Brown's theorem can generally be studied only by reverting to analysis of the originally given functor and is therefore of very limited use for purposes of computation.

The bulk of Stasheff's survey is devoted to the various alternative methods of classification, and we shall not give references here. In contrast to the general methods described in the previous two paragraphs, these alternative methods appear to be specific to particular types of fibration, or at least to require considerable reworking to be made applicable to varying types. Technically, with one exception, each such method involves at least one of the theory of simplicial sets, a combinatorial theory of cellular monoids, careful local pasting arguments, or the use of higher homotopies. The exception is Stasheff's original proof [32] of the classification theorem for fibrations with fibres of the homotopy type of a finite CW-complex.

Why are these results not adequate? First, fibrations with localized or completed spheres as fibres play a key role in Sullivan's beautiful proof of the Adams conjecture [39]. Such spaces are not finite CW-complexes, and one source of technical difficulty in Sullivan's argument is the absence of a good model for the relevant classifying spaces. Our Corollary 9.5 will rectify this, and we shall return to this point in [21] where a new theory of localization and completion of topological spaces will be given.

Second, spherical fibrations and bundles oriented with respect to an extraordinary cohomology theory are central to many applications, and there is no proof of a classification theorem for such structures in the literature. As we shall show in [19] and [20], Theorem 11.1 implies such a classification theorem, and its use allows easy derivations of some of the results of Adams on vector bundles and of Sullivan on topological sphere bundles.

Third, for the study of orientations, it is very convenient to have a variant of Stasheff's theorem in which fibrations are given with a cross-section which is a cofibration. Corollary 9.8 will give such a result.

Beyond these explicit applications, there is an evident need for a single coherent theory of fibrations and their classification which will simultaneously yield the various classification theorems desired in practice as special cases of one general result, or at least as consequences of one general pattern of proof. Moreover, such a theory should if possible avoid techniques, such as those listed a few paragraphs earlier, which, however great their interest within the theory of fibrations, are irrelevant to the actual computations based on the theory. Needless to say, our theory does meet these criteria.

We should say a bit more about two of the techniques we avoid. Much has been written about the inevitability of the appearance of higher homotopies in any complete theory of fibrations, and we freely admit that they are indeed implicitly present. Nevertheless, at each place where it is generally felt they ought to appear, we shall find that some conceptual trick leads to an equivalent solution with no such notion visible. Thus we shall classify principal G-fibrations for arbitrary grouplike topological monoids

G in Corollary 9.4, we shall show the independence of the choice of fibre in the classification of fibrations with fibres of a given homotopy type in section 12, and we shall classify such fibrations by use of associative transports (which are actions on fibres by the Moore loop space of the base) in section 14. In each case, the actual details are very much simpler than would be the case if higher homotopies were explicitly introduced.

Similarly, we have chosen to work with Hurewicz, rather than with Dold (or weak) fibrations, throughout. We freely admit that Dold fibrations have important technical advantages and are implicit in the notion of fibre homotopy equivalence. However, since the local pasting arguments for which they are essential are unnecessary in our work, their use would introduce considerable additional complexity while adding nothing of significance to our theory. Although the most important results concerning the local nature of fibrations are valid in our general context, local considerations will only play a role in those instances of our classification theorems which involve bundle theory.

The paper consists of fifteen sections, with logical interrelationships as indicated in the following chart:

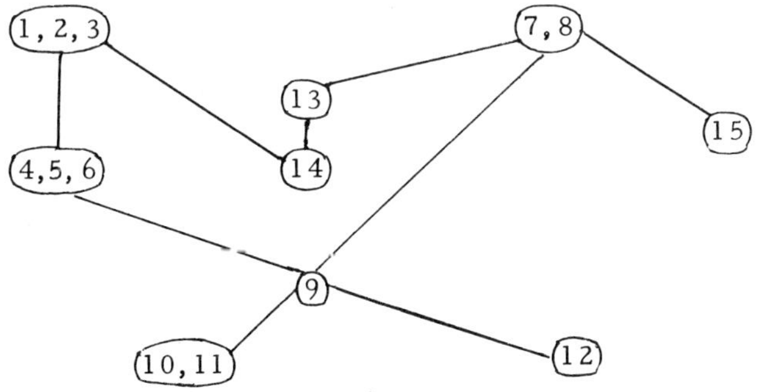

Introduction

The first three sections are devoted to a redevelopment of the theory of fibrations, including the basic theorems of Dold [7, 3.3 and 6.3] and Hurewicz [11], for fibrations with fibres constrained to lie in any preassigned category $\mathcal{F}$. Technically, the main point here is that the section extension property which Dold takes as fundamental does not generalize to our context, hence we have been forced to find alternative proofs. While these are still based on the ideas of Dold, they are shorter and may seem simpler even in the classical case.

The special properties required of $\mathcal{F}$ in order to classify $\mathcal{F}$-fibrations are discussed in sections 4 and 5, and examples of categories which satisfy these properties are given in section 6. The relevant properties ensure that associated principal fibrations can be constructed and that quasifibrations can be replaced by fibrations; for the latter, the point of interest is that the standard procedure is inadequate for the study of fibrations with cross-section.

In sections 7 and 8, we summarize the topological properties of the two-sided geometric bar construction. Most of the proofs have already been given, in a more general setting, in [17, §9-11 (which are independent of §1-8)] or [18, Appendix]. This construction is a straightforward generalization, implicit in Stasheff's paper [34], of the standard Milgram-Steenrod [23, 38] classifying space functor. The generalization, despite its simplicity, transforms the bar construction from an invariant of topological monoids to an extremely flexible tool in the theory of fibrations and their classification.

Introduction

We reach our basic classification theorems for fibrations and bundles in section 9. The method of proof is to write down an explicit universal fibration (or bundle) and to verify that it classifies by explicitly constructing a classifying map for any given fibration (or bundle). In section 12, we generalize the standard Segal [30] classifying space functor on small topological categories to a two-sided bar construction (technically, using both face and degeneracy operators). This generalization allows us to rework our basic theory, in favorable cases, so as not to give any particular choice of fibre a privileged role.

A general notion of additional global structure on a fibration or bundle is introduced in section 10. Special cases include reductions of the structural monoid, trivializations with respect to a coarser type of fibration, and orientations (of spherical fibrations or bundles) with respect to an extraordinary cohomology theory. We demonstrate in section 11 that the proof of our classification theorems directly generalizes to a proof of classification theorems for such fibrations or bundles with additional structure.

The last three sections are primarily concerned with homological properties of the geometric bar construction. In section 13, after generalizing results of Milgram [23] and Steenrod [38] concerning the cellular properties of classifying spaces, we obtain a technical result (Theorem 13.9) which relates the two-sided algebraic and geometric bar constructions by mixed use of singular and cellular chain groups. This result, which should be regarded as a generalization of a special case of a result of Stasheff [31], gives the Eilenberg-Moore and Rothenberg-Steenrod spectral sequences [26, 29], with their products, and shows that the two are in fact the same. In section 14, we introduce the notion of a transport, use it to prove a classification theorem for fibrations (suggested by a result due to Stasheff [33]), and combine it with Theorem 13.9 to give a novel derivation of the Serre spectral sequence, with its products. Finally, in section 15, we give a brief proof of the "group completion theorem", due to Barratt-Priddy [1] and Quillen [28], which analyzes the homological behavior of the natural map $G \to \Omega BG$ for appropriate non-connected topological monoids G. This result plays a fundamental role in the theory of infinite loop spaces and its application to algebraic K-theory [18].

Contents

1. $\mathcal{F}$-spaces and $\mathcal{F}$-maps

We take the position that types of fibrations ought to be specified by assigning structure to the fibres and that this is most sensibly done by specifying a category in which the fibres must lie. We here develop a framework in which to define such fibrations and generalize to this framework a theorem of Dold [7, 3.3] to the effect that a local fibre homotopy equivalence is a fibre homotopy equivalence.

We shall work in the category $\mathcal{U}$ of compactly generated weak Hausdorff spaces [37; 22, §2]; thus products, function spaces, etc. are always to be given the compactly generated topology. Throughout the first five sections $\mathcal{F}$ will denote a category with a faithful "underlying space" functor $\mathcal{F} \to \mathcal{U}$. Thus each object of $\mathcal{F}$ is a space and the set $\mathcal{F}(F, F')$ of morphisms $F \to F'$ in $\mathcal{F}$ is a subset of $\mathcal{U}(F, F')$. We agree either to insist that $\mathcal{F}$ contain with each $F \in \mathcal{F}$ the spaces $F \times *$ and $* \times F$ and the evident homeomorphisms between these spaces and F or to identify these spaces with F, where $*$ is any one-point space.

Definition 1.1. An $\mathcal{F}$-space is a map $\pi: E \to B$ in $\mathcal{U}$ such that $\pi^{-1}(b) \in \mathcal{F}$ for each $b \in B$; B and E are the base space and total space of π. An $\mathcal{F}$-map $(g, f): \nu \to \pi$ is a commutative diagram

$$\begin{array}{ccc} D & \xrightarrow{g} & E \\ {\scriptstyle\nu}\downarrow & & \downarrow{\scriptstyle\pi} \\ A & \xrightarrow{f} & B \end{array}$$

in $\mathcal{U}$ such that $g: \nu^{-1}(a) \to \pi^{-1}(f(a))$ is in $\mathcal{F}$ for each $a \in A$; if $A = B$ and f is the identity map, then g is said to be an $\mathcal{F}$-map over B. An $\mathcal{F}$-homotopy is an $\mathcal{F}$-map (H, h) of the form

$$\begin{array}{ccc} D \times I & \xrightarrow{H} & E \\ {\scriptstyle \nu \times 1}\downarrow & & \downarrow{\scriptstyle \pi} \\ A \times I & \xrightarrow{h} & B \end{array}$$

Thus it is required that each (H_s, h_s) be an $\mathcal{F}$-map, $H_s(d) = H(d, s)$; if $A = B$ and $h_s(b) = b$, then H is said to be an $\mathcal{F}$-homotopy over B. An $\mathcal{F}$-map $g: D \to E$ over B is an $\mathcal{F}$-homotopy equivalence if there is an $\mathcal{F}$-map $g': E \to D$ over B such that $g'g$ and gg' are $\mathcal{F}$-homotopic over B to the respective identity maps. An $\mathcal{F}$-space $\pi: E \to B$ is said to be $\mathcal{F}$-homotopy trivial if it is $\mathcal{F}$-homotopy equivalent to the projection $\pi_1: B \times F \to B$ for some $F \in \mathcal{F}$.

By restriction to one-point base spaces, the definition specializes to define $\mathcal{F}$-homotopies and $\mathcal{F}$-homotopy equivalences between spaces in $\mathcal{F}$.

We can form induced $\mathcal{F}$-spaces precisely as usual. The following lemma fixes notations.

<u>Lemma 1.2.</u> Let $\pi: E \to B$ be an $\mathcal{F}$-space and let $f: A \to B$ be a map in $\mathcal{U}$. Define a space f^*E and maps $f^*\pi: f^*E \to A$ and $\tilde{f}: f^*E \to E$ by

$$f^*E = \{(a, e) \mid f(a) = \pi(e)\} \subset A \times E, \quad f^*\pi(a, e) = a, \text{ and } \tilde{f}(a, e) = e.$$

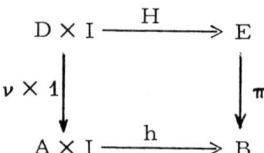

Then $f^*\pi$ is an $\mathcal{F}$-space and $(\tilde{f}, f)$ is an $\mathcal{F}$-map. Moreover, if $\nu: D \to A$ is an $\mathcal{F}$-space and $(g, f): \nu \to \pi$ is an $\mathcal{F}$-map, then the unique map $\bar{g}: D \to f^*E$ which makes the following diagram commutative, namely $\bar{g}(d) = (\nu(d), g(d))$, is an $\mathcal{F}$-map over A:

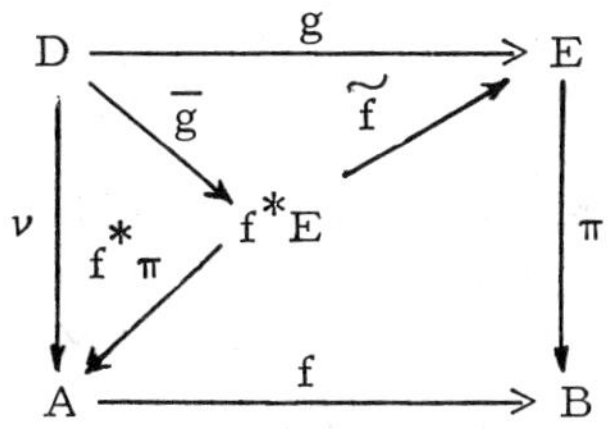

The remainder of this section will be devoted to the promised generalization of Dold's theorem. We assume given $\mathcal{F}$-spaces $\nu: D \to B$ and $\pi: E \to B$ and an $\mathcal{F}$-map $g: D \to E$ over B. We require some notations and a lemma.

Notations 1.3. Let G denote the subspace of $D \times E^I$ consisting of all pairs (d, ε) such that $g(d) = \varepsilon(0)$ and $\varepsilon(I) \subset \pi^{-1}(\nu(d))$. Define maps $\alpha: G \to D$ and $\beta_s: G \to E$ for $s \in I$ by $\alpha(d, \varepsilon) = d$ and $\beta_s(d, \varepsilon) = \varepsilon(s)$. By an $\mathcal{F}$-section of g, we understand a map $\sigma: E \to G$ such that $\beta_1 \circ \sigma = 1$ on E and such that, for all $b \in B$, each of the maps $\alpha\sigma: \pi^{-1}(b) \to \nu^{-1}(b)$ and $\beta_s \circ \sigma: \pi^{-1}(b) \to \pi^{-1}(b)$ is in $\mathcal{F}$. Thus an $\mathcal{F}$-section of g consists of an $\mathcal{F}$-map $g' = \alpha \circ \sigma: E \to D$ over B together with an $\mathcal{F}$-homotopy $H_s = \beta_s \circ \sigma$ over B from gg' to the identity map of E.

If A is a subspace of B, we let E_A, G_A, etc., denote the part of E, G, etc., over A; explicitly, $G_A = \{(d,\varepsilon) \mid \nu(d) \in A\} \subset G$.

Lemma 1.4. Assume that g is an $\mathcal{J}$-homotopy equivalence. Let $\phi: B \to I$ be a map, let $A = \phi^{-1}(1)$, and let $V = \phi^{-1}(0,1]$. Then, for any $\mathcal{J}$-section $\sigma: E_V \to G_V$ of g, there exists an $\mathcal{J}$-section $\rho: E \to G$ of g such that $\rho = \sigma$ on E_A.

Proof. Let $g': E \to D$ be an $\mathcal{J}$-homotopy inverse of g and let $H: gg' \simeq 1_E$ and $H': g'g \simeq 1_D$ be $\mathcal{J}$-homotopies over B. Define an $\mathcal{J}$-section $\tau: E \to G$ of g by $\tau(e) = (g'(e), H(e))$, where $H(e)(s) = H(e,s)$. Define a homotopy $L: G \times I \to G$ from the composite $\tau \circ \beta_1$ to the identity map of G by the formula

$$L((d,\varepsilon),t) = \begin{cases} (g'\varepsilon(1-2t),\ H\varepsilon(1-2t)) & \text{if } t \le 1/2 \\ (H'(d,2t-1),\ J(d,\varepsilon,2t-1)) & \text{if } 1/2 \le t \end{cases}$$

where $J(d,\varepsilon,t)(s) = jk(s,t)$ for some chosen retraction

$$k: I \times I \to (I \times 0) \cup (0 \times I) \cup (I \times 1)$$

and where $j(s,0) = H(\varepsilon(0),s)$, $j(0,t) = gH'(d,t)$, and $j(s,1) = \varepsilon(s)$. The desired $\mathcal{J}$-section ρ of g is then defined by the formula

$$\rho(e) = \begin{cases} \tau(e) & \text{if } \phi\pi(e) \le 1/2 \\ L(\sigma(e),\ 2\phi\pi(e)-1) & \text{if } 1/2 \le \phi\pi(e) \end{cases}$$

Recall that a cover $\mathcal{C}$ of a space B is said to be numerable if it is locally finite and if for each $U \in \mathcal{C}$ there is a map $\lambda_U : B \to I$ such that $U = \lambda_U^{-1}(0, 1]$. Recall too that a space $B \in \mathcal{U}$ is paracompact if and only if every open cover of B admits a numerable refinement and that any CW-complex is paracompact.

Theorem 1.5. Let $\nu : D \to B$ and $\pi : E \to B$ be $\mathcal{J}$-spaces. Let $g: D \to E$ be an $\mathcal{J}$-map over B such that g restricts to an $\mathcal{J}$-homotopy equivalence over each set of a numerable cover $\mathcal{C}$ of B. Then g is an $\mathcal{J}$-homotopy equivalence.

Proof. It suffices to construct an $\mathcal{J}$-section $\sigma : E \to G$ of g. Indeed, this will give a right $\mathcal{J}$-homotopy inverse g' of g. g' will restrict to an $\mathcal{J}$-homotopy equivalence over each $U \in \mathcal{C}$ since if f_U is an $\mathcal{J}$-homotopy inverse to g_U, then $f_U \simeq f_U g_U g'_U \simeq g'_U$ (where $\simeq$ means "is $\mathcal{J}$-homotopic to"). Therefore g' will itself have a right $\mathcal{J}$-homotopy inverse g", and $g \simeq gg'g" \simeq g"$. For $U \in \mathcal{C}$, choose a map $\lambda_U : B \to I$ such that $U = \lambda_U^{-1}(0, 1]$. For a union $V = \bigcup_{j \in J} U_j$ of sets $U_j \in \mathcal{C}$ define $\lambda_V = \sum_{j \in J} \lambda_{U_j}$, so that $V = \{x \mid \lambda_V(x) > 0\}$. We assume that $\mathcal{C}$ is irredundant, and then $V \subset W$ if and only if $\lambda_V \leq \lambda_W$. Let $\mathcal{A}$ denote the set of pairs (V, σ) such that V is a union of sets in $\mathcal{C}$ and $\sigma : E_V \to G_V$ is an $\mathcal{J}$-section of g. Partial order $\mathcal{A}$ by $(V, \sigma) < (W, \tau)$ if $V \subset W$ and $\sigma(e) = \tau(e)$ for all $e \in \pi^{-1}V$ such that $\lambda_V \pi(e) = \lambda_W \pi(e)$ (thus, $\sigma(e) \neq \tau(e)$

implies $\pi(e) \in U$ for some $U \in \mathcal{C}$ such that $U \subset W$ but $U \not\subset V$). Any totally ordered subset $\{(V_k, \sigma_k) \mid k \in K\}$ of $\mathcal{A}$ has the upper bound (V, σ) defined by $V = \bigcup_{k \in K} V_k$ and $\sigma(e) = \sigma_k(e)$ for $e \in \pi^{-1}V$ and all sufficiently large k. Here σ is well-defined and continuous and $(V, \sigma) > (V_k, \sigma_k)$ for all $k \in K$ since if $b \in V$ and $V(b)$ denotes the union of those $U \in \mathcal{C}$ such that $b \in U \subset V$, then $V(b) \subset V_{k(b)}$ for some $k(b) \in K$ (because $\mathcal{C}$ is locally finite) and therefore $\sigma_k = \sigma_{k(b)}$ over $V(b)$ for all $k > k(b)$. By Zorn's lemma, $\mathcal{A}$ contains a maximal element (V, σ). We claim that $V = B$. Suppose not; choose $U \in \mathcal{C}$ such that $U \not\subset V$ and let $W = U \cup V$. Define $\phi: W \to I$ by

$$\phi(b) = \begin{cases} 1 & \text{if } \lambda_U(b) \leq \lambda_V(b) \quad (\text{hence } \lambda_V(b) > 0) \\ \lambda_V(b)/\lambda_U(b) & \text{if } \lambda_U(b) \geq \lambda_V(b) \quad (\text{hence } \lambda_U(b) > 0) \end{cases}$$

$\phi(b) > 0$ if and only if $\lambda_V(b) > 0$, hence σ is defined over $\phi^{-1}(0, 1]$. By the lemma, there is an $\mathcal{J}$-section $\rho: E_U \to G_U$ of g such that $\rho = \sigma$ over $\phi^{-1}(1) \cap U$. Define $\tau: E_W \to G_W$ by

$$\tau(e) = \begin{cases} \sigma(e) & \text{if } \lambda_U\pi(e) \leq \lambda_V\pi(e) \\ \rho(e) & \text{if } \lambda_U\pi(e) \geq \lambda_V\tau(e) \end{cases}$$

Clearly $(W, \tau) > (V, \sigma)$, which is the desired contradiction.

Definition 2.1. An $\mathcal{F}$-space $\pi: E \to B$ is an $\mathcal{F}$-fibration if it satisfies the following $\mathcal{F}$-covering homotopy property (abbreviated $\mathcal{F}$-CHP): for every $\mathcal{F}$-space $\nu: D \to A$ and $\mathcal{F}$-map $(g, f): \nu \to \pi$ and every homotopy $h: A \times I \to B$ of f, there exists a homotopy $H: D \times I \to E$ of g such that the pair (H, h) is an $\mathcal{F}$-homotopy.

A $\mathcal{U}$-fibration is clearly just an ordinary (Hurewicz) fibration. We here generalize the elementary theory of fibrations to $\mathcal{F}$-fibrations and generalize Dold's theorem [7, 6.3] to the effect that a map of fibrations is a fibre homotopy equivalence if it restricts to a homotopy equivalence on each fibre. We observe first that the $\mathcal{F}$-homotopy (H, h) asserted to exist by the $\mathcal{F}$-CHP is itself unique up to $\mathcal{F}$-homotopy.

Lemma 2.2. Let (H, h), (H', h'), and (J, j) be $\mathcal{F}$-homotopies with domain $\nu \times 1: D \times I \to A \times I$ and range $\pi: E \to B$, where ν is an $\mathcal{F}$-space and π is an $\mathcal{F}$-fibration. Assume that (J, j) is an $\mathcal{F}$-homotopy from (H_0, h_0) to (H'_0, h'_0) and assume given $k: A \times I \times I \to B$ such that

$$k(a, s, 0) = h(a, s), \quad k(a, s, 1) = h'(a, s), \quad \text{and} \quad k(a, 0, t) = j(a, t).$$

Let $C = (I \times 0) \cup (I \times 1) \cup (0 \times I) \subset I \times I$ and define $g: D \times C \to E$ by

$$g(d, s, 0) = H(d, s), \quad g(d, s, 1) = H'(d, s), \quad \text{and} \quad g(d, 0, t) = J(d, t).$$

Then there exists $K: D \times I \times I \to E$ such that $K|D \times C = g$ and the pair (K, k) is an $\mathcal{F}$-homotopy.

Proof. (g, f) is an $\mathcal{F}$-map, where $f = k|A \times C$. Since the pairs $(I \times I, C)$ and $(I \times I, I \times 0)$ are homeomorphic, the conclusion follows directly from the $\mathcal{F}$-CHP.

The following result is an easy consequence of the lemma.

Proposition 2.3. An $\mathcal{F}$-fibration $\pi: E \to B$ determines a functor L from the fundamental groupoid of B to the homotopy category of $\mathcal{F}$ by $L(b) = \pi^{-1}(b)$ for $b \in B$ and $L[h] = [H_1]$ for a path $h: I \to B$, where $H: \pi^{-1}h(0) \times I \to E$ is any homotopy of the inclusion $\pi^{-1}h(0) \to E$ such that (H, h) is an $\mathcal{F}$-homotopy. In particular, if B is connected, then any two fibres of π have the same $\mathcal{F}$-homotopy type.

We shall find it convenient to compose paths in the reverse of the usual order; with this convention, the functor L is covariant.

We show next that induced $\mathcal{F}$-fibrations behave properly.

Lemma 2.4. Let $\pi: E \to B \times I$ be an $\mathcal{F}$-fibration and let $\pi^s: E^s \to B$ denote the part of π over $B \times \{s\}$. Then π^0 and π^1 are $\mathcal{F}$-homotopy equivalent.

Proof. Define $h: B \times I \times I \times I \to B \times I$ by $h(b, r, s, t) = (b, (1-t)r + ts)$. By the $\mathcal{F}$-CHP, there exists $H: E \times I \times I \to E$ such that $H(e, s, 0) = e$ and (H, h) is an $\mathcal{F}$-homotopy. Define $K: E \times I \to E$ by $K(e, s) = H(e, s, 1)$. Observe that if $\pi': E \to B$ and $\pi'': E \to I$ are defined by $\pi(e) = (\pi'(e), \pi''(e))$, then $\pi K(e, s) = (\pi'(e), s)$ and traversal of $H(e, \pi''(e), t)$, $0 \leq t \leq 1$, gives an $\mathcal{F}$-homotopy over $B \times I$ from the identity of E to the $\mathcal{F}$-map $k: E \to E$ over $B \times I$ defined by $k(e) = K(e, \pi''(e))$. Define $k^1: E^0 \to E^1$ and $k^0: E^1 \to E^0$ by $k^1(x) = K(x, 1)$ and $k^0(y) = K(y, 0)$. Via the homotopies $K(K(y, s), 1)$ and $K(K(x, 1-s), 0)$, $0 \leq s \leq 1$, the maps k^1k^0 and k^0k^1 are $\mathcal{F}$-homotopic over B to $kk|E^1$ and $kk|E^0$, respectively. Therefore k^1 and k^0 are inverse $\mathcal{F}$-homotopy equivalences.

Proposition 2.5. Let $\pi: E \to B$ be an $\mathcal{F}$-fibration. Then $f^*\pi: f^*E \to A$ is an $\mathcal{F}$-fibration for any map $f: A \to B$ and homotopic maps $A \to B$ induce $\mathcal{F}$-homotopy equivalent $\mathcal{F}$-fibrations over A. In particular, any $\mathcal{F}$-fibration over a contractible base space is $\mathcal{F}$-homotopy trivial.

Proof. The first half follows from Definition 2.1 and Lemma 1.2, and the second half follows by application of the previous lemma to $h^*\pi: h^*E \to A$ for any homotopy $h: A \times I \to B$.

Dold's theorem [7 , 6.3] now generalizes readily to our context.

Theorem 2.6. Let $\nu: D \to B$ and $\pi: E \to B$ be $\mathcal{F}$-fibrations. Let $g: D \to E$ be an $\mathcal{F}$-map over B such that $g: \nu^{-1}(b) \to \pi^{-1}(b)$ is an $\mathcal{F}$-homotopy equivalence for each $b \in B$. Assume that B admits a numerable cover $\mathcal{C}$ such that the inclusion map $U \to B$ is null-homotopic for each $U \in \mathcal{C}$. Then g is an $\mathcal{F}$-homotopy equivalence.

Proof. Let $U \in \mathcal{C}$ and let $h: U \times I \to B$ be a null-homotopy, $h_0(u) = u$ and $h_1(u) = b$. Define $\bar{g}: h^*D \to h^*E$ by the universal property of

h^*E and let $\overline{g}^s$ denote the restriction of $\overline{g}$ to the part of h^*D over $U \times \{s\}$. Then $\overline{g}^0 = g: \nu^{-1}U \to \pi^{-1}U$ and $\overline{g}^1 = 1 \times g: U \times \nu^{-1}(b) \to U \times \pi^{-1}(b)$. Construct maps K, k, k^1 and k^0 for $h^*\pi$ and J, j, j^1 and j^0 for $h^*\nu$ by the proof of Lemma 2.4. Then $\overline{g}^0$ is $\mathcal{J}$-homotopic over $U \times \{0\}$ to $k\overline{g}^0 j$ and, via the homotopy $K(\overline{g}^s J(x, s), 0)$, $0 \leq s \leq 1$, $k\overline{g}^0 j$ is $\mathcal{J}$-homotopic over $U \times \{0\}$ to the composite of $\mathcal{J}$-homotopy equivalences $k^0\overline{g}^1 j^1$. Thus g is an $\mathcal{J}$-homotopy equivalence over each $U \in \mathcal{C}$, and the result follows from Theorem 1.5.

Observe that the assumption on B is invariant under homotopy equivalence and is satisfied by spaces, such as CW-complexes, which are paracompact and locally contractible. In [7, 6.7], Dold has given a direct construction of a cover of the required type for any CW-complex B. Of course, if B is connected, the assumption on $g: \nu^{-1}(b) \to \pi^{-1}(b)$ will be satisfied for all $b \in B$ if it is satisfied for any one $b \in B$.

3. $\mathcal{J}$-lifting functions

We here generalize to our context the relationship between fibrations and lifting functions and Hurewicz's theorem [11] to the effect that a local fibration is a fibration. We first fix notations for Moore paths; use of such paths will simplify proofs here and will be essential in later sections.

Notations 3.1. For $B \in \mathcal{U}$, let ΠB denote the set of paths (β, s) $\beta: [0, s] \to B$. When convenient, we let $\beta(t) = \beta(s)$ for $t \geq s$; a point $(\beta, s) \in \Pi B$ is then specified by $\beta: [0, \infty] \to B$ and $s \in [0, \infty)$, and ΠB is topologized as a subspace of $\mathcal{U}([0, \infty], B) \times [0, \infty)$. Define the composite $(\alpha\beta, r+s)$ of paths (α, r) and (β, s) such that $\alpha(0) = \beta(s)$ by

$$(\alpha\beta)(t) = \beta(t) \text{ if } 0 \leq t \leq s \text{ and } (\alpha\beta)(t) = \alpha(t-s) \text{ if } s \leq t \leq r+s.$$

We shall abbreviate $\beta = (\beta, s)$, and we shall write $\ell(\beta) = s$ and $p(\beta) = \beta(s)$ for the length map and end-point projection. We shall consider B to be contained in ΠB as the subspace consisting of all paths of length zero.

Definition 3.2. Let $\pi: E \to B$ be an $\mathcal{J}$-space. Define a space ΓE and maps $\Gamma\pi: \Gamma E \to B$, $\eta: E \to \Gamma E$, and $\mu: \Gamma\Gamma E \to \Gamma E$ by

$$\Gamma E = \{(\beta, e) \mid \beta(0) = \pi(e)\} \subset \Pi B \times E \text{ and } \Gamma\pi(\beta, e) = p(\beta),$$

$$\eta(e) = (\pi(e), e) \quad \text{and} \quad \mu(\alpha, (\beta, e)) = (\alpha\beta, e).$$

An $\mathcal{J}$-lifting function ξ for π is a map $\xi: \Gamma E \to E$ such that $\pi \circ \xi = \Gamma\pi$, $\xi \circ \eta = 1$, and the map $\xi \circ \tilde{\beta}: \pi^{-1}\beta(0) \to \pi^{-1}p(\beta)$ is in $\mathcal{J}$ for each $\beta \in \Pi B$, where $\tilde{\beta}: \pi^{-1}\beta(0) \to (\Gamma\pi)^{-1}p(\beta)$ is given by $\tilde{\beta}(e) = (\beta, e)$. ξ is said to be transitive if, whenever $\beta(0) = \pi(e)$ and $\alpha(0) = p(\beta)$,

$$\xi(\alpha, \xi(\beta, e)) = \xi(\alpha\beta, e).$$

For example, μ is a transitive $\mathcal{U}$-lifting function for $\Gamma\pi: \Gamma E \to B$.

Lemma 3.3. If ξ is a (transitive) $\mathcal{J}$-lifting function for an $\mathcal{J}$-space $\pi: E \to B$ and $f: A \to B$ is a map, then $f^*\xi$ is a (transitive) $\mathcal{J}$-lifting function for $f^*\pi: f^*E \to B$, where $f^*\xi$ is defined by

$$(f^*\xi)(\alpha, (a, e)) = (p(\alpha), \xi(f \circ \alpha, e))$$

for $\alpha \in \Pi A$, $a \in A$, and $e \in E$ such that $\alpha(0) = a$ and $f(a) = \pi(e)$.

Proposition 3.4. An $\mathcal{J}$-space $\pi: E \to B$ is an $\mathcal{J}$-fibration if and only if π has an $\mathcal{J}$-lifting function ξ.

Proof. If $p_0: \Pi B \to B$ is the initial projection, then $\Gamma E = p_0^* E$ as a space and $(\tilde{p}_0, p_0): p_0^*\pi \to \pi$ is an $\mathcal{J}$-map. Define a homotopy $h: \Pi B \times [0, \infty] \to B$ of p_0 by $h(\beta, t) = \beta(t)$. If π is an $\mathcal{J}$-fibration, there is a homotopy $H: \Gamma E \times [0, \infty] \to E$ of $\tilde{p}_0$ such that (H, h) is an $\mathcal{J}$-homotopy, and an $\mathcal{J}$-lifting function ξ is then given by $\xi(\beta, e) = H((\beta, e), \ell(\beta))$. Conversely, assume given ξ. For an $\mathcal{J}$-map $(g, f): \nu \to \pi$, $\nu: D \to A$, and a homotopy h of f, let $h_t(a)$ denote the path of length t in B given by $h_t(a)(u) = h(a, u)$, $a \in A$, and define $H(d, t) = \xi(h_t \nu(d), g(d))$. Clearly H is a homotopy of g such that (H, h) is an $\mathcal{J}$-homotopy.

The following immediate consequence should be noted.

Corollary 3.5. Let $\mathcal{J} \to \mathcal{J}'$ be a functor over $\mathcal{U}$ (that is, the underlying space functor $\mathcal{J} \to \mathcal{U}$ is the composite $\mathcal{J} \to \mathcal{J}' \to \mathcal{U}$). Then an $\mathcal{J}$-fibration π is an $\mathcal{J}'$-fibration; in particular, π is a fibration.

When $\mathcal{J} = \mathcal{U}$, Definition 3.2 describes the standard procedure for replacing a map by a fibration. We note the following facts about this process.

Remarks 3.6. Let $\pi: E \to B$ be a map (that is, a $\mathcal{U}$-space).

(i) If ξ is a lifting function for π, then $1 \simeq \eta\xi$ over B via the homotopy $\gamma_t(\beta, e) = (\beta_t', \xi(\beta_t, e))$, where

$$\ell(\beta_t) = t\ell(\beta) \quad \text{and} \quad \beta_t(u) = \beta(u)$$

and

$$\ell(\beta_t') = (1-t)\ell(\beta) \quad \text{and} \quad \beta_t'(u) = \beta(u+t\ell(\beta)).$$

Therefore η and ξ are inverse fibre homotopy equivalences.

(ii) $\eta(E)$ is a strong deformation retract of ΓE via the homotopy $h_t(\beta, e) = (\beta_t, e)$. Thus η restricts to a weak homotopy equivalence on each fibre if π is a quasi-fibration (so that $\pi_*: \pi_i(E, \pi^{-1}b, e) \to \pi_i(B, b)$ is a bijection, $i \geq 1$, and $\pi_0(\pi^{-1}b, e) \to \pi_0(E, e) \to \pi_0(B, b) \to *$ is exact for all $b \in B$ and $e \in \pi^{-1}b$).

Remarks 3.7. For a $\mathcal{U}$-map $(g, f): \nu \to \pi$, $\nu: D \to A$ and $\pi: E \to B$, define a $\mathcal{U}$-map $\Gamma(g, f): \Gamma\nu \to \Gamma\pi$ by $\Gamma(g, f) = (\Gamma g, f)$, where $\Gamma g: \Gamma D \to \Gamma E$ is given by $\Gamma g(\alpha, d) = (f \circ \alpha, g(d))$. Then Γ is a functor from the category of $\mathcal{U}$-spaces to itself. The $\mathcal{U}$-maps $\eta: E \to \Gamma E$ and $\mu: \Gamma\Gamma E \to \Gamma E$ over B define natural transformations $\eta: 1 \to \Gamma$ and $\mu: \Gamma\Gamma \to \Gamma$ such that the following diagrams of $\mathcal{U}$-maps over B are commutative for each $\pi: E \to B$:

$$\begin{array}{ccccc} \Gamma E & \xrightarrow{\Gamma\eta} & \Gamma\Gamma E & \xleftarrow{\eta} & \Gamma E \\ & \diagdown\!\!\diagdown & \downarrow{\scriptstyle \mu} & \diagup\!\!\diagup & \\ & & \Gamma E & & \end{array} \qquad \text{and} \qquad \begin{array}{ccc} \Gamma\Gamma\Gamma E & \xrightarrow{\mu} & \Gamma\Gamma E \\ {\scriptstyle \Gamma\mu}\downarrow & & \downarrow{\scriptstyle \mu} \\ \Gamma\Gamma E & \xrightarrow{\mu} & \Gamma E \end{array}$$

In categorical language, (Γ, μ, η) is a monad in the category of $\mathcal{U}$-spaces [17, 2.1]. A transitive lifting function ξ for π is a $\mathcal{U}$-map $\xi: \Gamma E \to E$ over B such that the following diagrams of $\mathcal{U}$-maps over B are commutative:

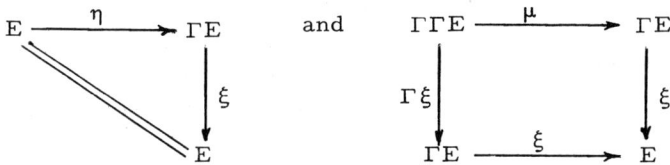

$$\begin{array}{ccc} E & \xrightarrow{\eta} & \Gamma E \\ & \diagdown\!\!\diagdown & \downarrow{\scriptstyle \xi} \\ & & E \end{array} \qquad \text{and} \qquad \begin{array}{ccc} \Gamma\Gamma E & \xrightarrow{\mu} & \Gamma E \\ {\scriptstyle \Gamma\xi}\downarrow & & \downarrow{\scriptstyle \xi} \\ \Gamma E & \xrightarrow{\xi} & E \end{array}$$

Thus the pair (π, ξ) is a Γ-algebra in the sense of [17, 2.2]. Moreover, by [17, 2.9], for any $\mathcal{U}$-space π, $(\Gamma\pi, \mu)$ is the free Γ-algebra generated by π.

The observation above was also noted by Malraison [15].

The rest of this section is devoted to the proof of the following generalization of Hurewicz's theorem [11].

<u>Theorem 3.8.</u> Let $\pi: E \to B$ be an $\mathcal{F}$-space and assume that B admits a numerable cover $\mathcal{C}$ such that $\pi: E_U \to U$ is an $\mathcal{F}$-fibration for each $U \in \mathcal{C}$. Then π is an $\mathcal{F}$-fibration. Therefore an $\mathcal{F}$-space

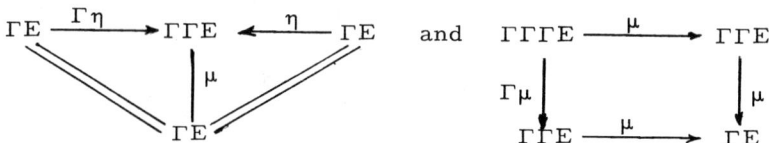

over a paracompact base space is an $\mathcal{J}$-fibration if and only if it is a local $\mathcal{J}$-fibration.

Proof. Our argument is a corrected version of Brown's modification [3] of Hurewicz's original proof. We shall construct an $\mathcal{J}$-lifting function ξ for π. For any finite ordered set $s = \{U_1, \ldots, U_n\}$ of not necessarily distinct sets in $\mathcal{C}$, define

$$W_s = \{\beta \mid \beta(t) \in U_i \text{ if } (i-1)\ell(\beta)/n \leq t \leq i\ell(\beta)/n\} \subset \Pi B.$$

By hypothesis, there exists an $\mathcal{J}$-lifting function $\xi_i: \Gamma E_{U_i} \to E_{U_i}$ for each i. For $0 \leq u < v \leq 1$ and a path $\beta \in \Pi B$, define the sub-path $\beta[u,v]$ of β by $\ell\,\beta[u,v] = (v-u)\ell(\beta)$ and

$$\beta[u,v](t) = \beta(t + u\ell(\beta)) .$$

Let $(i-1)/n \leq u < i/n$ and $(j-1)/n < v \leq j/n$ for integers $0 \leq i \leq j \leq n$. For $e \in \pi^{-1}\beta[u,v](0)$ and $\beta \in W_s$, define

$$(1) \quad \xi_s(\beta[u,v], e) = \xi_j(\beta[\tfrac{j-1}{n}, v], \xi_{j-1}(\beta[\tfrac{j-2}{n}, \tfrac{j-1}{n}], \ldots$$

$$\ldots \xi_{i+1}(\beta[\tfrac{i}{n}, \tfrac{i+1}{n}], \xi_i(\beta[u, \tfrac{i}{n}], e)) \ldots)) .$$

Let $\lambda_i: B \to I$ be such that $\lambda_i^{-1}(0,1] = U_i$ and define $\lambda_s: \Pi B \to I$ by

$$\lambda_s(\beta) = \inf\{\lambda_i \beta(t) \mid (i-1)\ell(\beta)/n \leq t \leq i\ell(\beta)/n \text{ and } 1 \leq i \leq n\}.$$

Then $W_s = \lambda_s^{-1}(0,1]$. $\{W_s\}$ is a cover of ΠB, but it is not locally finite. Let $c(s) = n$ if s has n elements and note that $\{W_s \mid c(s) < n\}$ is a locally finite set for each fixed n. Define θ_n on ΠB by $\theta_n(\beta) = \sum_{c(s)<n} \lambda_s(\beta)$. Then define γ_s on ΠB by

$$\gamma_s(\beta) = \max(0, \lambda_s(\beta) - n\theta_n(\beta)) \quad \text{if} \quad c(s) = n.$$

Define $V_s = \{\beta \mid \gamma_s(\beta) > 0\} \subset W_s$. It is easily verified that $\{V_s\}$ is a locally finite cover of ΠB. Total order the set $\mathcal{S}$ of finite ordered sets of sets in $\mathcal{C}$. For $(\beta, e) \in \Gamma E$, define

$$(2) \qquad \xi(\beta, e) = \xi_{s_q}(\beta[t_{q-1}, t_q], \xi_{s_{q-1}}(\beta[t_{q-2}, t_{q-1}], \ldots, \xi_{s_1}(\beta[t_0, t_1], e)\ldots)),$$

where $s_1 < \ldots < s_q$ are all elements $s \in \mathcal{S}$ such that $\beta \in V_s$ and where $t_j = \sum_{i=1}^{j} \gamma_{s_i}(\beta) / \sum_{i=1}^{q} \gamma_{s_i}(\beta)$.

ξ is the desired lifting function. It is clear from (1) and (2) that ξ restricts to a finite composite of maps in $\mathcal{F}$ for each fixed β.

4. Categories of fibres

In order to classify $\mathcal{F}$-fibrations, we must of course place severe restrictions on the category $\mathcal{F}$. We here define the notion of a "category of fibres," which is essentially a category with just enough structure to allow the development of a theory of associated principal fibrations. Such a theory is essential in our approach to classification theorems and is an obvious desideratum of any general theory of fibrations.

We topologize $\mathcal{F}(X, X')$ as a subspace of the function space $\mathcal{U}(X,X')$, with the (compactly generated) compact-open topology.

Definition 4.1. Let $\mathcal{F}$ have a distinguished object F. Then $(\mathcal{F}, F)$ is said to be a category of fibres if every map in $\mathcal{F}$ is a weak homotopy equivalence, $\mathcal{F}(F, X)$ is non-empty for each $X \in \mathcal{F}$, and composition with ϕ

$$\mathcal{F}(1, \phi): \mathcal{F}(F,F) \rightarrow \mathcal{F}(F, X)$$

is a weak homotopy equivalence for each $\phi \in \mathcal{F}(F, X)$. $\mathcal{F}$ is said to be a homogeneous category of fibres if $(\mathcal{F}, F)$ is a category of fibres for every object F.

Recall that a topological monoid is an associative H-space G with a two-sided identity element e and that a left G-space is a space X with an associative and unital action map $G \times X \rightarrow X$. G is said to be grouplike if $\pi_0 G$ is a group under the product induced by that of G. This holds if each right translation map $g: G \rightarrow G$ induces an isomorphism on $\pi_0 G$, and then translation by g on any (left or right) G-space is necessarily a homotopy equivalence.

Definition 4.2. A category of fibres $(\mathcal{G}, G)$ is said to be principal if G is a grouplike topological monoid, each object $Y \in \mathcal{G}$ is a (non-empty) right G-space, and the space $\mathcal{G}(Y, Y')$ coincides with the space of right G-maps from Y to Y'. Identify the space $\mathcal{G}(G, Y)$ with Y via $\phi \Longleftrightarrow \phi(e)$ and note that $\mathcal{G}(1, \phi): G \to Y$ is given by $g \to \phi(e)g$ and is required to be a weak homotopy equivalence. This condition already implies that all maps in $\mathcal{G}$ are weak homotopy equivalences (as one sees by composing a map $Y \to Y'$ in $\mathcal{G}$ with any map $G \to Y$ in $\mathcal{G}$).

Observe that to specify a principal category of fibres, we need only specify an appropriate collection of right G-spaces.

Definition 4.3. Let $(\mathcal{F}, F)$ be a category of fibres. Define the associated principal category of fibres $(\mathcal{G}, G)$ by letting $\mathcal{G}$ have objects $\mathcal{F}(F, X)$ for $X \in \mathcal{F}$, with $G = \mathcal{F}(F, F)$; the product on G and the action of G on $\mathcal{F}(F, X)$ are given by composition. For an $\mathcal{F}$-space $\pi: E \to B$, define a $\mathcal{G}$-space $P\pi: PE \to B$ by letting PE be the subspace of $\mathcal{U}(F, E)$ which consists of those maps $\psi: F \to E$ such that $\psi(F) \subset \pi^{-1}(b)$ for some $b \in B$ and $\psi: F \to \pi^{-1}(b)$ is a map in $\mathcal{F}$ and by letting $(P\pi)(\psi) = \pi\psi(F)$. For an $\mathcal{F}$-map $(g, f): \nu \to \pi$, $\nu: D \to A$, define a $\mathcal{G}$-map $P(g, f): P\nu \to P\pi$ by $P(g, f) = (Pg, f)$, where $(Pg)(\psi) = g \circ \psi$ for $\psi \in PD$. Then P is a functor from the category of $\mathcal{F}$-spaces to the category of $\mathcal{G}$-spaces.

The definition of P is based on ideas of Dold and Lashof [8], who called P Prin.

<u>Remark 4.4.</u> A principal category of fibres $(\mathcal{G}, G)$ may be identified with its associated principal category of fibres. Let $\pi: E \to B$ be a $\mathcal{G}$-space. Define a bijection of sets $\alpha: E \to PE$ by $\alpha(x)(g) = xg$ for $x \in E$ and $g \in G$; $\alpha^{-1}(\psi) = \psi(e)$ for $\psi: G \to E$ in PE. α^{-1} is always continuous and α is continuous provided that the given right actions of G on the fibres of π define a continuous function $E \times G \to E$. Henceforward, by a $\mathcal{G}$-space, we understand one for which $E \times G \to E$ is continuous; this is a reasonable restriction since, in the contrary case, we can retopologize E by requiring α to be a homeomorphism and so make the action continuous. With this convention, we can identify E and PE via α and regard P as the identity functor on $\mathcal{G}$-spaces.

The following pair of lemmas record obvious properties of the "associated principal $\mathcal{G}$-space" functor P for a fixed category of fibres $(\mathcal{F}, F)$.

<u>Lemma 4.5.</u> If $\pi: E \to B$ is an $\mathcal{F}$-space and $f: A \to B$ is a map, then there is a unique $\mathcal{G}$-map $\tau: Pf^*E \to f^*PE$ over A such that the following diagram is commutative:

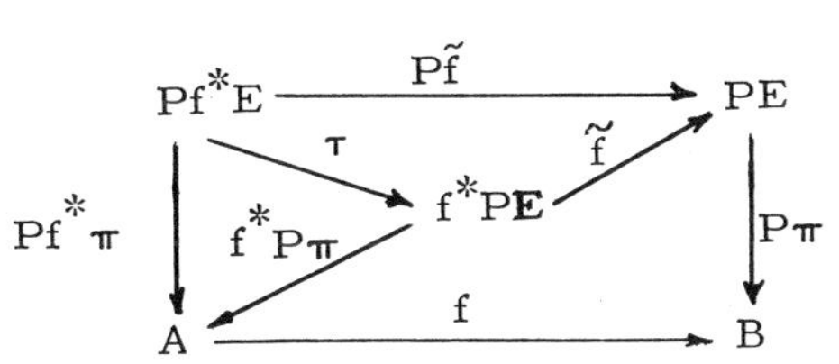

Lemma 4.6. Let $\pi: E \to B$ be an $\mathcal{J}$-fibration with (transitive) $\mathcal{J}$-lifting function $\xi: \Gamma E \to E$. Then $P\pi: PE \to B$ is a $\mathcal{G}$-fibration with (transitive) $\mathcal{G}$-lifting function $P\xi: \Gamma PE \to PE$ defined by

$$(P\xi)(\beta, \psi)(x) = \xi(\beta, \psi(x)) \text{ , hence } (P\xi)(\beta, \psi) = \xi \circ \tilde{\beta} \circ \psi: F \to \pi^{-1} p(\beta) \text{ ,}$$

for $x \in F$, $\beta \in \Pi B$, and $\psi \in PE$ such that $\beta(0) = (P\pi)(\psi)$.

5. $\mathcal{F}$-quasifibrations and based fibres

Our explicit classifying space constructions will yield universal quasifibrations. Since pullbacks of quasifibrations need not be quasifibrations, we shall sometimes have to use the functor Γ to replace $\mathcal{F}$-quasifibrations, by which we understand $\mathcal{F}$-spaces $\pi: E \to B$ such that π is a quasifibration, by $\mathcal{F}$-fibrations. The following definition records the minimum amount of information that will suffice for this purpose.

Definition 5.1. A category of fibres $(\mathcal{F}, F)$ is Γ**-complete in a full** subcategory $\mathcal{L}$ of $\mathcal{U}$ if $\mathcal{F} \subset \mathcal{L}$, $\mathcal{H} \subset \mathcal{L}$, and the following statements are valid for $\mathcal{F}$-quasifibrations $\pi: E \to B$ with B and E in $\mathcal{L}$.

(1) $\Gamma\pi: \Gamma E \to B$ is a $\mathcal{F}$-fibration with $\mathcal{F}$-lifting function μ.

(2) $\eta: E \to \Gamma E$ is an $\mathcal{F}$-map over B.

(3) Γ takes $\mathcal{F}$-maps between $\mathcal{F}$-quasifibrations in $\mathcal{L}$ to $\mathcal{F}$-maps.

Let $\mathcal{T}$ denote the category of nondegenerately based spaces in $\mathcal{U}$ and basepoint preserving maps. In some very important examples, the functor $\mathcal{F} \to \mathcal{U}$ factors through $\mathcal{T}$. In the definition just given, there is clearly no way to give the fibres of $\Gamma\pi$ basepoints such that each $\mu\tilde{\beta}: (\Gamma\pi)^{-1}\beta(0) \to (\Gamma\pi)^{-1}p(\beta)$ is basepoint preserving (since $\mu\tilde{\beta}(\alpha, e) = (\beta\alpha, e)$). We need a few more definitions in order to circumvent this difficulty.

Definition 5.2. When given a category $\mathcal{F}$ with a faithful functor $\mathcal{F} \to \mathcal{T}$, redefine an $\mathcal{F}$-space to be a map $\pi: E \to B$ such that not only is $\pi^{-1}(b)$ in $\mathcal{F}$ for all $b \in B$ but also the function $\sigma: B \to E$ specified by sending b to the basepoint of $\pi^{-1}(b)$ is continuous and is a fibrewise cofibration, in the sense that there exists a representation (h, u) of $(E, \sigma B)$ as an NDR-pair [17, A.1] such that $h: E \times I \to E$ is a $\mathcal{J}$-homotopy over B (so that (h, u) restricts to a representation of $(\pi^{-1}(b), \sigma b)$ as an NDR-pair). All other definitions and results obtained so far in this paper apply verbatim to these $\mathcal{F}$-spaces with a canonical cross-section (because $\mathcal{F}$-maps are automatically section preserving). We refer to a category of fibres $(\mathcal{F}, F)$ such that $\mathcal{F}$ maps faithfully to $\mathcal{J}$ as a category of based fibres.

Definition 5.3. Let $\pi: E \to B$ be a $\mathcal{J}$-space. Define a $\mathcal{J}$-space $\Gamma'\pi: \Gamma'E \to B$ and maps $\eta': E \to \Gamma'E$ and $\mu': \Gamma\Gamma'E \to \Gamma'E$ as follows. $\Gamma'E$ is obtained by growing a long whisker on each fibre of $\Gamma\pi$. Formally, $\Gamma'E$ is the quotient space obtained from the disjoint union of ΓE and $B \times [0, \infty]$ by identifying $(b, \sigma b) \in \Gamma E$ with $(b, 0) \in B \times [0, \infty]$ for each $b \in B$. $\Gamma'\pi$ coincides with $\Gamma\pi$ on ΓE and with the projection to the first coordinate on $B \times [0, \infty]$. The cross-section $\Gamma\sigma: B \to \Gamma'E$ is defined by $(\Gamma\sigma)(b) = (b, \infty)$ and is clearly a fibrewise cofibration. With (h, u) as in the previous definition, define η' by

$$\eta'(e) = \begin{cases} (\pi(e), 1/u(e) - 2) & \text{if } 0 \le u(e) \le 1/2 \\ (\pi(e), h(e, 2 - 2u(e))) & \text{if } 1/2 \le u(e) \le 1 \end{cases}$$

Then η' is a $\mathcal{J}$-map over B and a homotopy equivalence. Define μ' by $\mu' = \mu$ on $\Gamma\Gamma E \subset \Gamma\Gamma'E$ and by

$$\mu'(\beta, (b, t)) = \begin{cases} (\beta'_t, \sigma\beta'_t(0)) & \text{if } 0 \le t \le \ell(\beta) \\ (p(\beta), t - \ell(\beta)) & \text{if } \ell(\beta) \le t \le \infty \end{cases}$$

where $\beta(0) = b$, $\ell(\beta'_t) = \ell(\beta) - t$, and $\beta'_t(s) = \beta(s+t)$. Then μ' is easily verified to be a $\mathcal{J}$-lifting function for $\Gamma'\pi$. With the evident definition on $\mathcal{J}$-maps, Γ' becomes a functor from $\mathcal{J}$-spaces to $\mathcal{J}$-fibrations.

Definition 5.4. A category of based fibres $(\mathcal{F}, F)$ is Γ'-complete in a full subcategory $\mathcal{Q}$ of $\mathcal{U}$ if $\mathcal{F} \subset \mathcal{Q}$, $\mathcal{H} \subset \mathcal{Q}$, and the following statements are valid for $\mathcal{F}$-quasifibrations $\pi: E \to B$ with B and E in $\mathcal{Q}$.

(1) $\Gamma'\pi: \Gamma'E \to B$ is an $\mathcal{F}$-fibration with $\mathcal{F}$-lifting function μ'.

(2) $\eta': E \to \Gamma'E$ is an $\mathcal{F}$-map over B.

(3) Γ' takes $\mathcal{F}$-maps between $\mathcal{F}$-quasifibrations in $\mathcal{Q}$ to $\mathcal{F}$-maps.

A five lemma argument gives the following observation.

Lemma 5.5. Let $(\mathcal{F}, F)$ be Γ-complete (or Γ'-complete) in $\mathcal{Q}$ and let $\pi: E \to B$ be an $\mathcal{F}$-quasifibration with B and E in $\mathcal{Q}$. If $P\pi: PE \to B$ is again a quasifibration, then the $\mathcal{H}$-map $P\eta: PE \to P\Gamma E$ (or $P\eta': PE \to P\Gamma'E$) over B is a weak homotopy equivalence.

We record the following remarks for use in [19].

Remarks 5.6. Let $\nu: D \to A$ and $\pi: E \to B$ be $\mathcal{J}$-spaces. Define a $\mathcal{J}$-space $\nu \wedge \pi: D \wedge E \to A \times B$, the fibrewise smash product of ν and π, as follows. Let $D \wedge E = D \times E/(\approx)$, where the equivalence identifies the wedge $(\sigma a, \pi^{-1}b) \vee (\nu^{-1}a, \sigma b)$ to the point $(\sigma a, \sigma b)$ for each $(a, b) \in A \times B$, and let $\nu \wedge \pi$ be induced from $\nu \times \pi$; the cross-section of $\nu \wedge \pi$ is induced from $\sigma \times \sigma$. If ν and π are $\mathcal{J}$-fibrations, then so is $\nu \wedge \pi$ since $\nu \wedge \pi$ clearly inherits the $\mathcal{J}$-CHP from ν and π. There is a natural $\mathcal{J}$-map $g: \Gamma'D \wedge \Gamma'E \to \Gamma'(D \wedge E)$ over $A \times B$ specified for $(\alpha, d) \in \Gamma D$,

$(\beta, e) \in \Gamma E$, $(a, s) \in A \times [0,\infty]$ and $(b,t) \in B \times [0,\infty]$ by

$$g((\alpha, d) \wedge (\beta, e)) = (\alpha \times \beta, d \wedge e)$$

$$g((a, s) \wedge (b, t)) = (a \times b, \max(s,t))$$

$$g((\alpha, d) \wedge (b, t)) = \begin{cases} (\alpha'_t \times b, \sigma\alpha'_t(0) \wedge \sigma b) & \text{if } t \le \ell(\alpha) \\ (p\alpha \times b, t - \ell(\alpha)) & \text{if } \ell(\alpha) \le t \end{cases}$$

and

$$g((a,s) \wedge (\beta,e)) = \begin{cases} (a \times \beta'_s, \sigma a \wedge \sigma\beta'_s(0)) & \text{if } s \le \ell(\beta) \\ (a \times p\beta, s - \ell(\beta)) & \text{if } \ell(\beta) \le s \end{cases}$$

where $\ell(\alpha'_t) = \ell(\alpha) - t$, $\alpha'_t(u) = \alpha(t+u)$, $\ell(\beta'_s) = \ell(\beta) - s$, and $\beta'_s(u) = \beta(s+u)$. It is not hard to see that g restricts to a weak homotopy equivalence on each fibre if ν and π are quasifibrations with connected fibres.

6. Examples of categories of fibres

We here define the functors to which our classification theorem will apply (in favorable cases) and discuss various examples of categories of fibres.

Definition 6.1. Let $A \in \mathcal{U}$. Define $\mathcal{E}\mathcal{F}(A)$ to be the collection (assumed to be a set) of equivalence classes of $\mathcal{F}$-fibrations over A under the equivalence relation generated by the $\mathcal{F}$-maps over A. For a map $f: A \to A'$, define $f^*: \mathcal{E}\mathcal{F}(A') \to \mathcal{E}\mathcal{F}(A)$ by $f^*\{\nu\} = \{f^*\nu\}$, where $\{\nu\}$ denotes the equivalence class of ν. By Proposition 2.5, $\mathcal{E}\mathcal{F}$ is a contravariant functor from the homotopy category of $\mathcal{U}$ to the category of sets. By Lemmas 4.5 and 4.6, P induces a natural transformation $\mathcal{E}\mathcal{F} \to \mathcal{E}\mathcal{B}$ when $(\mathcal{F}, F)$ is a category of fibres. By Corollary 3.5, any functor $\mathcal{F} \to \mathcal{F}'$ over $\mathcal{U}$ induces a natural transformation $\mathcal{E}\mathcal{F} \to \mathcal{E}\mathcal{F}'$.

Note that the assumption that our equivalence relation leads to a set of equivalence classes is non-trivial. It will hold in our classification theorem because our constructive proof will display a set of $\mathcal{F}$-fibrations over A such that any given $\mathcal{F}$-fibration over A is equivalent to an element of the displayed set.

At first sight, our choice of equivalence relation may seem less natural than the obvious (and more restrictive) one of $\mathcal{F}$-homotopy equivalence. In the classical examples, every map in $\mathcal{F}$ is an $\mathcal{F}$-homotopy equivalence. In such cases, Theorem 2.6 ensures that our equivalence relation coincides with $\mathcal{F}$-homotopy equivalence (over good base spaces).

In the contrary case, it is very hard to verify that a given $\mathcal{J}$-map is in fact an $\mathcal{J}$-homotopy equivalence. Our equivalence relation allows us to ignore this problem and to freely use arbitrary $\mathcal{J}$-maps over A. It is this freedom which enables us to avoid both local pasting arguments and higher homotopies.

We now turn to examples. We first consider the principal case.

Examples 6.2. Let G be a grouplike topological monoid. Specify four successively smaller categories $\mathcal{Y}$ such that $(\mathcal{Y}, G)$ is a principal category of fibres by letting a right G-space Y be an object of $\mathcal{Y}$ if and only if the maps $\tilde{y}: G \to Y$ defined for $y \in Y$ by $\tilde{y}(g) = yg$ are all

(i) weak homotopy equivalences; write $(\mathcal{Y}, G) = G\mathcal{U}$.

(ii) homotopy equivalences; write $(\mathcal{Y}, G) = G\mathcal{W}$.

(iii) G-equivariant homotopy equivalences.

(iv) homeomorphisms, where G is a topological group.

Let $\mathcal{W}$ denote the full subcategory of $\mathcal{U}$ of spaces having the homotopy type of CW-complexes. Clearly, (i) and (ii) are appropriate to $\mathcal{U}$ and $\mathcal{W}$, respectively, but are conceptually similar. Case (iii) is most refractory, and we shall not study it. The point is that, in general, there is no effective way of telling when a G-equivariant map which is a homotopy equivalence is a G-equivariant homotopy equivalence. In particular, we have no analog of the following lemma.

Lemma 6.3. Let G be a grouplike topological monoid. Then $G\mathcal{U}$ is Γ-complete in $\mathcal{U}$ and, if $G \in \mathcal{W}$, $G\mathcal{W}$ is Γ-complete in $\mathcal{W}$.

Proof. Write $\mathcal{G}$ for $G\mathcal{U}$ or $G\mathcal{W}$ and let $\pi: E \to B$ be a $\mathcal{G}$-quasifibration, with B and E in $\mathcal{U}$ or $\mathcal{W}$. Via $(\beta, e)g = (\beta, eg)$, the right action of G on E induces a right action of G on ΓE such that $\eta: E \to \Gamma E$, $\mu: \Gamma\Gamma E \to \Gamma E$, and the $\tilde{\beta}$ are all maps of right G-spaces. We must show that $(\Gamma\pi)^{-1}(b) \in \mathcal{G}$ for all $b \in B$. Let $y = (\beta, x)$ be a typical point in $(\Gamma\pi)^{-1}(b)$, so that $\beta(0) = \pi(x)$ and $\rho(\beta) = b$. We must verify that the map $\tilde{y}: G \to (\Gamma\pi)^{-1}(b)$ is a weak homotopy equivalence; when $\mathcal{G} = G\mathcal{W}$ and $G \in \mathcal{W}$, the Whitehead theorem and the theorem of Stasheff [32] quoted just below will imply that $(\Gamma\pi)^{-1}(b) \in \mathcal{W}$ and will thus complete the proof. Consider the following commutative diagram:

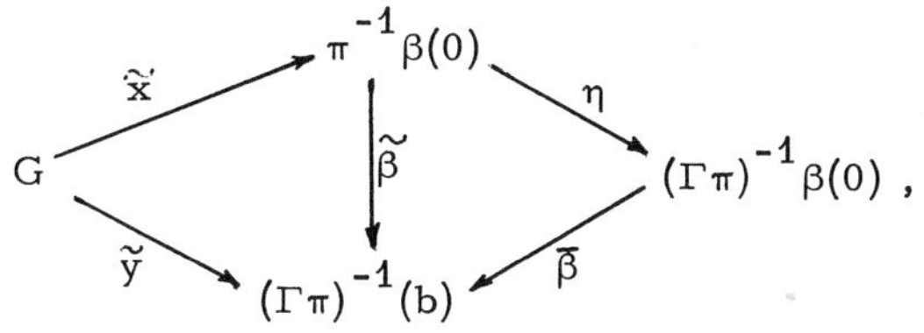

where $\bar{\beta}(\gamma, w) = (\beta\gamma, w)$ for $\gamma \in \Pi B$ and $w \in E$ such that $\gamma(0) = \pi(w)$ and $p\gamma = \beta(0)$. Since $\eta, \bar{\beta}$, and $\tilde{x}$ are weak homotopy equivalences by Remarks 3.6, Proposition 2.3, and hypothesis, $\tilde{\beta}$ and $\tilde{y}$ are also weak homotopy equivalences.

The required result of Stasheff[1] can be stated as follows.

Theorem 6.4. Let $\nu : D \to A$ be a fibration with $A \in \mathcal{W}$. Then
(i) $D \in \mathcal{W}$ if and only if $\nu^{-1}(a) \in \mathcal{W}$ for all $a \in A$; and
(ii) If $A' \in \mathcal{W}$, $f: A' \to A$ is a map, and $D \in \mathcal{W}$, then $f^*D \in \mathcal{W}$.

1. The proof in [32] is not correct, but can be patched.

For the next example, recall the following standard results about the relationship between $\mathcal{U}$ and $\mathcal{W}$ (e.g. [21, 2.9 and 3.10]). Let [X, Y] denote the set of homotopy classes of maps $X \to Y$.

Theorem 6.5. (i) If $\phi: Y \to Z$ is a weak homotopy equivalence and $X \in \mathcal{W}$, then $\phi_*: [X, Y] \to [X, Z]$ is an isomorphism.

(ii) There are a functor and natural transformation on the homotopy category of $\mathcal{U}$ which assign a CW-complex X' and a homotopy class of weak homotopy equivalences $X' \to X$ to a space X.

Example 6.6. Let $F \in \mathcal{W}$. Define two categories of fibres $F\mathcal{U}$ and $F\mathcal{W}$ with distinguished object F as follows.

(i) $X \in F\mathcal{U}$ if X is of the same weak homotopy type as F; the maps in $F\mathcal{U}$ are the weak homotopy equivalences $X \to X'$.

(ii) $X \in F\mathcal{W}$ if X is of the same homotopy type as F; the maps in $F\mathcal{W}$ are the homotopy equivalences $X \to X'$; thus $F\mathcal{W} = \mathcal{W} \cap F\mathcal{U}$.

In (i), $F\mathcal{U}(F, X)$ is non-empty by Theorem 6.5(ii) and the fact that $F \in \mathcal{W}$. For $\phi: F \to X$ in $F\mathcal{U}$ and any CW-complex K,

$$\phi_*: [K \times F, F] \to [K \times F, X]$$

is an isomorphism, by Theorem 6.5(i), and therefore

$$\mathcal{U}(1, \phi)_*: [K, \mathcal{U}(F, F)] \to [K, \mathcal{U}(F, X)]$$

is an isomorphism. Since $F\mathcal{U}(F, F)$ and $F\mathcal{U}(F, X)$ are unions of components of $\mathcal{U}(F, F)$ and $\mathcal{U}(F, X)$, because a map homotopic to a weak homotopy equivalence is a weak homotopy equivalence, it follows that $F\mathcal{U}(1, \phi)$ is a weak homotopy equivalence and thus that (i) and (ii) do indeed define categories of fibres.

Recall the following result of Milnor [25].

Theorem 6.7. If $X \in \mathcal{W}$ and $C \in \mathcal{U}$ is compact, then $\mathcal{U}(C, X) \in \mathcal{W}$.

For this reason, case (ii) is adequate when F is compact. When F is not compact, for example when F is a localization or completion of the n-sphere at a set of primes, we shall have to use fibres not in $\mathcal{W}$, as allowed in case (i).

Theorem 6.4 and a proof similar to, but simpler than, that of Lemma 6.3 give the following result.

Lemma 6.8. Let $F \in \mathcal{W}$. Then $F\mathcal{U}$ is Γ-complete in $\mathcal{U}$ and, if F is compact, $F\mathcal{W}$ is Γ-complete in $\mathcal{W}$.

There is a based variant of the preceding example. Let $\mathcal{V} = \mathcal{T} \cap \mathcal{W}$, the category of nondegenerately based spaces in $\mathcal{W}$, and recall that Theorems 6.5 and 6.7 remain valid if all maps and homotopies in sight are required to preserve basepoints.

Example 6.9. Let $F \in \mathcal{V}$. Define two categories of based fibres $F\mathcal{T}$ and $F\mathcal{V}$ with distinguished object F as follows.

(i) $X \in F\mathcal{T}$ if $X \in \mathcal{T}$ is of the same based weak homotopy type as F; the maps in $F\mathcal{T}$ are the based weak homotopy equivalences.

(ii) $X \in F\mathcal{V}$ if $X \in \mathcal{V}$ is of the same based homotopy type as F; the maps in $F\mathcal{V}$ are the based homotopy equivalences; thus $F\mathcal{V} = \mathcal{V} \cap F\mathcal{T}$.

Lemma 6.10. Let $F \in \mathcal{V}$. Then $F\mathcal{J}$ is Γ'-complete in $\mathcal{U}$ and, if F is compact, $F\mathcal{V}$ is Γ'-complete in $\mathcal{N}$.

Note that we impose basepoints only on fibres, not on base spaces or total spaces.

Example 6.11. Let G be a topological group and let F be a left G-space on which G acts effectively. Define a category $\mathcal{J}$ as follows. Let $\mathcal{J}$ have objects all pairs (X, x) such that X is a left G-space and $x: F \to X$ is a homeomorphism of left G-spaces. Let the set of morphisms from (X, x) to (X', x') be $\{x'gx^{-1} | g \in G\}$, with the evident operation of composition. $\mathcal{J}$ has the distinguished object $(F, 1)$, and we call $(\mathcal{J}, F)$ a category of bundle fibres. If $(\mathcal{G}, G)$ is the associated principal category of fibres, then G is the given group retopologized with its possibly coarser topology as a subspace of $\mathcal{U}(F, F)$; we insist that G, so topologized, again be a topological group. Of course, in practice, the two topologies usually agree. By Theorem 3.8, a Steenrod fibre bundle with group G (with either topology) and fibre F which is trivial over each set of a numerable cover of its base space is an $\mathcal{J}$-fibration. Following Dold [7], we say that such a bundle is numerable.

7. The geometric bar construction

We here review the definition and properties of the two-sided geometric bar construction introduced in [17, §9-11]. Let G be a topological monoid such that its identity element e is a strongly nondegenerate basepoint (in the sense that (G, e) is a strong NDR-pair [17, A.1]). Let X and Y be left and right G-spaces. Define a simplicial topological space $B_*(Y, G, X)$ by letting the space of j-simplices be $Y \times G^j \times X$, with typical elements written in the form $y[g_1, \ldots, g_j]x$, and letting the face and degeneracy operators be given by

$$\partial_i(y[g_1, \ldots, g_j]x) = \begin{cases} yg_1[g_2, \ldots, g_j]x & \text{if } i = 0 \\ y[g_1, \ldots, g_{i-1}, g_ig_{i+1}, g_{i+2}, \ldots, g_j]x & \text{if } 1 \leq i < j \\ y[g_1, \ldots, g_{j-1}]g_jx & \text{if } i = j \end{cases}$$

and $$s_i(y[g_1, \ldots, g_j]x) = y[g_1, \ldots, g_i, e, g_{i+1}, \ldots, g_j]x \quad .$$

Let $B(Y, G, X)$ denote the geometric realization of $B_*(Y, G, X)$, as defined in [17, 11.1]. Then B is a functor to $\mathcal{U}$ from the category $\mathcal{A}(\mathcal{U})$ of triples (Y, G, X); the morphisms of $\mathcal{A}(\mathcal{U})$ are triples $(k, f, j): (Y, G, X) \to (Y', G', X')$ where $f: G \to G'$ is a map of topological monoids and $j: X \to X'$ and $k: Y \to Y'$ are f-equivariant maps, $j(gx) = f(g)j(x)$ and $k(yg) = k(y)f(g)$. The functor B was first defined (implicitly) by Stasheff [34]. Let $*$ denote the one-point G-space and define

$$BG = B(*, G, *) \quad \text{and} \quad EG = B(*, G, G).$$

BG is the standard classifying space of G, namely the normalized version of the Dold-Lashof [8] construction, as defined by Stasheff [31, p. 289], exploited by Milgram [23], and analyzed in detail by Steenrod [38].

Many of the results of this section and the next are due to the authors cited above, but our explicit use of simplicial spaces simplifies nearly all of the proofs by reducing them to trivial verifications on the level of simplicial spaces followed by quotations of general results about geometric realization. The following series of propositions give the basic facts about the topological behavior of the functor B.

Proposition 7.1. $B_*(Y, G, X)$ is a proper simplicial space. $B(Y, G, X)$ is n-connected if G is (n-1)-connected and X and Y are n-connected.

Proof. The first statement means that $(Y, \emptyset) \times (G, e)^j \times (X, \emptyset)$ is a strong NDR-pair (where $\emptyset$ is the empty set) and holds by [17, A.3]. The second statement follows by [17, 11.12] (its extra hypothesis of strict propriety being unnecessary by [18, A.5]).

Now [18, A.6 and A.4] imply the following two results.

Proposition 7.2. If Y, G, and X are in $\mathcal{W}$, then so is $B(Y, G, X)$.

Proposition 7.3. Let $(k, f, j): (Y, G, X) \to (Y', G', X')$ be a morphism in $\mathcal{A}(\mathcal{U})$.

(i) If k, f, and j induce isomorphisms on integral homology, then so does $B(k, f, j)$.

(ii) If k, f, and j are homotopy equivalences, then so is B(k, f, j).

Note in (ii) that no equivariance conditions are required of the given homotopy inverses and homotopies.

Since B_* preserves products by [17, 10.1] and geometric realization preserves products by [17, 11.5], the following result holds.

Proposition 7.4. For (Y, G, X) and (Y', G', X') in $\mathcal{A}(\mathcal{U})$, the projections define a natural homeomorphism

$$B(Y \times Y', G \times G', X \times X') \to B(Y, G, X) \times B(Y', G', X').$$

We shall often write

$$\tau = \tau(\rho): Z \to B(Y, G, X) \quad \text{and} \quad \varepsilon = \varepsilon(\lambda): B(Y, G, X) \to Z$$

for the maps induced via [17, 9.2 and 11.8] from a map $\rho: Z \to Y \times X$ and from a map $\lambda : Y \times X \to Z$ such that $\lambda(yg, x) = \lambda(y, gx)$; the intended choice of ρ and λ should be clear from the context. Clearly ε factors through $Y \times_G X$, the quotient of $Y \times X$ by the equivalence relation generated by $(yg, x) \approx (y, gx)$. Note that B(G, G, X) is a left G-space (again, because realization preserves products). The following result is a consequence of [17, 9.8, 9.9, and 11.10].

Proposition 7.5. $\varepsilon : B(G, G, X) \to X$ is a map of left G-spaces and a strong deformation retraction (with right inverse τ). The symmetric conclusion holds for $\varepsilon : B(Y, G, G) \to Y$.

We shall always write

$p: B(Y,G,X) \to B(Y,G,*)$ and $q: B(Y,G,X) \to B(*,G,X)$

for the maps induced from the trivial G-maps $X \to *$ and $Y \to *$.

Theorem 7.6. If G is grouplike, then p and q are quasi-fibrations.

Proof. Consider p, the case q being handled symmetrically. As realizations of simplicial spaces, $B(Y,G,*)$ and $B(Y,G,X)$ are filtered spaces [17, 11.1] and $F_jB(Y,G,X) = p^{-1}F_jB(Y,G,*)$. Visibly,

$$F_0B(Y,G,X) = Y \times X \text{ and, if } j > 0,$$

$$F_j(B(Y,G,X) - F_{j-1}B(Y,G,X) = (F_jB(Y,G,*) - F_{j-1}B(Y,G,*)) \times X.$$

By [17, A.3 and A.4], any representations of (G,e) and $(\Delta_j, \partial\Delta_j)$ as strong NDR-pairs determine a representation (k,v) of $(G,e)^j \times (\Delta_j, \partial\Delta_j)$ as a strong NDR-pair. Together with the obvious representations of $(X, \emptyset)$ and $(Y, \emptyset)$ as strong NDR-pairs (namely, the constant homotopies and the trivial maps onto $\{1\} \subset I$), (k,v) determines representations (h,u) and (H, up) of

$$(F_jB(Y,G,*), F_{j-1}B(Y,G,*) \text{ and } (F_jB(Y,G,X), F_{j-1}B(Y,G,X))$$

as strong NDR-pairs such that H covers h. Let $U = u^{-1}[0,1)$. Then h restricts to a deformation of U onto $F_{j-1}B(Y,G,*)$. (It is for this that strong NDR-pairs, rather than just NDR-pairs, are needed.) By results of Dold and Thom [9] (as formulated in [17, 7.2]), it suffices to verify that, for all $z \in U$, $H_1: p^{-1}(z) \to p^{-1}h_1(z)$ is a weak homotopy equivalence. If $z \in F_{j-1}B(Y,G,*)$, H_1 is the identity. Thus let $z = |y[g_1, \ldots, g_j], a| \in U$, where $g_k \in G - \{e\}$ and $a \in \Delta_j - \partial\Delta_j$, and let $|y'[g'_1, \ldots, g'_i], a'|$, $i < j$, be the non-degenerate representative for $h_1(z)$ [17, 11.3]. Since G is grouplike, it suffices to show that there exists $g \in G$ such that the diagram

$$\begin{array}{ccc} X & \xrightarrow{\;g\;} & X \\ {\scriptstyle \iota}\downarrow & & \downarrow{\scriptstyle \iota'} \\ p^{-1}(z) & \xrightarrow{\;H_1\;} & p^{-1}h_1(z) \end{array}$$

commutes, where ι and ι' are the homeomorphisms

$$\iota(x) = |y[g_1, \ldots, g_j]x, a| \quad \text{and} \quad \iota'(x) = |y'[g_1', \ldots, g_i']x, a'|.$$

Let $k_1(g_1, \ldots, g_j, a) = (g_1'', \ldots, g_j'', a'')$. The reduction of the point $H_1\iota(x) = |y[g_1'', \ldots, g_j'']x, a''|$ to non-degenerate form by use of [17, 11.3] will yield a point $\iota'(gx)$, where g results from last face operators and is independent of x since the particular face and degeneracy operators required for the reduction are independent of x.

$p: EG \to BG$ should be thought of as the universal $G\mathcal{U}$-quasifibration. The following corollary asserts its essential uniqueness.

Corollary 7.7. Let G be grouplike and let $p': E' \to B'$ be a $G\mathcal{U}$-quasifibration such that E' is aspherical. Then the maps ε and q are weak homotopy equivalences in the following commutative diagram:

$$\begin{array}{ccccc} E' & \xleftarrow{\;\varepsilon\;} & B(E', G, G) & \xrightarrow{\;q\;} & EG \\ {\scriptstyle p'}\downarrow & & {\scriptstyle p}\downarrow & & \downarrow{\scriptstyle p} \\ B' & \xleftarrow{\;\varepsilon\;} & B(E', G, *) & \xrightarrow{\;q\;} & BG \end{array}$$

$p: B(Y, G, X) \to B(Y, G, *)$ should be thought of as the quasifibration with fibre X associated to the principal quasifibration $p: B(Y, G, G) \to B(Y, G, *)$. According to the following result, it can be thought of as classified by q.

Proposition 7.8. Let $(k, f, 1): (Z, H, X) \to (Y, G, X)$ be a morphism in $\mathcal{A}(\mathcal{U})$. Then the following diagrams are pullbacks:

$$\begin{array}{ccc} B(Z,H,X) & \xrightarrow{B(k,f,1)} & B(Y,G,X) \\ \downarrow p & & \downarrow p \\ B(Z,H,*) & \xrightarrow{B(k,f,1)} & B(Y,G,*) \end{array} \quad \text{and} \quad \begin{array}{ccc} B(Y,G,X) & \xrightarrow{q} & B(*,G,X) \\ \downarrow p & & \downarrow p \\ B(Y,G,*) & \xrightarrow{q} & BG \end{array}$$

Proof. The second diagram is the case $k: Y \to *$ and $f = 1$ of the first. Since geometric realization preserves pullbacks [17, 11.6], the result follows from the observation that the diagrams

$$\begin{array}{ccc} Z \times H^j \times X & \xrightarrow{k \times f^j \times 1} & Y \times G^j \times X \\ \downarrow p & & \downarrow p \\ Z \times H^j & \xrightarrow{k \times f^j} & Y \times G^j \end{array}$$

are pullbacks for all $j \geq 0$.

Proposition 7.9. If G is grouplike and Y is a right G-space, then

$$G \xrightarrow{\iota} Y \xrightarrow{\tau} B(Y, G, *) \xrightarrow{q} BG$$

is a quasifibration sequence, where $\iota(g) = y_0 g$ for any chosen $y_0 \in Y$.

Proof. We must show that τ is equivalent to a quasifibration, with ι equivalent to the inclusion of the fibre. Consider the following diagram, in which the maps τ with range $B(Y, G, G)$ are induced from the maps $G \to Y \times G$ and $Y \to Y \times G$ specified by $g \to (y_0, g)$ and $y \to (y, e)$:

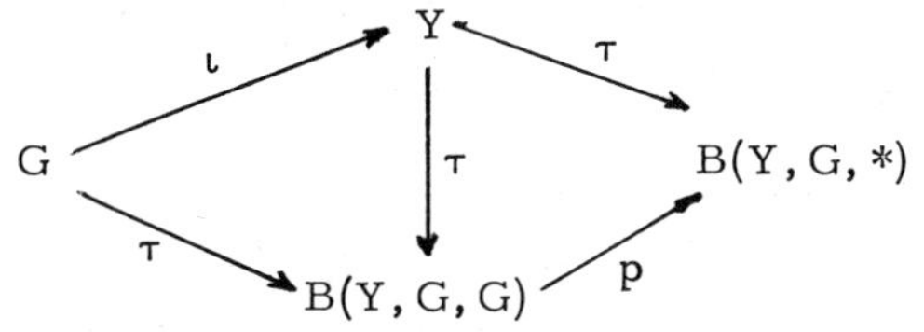

The right triangle commutes and the left triangle homotopy commutes via the homotopy $h(g, t) = |y_0[g]e, (t, 1-t)|$.

Our final result strengthens the analogy with bundle theory.

<u>Proposition 7.10.</u> Let $(\mathcal{F}, F)$ be a category of fibres with associated principal category of fibres $(\mathcal{G}, G)$. Let $f: H \to G$ be a map of topological monoids and let Y be a right H-space. Then there is a homeomorphism

$$\alpha: B(Y, H, G) \to PB(Y, H, F)$$

of $\mathcal{G}$-spaces over $B(Y, H, *)$. In particular, if H is grouplike, $Pp: PB(Y, H, F) \to B(Y, H, *)$ is a quasifibration.

<u>Proof.</u> $\alpha|y[h_1, \ldots, h_j]g, a|(f) = |y[h_1, \ldots, h_j]g(f), a|$ for $y \in Y$, $h_i \in H$, $g \in G = \mathcal{F}(F, F)$, $a \in \Delta_j$, and $f \in F$. α^{-1} is given by

$$\alpha^{-1}(\psi) = |y[h_1, \ldots, h_j]g, a|,$$

where $(Pp)(\psi) = |y[h_1, \ldots, h_j], a|$, in non-degenerate form, and where $g: F \to F$ is defined by $\psi(f) = |y[h_1, \ldots, h_j]g(f), a|$.

8. Groups, homogeneous spaces, and Abelian monoids

We here give special properties of BG and EG when G is a topological group or Abelian topological monoid. We also develop a generalized concept of "homogeneous space" for use in section 10.

The following theorem is due to Steenrod [38].

Theorem 8.1. Let G be a topological group. Then EG admits a natural structure of topological group such that the following statements hold.

(i) G is a closed subgroup of EG and the action $EG \times G \to EG$ agrees with the product in EG.

(ii) BG is the homogeneous space EG/G of right cosets and $p: EG \to BG$ is the natural projection.

(iii) The natural homeomorphism $E(G \times G') \to EG \times EG'$ is an isomorphism of groups when G and G' are groups.

Proof. A product-preserving functor D_* from spaces to simplicial spaces was constructed in [17, 10.2]; of course, D_* necessarily takes topological groups to simplicial topological groups. A homeomorphism $\alpha_*: E_*G \to D_*G$ of simplicial right G-spaces was defined in [17, 10.3]. Thus, by [17, 11.7], EG inherits a natural structure of topological group from $|D_*G|$. $G = D_0G$, and (i) and (ii) hold by inspection of [17, 10.2 and 10.3]. Part (iii) is clear.

Steenrod's construction of a group structure on EG is rather different from ours, and I have not tried to compare definitions. The following theorem is an improvement due to McCord [22, § 4] of a result due to Milgram and Steenrod [23, 38].

Theorem 8.2. Let G be a topological group (with identity element a nondegenerate basepoint). Then $p: EG \to BG$ is a numerable principal G-bundle.

Proof. Write $E = EG$ and $E_n = F_nEG$. The representation of (E_j, E_{j-1}) as an NDR-pair defined inthe proof of Theorem 7.6 is G-equivariant. As observed by Steenrod [38, 4.2], [37, 7.1 and 9.4] imply that each (E, E_n) also admits a representation, (h_n, u_n) say, as a G-NDR pair. For $n \geq 0$, define a G-map $\rho_n: E \to I$ (where G acts trivially on I) by

$$\rho_0(x) = 1-u_0(x) \text{ and, if } n > 0, \quad \rho_n(x) = (1-u_n(x))u_{n-1}(h_n(x, 1)).$$

Let $r_i: E \to E$ be the G-map defined by $r_i(x) = h_i(x, 1)$. Then

$$E_0 \subset \rho_0^{-1}(0, 1] \subset r_0^{-1}E_0 \text{ and, if } n > 0, \ E_n - E_{n-1} \subset \rho_n^{-1}(0, 1] \subset r_n^{-1}(E_n - E_{n-1}).$$

Define further G-maps $\pi_n: E \to I$, $n \geq 0$, by

$$\pi_n(x) = \max\left(0, \rho_n(x) - n\sum_{i=0}^{n-1} \rho_i(x)\right)$$

and define $W_n = \pi_n^{-1}(0, 1]$ and $V_n = pW_n \subset BG$. Then $\{V_n\}$ is a numerable open cover of BG. We have a G-homeomorphism

$$E_0 \cong F_0BG \times G \text{ or, if } n > 0, \quad E_n - E_{n-1} \cong (F_nBG - F_{n-1}BG) \times G,$$

and we define $\gamma_n: W_n \to G$ to be the composite of $r_0: W_0 \to E_0$ or, if $n > 0$, $r_n: W_n \to E_n - E_{n-1}$ and the second coordinate of this homeomorphism. Define $\xi_n: W_n \times G \to W_n$ by

$$\xi_n(y, g) = y\gamma_n(y)^{-1} g .$$

Then ξ_n induces a map $\zeta_n : V_n \times G \to W_n$, and ζ_n is a homeomorphism with inverse $p \times \gamma_n$ by direct calculation. Now $p : EG \to BG$ is a principal G-bundle by [36, 7.4], since the product structure on W_0 gives a local cross-section of G in EG, and the result is proven.

The theorem and Proposition 7.8 give the following result.

Corollary 8.3. For any right G-space Y, $p : B(Y, G, G) \to B(Y, G, *)$ is a principal G-bundle classified by $q : B(Y, G, *) \to BG$.

We can see the following complement in two ways.

Corollary 8.4. For any right G-space Y and left G-space F on which G acts effectively, $p : B(Y, G, F) \to B(Y, G, *)$ is the G-bundle with fibre F associated to $p : B(Y, G, G) \to B(Y, G, *)$.

Proof. On the one hand, it is evident that

$$B(Y, G, F) = B(Y, G, G) \times_G F .$$

On the other hand, if $H = G$ and $PB(Y, G, F)$ is retopologized as the classical associated principal bundle in Proposition 7.10, then the map α displayed there is an equivalence of principal G-bundles.

In a sense, every bundle arises in this fashion.

Proposition 8.5. Let G act principally from the right on Y and effectively from the left on F. Then the following diagram is a pullback in which the maps ε are weak homotopy equivalences (and $\delta : F \to *$ is the trivial map):

$$\begin{array}{ccc} B(Y,G,F) & \xrightarrow{\varepsilon} & Y \times_G F \\ \downarrow p & & \downarrow 1 \times_G \delta \\ B(Y,G,*) & \xrightarrow{\varepsilon} & Y \times_G * \end{array}$$

Proof. The bottom map ε is a weak homotopy equivalence by the case $F = G$ of the diagram and Proposition 7.5, and the diagram implies that the top map ε is also a weak homotopy equivalence. By [17, 9.2, 11.8, and 11.6], it remains to verify that the following diagram is a pullback for $j \geq 0$:

$$\begin{array}{ccc} Y \times G^j \times F & \xrightarrow{\varepsilon} & Y \times_G F \\ \downarrow p & & \downarrow 1 \times_G \delta \\ Y \times G^j & \xrightarrow{\varepsilon} & Y \times_G * \end{array}$$

Write $\{y, x\}$ for the image of $(y, x) \in Y \times F$ in $Y \times_G F$. The map from $Y \times G^j \times F$ into the fibred product of the bottom map ε and $1 \times_G \delta$ specified by

$$(y, g_1, \ldots, g_j, x) \to ((y, g_1, \ldots, g_j), \{y, g_1 \cdots g_j x\})$$

is a homeomorphism with inverse specified by

$$((y, g_1, \dots, g_j), \{y', x'\}) \to (y, g_1, \dots, g_j, g_j^{-1} \dots g_1^{-1} gx'),$$

where g is the unique element of G such that $y' = yg$.

When $Y = G'$, where G is a closed subgroup with a local cross-section in G', $G' \times_G *$ is the homogeneous space of right cosets of G in G'. With Stasheff [34], we define generalized homogeneous spaces as follows.

Definition 8.6. Let $f: H \to G$ be any map of topological monoids. Define

$$G/H = B(G, H, *) \quad \text{and} \quad H\backslash G = B(*, H, G),$$

where H acts on G (from the left and right) through f.

We shall compare G/H to the fibre of Bf, but we must first insert the standard comparison of G to ΩBG (where BG has the basepoint $* = |[\ ], (1)| = F_0 BG$). Write $\mathcal{J}(I, X)$ for the path space of a based space X and write $p: \mathcal{J}(I, X) \to X$ for the endpoint projection. Write χ for the standard inverse map $\Omega X \to \Omega X$. For a based map $k: Y \to X$, write Fk for the homotopy theoretic fibre of k,

$$Fk = \{(\beta, y) \mid \beta \in \mathcal{J}(I, X),\ y \in Y,\ \beta(1) = k(y)\},$$

and write $\iota: \Omega X \to Fk$ and $\pi: Fk \to Y$ for the natural inclusion and projection, $\iota(\beta) = (\beta, *)$ and $\pi(\beta, y) = y$. With these notations, the proofs of the following two propositions are straightforward verifications from the definition, [17, 11.1], of geometric realization and the form of the face and degeneracy operators on the relevant simplicial spaces.

Proposition 8.7. For a topological monoid G, define $\tilde{\zeta}: EG \to \mathcal{J}(I, BG)$ by

$$\tilde{\zeta}|[g_1, \dots, g_j]g_{j+1}, a|(t) = |[g_1, \dots, g_{j+1}], (ta, 1-t)|$$

for $g_i \in G$, $a \in \Delta_j$, and $t \in I$. Define $\zeta: G \to \Omega BG$ by

$$\zeta(g)(t) = |[g], (t, 1-t)|.$$

Then the following diagram is commutative, hence ζ is a weak homotopy equivalence if G is grouplike:

$$\begin{array}{ccccc} G & \xrightarrow{\tau} & EG & \xrightarrow{p} & BG \\ \downarrow{\scriptstyle\zeta} & & \downarrow{\scriptstyle\tilde{\zeta}} & & \| \\ \Omega BG & \xrightarrow{\subset} & \mathcal{J}(I, BG) & \xrightarrow{p} & BG \end{array}$$

The behavior of ζ when G is not grouplike will be studied in section 15.

Proposition 8.8. Let $f: H \to G$ be a map of topological monoids. Define $\psi: G/H \to FBf$ by $\psi(x) = (\beta(x), q(x))$, where

$$\beta|g[h_1, \dots, h_j], a|(t) = |[g, f(h_1), \dots, f(h_j)], (1-t, ta)|$$

for $g \in G$, $h_i \in H$, $a \in \Delta_j$, and $t \in I$. Then the following diagram is commutative, hence ψ is a weak homotopy equivalence if H and G are both grouplike:

$$\begin{array}{ccccccccc} H & \xrightarrow{f} & G & \xrightarrow{\tau} & G/H & \xrightarrow{q} & BH & \xrightarrow{Bf} & BG \\ \downarrow{\scriptstyle\zeta} & & \downarrow{\scriptstyle\zeta} & & \downarrow{\scriptstyle\psi} & & \| & & \| \\ \Omega BH & \xrightarrow{\Omega Bf} & \Omega BG & \xrightarrow{\iota\chi} & FBf & \xrightarrow{\pi} & BH & \xrightarrow{Bf} & BG \end{array}$$

By symmetry, an analogous result is valid for $H\backslash G$, and it follows that G/H and $H\backslash G$ are weakly homotopy equivalent when H and G are grouplike. We shall use the following observations in section 10.

<u>Remarks 8.9.</u> Let $f: H \to G$ be a morphism of monoids. Then the following two diagrams are commutative:

$$\begin{array}{ccccc}
G & \xrightarrow{\tau} & B(H\backslash G, G, G) & \xrightarrow{p} & B(H\backslash G, G, *) \\
\| & & \downarrow{\scriptstyle \varepsilon} & & \downarrow{\scriptstyle \varepsilon(p)} \\
G & \xrightarrow{\tau} & H\backslash G & \xrightarrow{p} & BH
\end{array}$$

and

$$\begin{array}{ccccc}
BH & \xleftarrow{\varepsilon(p)} & B(H\backslash G, G, *) & \xrightarrow{q} & BG \\
{\scriptstyle Bf}\downarrow & & \downarrow{\scriptstyle Bf} & & \| \\
BG & \xleftarrow{\varepsilon(p)} & B(G\backslash G, G, *) & \xrightarrow{q} & BG
\end{array}$$

where, in the middle, Bf is short for $B(B(1, f, 1), 1, 1)$. Let H and G be grouplike. Then, by Proposition 7.5 and Theorem 7.6, the first diagram shows that $\varepsilon(p): B(H\backslash G, G, *) \to BH$ is a weak homotopy equivalence. In the second diagram, $G\backslash G = EG$ and the bottom maps $\varepsilon(p)$ and q are weak homotopy equivalences by Corollary 7.7. The last step of the proof of Theorem 9.2 below will give that, for $A \in \mathcal{W}$, the automorphism $q_* \varepsilon(p)_*^{-1}$ of $[A, BG]$ is the identity. We conclude that, from the point of view of representable (or rather, represented) functors on $h\mathcal{W}$, the maps $q: B(H\backslash G, G, *) \to BG$ and $Bf: BH \to BG$ can be used interchangeably.

The following pair of remarks summarize properties of BG and EG when G is an Abelian topological monoid and relate the functors B and E to the infinite symmetric product . These results are due to Milgram [23].

Remarks 8.10. If G is Abelian, its product is a morphism of monoids and therefore EG and BG are Abelian topological monoids by Proposition 7.4 and naturality. If, in addition, G is a topological group, then its inverse map is also a morphism of monoids and EG and BG are topological groups by naturality. The group structure so defined on EG coincides with that obtained in Theorem 8.1 since a trivial verification shows that this is true on the level of simplicial spaces. Of course, BG is the quotient group EG/G when G is an Abelian group.

Remarks 8.11. Let NX denote the infinite symmetric product of a space $X \in \mathcal{T}$ and let $\eta : X \to NX$ denote the natural inclusion [9 or 17, §3]. Then there is a commutative diagram

$$\begin{array}{ccccc} X & \xrightarrow{\iota} & CX & \xrightarrow{\pi} & \Sigma X, \\ \downarrow{\scriptstyle\eta} & & \downarrow{\scriptstyle\tilde{\eta}} & & \downarrow{\scriptstyle\bar{\eta}} \\ NX & \xrightarrow{\tau} & ENX & \xrightarrow{p} & BNX \end{array}$$

where CX and ΣX are the (reduced) cone and suspension on X, ι and π are the natural inclusion and quotient map, $\bar{\eta}$ is determined by commutativity of the diagram, and

$$\tilde{\eta}(x, t) = |[\eta(x)]e, (t, 1-t)|$$

for $x \in X$ and $t \in I$; here the left square commutes since

$$\tau\eta(x) = |[\ \ ]\eta(x), (1)| = |[\eta(x)]e, (1,0)| = \tilde{\eta}\iota(x).$$

Since NX is the free Abelian topological monoid generated by X, there result maps $\phi(\tilde{\eta})$ and $\phi(\bar{\eta})$ of topological monoids such that the following diagram is commutative:

$$\begin{array}{ccccc} NX & \xrightarrow{N\iota} & NCX & \xrightarrow{N\pi} & N\Sigma X \\ \| & & \downarrow \phi(\tilde{\eta}) & & \downarrow \phi(\bar{\eta}) \\ NX & \xrightarrow{\tau} & ENX & \xrightarrow{p} & BNX \end{array}$$

As noted by Milgram [23 , p. 245], $\phi(\tilde{\eta})$ and $\phi(\bar{\eta})$ are in fact homeomorphisms .

9. The classification theorems

It is now an easy matter to use the bar construction to prove a general classification theorem for fibrations, and another for bundles. We shall only classify over base spaces in $\mathcal{W}$; greater generality would be useless for purposes of calculation. Nevertheless, for some important examples, we cannot insist that all spaces in sight be in $\mathcal{W}$; in such cases, we shall rely on the following consequences of the Whitehead theorem.

Remarks 9.1. Let $f: B \to A$ be a weak homotopy equivalence, where $A \in \mathcal{W}$. Since $f_*: [A, B] \to [A, A]$ is an isomorphism, there exists one and, up to homotopy, only one map $g: A \to B$ such that $fg \simeq 1$. Moreover, g is natural in the sense that, given a homotopy commutative diagram

$$\begin{array}{ccc} B & \xrightarrow{f} & A \\ {\scriptstyle k}\downarrow & & \downarrow{\scriptstyle j} \\ B' & \xrightarrow{f'} & A' \end{array}$$

in which $A, A' \in \mathcal{W}$ and f and f' are weak homotopy equivalences, the following diagram is also homotopy commutative:

$$\begin{array}{ccc} A & \xrightarrow{g} & B \\ {\scriptstyle j}\downarrow & & \downarrow{\scriptstyle k} \\ A' & \xrightarrow{g'} & B' \end{array}$$

(since $f'g'j \simeq j \simeq jfg \simeq f'kg$ and f'_* is an isomorphism).

To avoid cluttering up the statement and proof of the following theorem with minor technicalities, we tacitly assume that the identity elements of all monoids are (strongly) nondegenerate basepoints and that all cross-sections are fibrewise cofibrations. These assumptions will be discussed in Remarks 9.3 and 9.7 below.

Theorem 9.2. Assume one of the following hypotheses.

(a) $(\mathcal{F}, F)$ is a category of fibres which is either

(i) Γ-complete in $\mathcal{U}$ or

(ii) Γ-complete in $\mathcal{W}$.

(b) $(\mathcal{F}, F)$ is a category of based fibres which is either

(i) Γ'-complete in $\mathcal{U}$ or

(ii) Γ'-complete in $\mathcal{W}$.

(c) $(\mathcal{F}, F)$ is a category of bundle fibres.

Let $(\mathcal{G}, G)$ be the associated principal category of fibres of $(\mathcal{F}, F)$. Then, for $A \in \mathcal{W}$, the set $\mathcal{E}\mathcal{F}(A)$ of equivalence classes of $\mathcal{F}$-fibrations over A is naturally isomorphic to $[A, BG]$.

Proof. In the cases (ii), Theorems 6.4 and 6.7 and Proposition 7.2 will ensure that all spaces in sight are in $\mathcal{W}$. By abuse, let us agree to write (Γ, η) for (Γ', η') in case (b) and to write (Γ, η) for the identity functor and identity natural transformation in case (c). With this uniform notation, let

$$\pi = \Gamma p : \Gamma B(*, G, F) \rightarrow BG$$

in all cases. By Definitions 5.1 and 5.4 and Theorem 7.6 in cases (a) and (b) and by Example 6.11 and Theorem 8.2 in case (c), π is an $\mathcal{F}$-fibration. Define

$$\Psi : [A, BG] \rightarrow \mathcal{E}\mathcal{F}(A)$$

by $\Psi[f] = \{f^*\pi\}$. Ψ is well-defined and natural by Proposition 2.5. In the other direction, define

$$\Phi : \mathcal{E}\mathcal{F}(A) \rightarrow [A, BG]$$

as follows. Given an $\mathcal{F}$-fibration $\nu : D \rightarrow A$, consider the following commutative diagram, where $\gamma : PD \times G \rightarrow PD$ is given by composition:

$$\begin{array}{ccccc}
PD & \xleftarrow{\varepsilon(\gamma)} & B(PD, G, G) & \xrightarrow{q} & EG \\
\Big\downarrow{\scriptstyle P\nu} & & \Big\downarrow{\scriptstyle p} & & \Big\downarrow{\scriptstyle p} \\
A & \underset{g}{\overset{\varepsilon(P\nu)}{\rightleftarrows}} & B(PD, G, *) & \xrightarrow{q} & BG
\end{array}$$

$\varepsilon(\gamma)$ is a homotopy equivalence by Proposition 7.5 and it restricts to a weak homotopy equivalence on each fibre since $(\varepsilon(\gamma), \varepsilon(P\nu))$ is a $\mathcal{G}$-map. Since p and $P\nu$ are quasifibrations, $\varepsilon(P\nu)$ is a weak homotopy equivalence by the five lemma. Let g be a right homotopy inverse to $\varepsilon(P\nu)$ and define $\Phi\{\nu\} = [qg]$. Φ is well-defined and natural by the evident naturality of the diagram above, before insertion of g, and by Remarks 9.1. $\Psi\Phi$ is the identity transformation on $\mathcal{E}\mathcal{F}(A)$. Indeed, with the notations above, the following diagram displays a chain of $\mathcal{F}$-maps over A which connects ν to

$f^*\pi$, where $f = qg$, and thus displays an equivalence (in the sense of Definition 6.1) between these $\mathcal{J}$-fibrations over A:

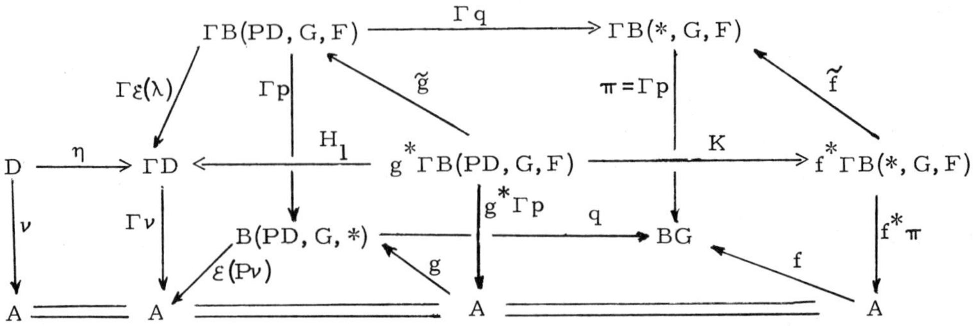

Here $\lambda: PD \times F \to D$ is the evaluation map, H_1 is obtained by application of the $\mathcal{J}$-CHP to the $\mathcal{J}$-map $(\Gamma\epsilon(\lambda)\circ\tilde{g}, \epsilon(P\nu)\circ g)$ and any homotopy $h: A \times I \to A$ from $\epsilon(P\nu)\circ g$ to the identity, and K is given by the universal property of $f^*\pi$. Finally, to analyze $\Phi\Psi$, assume given $f: A \to BG$ and consider the following diagram:

$$\begin{array}{ccccc}
A & \xleftarrow{\epsilon(Pf^*\pi)} & B(Pf^*\Gamma B(*,G,F),G,*) & \xrightarrow{q} & BG \\
{\scriptstyle f}\downarrow & \xrightarrow{g} & \downarrow{\scriptstyle B(P\tilde{f},1,1)} & & \| \\
BG & \xleftarrow{\epsilon(P\pi)} & B(P\Gamma B(*,G,F),G,*) & \xrightarrow{q} & BG
\end{array}$$

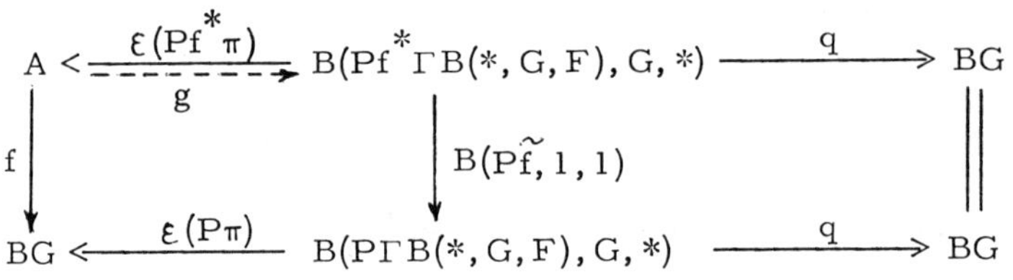

The argument used to define Φ demonstrates that $\epsilon(P\pi)$ is a weak homotopy equivalence (although it need not have a right inverse since BG need not be in $\mathcal{W}$). By Lemma 5.5 and Propositions 7.5 and 7.10, $P\Gamma B(*,G,F)$ is of the same weak homotopy type as EG and is thus aspherical. Since the bottom map q is a quasifibration with aspherical fibre, it is a weak homotopy equivalence. By Theorem 6.5 and the diagram, $\Phi\Psi$ is an automorphism of $[A, BG]$. But then Ψ is a bijection and $\Psi = (\Psi\Phi)\Psi = \Psi(\Phi\Psi)$, hence $\Phi\Psi$ is the identity transformation.

Remarks 9.3. In case (c), the requirement that (G, e) be an NDR-pair appears to be an essential hypothesis. In cases (a) and (b), such an hypothesis can be eliminated as follows. Let G' be the monoid obtained from G by growing a whisker from e [17,A.8]. Via the retraction $G' \to G$ (which is a homotopy equivalence) any left or right G-space is also a G'-space. Replace each $B(Y, G, X)$ in the statement and proof of the theorem by $B(Y, G', X)$. Then, with trivial modifications, the argument goes through to give $\mathcal{E}\mathcal{H}(A) \cong [A, BG']$. Note in particular that $B(Y, G', G')$ is homotopy equivalent to $B(Y, G', G)$ by Proposition 7.3 and that $B(Y, G', G)$ is homeomorphic to $PB(Y, G', F)$ by Proposition 7.10.

In all of the following corollaries, we agree to read BG' for BG if the basepoint of G happens to be degenerate.

In view of Example 6.2 and Lemma 6.3, we have the following classification theorem for principal fibrations.

Corollary 9.4. Let G be a grouplike topological monoid and let $A \in \mathcal{W}$.

(i) $\mathcal{E}G\mathcal{U}(A)$ is naturally isomorphic to $[A, BG]$.

(ii) If $G \in \mathcal{W}$, $\mathcal{E}G\mathcal{W}(A)$ is naturally isomorphic to $[A, BG]$, hence the natural map $\iota : \mathcal{E}G\mathcal{W}(A) \to \mathcal{E}G\mathcal{U}(A)$ is a bijection.

Similarly, in view of Example 6.6, Theorem 6.7, and Lemma 6.8, we have the following generalization of Stasheff's classification theorem [32] for fibrations with fibres of the homotopy type of a given finite CW-complex.

Corollary 9.5. Let $F \in \mathcal{W}$, let HF denote the topological monoid of homotopy equivalences of F, and let $A \in \mathcal{W}$.

(i) $\mathcal{E}F\mathcal{U}(A)$ is naturally isomorphic to $[A, BHF]$.

(ii) If F is compact, $\mathcal{E}F\mathcal{W}(A)$ is naturally isomorphic to $[A, BHF]$, hence the natural map $\iota: \mathcal{E}F\mathcal{W}(A) \to \mathcal{E}F\mathcal{U}(A)$ is a bijection.

Of course, in (ii), the equivalence relation used to define $\mathcal{E}F\mathcal{W}(A)$ coincides with fibre homotopy equivalence.

The compatibility of the previous two corollaries is immediate from our construction of classifying maps. Thus we have the following result.

Corollary 9.6. For $F \in \mathcal{W}$ and $A \in \mathcal{W}$, $P: \mathcal{E}F\mathcal{U}(A) \to \mathcal{E}HF\mathcal{U}(A)$ and, if F is compact, $P: \mathcal{E}F\mathcal{W}(A) \to \mathcal{E}HF\mathcal{W}(A)$ are bijections of sets.

Remarks 9.7. In case (b), we assumed in the proof of Theorem 9.2 that the cross-section of $p: B(Y,G,F) \to B(Y,G,*)$ is a cofibration for certain Y. Let F' be the G-space obtained from F by growing a whisker from the given basepoint and letting G act trivially on the whisker. The basepoint $1 \in F'$ is the endpoint of the whisker, and $(F', 1)$ is a G-equivariant NDR-pair. The cross-section of $p: B(Y,G,F') \to B(Y,G,*)$ is thus a fibrewise cofibration. Provided that $F' \in \mathcal{F}$ and the retraction $F' \to F$ is a map in $\mathcal{F}$, the proof of Theorem 9.2 goes through, with trivial modifications, with F replaced by F'.

In view of Example 6.9, Theorem 6.7, and Lemma 6.10, we have the following new variant of Stasheff's theorem. In [19], this result will play a key role in the study of E-oriented spherical fibrations for a commutative ring spectrum E.

<u>Corollary 9.8.</u> Let $F \in \mathcal{V}$, let JF denote the topological monoid of based homotopy equivalences of F, and let $A \in \mathcal{W}$.

(i) $\mathcal{E}F\mathcal{T}(A)$ is naturally isomorphic to [A, BJF].

(ii) If F is compact, $\mathcal{E}F\mathcal{V}(A)$ is naturally isomorphic to [A, BJF], hence the natural map $\iota : \mathcal{E}F\mathcal{V}(A) \to \mathcal{E}F\mathcal{T}(A)$ is a bijection.

In (ii), the equivalence relation used to define $\mathcal{E}\mathcal{J}\mathcal{V}(A)$ coincides with section preserving fibre homotopy equivalence, where homotopies are required to be section preserving for each parameter value $t \in I$.

<u>Corollary 9.9.</u> For $F \in \mathcal{V}$ and $A \in \mathcal{W}$, $P: \mathcal{E}F\mathcal{J}(A) \to \mathcal{E}JF\mathcal{U}(A)$ and, if F is compact, $P: \mathcal{E}F\mathcal{V}(A) \to \mathcal{E}JF\mathcal{W}(A)$ are bijections of sets.

<u>Theorem 9.10.</u> Let a topological group G act effectively from the left on a space F and let $\mathcal{B}\mathcal{J}(A)$ denote the set of equivalence classes of numerable G-bundles with fibre F over A. Assume that (G, e) is an NDR-pair. Then, for $A \in \mathcal{W}$, $\mathcal{B}\mathcal{J}(A)$ is naturally isomorphic to [A, BG].

Proof. Numerable Steenrod fibre bundles are known to satisfy the bundle-CHP (which is formulated precisely as was the $\mathcal{F}$-CHP but with $\mathcal{F}$-maps replaced by bundle maps) and the obvious analog of Lemma 2.4 [36, §11 and 7]. With P the classical associated principal bundle functor, the proof is formally identical to that of case (c) of Theorem 9.2.

Corollary 9.11. Let a topological group G act effectively from the left on a space F and let $(\mathcal{F}, F)$ denote the corresponding category of bundle fibres. For spaces $A \in \mathcal{W}$, let $\rho : \mathcal{B}\mathcal{F}(A) \to \mathcal{E}\mathcal{F}(A)$ denote the natural transformation obtained by regarding a G-bundle with fibre F as an $\mathcal{F}$-fibration. Then ρ is a bijection of sets provided that the identity map from G, with its given topology, to $\mathcal{F}(F, F)$, with the compact-open topology, is a weak homotopy equivalence and (G, e) is an NDR-pair in both topologies.

Proof. By our construction of classifying maps, the following diagram of natural transformations is commutative:

$$\begin{array}{ccc} \mathcal{B}\mathcal{F}(A) & \xrightarrow{\ \rho\ } & \mathcal{E}\mathcal{F}(A) \\ \Phi \downarrow & & \downarrow \Phi \\ [A, BG] & \xrightarrow{\ B(1)_*\ } & [A, B\mathcal{F}(F, F)] \end{array}$$

The conclusion follows, since $B(1)$ is a weak homotopy equivalence if 1 is (by a comparison of quasifibrations).

We thus have a precise comparison between bundle theory and fibration theory.

10. The definition and examples of Y-structures

Until otherwise specified, let $(\mathcal{F}, F)$ be a category of fibres which satisfies one of the hypotheses of Theorem 9.2, let $(\mathcal{G}, G)$ be its associated principal category of fibres, and let Y be any right G-space. Consider $q: B(Y, G, *) \to BG$. $B(Y, G, *)$ can be thought of as the classifying space for $\mathcal{F}$-fibrations together with a "Y-structure". In many important special cases, Y-structures can be described intrinsically, without reference to the classification theorem, and can then be proven to be classified by $B(Y, G, *)$. We give a general intrinsic definition and several examples in this section and prove such a classification theorem in the next. The motivating example of E-oriented spherical fibrations will be treated in [19]; it will in fact be a special case of Example 10.6 below.

Definition 10.1. Assume given an auxiliary space Z and an inclusion of Y in the function space $\mathcal{U}(F, Z)$ such that the right action of G on Y is induced by restriction from the action of $G = \mathcal{F}(F, F)$ on $\mathcal{U}(F, Z)$ given by composition. Define a Y-structure θ on an $\mathcal{F}$-space $\nu: D \to A$ to be a map $\theta: D \to Z$ such that the composite $\theta \circ \psi: F \to Z$ is an element of Y for every element $\psi: F \to D$ of PD. Define an $\mathcal{F}$-map $(\nu, \theta) \to (\nu', \theta')$ of $\mathcal{F}$-spaces with Y-structure to be an $\mathcal{F}$-map $(g, f): \nu \to \nu'$ such that $\theta' g$ is homotopic to θ via a homotopy $h: D \times I \to Z$ such that $h_t \psi: F \to Z$ is an element of Y for every $\psi \in PD$ and $t \in I$ (that is, via a homotopy through Y-structures). Define $\mathcal{E}\mathcal{F}(A; Y)$ to be the set of equivalence classes of

$\mathcal{F}$-fibrations with Y-structure under the equivalence relation generated by the $\mathcal{F}$-maps over A.

Our notions of $\mathcal{F}$-maps and of equivalence suggest that a Y-structure on a given $\mathcal{F}$-space should be reinterpreted as a homotopy class of Y-structures, and we adopt this terminology henceforward. Although the definition may seem artificial, at first sight, we shall see that it does satisfactorily account for the most important types of additional structure on $\mathcal{F}$-fibrations.

When $(\mathcal{F}, F)$ satisfies hypothesis (a) or (b) of Theorem 9.2, we shall need further conditions on Y and Z in order to ensure that $\mathcal{F}$-quasifibrations with Y-structure in $\mathcal{Q}$ ($\mathcal{Q} = \mathcal{U}$ or $\mathcal{Q} = \mathcal{W}$) can be replaced functorially by $\mathcal{F}$-fibrations with Y-structure in $\mathcal{Q}$. As in the proof of Theorem 9.2, we agree to write (Γ, η) for $(\Gamma, \eta), (\Gamma', \eta')$, or the identity functor and identity natural transformation according to whether $(\mathcal{F}, F)$ satisfies hypothesis (a), (b), or (c) of that theorem. The following definition should be compared with Definitions 5.1 and 5.4.

<u>Definition 10.2.</u> Let $(\mathcal{F}, F)$ be Γ-complete in $\mathcal{Q}$ [that is, Γ or Γ' complete], and let Y be a sub right G-space of $\mathcal{U}(F, Z)$. The pair (Y, Z) will be said to be admissible if $Y \in \mathcal{Q}$ and the following statements are valid for $\mathcal{F}$-quasifibrations $\pi: E \to B$ in $\mathcal{Q}$ with (homotopy class of) Y-structure $\theta: E \to Z$.

(1) $\Gamma\pi: \Gamma E \to \Gamma B$ admits a Y-structure $\Gamma\theta: \Gamma E \to Z$.

(2) $\eta: E \to \Gamma E$ defines an $\mathcal{F}$-map $(\pi, \theta) \to (\Gamma\pi, \Gamma\theta)$ over B.

(3) Γ takes $\mathcal{F}$-maps $(\pi, \theta) \to (\pi', \theta')$ to $\mathcal{F}$-maps $(\Gamma\pi, \Gamma\theta) \to (\Gamma\pi', \Gamma\theta')$.

If $(\mathcal{F}, F)$ is a category of bundle fibres, any pair (Y, Z) such that Y is a sub right G-space of $\mathcal{U}(F, Z)$ will be said to be admissible.

The following two examples give generalized versions of familiar types of Y-structures.

Example 10.3. Let $(\mathcal{F}', F)$ be a second category of fibres, with associated principal category $(\mathcal{G}', G')$, and let $j: \mathcal{F} \to \mathcal{F}'$ be a functor over $\mathcal{U}$ (or over $\mathcal{J}$ if $\mathcal{F}$ and $\mathcal{F}'$ are based). Then j defines a morphism of monoids $G \to G' = \mathcal{F}'(F, F)$. In Definition 10.1, set $Y = G'$ and $Z = F$. Then a G'-structure $\theta: D \to F$ on an $\mathcal{F}$-fibration $\nu: D \to A$ is just the second coordinate of an $\mathcal{F}'$-map $D \to A \times F$ over A (at least if $\alpha\beta \in \mathcal{F}'$ and $\beta \in \mathcal{F}'$ implies $\alpha \in \mathcal{F}'$). In other words, a G'-structure is precisely an $\mathcal{F}'$-trivialization of the $\mathcal{F}$-fibration ν. Of course, $B(G', G, *)$ is the generalized homogeneous space G'/G of Definition 8.6. In the interesting applications, $(\mathcal{F}, F)$ will be a category of bundle fibres, hence the question of admissibility will not arise.

Example 10.4. Let $f: H \to G$ be any morphism of monoids and set $Y = H\backslash G = B(*, H, G)$ and $Z = \Gamma B(*, H, F)$. Y is homeomorphic to $PB(*, H, F)$, by Proposition 7.10, and the inclusion of Y in $\mathcal{U}(F, Z)$ is the composite of this homeomorphism and the inclusion $P\eta$ of $PB(*, H, F)$ in PZ. Let $\nu: D \to A$ be an $\mathcal{F}$-space and let $\theta: D \to Z$ be a Y-structure. Because $\theta\psi$ is in Y for ψ in PD, θ is fibrewise with respect to ν and $\Gamma p: \Gamma B(*, H, F) \to BH$ (at least if every point of D is in the image of some ψ, as always holds in practice). We agree to strengthen the notion of an $H\backslash G$-structure by insisting that the induced function $A \to BH$ be continuous. Then $\Gamma\theta: \Gamma D \to \Gamma Z$ is defined, by Remarks 3.7, and the composite

$$\Gamma D \xrightarrow{\Gamma\theta} \Gamma Z = \Gamma\Gamma B(*, H, F) \xrightarrow{\mu} \Gamma B(*, H, F) = Z$$

is an $H\backslash G$-structure for $\Gamma\nu : \Gamma D \to A$ such that $\mu \circ \Gamma\theta \circ \eta = \theta$. Thus the pair (Y, Z) will always be admissible when $(\mathcal{F}, F)$ satisfies hypothesis (a) or (b), provided only that $H \in \mathcal{W}$ if $\mathcal{Q} = \mathcal{W}$. We call an $H\backslash G$-structure $\theta: D \to Z$ on an $\mathcal{F}$-fibration $\nu : D \to A$ a reduction of the structural monoid of ν to H. If H is a topological group and ν admits such a reduction θ, then ν is equivalent to the $\mathcal{F}$-fibration induced from $p: B(*, H, F) \to BH$ by the map $A \to BH$ derived from θ; of course, this $\mathcal{F}$-fibration is an H-bundle with fibre F if H acts effectively on F. As explained in Remarks 8.9, $B(H\backslash G, G, *)$ is weakly homotopy equivalent to BH (when H is group-like) in such a way that the maps $q: B(H\backslash G, G, *) \to BG$ and $Bf: BH \to BG$ are equivalent.

We also have the following generic types of Y-structures, the second of which will be central to [19].

<u>Example 10.5.</u> Let $F \in \mathcal{W}$ be compact, let $Z \in \mathcal{W}$, and let Y be the union of any set of components of $\mathcal{U}(F, Z)$ which is invariant under composition with homotopy equivalences of F. Then (Y, Z) is an admissible pair for $F\mathcal{W}$. Indeed, let $\pi: E \to B$ be an $F\mathcal{W}$-quasifibration with B and E in $\mathcal{W}$ and with Y-structure $\theta: E \to Z$. Choose a homotopy inverse $\zeta : \Gamma E \to E$ to η and define $\Gamma\theta = \theta\zeta : \Gamma E \to Z$. For $\psi: F \to (\Gamma\pi)^{-1}(b)$ in $P\Gamma E$ (that is,

a homotopy equivalence), consider the following diagram, in which $\eta(b)$ denotes the restriction of η to $\pi^{-1}(b)$ and $\zeta(b)$ is a chosen homotopy inverse to $\eta(b)$ (which need not be the restriction of ζ since ζ need not be fibrewise):

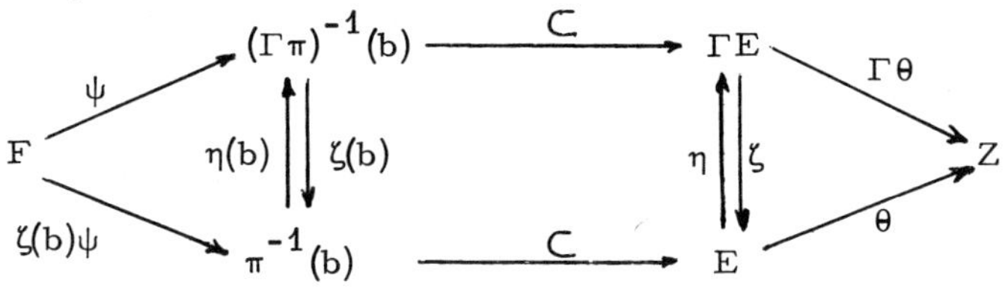

Here $\Gamma\theta\circ\psi \simeq \Gamma\theta\circ\eta(b)\circ\zeta(b)\circ\psi = \theta\circ\zeta\circ\eta\circ\zeta(b)\circ\psi \simeq \theta\circ\zeta(b)\circ\psi$ and, since $\zeta(b)\psi \in PD$, it follows that $\Gamma\theta\circ\psi$ is an element of Y. Clearly $\eta:(\pi,\theta)\to(\Gamma\pi,\Gamma\theta)$ is an $F\mathcal{W}$-map over B and Γ is functorial.

<u>Example 10.6.</u> Let $F\in\mathcal{V}$ be compact, let $Z\in\mathcal{V}$, and let Y be the union of any set of components of $\mathcal{J}(F, Z)$ which is invariant under composition with based homotopy equivalences of F. Then (Y, Z) is an admissible pair for $F\mathcal{V}$. Indeed, retaining the notations of the previous example (with (Γ,η) interpreted as (Γ',η')), we note that ζ can be chosen to be section preserving (although not fibre preserving) and that $\zeta\eta$ is then homotopic to the identity via a homotopy through section preserving maps (because the sections of π and $\Gamma\pi$ are cofibrations). These facts, and the nondegeneracy of the basepoints of the fibres of π and $\Gamma\pi$, allow the required use of based maps and homotopies in the verification that $\Gamma\theta\circ\psi$ is in Y for ψ in $P\Gamma E$. Observe that a Y-structure $\theta: D\to Z$ on an $F\mathcal{V}$-fibration $\nu: D\to A$ factors through the "Thom complex" $D/\sigma A$ since

$\theta\psi \in Y$ for $\psi \in PD$ implies that θ carries the basepoint of each fibre of ν to the basepoint of Z.

Of course, Definition 10.1 admits a bundle theoretic analog.

Definition 10.7. Let G be a topological group which acts effectively on a space F and let $\mathcal{F}$ be the derived category of bundle fibres (Example 6.11). Let Y be a right G-space and let $Y \to \mathcal{U}(F, Z)$ be a continuous one-to-one map under which the right action of G on Y agrees (as a function) with the right action of $\mathcal{F}(F,F)$ on $\mathcal{U}(F, Z)$ given by composition; the pair (Y, Z) is then said to be admissible. Define a Y-structure θ on a G-bundle $\nu: D \to A$ with fibre F to be a map $\theta: D \to Z$ such that the composite $\theta\psi: F \to Z$ is an element of Y for every element $\psi: F \to D$ of PD and such that the function $\tilde{\theta}: PF \to Y$ specified by $\tilde{\theta}(\psi) = \theta \circ \psi$ is continuous (where the associated principal bundle PD has its standard topology). Define a bundle map $(\nu, \theta) \to (\nu', \theta')$ of bundles with Y-structure to be a bundle map $(g, f): \nu \to \nu'$ such that $\theta' g$ is homotopic to θ by a homotopy through Y-structures. Define $\mathcal{B}\mathcal{F}(A; Y)$ to be the set of equivalence classes of G-bundles with fibre F and with Y-structure over A.

Example 10.3 applies directly to bundles, with $(\mathcal{F}, F)$ interpreted as the category of bundle fib res derived from F and G. There is also an obvious bundle theoretic analog of Example 10.3 in which G' is taken to be a group which contains G and also acts effectively on F.

In Example 10.4, interpreted bundle theoretically, if H is also a group and if $\nu : D \to A$ is a G-bundle with fibre F and $H\backslash G$-structure $\theta: D \to B(*, H, F)$, then θ determines an equivalence of G-bundles from D to the bundle $E \times_H F$, where E is the principal H-bundle induced from the universal bundle $EH \to BH$ by $A \to BH$. Thus our notion of a reduction of the group of a bundle agrees with the standard one. (Compare Lashof [12, §1], where the term lifting is used to emphasize that $H \to G$ is not assumed to be an inclusion.)

11. The classification of Y-structures

The following fundamental result should be regarded as an elaboration of Theorem 9.2 (to which it reduces when $Y = *$ and $Z = *$). We again tacitly assume that the identity elements of all monoids are strongly non-degenerate basepoints and that all cross-sections are fibrewise cofibrations. The discussions of these points in Remarks 9.3 and 9.7 apply verbatim to the present situation. We shall often abbreviate maps of the bar construction of the form $B(f, 1, 1)$ to Bf here.

Theorem 11.1. Let $(\mathcal{F}, F)$ be a category of fibres which satisfies one of the hypotheses (a), (b), or (c) of Theorem 9.2 and let $(\mathcal{G}, G)$ be its associated principal category of fibres. Let Y be a sub right G-space of $\mathcal{U}(F, Z)$ such that the pair (Y, Z) is admissible. Then, for $A \in \mathcal{W}$, the set $\mathcal{E}\mathcal{F}(A; Y)$ of equivalence classes of $\mathcal{F}$-fibrations with Y-structure over A is naturally isomorphic to $[A, B(Y, G, *)]$.

Proof. As usual, write (Γ, η) ambiguously for $(\Gamma, \eta), (\Gamma', \eta')$, and the identity functor and identity natural transformation in cases (a), (b), or (c). Let $\lambda : Y \times F \to Z$ be adjoint to the inclusion $Y \to \mathcal{U}(F, Z)$. Then $\varepsilon(\lambda): B(Y, G, F) \to Z$ is a Y-structure on $p: B(Y, G, F) \to B(Y, G, *)$. Define an $\mathcal{F}$-fibration with Y-structure (π, ω) by

$$\pi = \Gamma p : \Gamma B(Y, G, F) \to B(Y, G, *) \quad \text{and} \quad \omega = \Gamma \varepsilon(\lambda) : \Gamma B(Y, G, F) \to Z.$$

Now define $\Psi : [A, B(Y, G, *)] \to \mathcal{E}\mathcal{F}(A; Y)$ by $\Psi[f] = \{(f^*\pi, \omega\tilde{f})\}$. If $h: A \times I \to B(Y, G, *)$ is a homotopy from f to f', then the following composite is a homotopy through Y-structures from $\omega\tilde{f}$ to $\omega\tilde{f}'J_1$, where J is an $\mathcal{J}$-homotopy over A which starts at the identity map of $f^*\Gamma B(Y,G,F)$ and is obtained by application of the $\mathcal{J}$-CHP to the identity map of $A \times I$ (regarded as a homotopy):

$$f^*\Gamma B(Y,G,F) \times I \xrightarrow{J} h^*\Gamma B(Y,G,F) \xrightarrow{\tilde{h}} \Gamma B(Y,G,F) \xrightarrow{\omega} Z.$$

Therefore Ψ is well-defined. The same argument shows that $\mathcal{E}\mathcal{J}(A;Y)$ is in fact a functor of A, and Ψ is clearly natural. Define $\Phi: \mathcal{E}\mathcal{J}(A;Y) \to [A, B(Y,G,*)]$ as follows. Let $\nu: D \to A$ be an $\mathcal{J}$-fibration with Y-structure $\theta: D \to Z$ and let $\tilde{\theta}: PD \to Y$ be the map of right G-spaces specified by $\tilde{\theta}(\psi) = \theta \circ \psi$. Consider the maps

$$A \underset{g}{\overset{\varepsilon(P\nu)}{\rightleftarrows}} B(PD,G,*) \xrightarrow{B\tilde{\theta}} B(Y,G,*),$$

choose a right homotopy inverse g to $\varepsilon(P\nu)$ (as in the proof of Theorem 9.2), and define $\Phi\{(\nu,\theta)\} = [B\tilde{\theta} \circ g]$. Given (ν',θ'), an $\mathcal{J}$-map $k: D \to D'$ over A, and a homotopy $h: D \times I \to Z$ through Y-structures from θ to $\theta' k$, define $Ph: PD \times I \to Y$ by $(Ph)_t(\psi) = h_t \circ \psi$. For $\phi \in G$, $(Ph)_t(\psi \circ \phi) = (Ph)_t(\psi) \circ \phi$, hence Ph is a G-equivariant homotopy. It therefore induces a homotopy from $B\tilde{\theta}$ to $B\tilde{\theta} \circ BPk$. Thus Φ is well-defined, and a similar argument shows that Φ is natural. $\Psi\Phi$ is the identity transformation on $\mathcal{E}\mathcal{J}(A;Y)$. Indeed, with (ν,θ) as in the definition of Φ, set $f = B\tilde{\theta} \circ g$ and replace q by $B\tilde{\theta}$, BG by $B(Y,G,*)$, and $B(*,G,F)$ by $B(Y,G,F)$ in the diagram used for the corresponding step of the proof of Theorem 9.2. Then the resulting diagram displays an equivalence between ν and $f^*\pi$, and it is immediate from an argument like that used to prove that Ψ is well-defined and, in cases (a) and (b), from the functoriality of Γ such that η is an $\mathcal{F}$-map given by Definition 10.2 that the constructed equivalence is one of $\mathcal{F}$-fibrations with Y-structure. For the verification that $\Phi\Psi$ is an automorphism and therefore also the identity, construction of a diagram just like that used for the corresponding step of Theorem 9.2 shows that we need only check that

$$B\tilde{\omega}: B(P\Gamma B(Y,G,F), G, *) \to B(Y,G,*)$$

is a weak homotopy equivalence. By comparison of the quasifibrations q from the displayed spaces to BG, it suffices to check that $\tilde{\omega}: P\Gamma B(Y,G,F) \to Y$ is a weak homotopy equivalence, and this follows from Lemma 5.5, Propositions 7.5 and 7.10, and the fact that $\omega \circ \eta \simeq \varepsilon(\lambda)$, so that

$$\tilde{\omega} \circ P\eta \simeq \varepsilon : PB(Y,G,F) \cong B(Y,G,G) \to Y.$$

Under the hypotheses of the theorem, consider the quasifibration sequence

$$G \xrightarrow{\iota} Y \xrightarrow{\tau} B(Y,G,*) \xrightarrow{q} BG$$

obtained in Proposition 7.9. The following remarks interpret the corresponding sequence of represented functors on $\mathcal{N}$.

<u>Remarks 11.2.</u> (i) $q_*: [A, B(Y,G,*)] \to BG$ represents the forgetful transformation $\mathcal{EF}(A; Y) \to \mathcal{EF}(A)$ obtained by sending $\{(\nu, \theta)\}$ to $\{\nu\}$ since there is an $\mathcal{F}$-map j over $B(Y,G,*)$ such that the following diagram com-

mutes and since $\Phi\{\nu\} = [q]\Phi\{(\nu, \theta)\}$ by the proofs of Theorems 9.2 and 11.1:

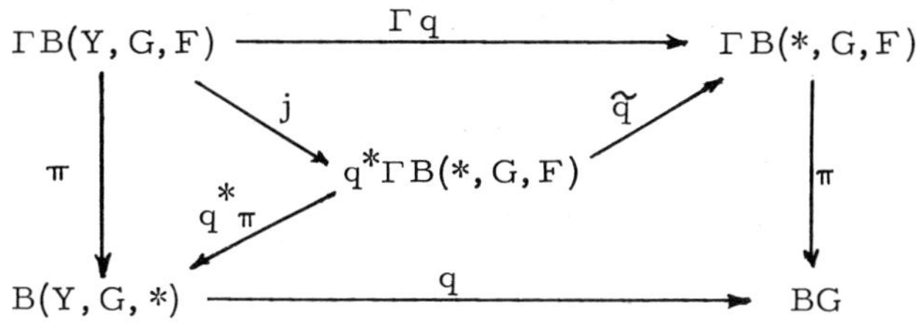

(ii) $[A, Y]$ is naturally isomorphic to the set of (homotopy classes of) Y-structures on the trivial $\mathcal{F}$-fibration $\varepsilon : A \times F \to A$. Indeed, given $f: A \to Y$, its adjoint $A \times F \to Z$ gives the corresponding Y-structure.

(iii) $\tau_*: [A, Y] \to [A, B(Y, G, *)]$ represents the transformation which sends a Y-structure θ on ε to the equivalence class $\{(\varepsilon, \theta)\}$, by inspection of the proof of Theorem 11.1.

(iv) $[A, G]$ is naturally isomorphic to the set of $\mathcal{F}$-homotopy classes of $\mathcal{F}$-maps over A from ε to itself. Indeed, given $f: A \to G$, its adjoint $A \times F \to F$ gives the second coordinate of the corresponding $\mathcal{F}$-map over A.

(v) $\iota_*: [A, G] \to [A, Y]$ represents the transformation which sends an $\mathcal{F}$-map $g: A \times F \to A \times F$ to the Y-structure $\theta_0 \circ g$, where $\theta_0 : A \times F \to Z$ is the Y-structure on ε with adjoint the trivial map $A \to \{y_0\} \in Y$, $y_0 = \iota(e)$.

Observe that if Y happens to admit a delooping, or classifying space, BY and if ι deloops to a map $B\iota : BG \to BY$ with fibre equivalent to $q: B(Y, G, *) \to BG$, then $B\iota$ defines the obstruction to the existence of a Y-structure on an $\mathcal{F}$-fibration ν; that is, ν admits a Y-structure if and only if $(B\iota)_* \Phi\{\nu\}$ is the trivial homotopy class. For example, when $Y = G'$

is as in Example 10.3, the quasifibration sequence above extends to

$$G \xrightarrow{j} G' \xrightarrow{\tau} G'/G \xrightarrow{q} BG \xrightarrow{Bj} BG'$$

by Proposition 8.8, and Bj defines the obstruction to the existence of an $\mathfrak{J}'$-trivialization of an $\mathfrak{J}$-fibration.

Remark 11.3. In the applications, one is often interested in two (or more) types of structure on $\mathfrak{J}$-fibrations. The theorem already handles such situations since, if Y and Y' are right G-spaces, then the square

$$\begin{array}{ccc} B(Y \times Y', G, *) & \xrightarrow{B\pi_1} & B(Y, G, *) \\ \downarrow{\scriptstyle B\pi_2} & & \downarrow{\scriptstyle q} \\ B(Y', G, *) & \xrightarrow{q} & BG \end{array}$$

is a pullback and since, in cases (a) or (b), if (Y, Z) and (Y', Z') are admissible pairs, then so also is $(Y \times Y', Z \times Z')$. When $Y' = H\backslash G$ for some morphism of monoids $f: H \to G$, the pullback above can be used interchangeably with the pullback

$$\begin{array}{ccc} B(Y, H, *) & \xrightarrow{B(1, f, 1)} & B(Y, G, *) \\ \downarrow{\scriptstyle q} & & \downarrow{\scriptstyle q} \\ BH & \xrightarrow{Bf} & BG \end{array}$$

in view of Remarks 8.9 and the following commutative diagram, in which all vertical arrows are weak homotopy equivalences:

$$
\begin{array}{ccccc}
B(H\backslash G, G, *) & \xrightarrow{q} & BG & \xleftarrow{q} & B(Y, G, *) \\
\Vert & & \uparrow{\scriptstyle q} & & \uparrow{\scriptstyle B\varepsilon} \\
B(H\backslash G, G, *) & \xrightarrow{Bf} & B(G\backslash G, G, *) & \xleftarrow{Bq} & B(B(Y, G, G), G, *) \\
{\scriptstyle \varepsilon(p)}\downarrow & & \downarrow{\scriptstyle \varepsilon(p)} & & \downarrow{\scriptstyle \varepsilon(p)} \\
BH & \xrightarrow{Bf} & BG & \xleftarrow{q} & B(Y, G, *)
\end{array}
$$

The proof of the following bundle theoretic analog is formally identical to the proof of Theorem 11.1. Recall Definition 10.7.

<u>Theorem 11.4.</u> Let a topological group G act effectively from the left on a space F and assume that (G, e) is an NDR-pair. Let the pair (Y, Z) be admissible. Then, for $A \in \mathcal{W}$, the set $\mathcal{B}\mathcal{J}(A; Y)$ is naturally isomorphic to $[A, B(Y, G, *)]$.

Of course, Remarks 11.3 apply verbatim to Theorem 11.4, and the obvious bundle theoretic analogs of Remarks 11.2 are valid.

We have an evident natural transformation $\zeta : \mathcal{B}\mathcal{J}(A; Y) \to \mathcal{E}\mathcal{J}(A; Y)$ where, on the right, Y has its topology as a subspace of $\mathcal{U}(F, Z)$. If the identity map from Y, with its given topology, to Y, with its function space topology, is a weak homotopy equivalence and if the hypotheses of Corollary 9.11 are satisfied, then ζ is a bijection of sets.

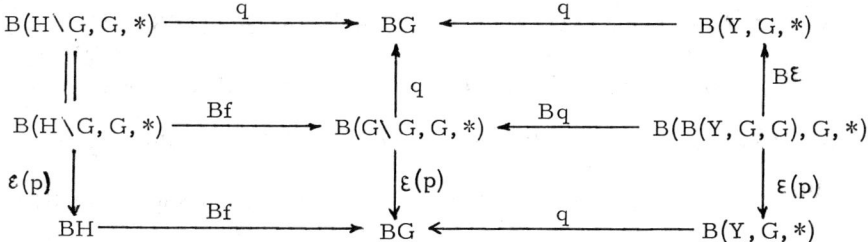

12. A categorical generalization of the bar construction

We here introduce an amusing categorical construction which allows us to generalize the material of section 9 to a context in which any set of fibres, rather than just a single fibre, is given a privileged role. This reworking of the theory yields an analysis of the effect of changing the choice of privileged fibre.

Let $\mathcal{O}$ be a fixed set (of objects) regarded for our purposes as a discrete topological space. Define an $\mathcal{O}$-graph to be a space $\mathcal{a}$ (of arrows) together with continuous maps $S: \mathcal{a} \to \mathcal{O}$ and $T: \mathcal{a} \to \mathcal{O}$ (called source and target). Let $\mathcal{O}Gr$ denote the category of $\mathcal{O}$-graphs; its morphisms are continuous maps $f: \mathcal{a} \to \mathcal{a}'$ such that $S \circ f = S$ and $T \circ f = T$. Regard $\mathcal{O}$ itself as that $\mathcal{O}$-graph with arrow space $\mathcal{O}$ and with S and T the identity map. Define the product over $\mathcal{O}$ of $\mathcal{O}$-graphs $\mathcal{a}$ and $\mathcal{a}'$ to be the $\mathcal{O}$-graph $\mathcal{a} \square \mathcal{a}'$ with arrow space $\{(a, a') \mid Sa = Ta'\} \subset \mathcal{a} \times \mathcal{a}'$ and with source and target defined by $S(a, a') = Sa'$ and $T(a, a') = Ta$. Clearly $\square$ is associative, up to the evident natural isomorphism, and is unital with respect to the natural isomorphisms $\lambda: \mathcal{a} \to \mathcal{O} \square \mathcal{a}$ and $\rho: \mathcal{a} \to \mathcal{a} \square \mathcal{O}$ specified by $\lambda(a) = (Ta, a)$ and $\rho(a) = (a, Sa)$ for $a \in \mathcal{a}$.

Thus $\mathcal{O}Gr$ is a monoidal category with product $\square$ and unit $\mathcal{O}$. We can therefore define the notion of a monoid $(\mathcal{G}, C, I)$ in $\mathcal{O}Gr$. Here $C: \mathcal{G} \square \mathcal{G} \to \mathcal{G}$ and $I: \mathcal{O} \to \mathcal{G}$ are maps of $\mathcal{O}$-graphs (called composition and identity) such that C is associative and I is a two-sided unit for C. In other words, $\mathcal{G}$ is just a small topological category with object space $\mathcal{O}$.

So far we have followed Mac Lane [14, p. 10 and 48], but we must now take cognizance of the assymmetry of $\square$. Define a right $\mathcal{O}$-graph to be a space $\mathcal{Y}$ together with a map $S: \mathcal{Y} \to \mathcal{O}$. Similarly, for a left $\mathcal{O}$-graph $\mathcal{X}$, only $T: \mathcal{X} \to \mathcal{O}$ is to be given. Observe that, for an $\mathcal{O}$-graph $\mathcal{A}$, we can define $\mathcal{Y} \square \mathcal{A}$ and $\mathcal{A} \square \mathcal{X}$ as right and left $\mathcal{O}$-graphs, and we can define $\mathcal{Y} \square \mathcal{X}$ as a space. Now let $\mathcal{G}$ be a monoid in $\mathcal{O}$Gr. Define a right $\mathcal{O}$-graph over $\mathcal{G}$ to be a right $\mathcal{O}$-graph $\mathcal{Y}$ together with a map $R: \mathcal{Y} \square \mathcal{G} \to \mathcal{Y}$ of right $\mathcal{O}$-graphs which satisfies the evident associativity and unit formulas $R(1 \square C) = R(R \square 1)$ and $R(1 \square I) = \rho^{-1}$. The notion of a left $\mathcal{O}$-graph over $\mathcal{G}$ is defined by symmetry.

At this point we can generalize the definition of the two-sided geometric bar construction to triples $(\mathcal{Y}, \mathcal{G}, \mathcal{X})$, where $\mathcal{G}$ is a monoid in $\mathcal{O}$Gr and $\mathcal{Y}$ and $\mathcal{X}$ are right and left $\mathcal{O}$-graphs over $\mathcal{G}$. Indeed, we need only replace $\times$ by $\square$ in the definition of section 7 to obtain a simplicial topological space $B_*(\mathcal{Y}, \mathcal{G}, \mathcal{X})$, and we define $B(\mathcal{Y}, \mathcal{G}, \mathcal{X})$ to be its geometric realization. To ensure that the construction has good topological properties, we insist that $(\mathcal{G}, I\mathcal{O})$ be a strong NDR-pair [17, A. 1]. When $\mathcal{O}$ is a singleton set, this two-sided bar construction reduces to that in section 7. The $\mathcal{O}$-graph $\mathcal{O}$ is itself a right and left $\mathcal{O}$-graph over $\mathcal{G}$ via

$$\mathcal{G} \square \mathcal{O} \xrightarrow{\rho^{-1}} \mathcal{G} \xrightarrow{T} \mathcal{O} \quad \text{and} \quad \mathcal{O} \square \mathcal{G} \xrightarrow{\lambda^{-1}} \mathcal{G} \xrightarrow{S} \mathcal{O}$$

$B\mathcal{G} = B(\mathcal{O}, \mathcal{G}, \mathcal{O})$ is the standard classifying space of the category $\mathcal{G}$ (e.g. Segal [30] or [18, §4]), and we write $E\mathcal{G} = B(\mathcal{O}, \mathcal{G}, \mathcal{G})$.

All of the results of section 7 and some of the results of section 8 generalize to $B(\mathcal{Y}, \mathcal{G}, \mathcal{X})$. Indeed, this is apparent from the fact that most of the proofs depended only on general properties of geometric realization and elementary properties of the simplicial bar construction. Note for Proposition 7.4 that the product $\mathcal{A} \times \mathcal{A}'$ of an $\mathcal{O}$-graph $\mathcal{A}$ and an $\mathcal{O}'$-graph $\mathcal{A}'$ is an $\mathcal{O} \times \mathcal{O}'$-graph and that there are evident natural isomorphisms of simplicial spaces

$$B_*(\mathcal{Y} \times \mathcal{Y}', \mathcal{G} \times \mathcal{G}', \mathcal{X} \times \mathcal{X}') \cong B_*(\mathcal{Y}, \mathcal{G}, \mathcal{X}) \times B_*(\mathcal{Y}', \mathcal{G}', \mathcal{X}')$$

for triples $(\mathcal{Y}, \mathcal{G}, \mathcal{X})$ over $\mathcal{O}$ and $(\mathcal{Y}', \mathcal{G}', \mathcal{X}')$ over $\mathcal{O}'$. For Theorem 7.6, we say that $\mathcal{G}$ is grouplike if its homotopy category is a groupoid (so that every homotopy class of maps in $\mathcal{G}$ is invertible); here p and q are induced from $T: \mathcal{X} \to \mathcal{O}$ and $S: \mathcal{Y} \to \mathcal{O}$ and are quasifibrations. Theorem 8.1 and Remarks 8.10 wholly fail to generalize; $E\mathcal{G}$ is clearly not a group (or groupoid) if $\mathcal{G}$ is a groupoid, and commutativity in $\mathcal{G}$ only makes sense on subcategories with one object.

Now suppose given a homogeneous category of fibres. Call it $\mathcal{G}$ rather than $\mathcal{F}$, to accord with our present emphasis on the morphism spaces rather than the object spaces, but continue to speak of $\mathcal{F}$-spaces and $\mathcal{F}$-fibrations. (Note that, in our previous notations, the object spaces of the associated principal category were certain of the morphism spaces of the original category.) Let $\mathcal{O}$ denote the collection of objects of $\mathcal{G}$, assume that $\mathcal{O}$ is a set, and give $\mathcal{O}$ the discrete topology; explicitly, $\mathcal{O}$ is to have one point $\{F\}$ for each object space F. Let $\mathcal{F}$ denote the left $\mathcal{O}$-graph over $\mathcal{G}$ which, as a space, is the disjoint union of the object spaces of $\mathcal{G}$; $T: \mathcal{F} \to \mathcal{O}$ is the map which collapses a space F of the disjoint union to the point $\{F\}$, and the left action $\mathcal{G} \square \mathcal{F} \to \mathcal{F}$ is defined by the evaluation maps $\mathcal{G}(F, F') \times F \to F'$.

Let $\nu: D \to A$ be an $\mathcal{F}$-space. Define a right $\mathcal{O}$-graph $\mathcal{P}D$ over $\mathcal{G}$ as follows. As a space, $\mathcal{P}D$ is the disjoint union over F of the subspaces of $\mathcal{U}(F, D)$ which consist of the maps in $\mathcal{G}$ from F to a fibre of ν. $S: \mathcal{P}D \to \mathcal{O}$ assigns the point $\{F\}$ to all $\psi: F \to D$ in $\mathcal{P}D$, and the right action $\mathcal{P}D \square \mathcal{G} \to \mathcal{G}$ is induced by the composition maps $\mathcal{U}(F', D) \times \mathcal{G}(F, F') \to \mathcal{U}(F, D)$.

We can now formally replace all bar constructions $B(Y, G, X)$ which occur in the proof of Theorem 9.2 by corresponding bar constructions $B(\mathcal{Y}, \mathcal{G}, \mathcal{X})$. In cases (a) and (b), the proof goes through without the slightest change. We must avoid case (c), because Theorem 8.2 is not available, but here all fibres are homeomorphic and the present elaboration would be uninteresting in any case. In practice, we must also avoid case (i) of (a) and (b), since here $\mathcal{F}$ is usually not homogeneous. This leaves case (ii), and here the categories $\mathcal{F}\mathcal{W}$ and $\mathcal{F}\mathcal{V}$ of Examples 6.6 and 6.9 are homogeneous.

We have ignored one difficulty: $\mathcal{O}$ was assumed to be a set, whereas the categories of interest are large. Probably the best solution to this problem is to restrict the constructions above to the various small full subcategories of $\mathcal{G}$. For example, Theorem 9.2 as originally developed is the case when $\mathcal{G}$ is replaced by its full subcategory with one object F.

Thus reinterpret $\mathcal{O}$ above to be a given set of objects of our original category. Now note that the bar construction is functorial in $\mathcal{O}$. Indeed, if $\mathcal{O}$ is a subset of $\mathcal{O}'$ and if $(\mathcal{Y}', \mathcal{G}', \mathcal{X}')$ is defined over $\mathcal{O}'$, then $(\mathcal{Y}, \mathcal{G}, \mathcal{X})$ is defined over $\mathcal{O}$, where

$$\mathcal{Y} = S^{-1}\mathcal{O} \subset \mathcal{Y}', \ \mathcal{G} = S^{-1}\mathcal{O} \cap T^{-1}\mathcal{O} \subset \mathcal{G}', \text{ and } \mathcal{X} = T^{-1}\mathcal{O} \subset \mathcal{X}'$$

and the inclusions induce a well-defined map

$$B(\mathcal{Y}, \mathcal{G}, \mathcal{X}) \to B(\mathcal{Y}', \mathcal{G}', \mathcal{X}').$$

In practice, by comparisons of quasifibrations, these maps will all be weak homotopy equivalent to BHX'.

For example, to show that BHX is homotopy equivalent to BHX' when X and X' are spaces of the homotopy type of a given compact space $F \in \mathcal{W}$, we need only map the quasifibrations EHX → BHX and EHX' → BHX' to the quasifibration $E\mathcal{G} \to B\mathcal{G}$, where $\mathcal{G}$ is the full subcategory of $F\mathcal{W}$ with objects X and X'. Indeed, by use of larger sets of objects, we actually obtain a coherent system of homotopy equivalences connecting the BHX as X ranges through the given homotopy type. Of course, these remarks apply equally well to any other homogeneous category of fibres and can easily be elaborated to a precise comparison of the natural transformations Φ and Ψ (of the proof of Theorem 9.2) obtained by use of different choices of privileged fibre.

13. The algebraic and geometric bar constructions

Let R be a commutative ring and take all homology with coefficients in R throughout the last three sections.

Just as the two-sided geometric bar construction is obtained by composing the simplicial bar construction on appropriate triples of spaces with geometric realization, so the two-sided algebraic bar construction is obtained by composing the simplicial bar construction B_* on appropriate triples of differential R-modules with the condensation functor C from simplicial differential R-modules to differential R-modules. The reader is referred to the Appendix of [10] for the details of the definition. Write $B(M, U, N)$ for the algebraic bar construction, where U is a differential R-algebra and M and N are right and left differential U-modules. Write $EU = B(R, U, U)$ and $BU = B(R, U, R)$, rather than the standard notations of [10], in conformity with the notations for the geometric bar construction.

In view of the similarity between their definitions, it is natural to expect there to be a close relationship between the geometric and algebraic bar constructions. In the case of BG, this relationship was first made precise by Stasheff [31], singularly, and later by Milgram [23], cellularly.

Let $\mathcal{C}$ denote the subcategory of $\mathcal{U}$ which consists of the CW-complexes and cellular maps. There is a derived subcategory $\mathcal{A}(\mathcal{C})$ of $\mathcal{A}(\mathcal{U})$, with objects those triples (Y, G, X) such that $Y, G, X \in \mathcal{C}$ and the product and unit of G and its actions on Y and X are cellular maps. The component maps of morphisms in $\mathcal{A}(\mathcal{C})$ are also required to be cellular.

We shall write $C_{\#}$ for cellular chains and C_* for normalized singular chains (both with coefficients in R). The product of CW-complexes X and Y is a CW-complex (since we are working in $\mathcal{U}$) with a natural cell structure such that $C_{\#}(X \times Y)$ can be identified with $C_{\#}X \otimes C_{\#}Y$.

In order to get the right signs, we agree to write simplices to the left, $|X| = \coprod \Delta_p \times X_p/(\approx)$, when forming the geometric realization of simplicial spaces.

Proposition 13.1. Let $(Y, G, X) \in \mathcal{Q}(\mathcal{C})$. Then $B(Y, G, X)$ is a CW-complex with a natural cell structure such that $C_{\#}B(Y, G, X)$ is naturally isomorphic to $B(C_{\#}Y, C_{\#}G, C_{\#}X)$.

Proof. Give $B_p(Y, G, X) = Y \times G^p \times X$ the product cell structure. Then, by [17, 11.4], $B(Y, G, X)$ is a CW-complex with one (n+p)-cell $\Delta_p \times \sigma$ for each n-cell $\sigma = \sigma_0 \times \sigma_1 \times \ldots \times \sigma_p \times \sigma_{p+1}$ of $Y \times (G - e)^p \times X$. Let $\Delta_p \times \sigma$ correspond to the element $\sigma_0[\sigma_1, \ldots, \sigma_p]\sigma_{p+1}$ of $B(C_{\#}Y, C_{\#}G, C_{\#}X)$. The pieces $\partial\Delta_p \times \sigma$ and $\Delta_p \times \partial\sigma$ of the boundary of $\Delta_p \times \sigma$ give rise to the simplicial (or external) and internal components of the differential $d = \sum_{i=0}^{p} (-1)^i \partial_i + (-1)^p \partial$ in the bar construction (see [10, A. 2]).

We require the following addendum to [17, 11.15] in order to describe the behavior of products under the isomorphism of chains given in the proposition.

Lemma 13.2. Let X and Y be simplicial objects in $\mathscr{C}$. Then the natural homeomorphism $f'\colon |X \times Y| \to |X| \times |Y|$ is naturally homotopic to a cellular map f and the natural homeomorphism $g = (f')^{-1}$ is itself cellular.

Proof. $f = (G_0 \times G_1)f'$, where the G_i are the Alexander-Whitney type homotopy equivalences specified in the proof of [17, 11.15], and the requisite verifications are the same as there.

We also write f and g for the maps of the geometric bar construction derived via the identifications (of [17, 10.1])

$$B_*(Y, G, X) \times B_*(Y', G', X') = B_*(Y \times Y', G \times G', X \times X').$$

In precise analogy, [10, A.3] gives natural chain homotopy equivalences, the Alexander-Whitney and shuffle maps,

$$\xi\colon C(M \otimes N) \to CM \otimes CN \quad \text{and} \quad \eta\colon CM \otimes CN \to C(M \otimes N)$$

for simplicial differential R-modules M and N. We also write ξ and η for the maps of the algebraic bar construction derived via the identifications (again, of [17, 10.1])

$$B_*(M, U, N) \otimes B_*(M', U', N') = B_*(M \otimes M', U \otimes U', N \otimes N').$$

Proposition 13.3. Let (Y, G, X) and (Y', G', X') be objects of $\mathscr{A}(\mathscr{C})$. Then, under the isomorphism of Proposition 13.1, $C_\# f$ coincides with ξ and $C_\# g$ coincides with η.

Proof. This is verified by explicit calculations. We omit the details (which really amount only to checks of signs) since the precise definitions of ξ and η were motivated by the present result.

The following results give special properties of BG and EG.

Proposition 13.4. If G is a cellular topological monoid, then BG admits a cellular diagonal approximation with respect to which $C_{\#}BG$ is naturally isomorphic to $BC_{\#}G$ as a differential coalgebra.

Proof. $f' \circ B\Delta = \Delta: BG \to BG \times BG$, and $f \circ B\Delta = (G_0 \times G_1)\Delta$ is the desired diagonal approximation; it is cellular by [17, 11.15]. The chain level statement does not follow by naturality from the previous result, since $B\Delta$ need not be cellular, but instead requires an easy direct calculation from the explicit coproduct

$$D[g_1, \ldots, g_p] = \sum_{i=0}^{p} (-1)^{(p-i)q_i}[g_1, \ldots, g_i] \otimes [g_{i+1}, \ldots, g_p] \ ,$$

$$q_i = \sum_{j=1}^{i} \deg g_j \ ,$$

on $BC_{\#}G$. Note that D is independent of any possible coproduct on $C_{\#}G$.

Proposition 13.5. If G is a cellular topological group, then so is EG.

Proof. This follows by naturality from the identification of EG with $|D_*G|$ in the proof of Theorem 8.1.

Proposition 13.6. If G is a cellular Abelian topological monoid, then so are BG and EG. $C_{\#}BG$ is naturally isomorphic to $BC_{\#}G$ as a differential Hopf algebra.

Proof. The first part is immediate by naturality and the second part follows from Propositions 13.4 and 13.3 since the product on $BC_{\#}G$ is derived from the shuffle map by naturality.

Remarks 13.7. Let π be an Abelian group regarded as a discrete CW-complex. As noted by Milgram [23], if we define $K(\pi, n) = B^n\pi$ by iteration, then $K(\pi, n)$ is a cellular Abelian topological group such that $C_{\#}K(\pi, n) \cong B^n(R\pi)$ as a differential Hopf algebra, where $R\pi$ denotes the group ring of π. This gives an alternative derivation to the classical one (given in detail in [10, Appendix]) of the geometric preliminaries necessary for Cartan's calculations of $H^*K(\pi, n)$.

In order to apply the above results in full generality, we note the following result. Although it surely ought to be well-known, it seems not to appear in the literature. Let S denote the total singular complex functor from spaces to simplicial sets, let T denote the geometric realization functor from simplicial sets to spaces, and let $\Phi : TSX \to X$ denote the natural weak homotopy equivalence, $\Phi|u, f| = f(u)$ for $u \in \Delta_p$ and $f: \Delta_p \to X$ (see e.g. [16]). Of course, T takes values in $\mathcal{C}$.

Proposition 13.8. The normalized singular chains C_*X are naturally isomorphic to the cellular chains $C_{\#}TSX$. The Alexander-Whitney and shuffle maps ξ and η agree under this isomorphism with the respective composites

$$C_{\#}TS(X\times Y) = C_{\#}T(SX\times SY) \xrightarrow{C_{\#}f} C_{\#}(TSX \times TSY) = C_{\#}TSX \otimes C_{\#}TSY$$

and

$$C_{\#}TSX \otimes C_{\#}TSY = C_{\#}(TSX\times TSY) \xrightarrow{C_{\#}g} C_{\#}T(SX \times SY) = C_{\#}TS(X\times Y) .$$

Proof. Let a non-degenerate simplex $f: \Delta_p \to X$ correspond to the cell $\Delta_p \times f \subset TSX$, where the p-simplex f is regarded as a vertex of the discrete CW-complex S_pX. This correspondence sets up the required isomorphism. (Note that, by the paragraph after [10, A.7], ξ and η on singular chains are special cases of the maps ξ and η of [10, A.3].)

Thus the geometry of simplices provides a rigorous construction of the Alexander-Whitney and shuffle maps on singular chains, rather than just the motivation for an algebraic definition.

The following theorem is the main technical result on the relationship between the geometric and algebraic bar constructions.

Theorem 13.9. For $(Y, G, X) \in \mathcal{A}(\mathcal{U})$, $B(TSY, TSG, TSX)$ admits a cellular diagonal approximation with respect to which $C_{\#}B(TSY, TSG, TSX)$ is naturally isomorphic to $B(C_*Y, C_*G, C_*X)$ as a differential coalgebra. Therefore $H^*B(Y, G, X)$ is naturally isomorphic as an algebra to the homology of the dual of $B(C_*Y, C_*G, C_*X)$.

Proof. On the level of differential R-modules, Propositions 13.1 and 13.8 establish the required isomorphism. For the statement about the diagonal, consider the following diagram (in which the isomorphisms are given by the results just cited):

$$
\begin{array}{ccc}
C_{\#}B(TSY, TSG, TSX) & \cong & B(C_*Y, C_*G, C_*X) \\
\downarrow{\scriptstyle C_{\#}B(TS\Delta, TS\Delta, TS\Delta)} & & \downarrow{\scriptstyle B(C_*\Delta, C_*\Delta, C_*\Delta)} \\
C_{\#}B(TS(Y\times Y), TS(G\times G), TS(X\times X)) & \cong & B(C_*(Y\times Y), C_*(G\times G), C_*(X\times X)) \\
\downarrow{\scriptstyle C_{\#}B(f, f, f)} & & \downarrow{\scriptstyle B(\xi, \xi, \xi)} \\
C_{\#}B(TSY\times TSY, TSG\times TSG, TSX\times TSX) & \cong & B(C_*Y\otimes C_*Y, C_*G\otimes C_*G, C_*X\otimes C_*X) \\
\downarrow{\scriptstyle C_{\#}f} & & \downarrow{\scriptstyle \xi} \\
C_{\#}B(TSY, TSG, TSX)\otimes C_{\#}B(TSY, TSG, TSX) & \cong & B(C_*Y, C_*G, C_*X)\otimes B(C_*Y, C_*G, C_*X)
\end{array}
$$

The fact that (ξ, ξ, ξ) is a morphism of triples in the appropriate category, so that $B(\xi, \xi, \xi)$ is defined, follows from the commutative diagram displayed in [10, A.3]. Since $\xi = C_{\#}f$ and $\eta = C_{\#}g$, the same combinatorial proof shows that the analogous diagram with condensation replaced by geometric realization also commutes, hence that (f, f, f) is a morphism in $\mathcal{A}(\mathcal{C})$. The upper two squares of our diagram commute by naturality (from Proposition 13.1), and the bottom square commutes by Proposition 13.3. The map $f\circ B(f\circ TS\Delta, f\circ TS\Delta, f\circ TS\Delta)$ is homotopic to the diagonal, and the coproduct on $B(C_*Y, C_*G, C_*X)$ is defined to be $\xi\circ B(\xi\circ C_*\Delta, \xi\circ C_*\Delta, \xi\circ C_*\Delta)$.

The last statement (which is given in cohomology solely in order to avoid flatness hypotheses for coproducts) follows since the map

$$B(\Phi, \Phi, \Phi): B(TSY, TSG, TSX) \to B(Y, G, X)$$

induces an isomorphism on homology by Proposition 7.3.

We complete this section with a discussion of the following theorem.

<u>Theorem 13.10.</u> Let $(Y, G, X) \in \mathcal{A}(\mathcal{U})$ and let R be a field. Then there is a natural spectral sequence of differential coalgebras which converges from the coalgebra $E^2 = \mathrm{Tor}^{H_*G}(H_*Y, H_*X)$ to the coalgebra $H_*B(Y, G, X)$.

R is taken to be a field to avoid awkward flatness hypotheses.

There are two conceptually different proofs. First, following Rothenberg and Steenrod [29], we can construct an exact couple by passage to homology from the filtration by successive cofibrations of the space $B(Y, G, X)$. This approach is analyzed in [17, 11.14], where the E^2-term is computed. Because of its geometric nature, this approach makes it simple to put Steenrod operations into the spectral sequence when $R = Z_p$ and is applicable to any homology theory with an appropriate Kunneth theorem.

Second, following Eilenberg and Moore [26], we can filter the algebra bar construction $B(M, U, N)$ by

$$F_pB(M, U, N) = \mathrm{Image} \sum_{i=0}^{p} B_i(M, U, N)$$

and observe that $E^1B(M, U, N) = B(HM, HU, HN)$, which is a suitable differential R-module (or R-coalgebra if HU is a Hopf algebra and HM and HN are right and left coalgebras over HU) for the computation of $\mathrm{Tor}^{HU}(HM, HN)$. This approach is analyzed in [10, A.9 and § 1, 2, 5]. Because of its algebraic nature, it gives maximal information on the internal structure of the spectral sequence.

Theorem 13.9 demonstrates the applicability of the second approach to the calculation of $H_*B(Y, G, X)$ and proves that both approaches yield the same spectral sequence. Indeed, the map $B(\Phi, \Phi, \Phi)$ certainly induces an isomorphism of Rothenberg-Steenrod spectral sequences, and the isomorphism between $C_\# B(TSY, TSG, TSX)$ and $B(C_*Y, C_*G, C_*X)$ implies that the Rothenberg-Steenrod spectral sequence for the triple (TSY, TSG, TSX) is isomorphic to the Eilenberg-Moore spectral sequence for the triple (C_*Y, C_*G, C_*X). Note that our chain level isomorphism yields a particularly conceptual proof that the obvious algebraic coproduct in E^2 converges to the correct geometric coproduct in the limit.

14. Transports and the Serre spectral sequence

As in section 1, assume given a category $\mathcal{F}$ with a faithful underlying space functor to $\mathcal{U}$. Recall that $\mathcal{J}$ denotes the category of nondegenerately based spaces in $\mathcal{U}$. For $A \in \mathcal{J}$, let ΛA and PA denote the Moore loop space and path space on A and let $p: PA \to A$ denote the end-point projection. The associated principal fibration functor previously denoted by P will be denoted by Prin here.

Definition 14.1. An $\mathcal{F}$-transport over a space $A \in \mathcal{J}$ is a space $X \in \mathcal{F}$ together with an (associative and unital) action $\tau: \Lambda A \times X \to X$ such that $\tau(\lambda, \ ): X \to X$ is a map in $\mathcal{F}$ for each $\lambda \in \Lambda A$. Define $\mathcal{J}\mathcal{F}(A)$ to be the collection, assumed to be a set, of equivalence classes of $\mathcal{F}$-transports over A under the equivalence relation generated by $\tau \approx \tau'$ if there exists a map $\phi: X \to X'$ in $\mathcal{F}$ such that the following diagram is commutative:

$$\begin{array}{ccc} \Lambda A \times X & \xrightarrow{\ \tau\ } & X \\ {\scriptstyle 1\times\phi}\downarrow & & \downarrow{\scriptstyle \phi} \\ \Lambda A \times X' & \xrightarrow{\ \tau'\ } & X' \end{array}$$

For a map $f: A' \to A$ in $\mathcal{J}$ and an $\mathcal{F}$-transport τ over A, define an $\mathcal{F}$-transport $f^*\tau$ over A' by $(f^*\tau)(\lambda', x) = \tau(f \circ \lambda', x)$ for $\lambda' \in \Lambda A'$ and $x \in X$. Then $\mathcal{J}\mathcal{F}$ is a contravariant functor from $\mathcal{J}$ to sets.

Observe that the adjoint $\Lambda A \to \mathcal{F}(X, X)$ of an $\mathcal{F}$-transport τ is a map of topological monoids and that, conversely, such a map has adjoint an $\mathcal{F}$-transport over A. Clearly $\mathcal{J}\mathcal{F}(A)$ is isomorphic to the set of equivalence classes of maps of monoids $\gamma: \Lambda A \to \mathcal{F}(X, X)$, $X \in \mathcal{F}$, under the equivalence relation generated by $\gamma \approx \gamma'$ if there exists a map $\phi: X \to X'$ in $\mathcal{J}$ such that $\phi(\gamma\lambda)(x) = (\gamma'\lambda)(\phi x)$ for all $\lambda \in \Lambda A$ and $x \in X$.

While the proof of the following theorem works somewhat more generally, we shall restrict attention to the categories of Example 6.6 for simplicity. The idea of the result, and the term "transport", are due to Stasheff [33].

Theorem 14.2. Let $F \in \mathcal{W}$ and let $\mathcal{J}$ be $F\mathcal{U}$ or, if F is compact, $F\mathcal{W}$. Then, for $A \in \mathcal{V}$, $\mathcal{E}\mathcal{J}(A)$ is naturally isomorphic to $\mathcal{T}\mathcal{J}(A)$.

Proof. $\mathcal{E}\mathcal{J}(A) \cong [A, BHF]$ is a well-defined set, and it will follow from the rest of the proof that $\mathcal{T}\mathcal{J}(A)$ is also well-defined. Define $\Phi: \mathcal{E}\mathcal{J}(A) \to \mathcal{T}\mathcal{J}(A)$ as follows. Given an $\mathcal{J}$-fibration $\nu: D \to A$, let $F\nu = (\Gamma\nu)^{-1}(*) \subset \Gamma D$ and let $\tau: \Lambda A \times F\nu \to F\nu$ be obtained by restriction from $\mu: \Gamma\Gamma D \to \Gamma D$. Then let $\Phi\{\nu\} = \{\tau\}$. Define $\Psi: \mathcal{T}\mathcal{J}(A) \to \mathcal{E}\mathcal{J}(A)$ as follows. Given an $\mathcal{J}$-transport $\tau: \Lambda A \times X \to X$, use it and the natural right action of ΛA on PA to construct the quotient space $PA \times_{\Lambda A} X$; it is weak Hausdorff because of the nondegeneracy of the basepoint $* \in A$. Define $\pi: PA \times_{\Lambda A} X \to A$ by $\pi(\beta, x) = p(\beta)$ and note that $\pi^{-1}(*) = X$. Define a transitive lifting function

$$\xi: \quad \Gamma(PA \times_{\Lambda A} X) \to PA \times_{\Lambda A} X$$

for π by $\xi(\alpha, (\beta, x)) = (\alpha\beta, x)$. Then define $\Psi\{\tau\} = \{\nu\}$. If τ is derived from $\nu: D \to A$ as in the definition of Φ, then

$$\mu: PA \times F\nu \subset \Gamma\Gamma D \to \Gamma D$$

induces an $\mathcal{J}$-map $PA \times_{\Lambda A} F\nu \to \Gamma D$ over A which restricts on $F\nu$ to the identity map. Since $\{\Gamma\nu\} = \{\nu\}$, it follows that $\Psi\Phi$ is the identity transformation of $\mathcal{E}\mathfrak{T}(A)$. Conversely, if π is derived as above from a $\mathcal{T}$-transport τ, so that

$$F\pi = \{(\alpha, (\beta, x)) \mid \alpha \in \prod A,\ (\beta, x) \in PA \times_{\Lambda A} F,\ \alpha(0) = p(\beta), p(\alpha) = *\},$$

define $\emptyset: F\pi \rightarrow X$ by $\emptyset(\alpha, (\beta, x)) = \tau(\alpha\beta, x)$. Then the diagram

$$\begin{array}{ccc} \Lambda A \times F\pi & \xrightarrow{\mu} & F\pi \\ {\scriptstyle 1\times\emptyset}\downarrow & & \downarrow{\scriptstyle \emptyset} \\ \Lambda A \times X & \xrightarrow{\tau} & X \end{array}$$

is commutative, because $\tau(\lambda, \tau(\alpha\beta, x)) = \tau(\lambda\alpha\beta, x)$, and therefore $\Phi\Psi$ is the identity transformation of $\mathcal{T}\mathfrak{T}(A)$.

Since, up to equivalence, a fibration determines and is determined by a transport, it is plausible that the Serre spectral sequence can be derived by use of a differential R-module which depends only on a transport. We shall show that this is the case. We first note the following fact.

Lemma 14.3. If $A \in \mathcal{T}$ is connected, then the diagram

$$A \xleftarrow{\varepsilon(p)} B(PA, \Lambda A, *) \xrightarrow{q} B\Lambda A$$

displays a weak homotopy equivalence between A and $B\Lambda A$.

Proof. Since A is connected, the fibration $p: PA \rightarrow A$ maps onto A and is thus a quasifibration. The result follows from Corollary 7.7.

Recall that homology and cohomology are to be taken with coefficients in R.

Theorem 14.4. Let $A \in \mathcal{J}$ be connected and let $\nu : D \to A$ be a quasifibration with $F = \nu^{-1}(*)$. Then there is a natural spectral sequence $\{E_r \nu\}$ of differential algebras such that $E_2\nu = H^*(A; \mathcal{H}^*(F))$ as an algebra and $\{E_r\nu\}$ converges to the algebra H^*D.

Proof. The result is stated in cohomology to avoid flatness hypotheses for coproducts, and we shall work in homology. Replacing ν by $\Gamma\nu : \Gamma D \to A$ if necessary, we may assume without loss of generality that ν has a transitive lifting function $\xi : \Gamma D \to D$. By restriction of ξ, we obtain

$$\xi : PA \times F \to D \quad \text{and} \quad \tau : \Lambda A \times F \to F .$$

The following diagram is commutative:

$$\begin{array}{ccccc}
D & \xleftarrow{\varepsilon(\xi)} & B(PA, \Lambda A, F) & \xrightarrow{q} & B(*, \Lambda A, F) \\
\downarrow{\scriptstyle \nu} & & \downarrow{\scriptstyle p} & & \downarrow{\scriptstyle p} \\
A & \xleftarrow{\varepsilon(p)} & B(PA, \Lambda A, *) & \xrightarrow{q} & B\Lambda A
\end{array}$$

$\varepsilon(\xi)$ restricts to the identity map on $F = p^{-1}(*)$, hence, by the lemma and Theorem 7.6, the upper row displays a weak homotopy equivalence between D and $B(*, \Lambda A, F)$. By Theorem 13.9, we conclude that $H_*D \cong HB(R, C_*\Lambda A, C_*F)$. Filter this algebraic bar construction by writing, additively,

$$B(R, C_*\Lambda A, C_*F) = BC_*\Lambda A \otimes C_*F$$

and then defining

$$F_pB(R, C_*\Lambda A, C_*F) = \sum_{i \leq p} B_iC_*\Lambda A \otimes C_*F .$$

Let $\{E^r\nu\}$ denote the derived spectral sequence. The degree i referred to in the filtration is the total degree, hence, visibly, the differential d^0 is just $1 \otimes d$. Thus, additively, $E^1_{pq}\nu = B_pC_*\Lambda A \otimes H_qF$ by the Künneth theorem. We may rewrite $E^1\nu = B(R, C_*\Lambda A, H_*F)$; this equality holds as differential R-modules, the point being that the last face operator depends on the action of $C_*\Lambda A$ on H_*F induced by the transport τ. Since $BC_*\Lambda A \cong C_\# B\Lambda A$, by Theorem 13.9 again, we conclude that $E^1\nu \cong C_\#(B\Lambda A; \mathcal{H}_*F)$. The reader is referred to Steenrod [36, §31] for a thorough treatment of cellular chains with local coefficients. Thus $E^2\nu \cong H_*(A; \mathcal{H}_*F)$. For the coproducts, merely note that Theorem 13.9 implies both that $B(R, C_*\Lambda A, C_*F)$ is a filtered differential coalgebra isomorphic to $C_\# B(*, \Lambda A, F)$ and that $E^1\nu$ is isomorphic to $C_\#(B\Lambda A; \mathcal{H}_*F)$ as a differential coalgebra.

A comparison to the standard construction of the Serre spectral sequence from a filtration on C_*D can easily be obtained by means of a chain level elaboration of the geometric diagram displayed in the proof. Our construction is similar in philosophy, but not in detail, to that given by Brown [3] in terms of twisted tensor products.

There is more than just a formal similarity between the diagram used in the previous argument and those used in the proof of Theorem 9.2. Indeed, with the notations of the proof just given, the following diagram is commutative (where $\tilde{\tau}$ and $\tilde{\xi}$ denote the adjoints of τ and ξ):

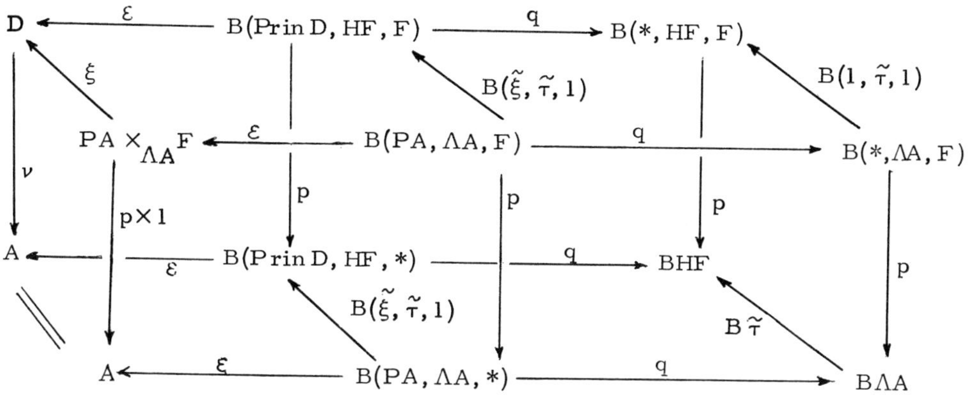

This diagram displays a connection between the classification theorems of Corollary 9.5 and Theorem 14.2. For $A \in \mathcal{V}$, the classifying map $A \to BHF$ for ν is transported to $B\tilde{\tau} : B\Lambda A \to BHF$ via the canonical homotopy equivalence between A and $B\Lambda A$. Of course, since Corollary 9.5 refers to a fixed chosen space $F \in \mathcal{F}$ whereas Theorem 14.2 refers to arbitrary spaces in $\mathcal{F}$, it would be necessary to use the theory of section 12 to make a more systematic comparison between the cited results.

15. The group completion theorem

We say that an H-space X is admissible if it is homotopy associative and if left translation by any given element is homotopic to right translation by the same element; the latter condition certainly holds if X is homotopy commutative.

Let $f: X \to Y$ be an H-map between admissible H-spaces, where Y is grouplike. We say that f is a group completion if $f_*: H_*X \to H_*Y$ is a localization of the ring H_*X at its multiplicative submonoid $\pi_0 X$ for every commutative coefficient ring R. This condition will hold if it does so for all prime fields $R = Z_p$ and for $R = Q$ [28 or 18, 1.4], hence we assume that R is a field below.

The purpose of this section is to prove the following result, which is a version due to Quillen [28] of a theorem of Barratt and Priddy [1].

Theorem 15.1. Let G be a topological monoid such that G and ΩBG are admissible H-spaces. Then the natural inclusion $\zeta : G \to \Omega BG$ is a group completion.

According to Quillen, the admissibility of ΩBG need not be assumed. The argument in [28] seemed unconvincing on this point, and Quillen subsequently obtained a quite different proof [private communication]. However, this hypothesis is usually satisfied in practice and allows the present technically simplified version of Quillen's original argument.

We begin the proof with the following lemma.

Lemma 15.2. $\pi_0\zeta: \pi_0 G \to \pi_0 \Omega BG$ is a group completion; that is, $\pi_0\zeta$ is universal with respect to morphisms of monoids from $\pi_0 G$ to groups.

Proof. $\pi_1 BG = H_1(BG; Z)$ since $\pi_0\Omega BG$ is commutative by the admissibility of ΩBG. By Theorem 13.9, we may use $BC_* G$ to compute $H_1 BG$. We find that it is the quotient of $H_0 G$ by the image of $H_0 G \otimes H_0 G$ under the map $d = \partial_0 - \partial_1 + \partial_2$, or

$$d(x \otimes y) = \varepsilon(x)y - xy + \varepsilon(y)x ,$$

where $\varepsilon: H_0 G \to Z$ is the augmentation and xy is the Pontryagin product. In other words, $H_1 BG$ is the quotient of the free Abelian group generated by the set $\pi_0 G$ by the subgroup generated by $\{x + y - xy \mid x, y \in \pi_0 G\}$. Clearly $\pi_0\zeta$ agrees with the natural map, and the conclusion follows.

To proceed further, we need Segal's (unpublished) analog for "special" simplicial spaces of the total singular complex. Let $\mathcal{S}\mathcal{J}$ denote the category of proper simplicial based spaces.

Definition 15.3. An object $Y \in \mathcal{S}\mathcal{J}$ is said to be reduced if $Y_0 = *$ and to be special if, in addition, the map $(\delta^0, \ldots, \delta^{p-1}): Y_p \to Y_1^p$ is a homotopy equivalence for all p, where $\delta^i = \partial_0 \cdots \partial_{i-1}\partial_{i+2} \cdots \partial_p$. For example, $B_* G$ is special for any topological monoid $G \in \mathcal{J}$. Let $\mathcal{S}^+\mathcal{J}$ denote the full subcategory of $\mathcal{S}\mathcal{J}$ whose objects are special. Let $T: \mathcal{S}^+\mathcal{J} \to \mathcal{J}$ denote the geometric realization functor. Define a right adjoint $S: \mathcal{J} \to \mathcal{S}^+\mathcal{J}$ to T by letting $S_p X$ be the space of maps

$(\Delta_p, \Delta_p^0) \to (X, *)$, where Δ_p^0 denotes the set of vertices of Δ_p, and letting the face and degeneracy operators be induced from those of Δ_p just as for the total singular complex [16, p. 2]. Obviously $S_1X = \Omega X$, and it is trivial to verify that SX is special. The adjunction

$$\mathcal{S}^+\mathcal{J}(Y, SX) \underset{\psi}{\overset{\phi}{\rightleftarrows}} \mathcal{J}(TY, X)$$

is given by $(\phi f)|y, u| = (fy)(u)$ and $(\psi g)(y)(u) = g|y, u|$ for $f: Y \to SX$ and $g: TY \to X$, where $y \in Y_p$ and $u \in \Delta_p$, precisely as in the classical case [16, p. 62]. Write

$$\Phi = \phi(1): TSX \to X \text{ and } \Psi = \psi(1): Y \to STY.$$

We shall need the following variant of Lemma 14.3. Its proof requires only trivial verifications and use of Proposition 8.7.

<u>Lemma 15.4.</u> Let $X \in \mathcal{J}$ and define $\xi: B\Lambda X \to X$ by the formula

$$\xi|[\lambda_1, \ldots, \lambda_p], u| = (\lambda_1 \cdots \lambda_p)(\sum_{i=1}^{p} u_i \ell(\lambda_i))$$

for $\lambda_i \in \Lambda X$ and $u = (t_0, \ldots, t_p) \in \Delta_p$, where $u_i = t_0 + \ldots + t_{i-1}$. Then the following composite is the identity map, hence ξ is a weak homotopy equivalence if X is connected:

$$\Omega X \subset \Lambda X \xrightarrow{\zeta} \Omega B\Lambda X \xrightarrow{\Omega\xi} \Omega X .$$

<u>Proposition 15.5.</u> $\Phi: TSX \to X$ is a weak homotopy equivalence for any connected space $X \in \mathcal{J}$ and $T\Psi: TY \to TSTY$ is a weak homotopy equivalence for any special simplicial space $Y \in \mathcal{S}^+\mathcal{J}$.

Proof. On purely categorical grounds, $\xi : B\Lambda X \to X$ is the composite

$$B\Lambda X = TB_*\Lambda X \xrightarrow{T\psi(\xi)} TSX \xrightarrow{\Phi} X .$$

A glance at the definitions shows that, on 1-simplices, $\psi_1(\xi): \Lambda X \to \Omega X$ agrees with the natural retraction. Since $B_*\Lambda X$ and SX are special, it follows that $\psi_p(\xi): B_p\Lambda X \to S_pX$ is a homotopy equivalence for all p. Therefore $T\psi(\xi)$ is a homotopy equivalence by [18, A.3]. This implies the first clause, and the second clause follows since the composite

$$TY \xrightarrow{T\Psi} TSTY \xrightarrow{\Phi} TY$$

is the identity map and since TY is connected by [17, 11.11].

As a final preliminary, we require the following variant of the comparison theorem (which is similar to Quillen's formulation [27, 3.8]).

Lemma 15.6. Let $\{f^r\}: \{E^r\} \to \{\overline{E}^r\}$ be a morphism of first quadrant spectral sequences. Assume that

(i) $E^\infty = E^0A$, $\overline{E}^\infty = E^0\overline{A}$, and $f^\infty = E^0f$ for filtered graded Abelian groups A and $\overline{A}$ and a filtration preserving isomorphism $f: A \to \overline{A}$ of graded Abelian groups;

(ii) $E^2_{0q} = 0$ and $\overline{E}^2_{0q} = 0$ for $q > 0$; and

(iii) For a given $n \geq 0$, f^2_{pq} is an isomorphism for $q < n$ and all p.

Then f^2_{1n} is an isomorphism and f^2_{2n} is an epimorphism.

Proof. By the argument in [13, p.356 and 357], (iii) implies

(iv) f^r_{pq} is a monomorphism if $q < n$ and an isomorphism if $q \leq n+1-r$.

By induction on q (for fixed p+q), (iv) and (i) imply

(v) f^{∞}_{pq} is an isomorphism if $q < n$ and an epimorphism if $q = n$.

(Use the exact sequences $0 \to F_{p-1}A \to F_pA \to E^{\infty}_{p*} \to 0$.) By downwards induction on r, (ii), (iv), and (v) imply

(vi) f^r_{pq}, $r \geq p$, is an isomorphism if $q < n$ and an epimorphism if $q = n$.

(Use the exact sequences $E^r_{p+r,q-r+1} \xrightarrow{d^r} E^r_{pq} \longrightarrow E^{r+1}_{pq} \longrightarrow 0$.) With $r = p = 2$, the second conclusion follows. Finally, E^r_{1n} is an isomorphism for $r \geq 2$ by an argument just like that at the bottom of [13, p. 357].

We can now use a variant of Quillen's argument [28, §4] to prove the theorem. For $Y \in \mathcal{S}^+\mathcal{J}$, consider the Segal spectral sequence $\{E^rY\}$ [30; 17, 11.14]. $\{E^rY\}$ converges to H_*TY. Since we are taking coefficients in a field R and Y is special, $E^1Y = BH_*Y_1$ and $E^2Y = \mathrm{Tor}^{H_*Y_1}(R, R)$. Clearly $E^2_{00}Y = R$ and $E^2_{0q}Y = 0$ for $q > 0$. From $\Psi : B_*G \to STB_*G = SBG$ we derive $\{E^r\Psi\} : \{E^rB_*G\} \to \{E^rSBG\}$. Since $T\Psi$ is a weak homotopy equivalence, by Proposition 15.5, hypothesis (i) of Lemma 15.6 is satisfied. Definition 15.3 shows that, on 1-simplices, $\Psi_1 = \zeta : G \to \Omega BG$. Therefore $E^2\Psi = \mathrm{Tor}^{\zeta_*}(1, 1)$.

For brevity of notation, we agree to write

$$g = \pi_0 G\,,\ \bar{g} = \pi_0 \Omega BG\,,\ A = H_*G\,,\ B = H_*\Omega BG\,.$$

Let $\iota : A \to \bar{A}$ denote the localization of A at its submonoid g. Note that $\bar{A}$ is A-flat (as a limit of free A-modules [18, 1.2]) and that $R = R \otimes_A \bar{A}$ (because the augmentation $\varepsilon : \bar{A} \to R$ takes the value one on elements of $\bar{g}$ regarded via Lemma 15.2 as elements of $\bar{A}$). Therefore

$$\mathrm{Tor}^{\iota}(1,1): \mathrm{Tor}^{A}(R,R) \to \mathrm{Tor}^{\overline{A}}(R,R)$$

is an isomorphism by [5, VI 4.1.1]. Thus, if $\zeta: \overline{A} \to B$ denotes the unique ring homomorphism such that $\overline{\zeta}\iota = \zeta_*$, then $E^2\Psi$ can be identified with

$$\mathrm{Tor}^{\overline{\zeta}}(1,1): \mathrm{Tor}^{\overline{A}}(R,R) \to \mathrm{Tor}^{B}(R,R) .$$

A, hence also $\overline{A}$, and B are Hopf algebras which, as coalgebras, are direct sums of connected coalgebras. Let $\overline{A}_e$ and B_e denote the components of the identity elements of $\overline{A}$ and B. If $R\overline{g}$ denotes the group ring of $\overline{g}$, then $A = \overline{A}_e \otimes R\overline{g}$ and $B = B_e \otimes R\overline{g}$ as R-algebras because $\overline{g}$ is central in both $\overline{A}$ and B by the admissibility of G and ΩBG. Clearly $\overline{\zeta} = \overline{\zeta}_e \otimes 1$, where $\overline{\zeta}_e$ is the restriction of $\overline{\zeta}$ to $\overline{A}_e$. By the Kunneth theorem for torsion products [5, XI 3.1],

$$\mathrm{Tor}^{\overline{A}}(R,R) = \mathrm{Tor}^{\overline{A}_e}(R,R) \otimes \mathrm{Tor}^{R\overline{g}}(R,R)$$

and similarly for B, hence $\mathrm{Tor}^{\overline{\zeta}}(1,1) = \mathrm{Tor}^{\overline{\zeta}_e}(1,1) \otimes 1$. We shall prove the following statement by induction on n.

$$P_n: \overline{\zeta}_e: \overline{A}_e \to B_e \text{ is an isomorphism in degrees } \leq n.$$

P_0 is trivial and we assume P_{n-1}. Then $E^2_{pq}\Psi$ is an isomorphism for $q < n$ and therefore, by Lemma 15.6, $E^2_{1n}\Psi$ is an isomorphism and $E^2_{2n}\Psi$ is an epimorphism. This means (e.g., by [40, § 7]) that $\overline{\zeta}_e$ induces a bijection between minimal sets of generators and a surjection between minimal sets of defining relations in degrees $\leq$ n and thus implies P_n. Alternatively, as in Quillen [28], P_n follows by application of the five lemma to exact sequences given by initial segments of the bar constructions on $\overline{A}_e$ and B_e.

Bibliography

1. M. G. Barratt and S. Priddy. On the homology of non-connected monoids and their associated groups. Comm. Math. Helv. 47 (1972), 1-14.

2. R. Bott, S. Gitler, and I. M. James. Lectures on Algebraic and Differential Geometry. Lecture Notes in Mathematics. Vol. 279, Springer-Verlag 1972.

3. E. H. Brown, Jr. Twisted tensor products I. Annals of Math. 69 (1959), 223-246.

4. E. H. Brown, Jr. Cohomology theories. Annals of Math. 75(1962), 467-484.

5. H. Cartan and S. Eilenberg. Homological Algebra. Princeton University Press 1956.

6. T. tom Dieck. Klassifikation numerierbarer Bündel. Arch. Math. (Basel) 17 (1966), 395-399.

7. A. Dold. Partitions of unity inthe theory of fibrations. Annals of Math. 78 (1963), 223-255.

8. A. Dold and R. Lashof. Principal quasi-fibrations and fibre homotopy equivalence of bundles. Illinois J. Math. 3 (1959), 285-305.

9. A. Dold and R. Thom. Quasifaserungen und unendliche symmetrische produkte. Annals of Math. 67 (1958), 239-281.

10. V.K.A.M. Gugenheim and J.P. May. On the Theory and Applications of Differential Torsion Products. Memoirs Amer. Math. Soc. No.142, 1974.

11. W. Hurewicz. On the concept of fibre space. Proceedings Nat. Acad. Sci. U.S.A. 41 (1955), 956-961.

12. R. Lashof. Poincaré duality and cobordism. Trans. Amer.Math. Soc. 109, 257-277.

13. S. MacLane. Homology. Springer-Verlag. 1963.

14. S. MacLane. Categories for the Working Mathematician . Springer-Verlag. 1971.

15. P. Malraison. Fibrations and triple algebras. J. Pure and Applied Algebra 3(1973), 287-293.

16. J.P. May. Simplicial Objects in Algebraic Topology. D. Van Nostrand 1967.

17. J.P. May. The Geometry of Iterated Loop Spaces. Lecture Notes in Mathematics Vol. 271. Springer-Verlag 1972.

18. J.P. May. E_∞ spaces, group completions, and permutative categories. London Mathematical Society Lecture Note Series 11, p. 61-94, 1974, Cambridge University Press.

19. J.P. May. $\mathcal{J}$-functors and orientation theory. To appear.

20. J.P. May. On k0-oriented bundle theories. To appear.

21. J.P. May. On duality and completions in homotopy theory. To appear.

22. M.C. McCord. Classifying spaces and infinite symmetric products. Trans. Amer. Math. Soc. 146 (1969), 273-298.

23. R.J. Milgram. The bar construction and Abelian H-spaces. Illinois J. Math. 11 (1967), 242-250.

24. J. Milnor. Construction of universal bundles, II. Annals of Math. 63(1956), 430-436.

25. J. Milnor. On spaces having the homotopy type of a CW-complex. Trans. Amer. Math. Soc. 90 (1959), 272-280.

26. J.C. Moore. Périodicité des Groupes d'Homotopie Stables des Groupes Classiques, d'apres Bott; expose 7. Seminaire Henri Cartan, 1959/60.

27. D.G. Quillen. An application of simplicial profinte groups. Comm. Math. Helv. 44 (1969), 45-60.

28. D.G. Quillen. On the group completion of a simplicial monoid. Preprint.

29. M. Rothenberg and N.E. Steenrod. The cohomology of classifying spaces of H-spaces. Bull. Amer. Math. Soc. 71 (1965), 872-875.

30. G. Segal. Classifying spaces and spectral sequences. Inst. Hautes Etudes Sci. Publ. Math. No.34 (1968), 105-112.

31. J.D. Stasheff. Homotopy associativity of H-spaces I, II. Trans. Amer. Math. Soc. 108 (1963), 275-312.

32. J.D. Stasheff. A classification theorem for fibre spaces. Topology 2 (1963), 239-246.

33. J.D. Stasheff. "Parallel" transport in fibre spaces. Bol. Soc. Mat. Mexicana 11 (1966), 68-84.

34. J.D. Stasheff. Associated fibre spaces. Michigan Math. J. 15 (1968), 457-470.

35. J.D. Stasheff. H-Spaces and classifying spaces: foundations and recent developments. Proceedings of Symposia in Pure Mathematics Vol. 22 Amer. Math. Soc. 1971, 247-272.

36. N.E. Steenrod. The Topology of Fibre Bundles. Princeton Univ. Press. 1951.

37. N.E. Steenrod. A convenient category of topological spaces. Michigan Math. J. 14 (1967), 133-152.

38. N.E. Steenrod. Milgram's classifying space of a topological group. Topology 7 (1968), 349-368.

39. D. Sullivan. Genetics of homotopy theory and the Adams conjecture. Mimeographed notes. M.I.T.

40. C.T.C. Wall. Generators and relations for the Steenrod algebra. Annals of Math. 72 (1960), p.429-444.

NUMBER 156

Sören Illman

Equivariant singular homology and cohomology I

MEMOIRS
OF THE AMERICAN MATHEMATICAL SOCIETY

VOLUME 1 · ISSUE 2 · NUMBER 156 (first of 3 numbers) · MARCH 1975 · CODEN: MAMCAU

Published bimonthly beginning in January, by the American Mathematical Society. Subscriptions for journals published by the American Mathematical Society should be addressed to American Mathematical Society, P.O. Box 1571, Annex Station, Providence, Rhode Island 02901.

Second-class postage permit pending at Providence, Rhode Island, and additional mailing offices.

Memoirs of the American Mathematical Society

Number 156

Sören Illman

Equivariant singular homology and cohomology I

Published by the

AMERICAN MATHEMATICAL SOCIETY

Providence, Rhode Island

VOLUME 1 · ISSUE 2 · NUMBER 156 (first of 3 numbers) · MARCH 1975

Abstract

Let G be a topological group. We construct an equivariant homology and equivariant cohomology theory, defined on the category of all G-pairs and G-maps, which both satisfy all seven equivariant Eilenberg-Steenrod axioms and have a given covariant and contravariant, respectively, coefficient system as coefficients. We also establish some further properties of these equivariant singular homology and cohomology theories, such as, a naturality property in the transformation group, transfer homomorphisms and a cup-product in equivariant singular cohomology with coefficients in a commutative ring coefficient system.

AMS (MOS) Subject Classifications (1970). Primary 55B99, 55B25, 57E99.

ISBN 0-8218-1856-2.

EQUIVARIANT SINGULAR HOMOLOGY AND COHOMOLOGY I

By Sören Illman

Let G be a topological group. By a G-space X we mean a topological space X together with a left action of G on X. A G-pair (X,A) consists of a G-space X and a G-subspace A of X. The notions G-map, G-homotopy, etc. have the usual meaning. In Chapter I of this paper we construct an equivariant homology and cohomology theory, defined on the category of all G-pairs and G-maps, which both satisfy all seven equivariant Eilenberg-Steenrod axioms and which have a given covariant coefficient system k and contravariant coefficient system m, respectively, as coefficients. See Definition 1.2. and Theorems 2.1. and 2.2. in Chapter I for precise statements. We call these equivariant homology and cohomology theories for "equivariant singular homology with coefficients in k" and "equivariant singular cohomology with coefficients in m". The construction of equivariant singular homology and cohomology is very much analogous to the construction of ordinary singular homology and cohomology. The ordinary singular theory in its present form is due to S. Eilenberg [5]. We have chosen the exposition in Eilenberg-Steenrod [6] as our model. This applies especially to the proof of the excision axiom.

For actions of discrete groups equivariant cohomology and homology theories which satisfy all seven equivariant Eilenberg-Steenrod axioms and have predescribed coefficients exist before, see G. Bredon [2], [3] and Th. Bröcker [4].

Received by the editors March 6, 1974.

In Chapter II we establish some further properties of equivariant singular homology and cohomology. We prove a naturality property in the transformation group and construct transfer homomorphisms in both equivariant singular homology and cohomology. Moreover we define a Kronecker index and also a cup-product in equivariant singular cohomology with coefficients in a commutative ring coefficient system. We conclude by proving that this cup-product is commutative.

This paper is a slightly extended and simplified version of chapter III and part of chapter IV of my thesis [8]. A geometrically more intuitive but technically more complicated construction of equivariant homology and cohomology theories which satisfy all seven equivariant Eilenberg-Steenrod axioms and have predescribed coefficients is given in [7], where also some other results from [8] can be found. This paper gives the details for everything stated in [9].

I. EQUIVARIANT SINGULAR THEORY

1. COEFFICIENT SYSTEMS

In the following G denotes an arbitrary topological group. Let R be a ring with identity element. All R-modules will be unitary.

DEFINITION 1.1. A family $\mathfrak{F}$ of subgroups of G is called an orbit type family for G if the following is true: if $H \in \mathfrak{F}$ and H' is conjugate to H, then $H \in \mathfrak{F}$.

Thus the family of all closed subgroups of G, and the family of all finite subgroups of G are examples of orbit type families for G. A more special example is the following. Let $G = O(n)$, and let $\mathfrak{F}$ be the family of all subgroups conjugate to $O(m)$ (standard inbedding) for some m, where $0 \leq m \leq n$.

DEFINITION 1.2. Let $\mathfrak{F}$ be an orbit type family for G. A covariant coefficient system k for $\mathfrak{F}$ over the ring R is a covariant functor from the category of G-spaces of the form G/H, where $H \in \mathfrak{F}$, and G-homotopy classes of G-maps to the category of <u>left</u> R-modules.

A contravariant coefficient system m for $\mathfrak{F}$ over the ring R is a contravariant functor from the category of G-spaces of the form G/H, where $H \in \mathfrak{F}$, and G-homotopy classes of G-maps to the category of <u>right</u> R-modules.

If $\alpha: G/H \to G/K$ is a G-map, and $H,K \in \mathfrak{F}$, we denote

$$k(\alpha) = \alpha_* : k(G/H) \to k(G/K),$$

and

$$m(\alpha) = \alpha^* : m(G/K) \to m(G/H).$$

Let k and k' be covariant coefficient systems for $\mathfrak{F}$. A natural transformation

$$\theta : k \to k'$$

will be called a homomorphism of covariant coefficient systems. If θ is natural equivalence, we vall θ an isomorphism. Similarly for contravariant coefficient systems.

2. THE EXISTENCE THEOREMS FOR EQUIVARIANT SINGULAR HOMOLOGY AND COHOMOLOGY

THEOREM 2.1. Let G be a topological group. Let $\mathfrak{F}$ be an orbit type family for G, and let k be a covariant coefficient system for $\mathfrak{F}$ over the ring R.

Then there exists an equivariant homology theory $H_*^G(\ ;k)$, defined on the category of all G-pairs and all G-maps, and with values in the category of left R-modules, which satisfies all seven equivariant Eilenberg-Steenrod axioms and which has the given coefficient system k as coefficients.

This means:

For each G-pair (X,A) we have a left R-module $H_n^G(X,A;k)$ for every integer n.
Each G-map $f : (X,A) \to (Y,B)$ induces a homomorphism

$$f_* : H_n^G(X,A;k) \to H_n^G(Y,B;k)$$

for every integer n.
Each G-pair (X,A) determines a boundary homomorphism

$$\partial : H_n^G(X,A;k) \to H_{n-1}^G(A;k)$$

for every integer n.

In addition the following axioms are satisfied.

A.1. If f = identity, then f_* = identity.

A.2. $f : (X,A) \to (Y,B)$ and $f' = (Y,B) \to (Z,C)$ are G-maps, then

$$(f'f)_* = f'_* f_* .$$

A.3. For any G-map $f : (X,A) \to (Y,B)$ we have

$$\partial f_* = (f|A)_* \, \partial .$$

A.4. (Exactness axiom). Any G-pair (X,A) gives rise to an exact homology sequence

$$\cdots \overset{i_*}{\leftarrow} H^G_{n-1}(A;k) \overset{\partial}{\leftarrow} H^G_n(X,A;k) \overset{j_*}{\leftarrow} H^G_n(X;k) \overset{i_*}{\leftarrow} H^G_n(A;k) \overset{\partial}{\leftarrow} \cdots$$

A.5. (Homotopy axiom). If $f_0, f_1 : (X,A) \to (Y,B)$ are G-homotopic, then

$$(f_0)_* = (f_1)_* .$$

A.6. (Excision axiom). An inclusion of the form

$$i : (X - U, A - U) \to (X,A)$$

where $\overline{U} \subset A^{\circ}$ (U and A are G-subsets) induces an isomorphism

$$i_* : H^G_n(X - U, A - U; k) \overset{\cong}{\to} H^G_n(X,A; k)$$

for every integer n.

A.7. (Dimension axiom). If $H \varepsilon \mathfrak{F}$, then

$$H^G_m(G/H;k) = 0 \quad \text{for all } m \neq 0.$$

Moreover, for every $H \varepsilon \mathfrak{F}$ we have an isomorphism

$$\gamma : H^G_0(G/H;k) \overset{\cong}{\to} k(G/H)$$

such that if also $K \varepsilon \mathfrak{F}$ and $\alpha : G/H \to G/K$ is a G-map, then the diagram

$$\begin{array}{ccc} H_0^G(G/H;k) & \xrightarrow{\gamma} & k(G/H) \\ \alpha_* \downarrow & & \downarrow \alpha_* \\ H_0^G(G/K;k) & \xrightarrow{\gamma} & k(G/H) \end{array}$$

commutes.

Moreover, this equivariant homology theory has no "negative homology", that is, for any G-pair (X,A) we have

$$H_m^G(X,A;k) = 0 \quad \text{if } m < 0.$$

We call this equivariant homology theory $H_*^G(\ ;k)$ for "equivariant singular homology with coefficients in k".

THEOREM 2.2. Let G be a topological group. Let $\mathfrak{F}$ be an orbit type family for G, and let m be a contravariant coefficient system for $\mathfrak{F}$ over the ring R.

Then there exists an equivariant cohomology theory $H_G^*(\ ;m)$ defined on the category of all G-maps, and with values in the category of right R-modules, which satisfies all seven equivariant Eilenberg-Steenrod axioms, and which has the given coefficient system m as coefficients.

That $H_G^*(\ ;m)$ satisfies the dimension axiom means the following. If $H \varepsilon \mathfrak{F}$, then

$$H_G^p(G/H;m) = 0 \quad \text{for all } p \neq 0,$$

and there is an isomorphism

$$\xi : H_G^0(G/H;m) \xrightarrow{\cong} m\ (G/H),$$

such that if also $K \varepsilon \mathfrak{F}$ and $\alpha : G/H \to G/K$ is a G-map, then the diagram

$$\begin{array}{ccc} H_G^0(G/K;m) & \xrightarrow{\xi} & m(G/K) \\ \downarrow \alpha^* & & \downarrow \alpha^* \\ H_G^0(G/H;m) & \xrightarrow{\xi} & m(G/H) \end{array}$$

commutes. The meaning of the rest of Theorem 2.2. is clear. Let us point out that also here the excision axiom is satisfied in the strong form that the G-subset U need not be open. We call the equivariant cohomology theory $H_G^*(\ ;m)$ for "equivariant singular cohomology with coefficients in m". We have $H_G^p(X,A;m) = 0$ if $p < 0$, for every G-pair (X,A).

EXAMPLE. As a simple illustration we determine the equivariant singular homology and cohomology of the following example. Let $G = S^1$ the circle group and $X = S^2$ the two-sphere. Assume that S^1 acts on S^2 by the standard rotation leaving the north and south poles fixed, and acting freely elsewhere. Let X_1 and X_2 denote the northern and southern hemispheres, respectively, and $X_0 = X_1 \cap X_2$ the equator.

Now assume that $\mathfrak{F}$ is an orbit type family, such that, both $G \varepsilon \mathfrak{F}$ and $\{e\} \varepsilon \mathfrak{F}$. Let, as before, R be a ring with identity element and k a covariant coefficient system for $\mathfrak{F}$ over R. It is a formal consequence of the axioms that we in this situation have the following exact Mayer-Vietoris sequence

$$0 \longleftarrow H_0^G(X;k) \xleftarrow{j_{1*}+j_{2*}} H_0^G(X_1;k) \oplus H_0^G(X_2;k) \xleftarrow{(i_{1*},-i_{2*})} H_0^G(X_0;k) \xleftarrow{\partial} H_1^G(X;k) \leftarrow 0.$$

Since both X_1 and X_2 are G-homotopy equivalent to a point and $X_0 \cong G$ as G-spaces, it follows that the above exact sequence equals

$$0 \leftarrow H_0^G(X;k) \leftarrow k(G/G) \oplus k(G/G) \xleftarrow{(p_*,-p_*)} k(G/\{e\}) \leftarrow H_1^G(X;k) \leftarrow 0$$

where $p_* : k(G/\{e\}) \to k(G/G)$ is induced by the G-map $p : G \to G/G$. Thus

$$H_0^G(X;k) \cong (k(G/G) \oplus k(G/G))/\{(p_*(a), -p_*(a)) | a \in k(G)\}$$

$$H_1^G(X;k) \cong \ker (p_* : k(G) \to k(G/G))$$

$$H_m^G(X;k) = 0 \qquad \text{for } m \neq 0,1.$$

Let us now consider this result for some specific convariant coefficient systems. Let the orbit type family $\mathfrak{F}$ be, for example, the family of all closed subgroups of $G = S^1$, and let R be the ring of integers Z.

1. Define a convariant coefficient system k_1 as follows. Let $k_1(G/H) = Z$ if $H \neq G$ and $k(G/G) = Z_2$, and let $p : G/H \to G/G$, where $H \neq G$, induce the natural projection $Z \to Z_2$ and let all other induced homomorphisms on k_1 be the identity on Z. Then

$$H_0^G(X;k_1) \cong Z_2$$

$$H_1^G(X;k_1) \cong Z$$

$$H_m^G(X;k_1) = 0 \quad \text{for } m \neq 0,1.$$

2. Define k_2 by : $k_2(G/\{e\}) = Z$, and $k_2(G/H) = 0$ for $H \neq \{e\}$. Then

$$H_0^G(X;k_2) = 0$$

$$H_1^G(X;k_2) \cong Z$$

$$H_m^G(X;k_2) = 0 \quad \text{for } m \neq 0,1.$$

3. Define k_3 by : $k_3(G/H) = 0$ for $H = G$, and $k_3(G/G) = Z$. Then

$$H_0^G(X;1_3) = Z \oplus Z$$

$$H_p^G(X;k_3) = 0 \qquad \text{for } p \neq 0.$$

Observe that this equals the ordinary singular homology of the fixed point set X^G.

4. Define k_4 by : $k_4(G/H) = Z$ for every closed subgroup H of G and all induced homomorphisms are the identity on Z. Then

$$H_0^G(X;k_4) \cong Z$$

$$H_p^G(X;k_4) = 0 \qquad \text{for } p \neq 0.$$

Observe that this equals the ordinary singular homology of the orbit space $G\backslash X$.

5. Define k_5 by : $k_5(G/H) = Z$ for every closed subgroup H of G, and every G-map $\alpha : G/H \to G/K$, where $H \subset K$ but $H \neq K$, induces the zero homomorphism, and every G-map $\beta : G/H \to G/H$ induces the identity on Z. Then

$$H_0^G(X;k_5) \cong Z \oplus Z$$

$$H_1^G(X;k_5) \cong Z$$

$$H_m^G(X;k_5) = 0 \qquad \text{for } m \neq 0,1.$$

To determine the equivariant singular cohomology of the G-space X we use the analogous exact Mayer-Vietoris sequence for cohomology

$$0 \to H_G^0(X;m) \xrightarrow{(j_1^*,j_2^*)} H_G^0(X_1;m) \oplus H_G^0(X_2;m) \xrightarrow{i_1^*-i_2^*} H_G^0(X_0;m) \xrightarrow{\partial} H_G^1(X;m) \to 0.$$

In the same way as above we see that this exact sequence equals

$$0 \to H^0_G(X;m) \to m(G/G) \oplus m(G/G) \xrightarrow{p^*(\pi_1 - \pi_2)} m(G/\{e\}) \to H^1_G(X;m) \to 0,$$

where $\pi_i : m(G/G) \oplus m(G/G) \to m(G/G)$ denotes the projection onto the i:th factor, $i = 1,2$. Thus

$$H^0_G(X;m) \cong \ker \ (p^* \ (\pi_1 - \pi_2) : m(G/G) \oplus m(G/G) \to m(G/\{e\})$$

$$H^1_G(X;m) \cong m(G/\{e\})/\mathrm{im}\ p^*(\pi_1 - \pi_2)$$

$$H^q_G(X;m) = 0 \qquad \text{for } q \neq 0,1.$$

3. CONSTRUCTION OF EQUIVARIANT SINGULAR HOMOLOGY

Let Δ_n be the standard n-simplex, that is, $\Delta_n = \{(x_0,\ldots,x_n) \varepsilon R^{n+1} \mid \sum_{i=0}^{n} x_i = 1,\ x_i \geq 0\}$. We have the face maps

$$e^i_n : \Delta_{n-1} \to \Delta_n \qquad i = 0,\ldots,n,$$

defined by $e^i_n(x_0,\ldots,x_{n-1}) = (x_0,\ldots,x_{i-1},0,x_i,\ldots,x_{n-1})$. The identity

$$e^i_n e^j_{n-1} = e^j_n e^{i-1}_{n-1}, \quad \text{where } 0 \leq j < i \leq n, \quad \text{is valid.}$$

Now let K be a subgroup of G. We call the G-space $\Delta_n \times G/K$ for the standard equivariant n-simplex of type K. We have the face maps

$$e^i_n \times \mathrm{id} : \Delta_{n-1} \times G/K \to \Delta_n \times G/K \qquad i = 0,\ldots,n.$$

DEFINITION 3.1. A G-map

$$T : \Delta_n \times G/K \to X$$

is called an equivariant singular n-simplex in X. We call K for the type of T and denote

$$t(T) = K.$$

The equivariant singular (n-1)-simplex

$$T^{(i)} = T(e_n^i \times id) : \Delta_{n-1} \times G/K \to X$$

is called the i:th face of T, $i = 0,\dots,n$.

DEFINITION 3.2. Let $\mathfrak{F}$ be an orbit type family for G. We say that the equivariant n-simplex $T : \Delta_n \times G/K \to X$ belongs to $\mathfrak{F}$ if $K \in \mathfrak{F}$.

Given an equivariant singular n-simplex $T : \Delta_n \times G/K \to X$ belongs to $\mathfrak{F}$, we form

$$Z_T \otimes k(G/t(T)) = Z_T \otimes k(G/K).$$

Here Z_T denotes the infinite cyclic group on the generator T, and the tensor product is over the integers. The left R-module structure on $k(G/t(T))$ makes $Z_T \otimes k(G/t(T))$ into a left R-module such that the map $i : k(G/t(T)) \to Z_T \otimes k(G/t(T))$ defined by $i(a) = T \otimes a$ is an isomorphism of left R-modules.

DEFINITION 3.3. We define

$$\hat{C}_n^G(X;k) = \sum_{t(T)\in\mathfrak{F}} \oplus\, (Z_T \otimes k(G/t(T))$$

where the direct sum is over all equivariant singular n-simplexes in X, which belong to $\mathfrak{F}$. Thus for $n < 0$ we have $\hat{C}_n^G(X;k) = 0$.

The boundary homomorphism

$$\hat{\partial}_n : \hat{C}_n^G(X;k) \to \hat{C}_{n-1}^G(X;k)$$

is defined in the usual way, that is, for $n \leq 0$ we define $\hat{\partial}_n = 0$, and if $n > 0$ and T is an equivariant singular n-simplex in X and $a \in k(G/t(T))$, we define

$$\hat{\partial}_n(T \otimes a) = \sum_{i=0}^{n} (-1)^{(i)} T^{(i)} \otimes a.$$

The standard calculation then shows that $\hat{\partial}_{n-1}\hat{\partial}_n = 0$.

Thus we get the chain complex

$$\hat{S}^G(X;k) = \{\hat{C}_n^G(X;k), \hat{\partial}_n\}.$$

Our main interest is not in the chain complex $\hat{S}^G(X;k)$, but in a quotient of it. We now proceed to define this quotient.

Let

$$h : \Delta_n \times G/K \to \Delta_n \times G/K'$$

be a G-map which covers $id : \Delta_n \to \Delta_n$. Every $x \in \Delta_n$ gives rise to a G-map

$$h_x : G/K \to G/K'$$

defined by $h_x(gK) = pr_2\, h(x,gK)$, where $pr_2 : \Delta_n \times G/K' \to G/K'$ is projection onto the second factor.

LEMMA 3.4. Let the notation be as above and let $x,y \in \Delta_n$. Then the G-maps $h_x, h_y : G/K \to G/K'$ are G-homotopic.

PROOF. Define $F: I \times G/K \to G/K'$ by $F(t,gK) = pr_2\, h((1-t)x + ty, gK)$. Then F is a G-homotopy from h_x to h_y. q.e.d.

Thus, if $K,K' \in \mathfrak{F}$, it follows that $(h_x)_* = (h_y)_* : k(G/K) \to k(G/K')$, that is, the G-map h induces in this way a unique homomorphism from $k(G/K)$ to $k(G/K')$. We denote this homomorphism by

$h_* : k(G/K) \to k(G/K')$.

Let for the moment $\mathcal{G}_n \subset \hat{C}^G_n(X;k)$ denote the set of all elements in $\hat{C}^G_n(X;k)$ that have at most one coordinate $\neq 0$. Every element in $\mathcal{G}_n$ has a unique expression of the form $T \otimes a$, where T is some equivariant singular n-simplex belonging to $\mathcal{F}$ in X, and $a \in k(G/t(T))$.

We define a relation $\sim$ in $\mathcal{G}_n$ in the following way. Let $T \otimes a$ and $T' \otimes a'$ be two arbitrary elements in $\mathcal{G}_n$, where $T : \Delta_n \times G/K \to X$ and $T' : \Delta_n \times \Delta_n \times G/K' \to X$ are equivariant singular n-simplexes belonging to $\mathcal{F}$ in X, and $a \in k(G/K_n)$, $a' \in k(G/K')$. We how define

$$T \otimes a \sim T' \otimes a' \Leftrightarrow \begin{cases} \text{there exists a G-map} \\ h : \Delta_n \times G/K \to \Delta_n \times G/K' \text{ covering} \\ id : \Delta_n \to \Delta_n \text{ such that } T = T'h \text{ and} \\ h_*(a) = a'. \end{cases}$$

DEFINITION 3.5. Let the notation be as above. We define

$$\overline{C}^G_n(X;k) \subset \hat{C}^G_n(X;k)$$

to be the submodule of $\hat{C}^G_n(X;k)$ consisting of all elements of the form

$$\sum_{i=1}^{s} (T_i \otimes a_i \sim T_i' \otimes a_i')$$

where $T_i \otimes a_i \sim T_i' \otimes a_i'$ or $T_i' \otimes a_i' \sim T_i \otimes a_i$, for $i = 1,\dots,s$.

DEFINITION 3.6. We define the left R-module $C^G_n(X;k)$ by

$$C^G_n(X;k) = \hat{C}^G_n(X;k)/\overline{C}^G_n(X;k).$$

Now observe that if $T \otimes a \sim T' \otimes a'$, then also $T^{(i)} \otimes a \sim (T')^{(i)} \otimes a'$, $i = 0,\dots,n$. It follows that the boundary homomorphism $\hat{\partial}_n : \hat{C}^G_n(X;k) \to \hat{C}^G_{n-1}(X;k)$

restricts to $\overline{\partial}_n : \overline{C}^G_n(X;k) \to \overline{C}^G_{n-1}(X;k)$, and thus induces a boundary homomorphism

$$\partial_n : C^G_n(X;k) \to C^G_{n-1}(X;k).$$

Since $\hat{\partial}_{n-1}\hat{\partial}_n = 0$ it follows that $\overline{\partial}_{n-1}\overline{\partial}_n = 0$ and $\partial_{n-1}\partial_n = 0$. Thus we have the chain complexes

$$\overline{S}^G(X;k) = \{\overline{C}^G_n(X;k), \overline{\partial}_n\}$$

$$S^G(X;k) = \{C^G_n(X;k), \partial_n\}.$$

It is the chain complex $S^G(X;k)$ that gives us the equivariant singular homology groups with coefficients in k of X. We shall now consider the relative case.

Let (X,A) be a G-pair. The inclusion $i : A \to X$ induces a monomorphism of chain complexes

$$\hat{i} : \hat{S}^G(A;k) \to \hat{S}^G(X;k).$$

Moreover, the image $\hat{i}(\hat{C}^G_n(A;k))$ is a direct summand in $\hat{C}^G_n(X;k)$, for each n. We identify $\hat{C}^G_n(A;k)$ with $\hat{i}(\hat{C}^G_n(A;k))$, that is, we consider $\hat{S}^G(A;k)$ as a subcomplex of $\hat{S}^G(X;k)$ through the monomorphism $\hat{i}$. We define

$$\hat{C}^G_n(X,A;k) = \hat{C}^G_n(X;k)/\hat{C}^G_n(A;k)$$

and denote the corresponding chain complex by $\hat{S}^G(X,A;k)$. We have the short exact sequence of chain complexes

$$0 \to \hat{S}^G(A;k) \to \hat{S}^G(X;k) \to \hat{S}^G(X,A;k) \to 0.$$

Clearly $\hat{i}$ restricts to $\overline{i}$: $\overline{S}^G(A;k) \to \overline{S}^G(X;k)$ and hence $\hat{i}$ induces

$$i : S^G(A;k) \to S^G(X;k).$$

LEMMA 3.7. The homomorphism $i : S^G(A;k) \to S^G(X;k)$ induced by $\hat{i}$ is a monomorphism. Moreover, $i(C_n^G(A;k))$ is a direct summand in $C_n^G(X;k)$ for each n.

PROOF. Define a homomorphism

$$\hat{\alpha} : \hat{C}_n^G(X;k) \to \hat{C}_n^G(A;k)$$

by

$$\hat{\alpha}(T \otimes a) = \begin{cases} T \otimes a & \text{if } \operatorname{Im}(T) \subset A \\ 0 & \text{if } \operatorname{Im}(T) \cap (X-A) \neq \emptyset. \end{cases}$$

Thus $\hat{\alpha}$ is a left inverse to $\hat{i}$. If $T \otimes a \sim T' \otimes a'$ it follows that $\operatorname{Im}(T) = \operatorname{Im}(T')$. Therefore $\hat{\alpha}$ restricts to $\bar{\alpha} : \bar{C}_n^G(X;k) \to \bar{C}_n^G(A;k)$, and hence $\hat{\alpha}$ induces a homomorphism

$$\alpha : C_n^G(X;k) \to C_n^G(A;k)$$

which is a left inverse to i. q.e.d.

We define

$$C_n^G(X,A;k) = C_n^G(X;k)/C_n^G(A;k),$$

and denote the corresponding chain complex by

$$S^G(X,A;k) = \{C_n^G(X,A;k), \partial_n\}.$$

DEFINITION 3.8. We define

$$H_n^G(X,A;k)$$

to be the n:th homology module of the chain complex $S^G(X,A;k)$.

By Lemma 3.7 and by definition we have the short exact sequence

$$0 \to S^G(A;k) \to S^G(X;k) \to S^G(X,A;k) \to 0.$$

This gives us the boundary homomorphism

$$\partial : H_n^G(X,A;k) \to H_{n-1}^G(A;k)$$

and the exact homology sequence of a G-pair (X,A) in the standard way.

More or less as a side remark let us point out the following. Define the chain complex $\overline{S}^G(X,A;k)$ to be the quotient of $\overline{S}^G(X;k)$ by $\overline{S}^G(A;k)$. Then the sequence

$$0 \to \overline{S}^G(X,A;k) \to \hat{S}^G(X,A;k) \to S^G(X,A;k) \to 0$$

is exact. This can be seen "directly" or by drawing the obvious commutative 3×3 diagram and applying the 3×3 lemma.

We denote the homology groups of the chain complexes $\overline{S}^G(X,A;k)$ and $\hat{S}^G(X,A;k)$ by $\overline{H}_*^G(X,A;k)$ and $\hat{H}_*^G(X,A;k)$, respectively. Thus we get a long exact sequence

$$\cdots \leftarrow \overline{H}_{n-1}^G(X,A;k) \overset{\partial}{\leftarrow} H_n^G(X,A;k) \leftarrow \hat{H}_n^G(X,A;k) \leftarrow \overline{H}_n^G(X,A;k) \leftarrow \cdots$$

Our main interest is in $H_*^G(\ ;k)$. But in the process of the proof of the fact that $H_*^G(\ ;k)$ satisfies all seven equivariant Eilenberg-Steenrod axioms it will be shown that both $\overline{H}_*^G(\ ;k)$ and $\hat{H}_*^G(\ ;k)$ satisfy the first six axioms.

Let (X,A) and (Y,B) be G-pairs and let $f : (X,A) \to (Y,B)$ be a G-map. If $T : \Delta_n \times G/K \to X$ is an equivariant singular n-simplex belonging to $\mathfrak{F}$ in X, then $fT : \Delta_n \times G/K \to Y$ is an equivariant singular n-simplex belonging to $\mathfrak{F}$ in Y. Thus we get a homomorphism

$$\hat{f}_{\#} : \hat{C}_n^G(X,A;k) \to \hat{C}_n^G(Y,B;k)$$

by defining $\hat{f}_{\#}(T \otimes a) = (fT) \otimes a$. Since $(fT)^{(i)} = fT^{(i)}$, for $i = 0,\ldots,n$, it follows that the homomorphisms $\hat{f}_{\#}$ form a chain homomorphism. If $T \otimes a \sim T' \otimes a'$, then $(fT) \otimes a \sim (fT') \otimes a'$, and hence $f_{\#}$ restricts to $\overline{f}_{\#} : \overline{S}^G(X,A;k) \to \overline{S}^G(Y,B;k)$ and hence $\hat{f}_{\#}$ induces a chain homomorphism

$$f_{\#} : S^G(X,A;k) \to S^G(Y,B;k).$$

Now $f_{\#}$ induces a homomorphism $f_* : H_n^G(X,A;k) \to H_n^G(Y,B;k)$ for every n. It is now clear that we have proved everything up to the exactness axiom in the statement of Theorem 2.1.

In the next section we construct the equivariant singular cohomology theory and establish at the same time everything up to the exactness axiom in the statement of Theorem 2.2. The homotopy, excision, and dimension axioms will be proved simultaneously for homology and cohomology in sections 5, 6, and 7.

4. CONSTRUCTION OF EQUIVARIANT SINGULAR COHOMOLOGY

Let G, $\mathcal{F}$ and R be as before. Let m be a contravariant coefficient system for $\mathcal{F}$ over the ring R. Recall that each $m(G/K)$, where $K \varepsilon \mathcal{F}$, is a right R-module.

Let X be a G-space. We denote

$$\hat{C}_n^G(X) = \sum_{t(T)\varepsilon\mathcal{F}} \oplus Z_T$$

where the direct sum is over all equivariant singular n-simplexes belonging to $\mathcal{F}$ in X. That is, $\hat{C}^G_n(X)$ is the free abelian group on all equivariant singular n-simplexes belonging to $\mathcal{F}$ in X. The boundary homomorphism

$$\hat{\partial}_n : \hat{C}^G_n(X) \to \hat{C}^G_{n-1}(X)$$

is defined by

$$\hat{\partial}_n(T) = \sum_{i=0}^{n} (-1)^i T^{(i)}.$$

Then $\hat{\partial}_{n-1}\hat{\partial} = 0$, and we have the chain complex

$$\hat{S}^G(X) = \{\hat{C}^G_n(X), \hat{\partial}_n\}.$$

That is

$$\hat{S}^G(X) = \hat{S}^G(\ ;Z)$$

where Z denotes the covariant coefficient for which Z(G/K) = Z for every K ε $\mathcal{F}$, (and all the induced homomorphisms are the identity on Z).

Denote

$$M = \sum_{K \varepsilon \mathcal{F}} \oplus\, m(G/K),$$

where the direct sum is over all subgroups belonging to $\mathcal{F}$. By $\mathrm{Hom}_Z(\hat{C}^G_n(X),M)$ we denote the set of all homomorphisms of abelian groups from $\hat{C}^G_n(X)$ to M. The right R-module structure on M makes $\mathrm{Hom}_Z(\hat{C}^G_n(X),M)$ into a right R-module.

DEFINITION 4.1. We define the right R-module $\hat{C}^n_G(X;m)$ by

$$\hat{C}^n_G(X;m) = \mathrm{Hom}_t(\hat{C}^G_n(X),M).$$

Here $\mathrm{Hom}_t(\hat{C}^G_n(X),M)$ consists of all homomorphisms of abelian groups $c : \hat{C}^G_n(X) \to M$ which satisfy the condition

$$c(T) \in m(G/t(T))$$

for every equivariant singular n-simplex T belonging to $\mathfrak{F}$ in X. Thus $\hat{C}^n_G(X;m)$ is a submodule of the right R-module $\mathrm{Hom}_Z(\hat{C}^G_n(X),M)$.

For any homomorphism $\hat{\alpha} : \hat{C}^G_n(Y)$ we have the dual homomorphism

$$\hat{\alpha}^* : \mathrm{Hom}_Z(\hat{C}^G_m(Y),M) \to \mathrm{Hom}_Z(\hat{C}^G_n(X),M)$$

defined by $\hat{\alpha}^*(c) = c\,\hat{\alpha}$, $c \in \mathrm{Hom}_Z(\hat{C}^G_m(Y),M)$. Observe that $\hat{\alpha}^*$ is a homomorphism of right R-modules.

DEFINITION 4.2. We call a homomorphism

$$\hat{\alpha} : \hat{C}^G_n(X) \to \hat{C}^G_m(Y)$$

for "type preserving" if the following condition is satisfied. For every equivariant singular n-simplex T, belonging to $\mathfrak{F}$, in X we have

$$\hat{\alpha}(T) = \sum_{j=0}^{q} r_j S_j, \quad r_j \in Z,$$

with $t(S_j) = t(T)$, for $j = 0,\ldots,q$. (Each S_j is an equivariant singular m-simplex, belonging to $\mathfrak{F}$, in Y).

Clearly the dual $\hat{\alpha}^*$ of a "type preserving" homomorphism $\hat{\alpha} : \hat{C}^G_n(X) \to \hat{C}^G_m(Y)$ restricts to give a homomorphism

$$\hat{\alpha}^{\#} : \hat{C}^m_G(Y;m) \to \hat{C}^n_G(X;m)$$

which we again call the dual of $\hat{\alpha}$.

The boundary homomorphism $\hat{\partial}_n$ is "type preserving". We denote its dual by

$$\hat{\delta}^{n-1} : \hat{C}^{n-1}_G(X;m) \to \hat{C}^n_G(X;m)$$

and call it the coboundary homomorphism. Then $\hat{\delta}^n\hat{\delta}^{n-1} = 0$, and we have the cochain complex

$$\hat{S}_G(X;m) = \{\hat{C}^n_G(X;m),\hat{\delta}^n\}.$$

Let (X,A) be a G-pair, and let $i : A \to X$ be the inclusion. Both the monomorphism $\hat{i}_{\#} : \hat{C}^G_n(A) \to \hat{C}^G_n(X)$ and the homomorphism $\hat{\alpha} : \hat{C}^G_n(X) \to \hat{C}^G(A)$, which is a left inverse to $\hat{i}_{\#}$, are "type preserving" (see the proof of Lemma 3.7). We denote the dual of $\hat{i}_{\#}$ by

$$\hat{i}^{\#} : \hat{C}^n_G(X;m) \to \hat{C}^n_G(A;m).$$

Then $\hat{\alpha}^{\#}$ is a right inverse to $\hat{i}^{\#}$, and it follows in particular that $\hat{i}^{\#}$ is onto.

Define $\hat{C}^n_G(X,A;m)$ to be the submodule of $\hat{C}^n_G(X;m) = \mathrm{Hom}_t(\hat{C}^G_n(X),M)$ consisting of all the homomorphisms that vanish on $\hat{C}^G_n(A)$. That is, we have a short exact sequence

$$0 \to \hat{C}^n_G(X,A;m) \to \hat{C}^n_G(X;m) \xrightarrow{\hat{i}^{\#}} \hat{C}^n_G(A;m) \to 0.$$

Since $\hat{i}^{\#}$ has a right inverse $\hat{\alpha}^{\#}$ it follows that the above sequence splits.

We have the corresponding short exact sequence of cochain complexes

$$0 \to \hat{S}_G(X,A;m) \to \hat{S}_G(X;m) \xrightarrow{\hat{i}^{\#}} \hat{S}_G(A;m) \to 0.$$

Let $f : (X,A) \to (Y,B)$ be a G-map. The induced homomorphism $\hat{f}_{\#} : \hat{C}^G_n(X) \to \hat{C}^G_n(Y)$ is "type preserving". We denote its dual by $\hat{f}^{\#} : \hat{C}^n_G(Y;m) \to \hat{C}^n_G(X;m)$. These homomorphisms commute with the coboundary homomorphisms and also restrict to the corresponding relative cochain groups. Thus we have a homomorphism of cochain complexes

$$\hat{f}^{\#}: \hat{S}_G(Y,B;m) \to \hat{S}_G(X,A;m).$$

In constructing equivariant singular homology we took a quotient of the "roof" chain complex. Here, dually, in constructing equivariant singular cohomology we shall consider an appropriate subcomplex of $\hat{S}_G(X;m)$. We now define this one.

DEFINITION 4.3. We define $C_G^n(X;m)$ to be the submodule of $\hat{C}_G^n(X;m) = \mathrm{Hom}_t(\hat{C}_n^G(X),M)$ consisting of all $c \in \mathrm{Hom}_t(\hat{C}_n^G(X),M)$ which satisfy the following condition. If $T : \Delta_n \times G/K \to X$ and $T' : \Delta_n \times G/K' \to X$ are equivariant singular n-simplexes belonging to $\mathcal{F}$ in X, and $h : \Delta_n \times G/K \to \Delta_n \times G/K'$ is a G-map, which covers $id : \Delta_n \to \Delta_n$, such that $T = T'h$, then

$$c(T) = h^* c(T') \in m(G/K).$$

Here $h^* : m(G/K') \to m(G/K)$ is the homomorphism induced by h, (see Lemma 3.4.).

DEFINITION 4.4. We say that a "type preserving" homomorphism $\hat{\alpha} : \hat{C}_n^G(X) \to \hat{C}_m^G(Y)$ "preserves the relation $\sim$" if $\hat{\alpha}$ besides being a homomorphism also determines the following extra structure. First, there exists a natural number q, and $q + 1$ integers r_j, $j = 0,\ldots,q$, such that, for any equivariant singular n-simplexes T and T' belonging to $\mathcal{F}$ in X we have

$$\hat{\alpha}(T) = \sum_{j=0}^{q} r_j S_j, \quad \text{and} \quad \hat{\alpha}(T') = \sum_{j=0}^{q} r_j S'_j,$$

where S_j and S'_j denote equivariant singular m-simplexes belonging to $\mathcal{F}$ in Y. Secondly, if $h : \Delta_n \times G/K \to \Delta_n \times G/K'$ is a G-map, which covers $id : \Delta_n \to \Delta_n$, then $\hat{\alpha}$ determines $q + 1$ G-maps

$$h_j : \Delta_m \to G/K \to \Delta_m \times G/K', \ j = 0,\ldots,q,$$

which cover $id : \Delta_m \to \Delta_m$, such that

$$[h_j] = [h] : G/K \to G/K', \quad j = 0,\ldots,q,$$

where $[h_j]$ and $[h]$ denote the G-homotopy classes determined by h_j and h, respectively (see Lemma 3.4), and such that if $T = T'h$ then

$$S_j = S_j'h_j \qquad j = 0,\ldots,q.$$

LEMMA 4.5. Assume that $\hat{\alpha} : \hat{C}_n^G(X) \to \hat{C}_m^G(Y)$ "preserves the relation $\sim$". Then its dual $\hat{\alpha}^{\#} : \hat{C}_G^m(Y;m) \to \hat{C}_G^n(X;m)$ restricts to a homomorphism

$$\alpha^{\#} : C_G^m(Y;m) \to C_G^n(X;m).$$

PROOF. Let $c \in C_G^m(Y;m)$. We claim that then $\hat{\alpha}^{\#}(c) \in C_G^n(X;m)$. Let $T : \Delta_n \times G/K \to X$, and $T' : \Delta_n \times G/K' \to X$, and let $h : \Delta_n \times G/K \to \Delta_n \times G/K'$ be a G-map, which covers $id : \Delta_n \to \Delta_n$, such that $T = T'h$. Preserving the same notation as in Definition 4.4 we then have

$$(\hat{\alpha}^{\#}(c))(T) = c(\hat{\alpha}(T)) = \sum_{j=0}^{q} r_j\, c\,(S_j) = \sum_{j=0}^{q} r_j\, c\,(S_j'h_j) =$$

$$h^*(\sum_{j=0}^{q} r_j\, c\,(S_j')) = h^*(c(\hat{\alpha}(T'))) = h^*(\hat{\alpha}^{\#}(c)(T')). \qquad \text{q.e.d.}$$

We also call $\alpha^{\#}: C_G^m(Y;m) \to C_G^n(X;m)$ for the dual of $\hat{\alpha} : \hat{C}_n^G(X) \to \hat{C}_m^G(Y)$.

The boundary homomorphism $\hat{\partial}_n : \hat{C}_n^G(X) \to \hat{C}_{n-1}^G(X)$ "preserves the relation $\sim$". To see that the conditions of Definition 4.4 are satisfied we simply take $q = n$, and $r_j = (-1)^j$, $j = 0,\ldots,n$, and given h the G-map h_j is the restriction of h to the j:th face, that is $h_j = h| : e_n^j(\Delta_{n-1}) \times G/K \to e_n^j(\Delta_{n-1}) \times G/K'$, $j = 0,\ldots,n$. Thus the coboundary $\hat{\delta}^{n-1} : C_G^{n-1}(X;m) \to \hat{C}_G^n(X;m)$ restricts to

$$\delta^{n-1} : C_G^{n-1}(X;m) \to C_G^n(X;m).$$

Then $\delta^n\delta^{n-1} = 0$, and we have the cochain complex

$$0 \to S_G(X,A;m) \to S_G(X;m) \overset{i^\#}{\to} S_G(A;m) \to 0,$$

where by definition $S_G(X,A;m) = \ker i^\#$. In each degree the above short exact sequence splits as a sequence of right R-modules. We now define the equivariant singular cohomology groups.

DEFINITION 4.6. We define

$$H^n_G(X,A;m)$$

to be the n:th homology module of the cochain complex $S_G(X,A;m)$.

Let $f : (X,A) \to (Y,B)$ be a G-map. The induced homomorphism $\hat{f}_\# : \hat{C}^G_n(X) \to \hat{C}^G_n(Y)$ clearly "preserves the relation ~". It follows that the dual $f^\#$ of $\hat{f}_\#$ gives us a homomorphism of cochain complexes

$$f^\# : S_G(Y,B;m) \to S_G(X,A;m)$$

and hence the induced homomorphisms $f^* : H^n_G(Y,B;m) \to H^n_G(X,A;m)$. It is now clear that so far we have proved everything up to the exactness axiom in the statement of Theorem 2.2.

We can also define a cochain complex $\overline{S}_G(X;m)$ by

$$\overline{S}_G(X;m) = \hat{S}_G(X;m)/S_G(X;m).$$

Both $\hat{i}^\# : \hat{C}^n_G(X;m) \to \hat{C}^n_G(A;m)$ and $\hat{\alpha}^\# : \hat{C}^n_G(A;m) \to \hat{C}^n_G(X;m)$ induce homomorphisms $\overline{i}^\#: \overline{C}^n_G(X;m) \to \overline{C}^n_G(A;m)$ and $\overline{\alpha}^\#: C^n_G(A;m) \to C^n_G(X;m)$, and $\overline{\alpha}^\#$ is a right inverse to $\overline{i}^\#$. Thus we have a short exact sequence of cochain complexes

$$0 \to S_G(X,A;m) \to \overline{S}_G(X;m) \overset{\overline{i}^\#}{\to} \overline{S}_G(A;m) \to 0$$

where by definition $\overline{S}_G(X,A;m) = \ker \overline{i}^\#$, which in each degree splits as a sequence of right R-modules.

Applying the 3 × 3-lemma we now see that we have the short exact sequence of cochain complexes

$$0 \to S_G(X,A;m) \to \hat{S}_G(X,A;m) \to \overline{S}_G(X,A;m) \to 0.$$

Define $\hat{H}^n_G(X,A;m)$ and $\overline{H}^n_G(X,A;m)$ to be the n:th homology modules of the cochain complexes $\hat{S}_G(X,A;m)$ and $\overline{S}_G(X,A;m)$, respectively. Thus we get the long exact sequence

$$\dots \to H^n_G(X,A;m) \to \hat{H}^n_G(X,A;m) \to \overline{H}^n_G(X,A;m) \overset{\delta}{\to} H^{n+1}_G(X,A;m) \to \dots$$

In the process of showing that $H^*_G(\ ;m)$ satisfies all seven equivariant Eilenberg-Steenrod axioms it will be shown that both $\hat{H}^*_G(\ ;m)$ and $\overline{H}^*_G(\ ;m)$ satisfy the first six axioms.

5. THE HOMOTOPY AXIOM

In this section we prove the homotopy axiom for both equivariant singular homology and cohomology.

Let V be a convex set in some euclidean space R^q, and let $v^0,\dots,v^n$ be $n + 1$ points in V. Denote $d^i = (0,\dots,1,\dots,0) \in \Delta_n$, $0 \leq i \leq n$, where the 1 occurs in the i-coordinate (recall that we index the coordinates such that a point in Δ_n is denoted by $(x_0,\dots,x_n)$). We use the notation

$$v^0\dots v^n : \Delta_n \to V$$

to denote the linear map from Δ_n into V, which is uniquely determined by the condition that it takes d^i into v^i, $i = 0,\dots,n$. We have

$(x_0,\dots,x_n) = \sum_{i=0}^{n} x_i d^i$, and thus $v^0\dots v^n(x_0,\dots,x_n) = \sum_{i=0}^{n} x_i v^i \in V$, where $(x_0,\dots,x_n) \in \Delta_n$. The map $v^0\dots v^n$ is a singular n-simplex in V, and its j:th face is the map $v^0\dots\hat{v}^j\dots v^n : \Delta_{n-1} \to V$. Naturally $v^0\dots v^n$ is called a linear n-simplex in V. We are now ready to begin the proof of the homotopy axiom.

Let $f_0, f_1 : (X,A) \to (Y,B)$ be two G-homotopic G-maps, and let $F : I \times (X,A) \to (Y,B)$ be a G-homotopy such that $F(0,x) = f_0(x)$ and $F(1,x) = f_1(x)$, for every $x \in X$. Using this specific G-homotopy F we now construct homomorphisms

$$\hat{D}_n : \hat{C}_n^G(X;k) \to \hat{C}_{n+1}^G(Y;k)$$

for all n, which form a chain homotopy from $(f_1)_\#$ to $(f_0)_\#$. We shall also show that this chain homotopy induces the other chain homotopies we need.

Let $T : \Delta_n \times G/K \to X$ be an equivariant singular n-simplex belonging to $\mathcal{F}$ in X. Composing $id_I \times T$ with the G-homotopy F we get the G-map $F(id_I \times T) : I \times \Delta_n \times G/K \to Y$. Now consider linear (n+1)-simplexes in $I \times \Delta_n$ of the form

$$(0,d^0)\cdots(0,d^i)(1,d^i)\cdots(1,d^n) : \Delta_{n+1} \to I \times \Delta_n, \quad 0 \le i \le n.$$

To shorten our notation we denote

$$\tau_{n+1}^i = (0,d^0)\cdots(0,d^i)(1,d^i)\cdots(1,d^n), \quad 0 \le i \le n.$$

We shall also have use of the following notation

$$\omega_n^i = (0,d^0)\cdots(0,d^{i-1})(1,d^i)\cdots(1,d^n) : \Delta_n \to I \times \Delta_n, \quad 0 \le i \le n+1.$$

Observe that $\omega_n^0 = (1,d^0)\cdots(1,d^n)$ and $\omega_n^{n+1} = (0,d^0)\cdots(0,d^n)$. Now consider G-maps of the form

$$F(\mathrm{id}_I \times T)(\tau_{n+1}^i \times \mathrm{id}_{G/K}) : \Delta_{n+1} \times G/K \to Y.$$

We define

$$\hat{D}_n(T \otimes a) = \sum_{i=0}^{n} (-1)^n[F(\mathrm{id} \times T)(\tau_{n+1}^i \times \mathrm{id})] \otimes a,$$

where $a \in k(G/K)$. (From now on we omit the subscripts on the identity maps. The symbol id denotes the identity map on the unit interval I if it appears immediately to the left of a product sign $\times$, and id denotes the identity map on a space of the form $G/t(T)$ whenever it appears immediately to the right of a product sign $\times$.) This defines the homomorphism $\hat{D}_n$: $\hat{C}_n^G(X;k) \to \hat{C}_{n+1}^G(X;k)$. We shall show that

$$\hat{\partial}_{n+1}\hat{D}_n + \hat{D}_{n-1}\hat{\partial}_n = (\hat{f}_1)_\# - (\hat{f}_0)_\#.$$

It is immediately seen that the identities

$$\tau_{n+1}^i e_{n+1}^j = \begin{cases} (\mathrm{id} \times e_n^j)\tau_{n-1}^{i-1}, & 0 \le j \le i-1 \le n-1 \\ \omega_n^i & , \quad 0 \le j = i \le n \\ \omega_n^{i+1} & , \quad 1 \le i+1 = j \le n+1 \\ (\mathrm{id} \times e_n^{j-1})\tau_n^i, & 2 \le i+2 \le j \le n+1 \end{cases}$$

are valid. Using these identities and the fact that, by definition, $T^{(j)} = T(e_n^j \times \mathrm{id})$, $0 \le j \le n$, we have (we omit the coefficient element $a \in k(G/t(T)$ in the calculation below).

$$\begin{aligned}\hat{\partial}_{n+1}\hat{D}_n(T) = &\sum_{0 \le j \le i-1 \le n-1} (-1)^{i+j}F(\mathrm{id} \times T^{(j)})(\tau_n^{i-1} \times \mathrm{id}) + \\ &\sum_{0 \le i \le n} F(\mathrm{id} \times T)\omega_n^i - \sum_{0 \le i \le n} F(\mathrm{id} \times T)\omega_n^{i+1} \\ &\sum_{2 \le i+2 \le j \le n+1} (-1)^{i+j}F(\mathrm{id} \times T^{(j-1)})(\tau_n^i \times \mathrm{id}).\end{aligned}$$

The second line of the above sum equals

$$F(\mathrm{id} \times T)\omega_n^0 - F(\mathrm{id} \times T)\omega_n^{n+1} = f_1T - f_0T.$$

Changing the index i to $i + 1$ on the first line of the sum and the index j to $j + 1$ on the third line of the sum and summing over the index i, we see that the sum of the first and third line in the above sum equals

$$- \sum_{0 \leq j \leq n} \hat{D}_{n-1}((-1)^{(j)}T^{(j)}) = - \hat{D}_{n-1}\hat{\partial}_n(T).$$

Thus $\hat{\partial}_{n+1}\hat{D}_n(T) = (\hat{f}_1)_{\#}(T) - (\hat{f}_0)_{\#}(T) - \hat{D}_{n-1}\hat{\partial}_n(T)$, which is exactly what we wanted to prove.

PROPOSITION 5.1. Let the notation be as above. The homomorphism $\hat{D}_n : \hat{C}_n^G(X;k) \to \hat{C}_{n+1}^G(Y;k)$ restricts to $\bar{D}_n : \bar{C}_n^G(X;k) \to \bar{C}_{n+1}^G(Y;k)$ and hence induces a homomorphism $D_n : C_n^G(X;k) \to C_{n+1}^G(Y;k)$. In fact the homomorphism $\hat{D}_n : \hat{C}_n^G(X) \to \hat{C}_{n+1}^G(Y)$ "preserves the relation $\sim$" and hence its dual $\hat{D}^{n+1}$: $\hat{C}_G^{n+1}(Y;m) \to \hat{C}_G^n(X;m)$ restricts to $D^{n+1} : C_G^{n+1}(Y;m) \to C_G^n(X;m)$ and thus also induces $\bar{D}^{n+1} : \bar{C}_G^{n+1}(Y;m) \to \bar{C}_G^n(X;m)$. All these homomorphisms induce homomorphisms on the corresponding relative versions.

PROOF. We shall show that the homomorphisms $\hat{D}_n : \hat{C}_n^G(X) \to \hat{C}_{n+1}^G(Y)$ "preverses the relation $\sim$". The conditions for this, given in Definition 4.4, are seen to be satisfied as follows. First, choose the integers r_i by $r_i = (-1)^i$, $i = 0,\ldots,n$. Secondly, if $h : \Delta_n \times G/K \to \Delta_n \times G/K'$ is a G-map, which covers $\mathrm{id} : \Delta_n \to \Delta_n$, we define the G-map $h_i : \Delta_{n+1} \times G/K \to \Delta_{n+1} \times G/K'$, $i = 0,\ldots,n$, by the condition that the diagram

$$\begin{array}{ccc}
\Delta_{n+1} \times G/K & \xrightarrow{\tau_{n+1}^i \times \mathrm{id}} & I \times \Delta_n \times G/K \\
\Big\downarrow h_i & & \Big\downarrow \mathrm{id} \times h \\
\Delta_{n+1} \times G/K' & \xrightarrow{\tau_{n+1}^i \times \mathrm{id}} & I \times \Delta_n \times G/K'
\end{array}$$

commutes. That such a G-map h_i exists and is unique follows immediately from the fact that the linear map $\tau^i_{n+1} : \Delta_{n+1} \to I \times \Delta_n$ is an imbedding. Also observe that each $h_i : \Delta_{n+1} \times G/K \to \Delta_{n+1} \times G/K'$ covers $id : \Delta_{n+1} \to \Delta_{n+1}$, and that each h_i determines the same G-homotopy class from G/K to G/K' as h does.

If now $T = T'h$, where T and T' are equivariant singular n-simplexes in X, then $F(id \times T)(\tau^i_{n+1} \times id) = F(id \times T')(\tau^i_{n+1} \times id)h_i$, $i = 0,\dots,n$. We have shown that the homomorphism $\hat{D}_n : \hat{C}^G_n(X) \to \hat{C}^G_{n+1}(Y)$ "preserves the relation $\sim$". At the same time we have shown that if $T \boxtimes a \sim T' \boxtimes a'$ then $\hat{D}_n(T \boxtimes a - T' \boxtimes a') \in \overline{C}^G_{n+1}(Y;k)$ and thus $\hat{D}_n : \hat{C}^G_n(X;k) \to \hat{C}^G_{n+1}(Y;k)$ restricts to $\overline{C}^G_n(X;k) \to \overline{C}^G_{n+1}(Y;k)$ and hence induces $D_n : C^G_n(X;k) \to C^G_{n+1}(Y;k)$.

Since $F : I \times (X,A) \to (Y,B)$ it follows immediately from the definition of the homomorphism $\hat{D}_n$ that all the homomorphisms D_n, $\overline{D}_n$, etc. induce homomorphisms on the corresponding relative versions. q.e.d.

COROLLARY 5.2. Let $f_0, f_1 : (X,A) \to (Y,B)$ be two G-homotopic G-maps. Then the induced homomorphisms

$$(f_0)_\#, (f_1)_\# \; : S^G(X,A;k) \to S^G(Y,B;k)$$

are chain homotopic, and the same is true for the homomorphisms

$$(f_0)^\#, (f_1)^\# \; : S_G(Y,B;m) \to S_G(X,A;m).$$

Moreover, the same conclusion holds for both the "roof" and "bar" versions.

q.e.d.

This result establishes the homotopy axiom for equivariant singular homology $H^G_*(\;;k)$, and equivariant singular cohomology $H^*_G(\;;m)$, as well as for the theories $\hat{H}^G_*(\;;k)$, $\overline{H}^G_*(\;;k)$, $\hat{H}^*_G(\;;m)$ and $\overline{H}^*_G(\;;m)$.

6. THE EXCISION AXIOM

In this section we prove the excision axiom for both equivariant singular homology and cohomology.

Consider the G-space $\Delta_n \times G/K$. An equivariant linear q-simplex in $\Delta_n \times G/K$ is a map of the form

$$v^0 \dots v^q \times \mathrm{id} : \Delta_q \times G/K \to \Delta_n \times G/K$$

where $v^0 \dots v^q$ is a linear q-simplex in Δ_n, and id denotes the identity mapping on G/K. Now assume that $K \varepsilon \mathcal{F}$, the fixed orbit type family under consideration. The equivariant linear q-simplexes in $\Delta_n \times G/K$ "generate" a submodule, which we denote by $\hat{C}^G_q Q(\Delta_n \times G/K;k)$, of $\hat{C}^G_q(\Delta_n \times G/K;k)$. Here "generate" means that $\hat{C}^G_q Q(\Delta_n \times G/K;k)$ is the submodule of $\hat{C}^G_q(\Delta_n \times G/K;k)$ generated by all elements of the form $T \otimes a$, where $a \varepsilon k(G/K)$ and $T = v^0 \dots v^q \times \mathrm{id}$ is an equivariant linear q-simplex in $\Delta_n \times G/K$. The boundary $\hat{\partial}_q$ maps $\hat{C}^G_q Q(\Delta_n \times G/K;k)$ into $\hat{C}^G_{q-1} Q(\Delta_n \times G/K;k)$ and we have the corresponding subcomplex $\hat{S}^G Q(\Delta_n \times G/K;k)$ of $\hat{S}^G(\Delta_n \times G/K;k)$.

Let $v \varepsilon \Delta_n$. Define homomorphisms

$$v\cdot : \hat{C}^G_q Q(\Delta_n \times G/K;k) \to \hat{C}^G_{q+1} Q(\Delta_n \times G/K;k)$$

by $v\cdot [(v^0 \dots v^q \times \mathrm{id}) \otimes a] = (vv^0 \dots v^q \times \mathrm{id}) \otimes a$, where $a \varepsilon k(G/K)$. Direct calculation shows that

$$\hat{\partial}_{q+1}(v\cdot\sigma) = \sigma - v\cdot(\hat{\partial}_q(\sigma)), \quad , \quad q \geq 1,$$

$$\hat{\partial}_1(v\cdot\sigma) - \sigma - (v \times \mathrm{id}) \otimes \mathrm{In}(\sigma), \quad q = 0,$$

where $\sigma \varepsilon \hat{C}^G_q Q(\Delta_n \times G/K;k)$ and $v \times \mathrm{id} : \Delta_0 \times G/K \to \Delta_n \times G/K$ is the equivariant linear 0-simplex in $\Delta_n \times G/K$ determined by the point $v \varepsilon \Delta_n$, and

In : $\hat{C}_0^G Q(\Delta_n \times G/K;k) \to k(G/K)$ is the homomorphism defined by $In[(v^0 \times id) \otimes a] = a$.

We now inductively define homomorphisms

$$\hat{Sd}_q : \hat{C}_q^G Q(\Delta_n \times G/K;k) \to \hat{C}_q^G Q(\Delta_n \times G/K;k)$$

$$\hat{R}_q : \hat{C}_q^G Q(\Delta_n \times G/K;k) \to \hat{C}_{q+1}^G Q(\Delta_n \times G/K;k)$$

in the following way. We set $\hat{Sd}_0 = id$ and $\hat{R}_0 = 0$. If $\omega = v^0 \ldots v^q \times id$ is an equivariant linear q-simplex in $\Delta_n \times G/K$ and $a \in k(G/K)$ we define

$$\hat{Sd}_q(\omega \otimes a) = b_\omega \cdot \hat{Sd}_{q-1}(\hat{\partial}_q(\omega \otimes a))$$

$$\hat{R}_q(\omega \otimes a) = b_\omega \cdot (\omega \otimes a - \hat{Sd}_q(\omega \otimes a) - \hat{R}_{q-1}(\hat{\partial}_q(\omega \otimes a))).$$

Here $b_\omega \in \Delta_n$ denotes the barycenter of ω, that is, the point $b_\omega = \frac{1}{q+1} v^0 + \ldots + \frac{1}{q+1} v^q$.

By induction with respect to q it is easy to prove that

$$\hat{\partial}_q \hat{Sd}_q = \hat{Sd}_{q-1} \hat{\partial}_q$$

$$\hat{\partial}_{q+1} \hat{R}_q + \hat{R}_{q-1} \hat{\partial}_q = id - \hat{Sd}_q,$$

that is, the homomorphisms $\hat{Sd}_q$ form a chain map $\hat{Sd}$, and the homomorphisms $\hat{R}_q$ form a chain homotopy $\hat{R}$ from id to $\hat{Sd}$.

Let X be an arbitrary G-space. We now define homomorphisms

$$\hat{Sd}_n : \hat{C}_n^G(X;k) \to \hat{C}_n^G(X;k)$$

$$\hat{R}_n : \hat{C}_n^G(X;k) \to \hat{C}_{n+1}^G(X;k)$$

in the following way. Let $T : \Delta_n \times G/K \to X$ be an equivariant singular n-simplex belonging to $\mathcal{F}$ in X, and let $a \in k(G/K)$. We define

$$\hat{Sd}_n(T \otimes a) = \hat{T}_{\#}\hat{Sd}_n((d^0 \ldots d^n \times id) \otimes a),$$

$$\hat{R}_n(T \otimes a) = \hat{T}_{\#}\hat{R}_n((d^0 \ldots d^n \times id) \otimes a).$$

It is easy to see that these homomorphisms $\hat{Sd}_n$ again form a chain map $\hat{Sd}$ and that these homomorphisms $\hat{R}_n$ form a chain homotopy from id to $\hat{Sd}$. The proof of this is a formal computation using the fact that both $\hat{Sd}_q$ and $\hat{R}_q$, when defined on the equivariant linear chain complexes, commute with the homomorphisms induced by the face maps $e_n^i \times id : \Delta_{n-1} \times G/K \to \Delta_n \times G/K$, and the fact that the homomorphisms $\hat{Sd}_q$ and $\hat{R}_q$ defined on the equivariant linear chain complexes already have the desired properties.

PROPOSITION 6.1. The homomorphism $Sd_n : \hat{C}_n^G(X;k) \to \hat{C}_n^G(X;k)$ restricts to $\overline{Sd} : \overline{C}_n^G(X;k) \to \overline{C}_n^G(X;k)$ and hence induces a homomorphism $Sd_n : C_n^G(X;k) \to C_n^G(X;k)$. In fact the homomorphism $\hat{Sd}_n : \hat{C}_n^G(X) \to \hat{C}_n^G(X)$ "preverses the relation $\sim$" and hence its dual $\hat{Sd}^n : \hat{C}_G^n(X;m) \to \hat{C}_G^n(X;m)$ restricts to $Sd^n : C_G^n(X;m) \to C_G^n(X;m)$ and thus also induces $\overline{Sd} : \overline{C}_G^n(X;m) \to \overline{C}_G^n(X;m)$. All the corresponding statements hold for the homomorphism $\hat{R}_n$.

PROOF. We first prove that the homomorphism $\hat{Sd}_n : \hat{C}_n^G(X) \to \hat{C}_n^G(X)$ "preserves the relation $\sim$". Consider $\hat{Sd}_n : \hat{C}_n^G Q(\Delta_n \times G/K) \to \hat{C}_n^G Q(\Delta_n \times G/K)$. We have $Sd_q(d^0 \ldots d^n \times id) = \sum_{j=1}^{N} \pm \sigma_j \times id$, where each σ_j is a linear n-simplex in Δ_n. Also observe that each expression $\pm \sigma_j$ and the integer N (in fact $N = (n + 1)!$) are independent of the subgroup K. Moreover it follows immediately from the recursive definition of Sd_q that each $\sigma_j : \Delta_n \to \Delta_n$ is an imbedding. If now $h : \Delta_n \times G/K \to \Delta_n \times G/K'$ is a G-map, which covers $id : \Delta_n \to \Delta_n$, we define the G-map $h_j : \Delta_n \times G/K \to \Delta_n \times G/K'$ by the condition that the diagram

$$\begin{array}{ccc} \Delta_n \times G/K & \xrightarrow{\sigma_j \times id} & \Delta_n \times G/K \\ h_j \downarrow & & \downarrow h \\ \Delta_n \times G/K' & \xrightarrow{\sigma_j \times id} & \Delta_n \times G/K' \end{array}$$

commutes. Observe that h_j covers $id : \Delta_n \to \Delta_n$ and that h_j determines the same G-homotopy class of G-maps from G/K to G/K' as h does.

Now recall that for any equivariant singular n-simplex, belonging to , $T : \Delta_n \times G/K \to X$ we have by definition, $\hat{Sd}_n(T) = \hat{T}_{\#}Sd_n(d^0 \ldots d^n \times id) = \sum_{j=1}^{N} \pm T(\sigma_j \times id)$. If now $T = T'h$, it follows that $T(\sigma_j \times id) = T'h(\sigma_j \times id) = T'h(\sigma_j \times id) = T'(\sigma_j \times id)h_j$. We have shown that $\hat{Sd}_n : \hat{C}_n^G(X) \to \hat{C}_n^G(X)$ "preserves the relation $\sim$".

To prove that the homomorphism $\hat{R}_n : \hat{C}_n^G(X) \to \hat{C}_{n+1}^G(X)$ "preserves the relation $\sim$" we proceed in a completely analogous way as above. First consider the homomorphism $\hat{R}_n : \hat{C}_n^G Q(\Delta_n \times G/K) \to \hat{C}_{n+1}^G Q(\Delta_n \times G/K)$, and observe that $\hat{R}_n(d^0 \ldots d^n \times id) = \sum_{i=1}^{M} \pm \tau_j \times id$, where each τ_j is a linear $(n + 1)$-simplex in Δ_n. Moreover, the expression $\pm \tau_j$ and the integer M are independent of the subgroup K. If now $\bar{h} : \Delta_n \times G/K \to \Delta_n \times G/K'$ is a G-map, which covers $id : \Delta_n \to \Delta_n$, we define the G-map $\bar{h}_j : \Delta_{n+1} \times G/K \to \Delta_{n+1} \times G/K'$ by requiring that $\bar{h}_j$ covers $id : \Delta_{n+1} \to \Delta_{n+1}$ and that the diagram

$$\begin{array}{ccc} \Delta_{n+1} \times G/K & \xrightarrow{\tau_j \times id} & \Delta_n \times G/K \\ h_j \downarrow & & \downarrow \bar{h} \\ \Delta_{n+1} \times G/K' & \xrightarrow{\tau_j \times id} & \Delta_n \times G/K' \end{array}$$

commutes. (Thus $\bar{h}_j(x,gK) = (x,pr_2\bar{h}(\tau_j(x),gK))$, where pr_2 denotes projection onto the second factor).

If $T = T'\bar{h}$, it follows that $T(\tau_j \times id) = T'(\tau_j \times id)\bar{h}_j$. Hence $\hat{R}_n(T) = \sum_{j=1}^{M} \pm T(\tau_j \times id) = \sum_{j=1}^{M} \pm T'(\tau_j \times id)\bar{h}_j$. This shows that $\hat{R}_n : \hat{C}_n^G(X) \to \hat{C}_{n+1}^G(X)$ "preserves the relation $\sim$". q.e.d.

Exactly as in the case of ordinary singular theory the subdivision chain map $Sd : S^G(X;k) \to S^G(X;k)$ and the chain homotopy $R : S^G(X;k) \to S^G(X;k)$ are the crucial ingredients for the proof of the excision axiom. We proceed to give the remaining details.

DEFINITION 6.2. Let V be a family of G-subsets of the G-space X. An equivariant singular n-simplex $T : \Delta_n \times G/K \to X$ is said to be in V if $T(\Delta_n \times G/K)$ is contained in at least one of the sets in V.

Clearly all equivariant singular n-simplexes belonging to $\mathfrak{F}$ in X which are in V "generate" a submodule $\hat{C}_n^G(X;k;V)$ of $\hat{C}_n^G(X;k)$. For the case $k = Z$, i.e. the coefficient system defined by $Z(G/K) = Z$, for every $K \varepsilon \mathfrak{F}$, and all the induced homomorphisms are the identity on Z, we use the simplified notation $\hat{C}_n^G(X;V)$, in complete analogy with our earlier notation. We denote the inclusion by

$$\hat{\eta} : \hat{C}_n^G(X;k,V) \to \hat{C}_n^G(X;k).$$

Now observe that if $T \otimes a \,\varepsilon\, \hat{C}_n^G(X;k;V)$ and $T \otimes a \sim T' \otimes a'$, where $T' \otimes a' \,\varepsilon\, \hat{C}_n^G(X;k)$, then it follows that also $T' \otimes a' \,\varepsilon\, \hat{C}_n^G(X;k;V)$. That is, the relation $\sim$ restricts to $\hat{C}_n^G(X;k;V)$ in this way. This fact allows us to use these new modules $\hat{C}_n^G(X;k;V)$ in a way completely analogous to the way we have used the modules $\hat{C}_n^G(X;k)$. To be more specific we mean the following. We define $\bar{C}_n^G(X;k;V)$ and $C_n^G(X;k;V)$ by complete analogy to the definitions of $\bar{C}_n^G(X;k)$

and $C_n^G(X;k)$. We can use the notion of a homomorphism $\hat{\alpha}$ which "preserves the relation ~" (see Definition 4.4.) also when the range or domain (or both) of the homomorphisms $\hat{\alpha}$ is one of the modules $\hat{C}_n^G(X;V)$. We can use the same kind of duals of $\hat{C}_n^G(X;V)$ as of $\hat{C}_n^G(X)$, that is, we define $\hat{C}_G^n(X;m;V) = \mathrm{Hom}_t(\hat{C}_n^G(X;V),M)$, the right R-module consisting of all homomorphisms of abelian groups $c : \hat{C}_n^G(X;V) \to M$, which satisfy the condition $c(T) \in m(G/t(T))$, for every equivariant singular n-simplex T, belonging to $\mathcal{F}$ in X, which is in V. (Recall that M denotes the right R-module $M = \sum_{K\in\mathcal{F}} \oplus\, m(G/K))$. Then $C_G^n(X;m;V)$ is defined to be the submodule of $\hat{C}_G^n(X;m;V)$ consisting of all homomorphisms $c : \hat{C}_n^G(X;V) \to M$ which satisfy the condition in Definition 4.3. If now for example $\hat{\alpha} : \hat{C}_n^G(X) \to \hat{C}_n^G(X;V)$ is a homomorphism which "preserves the relation ~" it follows that $\hat{\alpha}$ has a dual $\hat{\alpha}^{\#} : \hat{C}_G^n(X;m;V) \to \hat{C}_G^n(X;m)$ which restricts to $\alpha^{\#} : C_G^n(X;m;V) \to C_G^n(X;m)$, (and $\alpha^{\#}$ is again called the dual of $\hat{\alpha}$).

LEMMA 6.3. Let V be a family of G-subsets of X such that $X = \bigcup_{B\in V} B^{\circ}$. Let $T : \Delta_n \times G/K \to X$ be an equivariant singular n-simplex belonging to $\mathcal{F}$ in X, and $a \in k(G/K)$. Then there exists an integer m such that $\hat{Sd}^m(T \otimes a) \in \hat{S}^G(X;k;V)$.

PROOF. Consider the (ordinary) singular n-simplex $T| : \Delta_n \to X$, where $(T|)(x) = T(x,eK)$, $x \in \Delta_n$. From the corresponding result in ordinary singular theory we know that there exists m such that $Sd^m(T|) \in S(X;V)$ (see Eilenberg-Steenrod [6], p. 198-199.) Here $Sd : S(X) \to S(X)$ is the subdivision chain map on the ordinary singular chain complex of X. But since V is a family of G-subsets, it now follows from the way our Sd is defined that we have $\hat{Sd}^m(T \otimes a) \in \hat{S}^G(X;k;V)$. q.e.d.

For any equivariant singular simplex T belonging to $\mathcal{F}$ in X we denote by $m(T)$ the smallest integer for which $\hat{Sd}^{m(T)}(T \otimes a) \in \hat{S}^G(X;k;V)$. The coefficient element $a \in k(G/t(T))$ does not affect this situation at all. Clearly we have $m(T^{(i)}) \leq m(T)$. If $T \otimes a \in \hat{S}^G(X;k;V)$ then $m(T) = 0$, and if $T \otimes a \in \hat{S}^G(A;k)$ then also $\hat{Sd}^{m(T)}(T \otimes a) \in \hat{S}^G(A;k;V)$. The following proposition corresponds to Theorem 8.2. on page 197 in Eilenberg-Steenrod [6]. The proof we give follows the proof they give in the Notes at the end of Chapter III, and not the proof they give in the text. Note the remark on page 207 in Eilenberg-Steenrod [6].

PROPOSITION 6.4. Let V be a family of G-subsets of X such that $X = \bigcup_{B \in V} B^{\circ}$. Then, for any G-subset A of X, the inclusion

$$\eta : S^G(X,A;k;V) \to S^G(X,A;k)$$

is a homotopy equivalence, and the same is true for the corresponding $\hat{\eta}$ and $\bar{\eta}$. The dual

$$\eta^{\#} : S_G(X,A;m) \to S_G(X,A;m;V)$$

is also a homotopy equivalence, and the same is true for the corresponding $\hat{\eta}^{\#}$ and $\bar{\eta}^{\#}$.

PROOF. Let T be an equivariant singular n-simplex belonging to $\mathcal{F}$ in X, and $a \in k(G/t(T))$.

Define

$$\hat{\tau}(T \otimes a) = \hat{Sd}^{m(T)}(T \otimes a) + \sum_{i=0}^{n} (-1)^i \sum_{j=m(T^{(i)})}^{m(T)-1} \hat{R}\, \hat{Sd}^j(T^{(i)} \otimes a)$$

$$\hat{D}(T \otimes a) = \sum_{j=0}^{m(T)-1} \hat{R}\, \hat{Sd}^j(T \otimes a).$$

Observe that $\hat{\tau}(T \otimes a) \in \hat{C}^G_n(X;k;V)$ and that $\hat{D}(T \otimes a) \in \hat{C}^G_{n+1}(X;k)$.

This defines homomorphisms

$$\hat{\tau}_n : \hat{C}^G_n(X,A;k) \to \hat{C}^G_n(X,A;k;V)$$

$$\hat{D}_n : \hat{C}^G_n(X;A;k) \to \hat{C}^G_{n+1}(X,A;k).$$

A formal computation shows that

$$\hat{\partial}_{n+1}\hat{D}_n + \hat{D}_{n-1}\hat{\partial}_n = id - \hat{\eta}_n\hat{\tau}_n$$

for all n. From this it follows that $\hat{\partial}_n\hat{\eta}_n\hat{\tau}_n = \hat{\eta}_{n-1}\hat{\tau}_{n-1}\hat{\partial}_n$. Since $\hat{\eta}$: $\hat{S}^G(X,A;k;V) \to \hat{S}^G(X,A;k)$ is a chain map and an inclusion it follows that $\hat{\partial}_n\hat{\tau}_n = \hat{\tau}_{n-1}\hat{\partial}_n$, that is, the homomorphism $\hat{\tau}_n$ form a chain map $\hat{\tau} : \hat{S}(X,A;k;V) \to S^G(X,A;k)$. Since $\hat{\tau}\hat{\eta} = id$ and the above formula tells us that $\hat{\eta}\hat{\tau}$ is chain homotopic to the identity map, it follows that $\hat{\eta}$ is a homotopy equivalence, and in fact $\hat{\tau}$ is a homotopy inverse to $\hat{\eta}$.

By Proposition 6.1. the maps $\hat{Sd}$ and $\hat{R}$ restrict to maps $\overline{Sd}$ and $\overline{R}$, and hence induce Sd and R. Since $m(T) = m(T')$, if $T \otimes a \sim T' \otimes a'$, it follows that the maps $\hat{\tau}$ and $\hat{D}$ restrict to corresponding maps $\overline{\tau}$ and $\overline{D}$ and therefore induce τ and D. Thus τ and $\overline{\tau}$ are homotopy inverses to η and $\overline{\eta}$, respectively. In the same way it follows from Proposition 6.1. that the homomorphisms $\hat{\tau}_n : \hat{C}^G_n(X) \to \hat{C}^G_n(X;V)$ and $\hat{D}_n : \hat{C}^G_n(X) \to \hat{C}^G_{n+1}(X)$ "preverse the relation $\sim$" and thus the duals $\hat{\tau}^{\#}$, $\tau^{\#}$, and $\overline{\tau}^{\#}$ are homotopy inverses to $\hat{\eta}^{\#}$, $\eta^{\#}$, and $\overline{\eta}^{\#}$ respectively. q.e.d.

COROLLARY 6.5. Let (X,A) be a G-pair and let U a G-subset of X such that $\overline{U} \subset A^{\circ}$. Then the inclusion

$$i : (X - U, A - U) \to (X,A)$$

induces a homotopy equivalence

$$i_{\#} : S^G(X - U, A - U;k) \to S^G(X,A;k).$$

The corresponding $\hat{i}_{\#}$ and $\bar{i}_{\#}$ are also homotopy equivalences, and the same is true for the duals $\hat{i}^{\#}$, $i^{\#}$, and $\bar{i}^{\#}$.

PROOF. Let V denote the family consisting of the G-subsets A and $X - U$. Since $\bar{U} \subset A^{\circ}$ it follows that $X = (X - U)^{\circ} \cup A^{\circ}$, that is, the family V satisfies the dondition of Proposition 6.4. Since

$$\hat{S}^G(X;k;V) = \hat{S}^G(X - U;k) + \hat{S}^G(A;k)$$

$$\hat{S}^G(A;k;V) = \hat{S}^G(A;k) \quad \text{and}$$

$$\hat{S}^G(X - U;k) \cap \hat{S}^G(A;k) = \hat{S}^G(A - U;k)$$

it follows (by the Noether isomorphism theorem) that

$$\hat{j} : \hat{S}^G(X - U, A - U;k) \to \hat{S}^G(X,A;k;V)$$

is an isomorphism. Since $\hat{i}_{\#} = \hat{\eta}\hat{j}$, and $\hat{\eta} : \hat{S}^G(X,A;k;V) \to \hat{S}^G(X,A;k)$ is a homotopy equivalence by Proposition 6.4., it follows that $\hat{i}_{\#}$ is a homotopy equivalence. Since also the maps $\bar{j}$ and j corresponding to $\hat{j}$, and the duals $\hat{j}^{\#}$, $j^{\#}$, and $\bar{j}^{\#}$ are isomorphisms, it follows that the maps $\bar{i}_{\#}$ and $i_{\#}$, as well as the maps $\hat{i}^{\#}$, $i^{\#}$, and $\bar{i}^{\#}$ are homotopy equivalences. q.e.d.

This result proves the excision axiom for equivariant singular homology $H_*^G(\ ;k)$, and equivariant singular cohomology $H_G^*(\ ;m)$, as well as for the theories $H_*^G(\ ;k)$, $\bar{H}_*^G(\ ;k)$, $\hat{H}_G^*(\ ;m)$, and $\bar{H}_G^*(\ ;m)$,

7. THE DIMENSION AXIOM

Let $H \in \mathcal{F}$. We shall determine the R-modules $H_n^G(G/H;k)$, for every n. Let $\pi_n : \Delta_n \times G/H \to G/H$ be the projection onto the second factor. The map π_n is an equivariant singular n-simplex belonging to $\mathcal{F}$ in X. We have $\partial_n(\pi_n \otimes b) = \sum_{i=0}^{n} (-1)^i \pi_{n-1} \otimes b$, that is,

$$\hat{\partial}_n(\pi_n \otimes b) = \begin{cases} \pi_{n-1} \otimes b & , \text{ for } n \text{ even, and } n \geq 2 \\ 0 & , \text{ for } n \text{ odd,} \end{cases}$$

where $b \in k(G/H)$. Let $S^G\text{spe.}(G/H;k)$ denote the chain complex which in degree n is $C_n^G\text{spe.}(G/H;k) = Z_{\pi_n} \otimes k(G/H)$ and the boundary homomorphism is the standard one i.e. the one given above. Clearly the homology of the chain complex $S^G\text{spe.}(G/H;k)$ is given by

$$H_0(S^G\text{spe.}(G/H;k)) = C_0^G\text{spe.}(G/H;k) \cong k(G/H),$$

$$H_m(S^G\text{spe.}(G/H;k)) = 0, \quad \text{for } m \neq 0.$$

We shall now establish an isomorphism of chain complexes

$$\beta : S^G(G/H;k) \xrightarrow{\cong} S^G\text{spe.}(G/H;k).$$

First we define a chain map

$$\hat{\beta} : \hat{S}^G(G/H;k) \to S^G\text{spe.}(G/H;k)$$

as follows. Let $T : \Delta_n \times G/K \to G/H$ be any equivariant singular n-simplex belonging to $\mathcal{F}$ in G/H. Then define $\overline{T} : \Delta_n \times G/K \to \Delta_n \to G/H$, by $\overline{T}(x,gK) = (x,T(x,gK))$. Observe that $T = \pi_n\overline{T}$, and $T \otimes a \sim \pi_n \otimes (\overline{T})_*(a)$, where

$a \in k(G/K)$. Now define

$$\hat{\beta}(T \otimes a) = \pi_n \otimes (\overline{T})_*(a) \in C_n^G spe.(G/H;k).$$

Since $\hat{\beta}((-1)^i T^{(i)} \otimes a) = (-1)^i \pi_{n-1} \otimes (\overline{T^{(i)}})_*(a) = (-1)^i \pi_{n-1} \otimes (\overline{T})_*(a)$, $i = 0,\ldots,n$, it follows that $\hat{\beta}$ is a chain map. Now assume that $T \otimes a \sim T' \otimes a'$. Let $h : \Delta_n \times G/t(T) \to \Delta_n \times G/t(T')$ be a G-map which covers $id : \Delta_n \to \Delta_n$, such that $T = T'h$ and $h_*(a) = a'$. Then $\overline{T} = \overline{T}'h$, and hence $(\overline{T})_*(a) = (\overline{T'})_* h_*(a) = (\overline{T'})_*(a')$. Thus $\hat{\beta}(T \otimes a) = (T' \otimes a')$, and it follows that $\hat{\beta}$ induces the chain map β. An inverse $\alpha : S^G spe.(G/H;k) \to S^G(G/H;k)$ to β is defined as follows. Let $\pi_n \otimes b \in C_n^G spe.(G/H;k)$ and define $\alpha(\pi_n \otimes b) = \{\pi_n \otimes b\}$, where $\{\pi_n \otimes b\}$ denotes the image of the element $\pi_n \otimes b \in \hat{C}_n^G(G/H;k)$ under the natural projection $\hat{C}_n^G(G/H;k) \to C_n^G(G/H;k)$. Then $\beta\alpha(\pi_n \otimes b) = \beta\{\pi_n \otimes b\} = \pi_n \otimes (\overline{\pi}_n)_*(b) = \pi_n \otimes b$, since $\overline{\pi}_n = id : \Delta_n \times G/H \to \Delta_n \times G/H$. Also $\alpha\beta(\{T \otimes a\}) = \alpha(\pi_n \otimes (\overline{T})_*(a)) = \{\pi_n \otimes (\overline{T})_*(a)\} = \{T \otimes a\}$ since, as we already noted before, $T \otimes a \sim \pi_n \otimes (\overline{T})_*(a)$. Thus α is an inverse to β and we have shown that $\beta : S^G(G/H;k) \to S^G spe.(G/H;k)$ is an isomorphism of chain complexes. It follows that the homology R-modules of the chain complex $S^G(G/H;k)$, that is, the equivariant singular homology R-modules $H_p^G(G/H;k)$ are given by

$$H_0^G(G/H;k) \cong k(G/H) \quad ,$$

$$H_m^G(G/H;k) = 0 \quad , \text{ for } m \neq 0.$$

The explicit isomorphism

$$\alpha : H_0^G(G/H;k) \xrightarrow{\cong} k(G/H)$$

is described as follows. Since every element in $C_G^0(G/H;k)$ is a cycle and only the zero element is a boundary it follows that $H_0^G(G/H;k) = C_0^G(G/H;k)$.

Then $\alpha : H_0^G(G/H;k) = C_0^G(G/H;k) \to k(G/H)$ is defined by the following. If $T : \Delta_0 \times G/K = G/K \to G/H$ is any equivariant singular 0-simplex belonging to $\mathcal{F}$ in G/H, we have $\alpha(\{T \otimes a\}) = T_*(a) \in k(G/H)$, where $a \in k(G/K)$ and $T_* : k(G/K) \to k(G/H)$ is the homomorphism induced by the G-map $T : G/K \to G/H$. From this description of the isomorphism α it also follows immediately that α has the naturality property described in the statement of the dimension axiom in Theorem 2.1. This concludes the proof of the dimension axiom for equivariant singular homology.

Let us now prove the dimension axiom for equivariant singular cohomology. Denote $C_G^n\text{spe.}(G/H;m) = \text{Hom}_Z(Z_{\pi_n}, m(G/H))$.

We have

$$(\hat{\delta}^n(d))(\pi_{n+1}) = d(\hat{\partial}_{n+1}(\pi_{n+1})) = \begin{cases} d(\pi_n), & \text{for } n \text{ odd, } n \geq 1 \\ 0, & \text{for } n \text{ even,} \end{cases}$$

where $d \in C_G^n\text{spe.}(G/H;m)$. Thus $\hat{\delta}^n : C_G^n\text{spe.}(G/H;m) \to C_G^{n+1}\text{spe.}(G/H;m)$ is the zero homomorphism if n is even, and $\hat{\delta}^n$ is an isomorphism if n is odd and $n \geq 1$. Let $S_G\text{spe.}(G/H;m)$ denote the corresponding cochain complex. Clearly the homology of the cochain complex $S_G\text{spe.}(G/H;m)$ is given by

$$H_0(S_G\text{spe.}(G/H;m)) = C_G^0\text{spe.}(G/H;m) = \text{Hom}_Z(Z_{\pi_n}, m(G/H)) \cong m(G/H)$$

$$H_m(S_G\text{spe.}(G/H;m)) = 0.$$

We shall now define an isomorphism of cochain complexes

$$\alpha' : S_G(G/H;m) \to S_G\text{spe.}(G/H;m).$$

Let $c \in C_G^n(G/H;m)$, we then define $\alpha'(c) \in C_G^n\text{spe.}(G/H;m)$ by $(\alpha'(c)(\pi_n) =$

$c(\pi_n)$. Clearly α' is a cochain map. To see that α' is an isomorphism we define a cochain map

$$\beta' : S_G\text{spe.}(G/H;m) \to S_G(G/H;m),$$

and show that β' is the inverse to α'. We first define $\hat{\beta}' : S_G\text{spe.}(G/H;m) \to \hat{S}_G(G/H;m)$ as follows. If $d \in C^n_G\text{spe.}(G/H;m)$ we define $\hat{\beta}'(d)$ by $(\hat{\beta}'(d))(T) = (\overline{T})^*d(\pi_n) \in m(G/t(T))$, for any equivariant singular n-simplex T belonging to $\mathcal{F}$ in G/H. It is easy to see that $\hat{\beta}'$ is a cochain map. If now $T = T'h$, where $h : \Delta_n \times G/t(T) \to \Delta_n \times G/t(T')$ covers $id : \Delta_n \to \Delta_n$, then $(\hat{\beta}'(d))(T) = (\overline{T})^*d(\pi_n) = h^*(\overline{T'})^*d(\pi_n) = h^*(\hat{\beta}'(d))(T')$. Thus the values of $\hat{\beta}'$ in fact lie in $S_G(G/H;m)$. This defines the cochain map β'. For any equivariant singular n-simplex T belonging to in G/H we have $T = \pi_n\overline{T}$, and hence $c(T) = (\overline{T})^*c(\pi_n)$, for every $c \in C^n_G(G/H;m)$. Thus we have $(\beta'\alpha'(c)(T) = (\overline{T})^*\alpha'(C)(\pi_n) = (\overline{T})^*c(\pi_n) = c(T)$, that is $\beta'\alpha' = id$. Moreover $(\alpha'\beta'(d)(\pi_n) = (\beta'(d)(\pi_n) = (\overline{\pi}_n)^*d(\pi_n) = d(\pi_n)$, since $\overline{\pi}_n = id$. We have shown that β' is the inverse to α' and hence that α' is an isomorphism. It follows that the homology R-modules of the cochain complex $S_G(G/H;m)$, that is, the equivariant singular cohomology R-modules $H^P_G(G/H;m)$ are given by

$$H^0_G(G/H;m) \cong m(G/H),$$

$$H^q_G(G/H;m) = 0 \qquad , \text{ for } q \neq 0.$$

The explicit isomorphism

$$\xi : H^0_G(G/H;m) \xrightarrow{\cong} m(G/H)$$

is described as follows. First we have $H^0_G(G/H;m) = C^0_G(G/H;m)$, and then ξ is defined in the following way. Let $c \in C^0_G(G/H;m)$, then $\xi(c) = c(\pi_0)$, where $\pi_0 : \Delta_0 \times G/H = G/H \to G/H$ equals the identity map. Using this description of the isomorphism one easily shows that ξ has the naturality property described in the statement of the dimension axiom in Theorem 2.2. This concludes the proof of the dimension axiom for equivariant singular cohomology.

We have now completed the proofs of both Theorem 2.1. and Theorem 2.2.

8. ADDITIVITY PROPERTIES

Assume that the G-space X is the topological sum of the G-spaces X_j, $j \in J$. We denote this by $X = \biguplus_{j\in J} X_j$. Let A be a G-subset of X and denote $A_j = A \cap X_j$. Then also $A = \biguplus_{j\in J} A_j$. By $i_j : (X_j,A_j) \to (X,A)$ we denote the natural inclusion.

PROPOSITION 8.1. The homomorphisms

$$\sum_{j\in J} \oplus (i_j)_* : \sum_{j\in J} \oplus H_n^G(X_j,A_j;k) \to H_n^G(X,A;k)$$

$$\prod_{j\in J} (i_j)^* : H_G^n(X,A;m) \to \prod_{j\in J} H_G^n(X_j,A_j;m)$$

are isomorphisms for every n.

PROOF. Follows immediately using standard properties of direct sums and products from the way we have defined the equivariant singular homology and cohomology modules. q.e.d.

II. FURTHER PROPERTIES OF EQUIVARIANT SINGULAR HOMOLOGY AND COHOMOLOGY

1. NATURALITY IN THE GROUP

Let P and G be topological groups, and let $\mathcal{F}$ and $\mathcal{F}'$ be orbit type families for P and G, respectively. Assume that $\phi : P \to G$ is a continuous homomorphism, such that, if $Q \in \mathcal{F}$ then $\phi(Q) \in \mathcal{F}'$.

Let X be a P-space and Y a G-space, and let $f : X \to Y$ be a ϕ-map. Thus $f(px) = \phi(p)f(x)$, for every $p \in P$ and $x \in X$. Make Y into a P-space through the homomorphism $\phi : P \to G$. That is, P acts on Y by $py = \phi(p)y$. We denote this P-space by Y'. Now observe that we have the commutative diagram

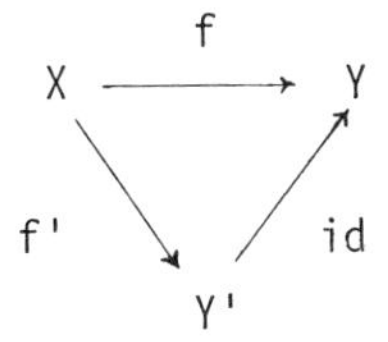

where the map f' equals f as a map of topological spaces, and id is the identity on the underlying topological spaces. The map $f' : X \to Y'$ is a P-map, and $id : Y' \to Y$ is a ϕ-map.

We shall define induced homomorphisms on equivariant singular homology and cohomology of the ϕ-map $f : X \to Y$. Due to the above commutative diagram it is enough to consider the ϕ-map

$$id : Y' \to Y,$$

and define the homomorphisms it induces on equivariant singular homology and cohomology.

Let

$$\alpha : P/Q \to P/N$$

be a P-map (Q and N are subgroups of P). Denote $\alpha(eQ) = p_0N$. Thus $\alpha(pQ) = pp_0N$. We have $Q \subset p_0Np_0^{-1}$, and hence $\phi(Q) \subset \phi(p_0)\phi(N)\phi(p_0)^{-1}$. Therefore we can define a G-map

$$\phi(\alpha) : G/\phi(Q) \to G/\phi(N)$$

by the condition $\phi(\alpha)(e\phi(Q)) = \phi(p_0)\phi(N)$. We then have $\phi(\alpha)(g\phi(Q)) = g\phi(p_0)\phi(N)$, $g \in G$.

Now let k' and k be covariant coefficient systems for $\mathfrak{F}'$ and $\mathfrak{F}$, respectively, over the ring R. Let

$$\Phi : k' \to k$$

be a natural transformation with respect to the homomorphism $\phi : P \to G$. By this we mean that for each $Q \in \mathfrak{F}$ we have a homomorphism of left R-modules

$$\Phi : k'(P/Q) \to k(G/\phi(Q)),$$

such that if $\alpha : P/Q \to P/N$ is a P-map, where also $N \in \mathfrak{F}'$, then the diagram

$$\begin{array}{ccc} k'(P/Q) & \xrightarrow{\Phi} & k(G/\phi(Q)) \\ \alpha_* \downarrow & & \downarrow (\phi(\alpha))_* \\ k'(P/N) & \xrightarrow{\Phi} & k(G/\phi(N)) \end{array}$$

commutes.

PROPOSITION 1.1. Let the homomorphism $\phi : P \to G$, and the natural transformation $\Phi : k' \to k$ be as above. Let (Y,B) be a G-pair and make it into a P-pair (Y',B') through the homomorphism ϕ. Then the ϕ-map $id : (Y',B') \to (Y,B)$ induces homomorphisms

$$(\phi,\Phi)_* : H_n^P(Y',B';k') \to H_n^G(Y,B;k),$$

for all n, with the following properties.

1. If $\phi = \mathrm{id}$ and $\Phi = \mathrm{id}$, then $(\phi,\Phi)_* = \mathrm{id}$.
2. $(\phi,\Phi)_*$ commutes with the boundary homomorphisms.
3. If $s : (Y,B) \to (\tilde{Y},\tilde{B})$ is a G-map, and $s' = s : (Y',B') \to (\tilde{Y}',\tilde{B}')$ is the corresponding P-map, we have $s_*(\phi,\Phi)_* = (\phi,\Phi)_* s'_*$.

PROOF. We define a chain map

$$\widehat{(\phi,\Phi)}_\# = \hat{S}^P(Y',B';k') \to \hat{S}^G(Y,B;k)$$

as follows. Let

$$T_P : \Delta_n \times P/Q \to Y'$$

be any P-equivariant singular n-simplex belonging to $\mathfrak{F}'$ in Y'. We define the corresponding G-equivaraint singular n-simplex in Y

$$T_G : \Delta_n \times G/\phi(Q) \to Y$$

by $T_G(x,g\phi(Q)) = g\,T_P(x,eQ)$. Since the point $T_P(x,eQ) \in Y'$ is fixed under the subgroup Q, it follows that the same point $T_P(x,eQ) \in Y$, when considered as a point in the G-space Y, is fixed under $\phi(Q)$. It follows that T_G is a well-defined G-map. Since we assumed that $Q \in \mathfrak{F}'$, it follows that $\phi(Q) \in \mathfrak{F}$, that is, that T_G belongs to $\mathfrak{F}$. Now put

$$\widehat{(\phi,\Phi)}_\#(T_P \otimes b) = T_G \otimes \Phi(b),$$

where $b \in k'(P/t(T_P))$, and $\Phi : k'(P/t(T_P)) \to k(G/t(T_G))$ (we have $t(T_G) = \phi(t(T_P)))$. This defines the homomorphism $\widehat{(\phi,\Phi)}_n : \hat{C}_n^P(Y';k') \to \hat{C}_n^G(Y;k)$. It is clear that $\widehat{(\phi,\Phi)}_n$ maps $\hat{C}_n^P(B';k')$ into $\hat{C}_n^G(B;k)$, and that the homomorphisms

$\widehat{(\phi,\Phi)}_n$ commute with the boundary. We have constructed the chain map $\widehat{(\phi,\Phi)}_\#$.

It remains to show that $\widehat{(\phi,\Phi)}_n$ restricts to $\overline{(\phi,\Phi)}_n : \overline{C}^P_n(Y';k') \to \overline{C}^G_n(Y;k)$, and therefore induces $(\phi,\Phi)_n : C^P_n(Y';k') \to C^G_n(Y;k)$. Assume that $T_P \otimes b \sim T'_P \otimes b'$, where $T_P : \Delta_n \times P/Q \to Y'$, $T'_P : \Delta_n \times P/Q' \to Y'$, and $b \in k'(P/Q)$, $b' \in k'(P/Q')$. Let $h_P : \Delta_n \times P/Q \to \Delta_n \times P/Q'$ be a P-map, which covers $id : \Delta_n \to \Delta_n$, such that $T_P = T'_P h_P$ and $(h_P)_*(b) = b'$. Consider the diagram

$$\begin{array}{ccc}
\Delta_n \times P/Q & \xrightarrow{\ \Delta(\phi)\ } & \Delta_n \times G/\phi(Q) \\
{\scriptstyle h_P}\downarrow & & \downarrow{\scriptstyle h_G} \\
\Delta_n \times P/Q' & \xrightarrow{\ \Delta(\phi)'\ } & \Delta_n \times G/\phi(Q')
\end{array}$$

where $\Delta(\phi)$ is defined by $\Delta(\phi)(x,pQ) = (x,\phi(p)\phi(Q))$, $x \in \Delta_n$, $p \in P$, and $\Delta(\phi)'$ is defined analogously. It is immediately seen that $\Delta(\phi)$ and $\Delta(\phi)'$ are well-defined ϕ-maps. Define the map h_G by $h_G(x,g\phi(Q)) = g(\Delta(\phi)'h_P(x,eQ))$, $g \in G$. The point $\Delta(\phi)'h_P(x,eQ)$ is fixed under the subgroup $\phi(Q)$ of G, since the composite map $\Delta(\phi)'h_P$ is a ϕ-map. It follows that h_G is a well-defined G-map. It is now easy to see that $T'_G h_G = T_G$.

We claim that $(h_G)_*\Phi(b) = \Phi(b')$. This is seen as follows. Restricting the maps h_P and h_G to, for example, the orbit over $d^0 \in \Delta_n$ gives us the P-map $(h_P)_0 : P/Q \to P/Q'$ and the G-map $(h_G)_0 : G/\phi(Q) \to G/\phi(Q')$. It is easily seen that $(h_G)_0 = \phi((h_P)_0)$. It follows that $(h_G)_*\Phi = \Phi(h_P)_*$. Thus we have $(h_G)_*\Phi(b) = \Phi(h_P)_*(b) = \Phi(b')$. We have now shown that $T_G \otimes \Phi(b) \sim T'_G \otimes \Phi(b')$.

Thus the chain map $\widehat{(\phi,\Phi)}_\#$ restricts to $\overline{(\phi,\Phi)}_\#$, and hence induces a chain map $(\phi,\Phi)_\#$. This chain map induces homomorphisms

$$(\phi,\Phi)_* : H_n^P(Y',B';k') \to H_n^G(Y,B;k),$$

for all n. It is clear that the properties 1-3 are satisfied. q.e.d.

Let P_1 be another topological group and let $\mathfrak{F}_1$ be an orbit type family for P_1, and let k_1 be a covariant coefficient system for $\mathfrak{F}_1$ over R. Assume that $\phi_1 : P_1 \to P$ is a continuous homomorphism, such that, $\phi_1(Q_1) \in \mathfrak{F}'$ if $Q_1 \in \mathfrak{F}_1$, and let $\Phi_1 : k_1 \to k'$ be a natural transformation with respect to ϕ_1. It is immediately seen that for any P_1-map $\alpha_1 : P_1/Q_1 \to P_1/N_1$, we have $\phi((\phi_1)(\alpha)) = (\phi\phi_1)(\alpha) : G/\phi\phi_1(Q_1) \to G/\phi\phi_1(N_1)$ and that $\Phi\Phi_1 : k_1 \to k$ is a natural transformation with respect to the homomorphism $\phi\phi_1 : P_1 \to G$.

PROPOSITION 1.2. Let the notation be as above. We then have $(\phi\phi_1,\Phi\Phi_1)_* = (\phi,\Phi)_*(\phi_1,\Phi_1)_*$. q.e.d.

Let us now return to the situation where we are given a ϕ-map $f : (X,A) \to (Y,B)$ from the P-pair (X,A) to the G-pair (Y,B). (All notation and terminology will be as above). The ϕ-map f induces homomorphisms

$$(f,\phi,\Phi)_* : H_n^P(X,A;k') \to H_n^G(Y,B;k)$$

which by definition, are given by $(f,\phi,\Phi)_* = (\phi,\Phi)_* f'_*$, where $f'_* : H_n^P(X,A;k') \to H^P(Y',B';k')$ is the homomorphism induced by the P-map $f' : (X,A) \to (Y',B')$.

COROLLARY 1.3. Let all the notation be as above. The ϕ-map $f : (X,A) \to (Y,B)$ induces homomorphism

$$(f,\phi,\Phi) : H_n^P(X,A;k') \to H_n^G(Y,B;k),$$

for all n, with the following properties.

1. If $f = \mathrm{id}$, $\phi = \mathrm{id}$ and $\Phi = \mathrm{id}$, then $(f,\phi,\Phi)_* = \mathrm{id}$.
2. $(f,\phi,\Phi)_*$ commutes with the boundary homomorphism.
3. Let $s : (Y,B) \to (\tilde{Y},\tilde{B})$ be a G-map, then $(sf,\phi,\Phi)_* = s_*(f,\phi,\Phi)_*$.

4. Let $r : (\tilde{X},\tilde{A}) \to (X,A)$ be a P-map, then $(fr,\phi,\Phi)_* = (f,\phi,\Phi)_* r_*$.

5. Let

$$\begin{array}{ccc} (X,A) & \xrightarrow{\quad f \quad} & (Y,B) \\ r\downarrow & & \downarrow s \\ (\tilde{X},\tilde{A}) & \xrightarrow{\quad \tilde{f} \quad} & (\tilde{Y},\tilde{B}) \end{array}$$

be a commutative diagram where s is a G-map, r is a P-map, and both f and $\tilde{f}$ are ϕ-maps. Then we have $(\tilde{f},\phi,\Phi)_* r_* = s_*(f,\phi,\Phi)_*$

6. If $h : (U,C) \to (X,A)$ is a ϕ_1-map, we have

$$(fh,\phi\phi_1,\Phi\Phi_1)_* = (f,\phi,\Phi)_*(h,\phi_1,\Phi_1)_*.$$

PROOF. The definition of the induced homomorphism $(f,\phi,\Phi)_*$ was already given above, and the properties 1. and 2. are then immediate consequences of the corresponding properties in Proposition 1.1. We shall next show that property 3. is valid. Let $s : (Y,B) \to (\tilde{Y},\tilde{B})$ be a G-map and $s' : (Y',B') \to (\tilde{Y}',\tilde{B}')$ the corresponding P-map. Since $s_*(\phi,\Phi)_* = (\phi,\Phi)_* s'_*$, by property 3. in Proposition 1.1., it follows that $(sf,\phi,\Phi)_* = (\phi,\Phi)_* (sf)'_* = (\phi,\Phi)_* s'_* f'_* = s_*(\phi,\Phi)_* f'_* = s_*(f,\phi,\Phi)_*$. The property 4. is an immediate consequence of the definition, since $(fr,\phi,\Phi)_* = (\phi,\Phi)_*(fr)'_* = (\phi,\Phi)_* f' r_* = (f,\phi,\Phi)_* r_*$. The property 5. is a consequence of properties 3. and 4.

Finally we prove property 6. The notation will be as follows. Let $h'' : (U,C) \to (X'',A'')$ be the P_1-map corresponding to the ϕ_1-map $h : (U,C) \to (X,A)$. Let $f' : (X,A) \to (Y',B')$ be the P-map corresponding to the ϕ-map $f : (X,A) \to (Y,B)$, and let then $f'' : (X'',A'') \to (Y'',B'')$ be the P_1-map corresponding to the P-map f'. By property 3. in Proposition 1.1. we have $f'_*(\phi_1,\Phi_1)_* = (\phi_1,\Phi_1)_* f''_*$. By Proposition 1.2. we have $(\phi\phi_1,\Phi\Phi_1)_* = (\phi,\Phi)_*(\phi_1\Phi_1)_*$.

Thus, using these two results, we have $(fh,\phi\phi_1,\Phi\Phi_1)_* = (\phi\phi_1,\Phi\Phi_1)_*f''_*h''_* =$ $(\phi,\Phi)_*(\phi_1,\Phi_1)_*f''_*h''_* = (\phi,\Phi)_*f'_*(\phi_1,\Phi_1)_*h''_* = (f,\phi,\Phi)_*(h,\phi_1,\Phi_1)_*$.

Let us now consider the cohomology version of Proposition 1.1. Let the continuous homomorphism $\phi : P \to G$ and the orbit type families $\mathfrak{F}'$ and $\mathfrak{F}$ for P and G, respectively, be as before. Let m' and m be contravariant coefficient systems for $\mathfrak{F}'$ and $\mathfrak{F}$, respectively, over the ring. Let

$$\Psi : m \to m'$$

be a natural transformation with respect to the homomorphism $\phi : P \to G$. By this we mean that for any $Q \in \mathfrak{F}'$ we have a homomorphism of R-modules

$$\Psi : m(G/\phi(Q)) \to m'(P/Q),$$

such that if $\alpha : P/Q \to G/N$ is a P-map, where also $N \in \mathfrak{F}'$, then the diagram

$$\begin{array}{ccc} m(G/\phi(N)) & \xrightarrow{\Psi} & m'(P/N) \\ \phi(\alpha)^* \downarrow & & \downarrow \alpha^* \\ m(G/\phi(Q)) & \xrightarrow{\Psi} & m'(P/Q) \end{array}$$

commutes.

PROPOSITION 1.4. Let the homomorphism $\phi : P \to G$, and the natural transformation $\Psi : m \to m'$ be as above. Let (Y,B) be a G-pair and make it into a P-pair (Y',B') through the homomorphism ϕ. Then the ϕ-map $id :$ $(Y',B') \to (Y,B)$ induces homomorphisms

$$(\phi,\Psi)^* : H^n_G(Y,B;m) \to H^n_P(Y',B';m'),$$

for all n, and the contravariant versions of the properties 1.-3.

in Proposition 1.1. are valid.

PROOF. Define a cochain map

$$\widehat{(\phi,\Psi)}^{\#} : \hat{S}_G(Y,B;m) \to \hat{S}_P(Y',B';m')$$

as follows. If $T_P : \Delta_n \times P/Q \to Y'$ is a P-equivariant singular n-simplex belonging to $\mathcal{F}'$ in Y' we let $T_G : \Delta_n \times G/\phi(Q) \to Y$ be the corresponding G-equivariant singular n-simplex belonging to $\mathcal{F}$ in Y, as defined in the proof of Proposition 1.1. Now let $\hat{c} \in \hat{C}^n_G(Y,B;m)$ and define $\widehat{(\phi,\Psi)}^{\#}(\hat{c})$ by,

$$((\widehat{(\phi,\Psi)}^{\#}(\hat{c}))(T_P) = \Psi(\hat{c}(T_G)) \in m'(P/Q),$$

where $\Psi : m(G/\phi(Q)) \to m'(P/Q)$. This defines the homomorphism $\widehat{(\phi,\Psi)}^{\#}$, and it is immediately seen that $\widehat{(\phi,\Psi)}^{\#}$ is a cochain map.

It remains to show that $\widehat{(\phi,\Psi)}^{\#}$ restricts to $(\phi,\Psi)^{\#} : S_G(Y,B;m) \to S_P(Y',B';m')$. Let $c \in C^n_G(Y,B;m)$. We shall use the same notation as in the proof of Proposition 1.1. Assume that $T_P = T'_P h_P$, and recall that then $T_G = T'_G h_G$. Moreover $(h_G)_0 = \phi((h_P)_0)$, and hence $\Psi(h_G)^* = (h_P)^*\Psi$. Thus we have $((\widehat{(\phi,\Psi)}^{\#}(c))(T_P) = \Psi(c(T_G)) = \Psi((h_G)^*c(T'_G)) = (h_P)^*\Psi(c(T'_G)) = (h_P)^*((\widehat{(\phi,\Phi)}^{\#}(c)(T'_P)$. Hence $\widehat{(\phi,\Psi)}^{\#}(c) \in C^n_P(Y',B';m)$. This completes the proof. q.e.d.

REMARK. Observe that the expression $(\phi,\Psi)^*$ is contravariant in ϕ but covariant in Ψ. Thus, if $\Psi_1 : m' \to m_1$ is a natural transformation (between contravariant coefficient systems m' and m_1) with respect to the continuous homomorphism $\phi_1 : P_1 \to P$, the cohomology analogue of Proposition 1.2. reads

$$(\phi\phi_1,\Psi_1\Psi)^* = (\phi_1,\Psi_1)^*(\phi,\Psi)^*.$$

Returning to the ϕ-map $f : (X,A) \to (Y,B)$, from the P-pair (X,A) to the G-pair (Y,B), we now define its induced homomorphisms

$$(f,\phi,\Psi)^* : H_G^n(Y,B;m) \to H_P^n(X,A;m')$$

by $(f,\phi,\Psi)^* = (f')^*(\phi,\Psi)^*$, where $(f')^* : H_P^n(Y',B';m') \to H_P^n(X,A;m')$ is the homomorphism induced by the P-map $f' : (X,A) \to (Y',B')$. The following corollary follows from Proposition 1.4. and the above remark is exactly the same way as Corollary 1.3. followed from Propositions 1.1. and 1.2.

COROLLARY 1.5. The ϕ-map $f : (X,A) \to (Y,B)$ induces homomorphisms

$$(f,\phi,\Psi)^* : H_G^n(Y,B;m) \to H_P^n(X,A;m')$$

for all n, and the cohomology versions of the properties 1.-6. in Corollary 1.3. are valid. q.e.d.

We conclude this section by determining the induced homomorphisms of the ϕ-map $f_\phi : P/Q \to G/\phi(Q)$, where $Q \in \mathcal{F}'$ and $f_\phi(pQ) = \phi(p)\phi(Q)$. Since $Q \in \mathcal{F}'$ and $\phi(Q) \in \mathcal{F}$, it follows by the dimension axiom that the P-space P/Q has only 0-dimensional P-equivariant singular homology and cohomology, and that the G-space $G/\phi(Q)$ has only 0-dimensional G-equivariant singular homology and cohomology. The following proposition thus gives the complete answer.

PROPOSITION 1.6. Consider the ϕ-map $f_\phi : P/Q \to G/\phi(Q)$, where $Q \in \mathcal{F}'$ and $f_\phi(pQ) = \phi(p)\phi(Q)$. We have the commutative diagrams

$$\begin{array}{ccc} H_0^P(P/Q;k') & \xrightarrow{(f_\phi,\phi,\Phi)^*} & H_0^G(G/\phi(Q);k) \\ \gamma_P \downarrow \cong & & \cong \downarrow \gamma_G \\ k'(P/Q) & \xrightarrow{\Phi} & k(G/\phi(Q)) \end{array}$$

$$\begin{array}{ccc} H^0_G(G/\phi(G);m) & \xrightarrow{(f_\phi,\phi,\Psi)^*} & H^0_P(P/Q;m') \\ \xi_G \downarrow \cong & & \cong \downarrow \xi_P \\ m(G/\phi(Q)) & \xrightarrow{\Psi} & m'(P/Q) \end{array}$$

Here the vertical arrows denote isomorphisms given by the dimension axiom.

PROOF. Let $T \otimes b \in C^P_0(P/Q;k') = H^P(P/Q;k')$, where $T : \Delta_0 \times P/N = P/N \to P/Q$ is a P-equivariant singular 0-simplex belonging to $\mathfrak{F}'$ in P/Q, and $b \in k'(P/N)$. We have $(f_\phi,\phi,\Phi)_*(T \otimes b) = (\phi,\Phi)_*(f'_\phi)_*(T \otimes b) = (\phi,\Phi)_*((f'_\phi T) \otimes b) = (f'_\phi T)_G \otimes \Phi(b)$. Here $(f'_\phi T)_G : G/\phi(N) \to G/\ (Q)$ denotes the G-equivariant singular 0-simplex in $G/\phi(Q)$ corresponding to the P-equivariant singular 0-simplex $f'_\phi T : P/N \to G/\phi(N)'$, where $(G/\phi(Q))'$ denotes the P-space obtained by making the G-space $G/\phi(Q)$ into a P-space through the homomorphism $\phi : P \to G$. It is immediately seen that $(f'_\phi T)_G = \phi(T) : G/\phi(N) \to G/\phi(Q)$ the G-map corresponding to the P-map $T : P/N \to P/Q$. Since $\phi(T)_*\Phi = \Phi T_*$ it follows that $\gamma_G(f_\phi,\phi,\Phi)_*(T \otimes b) = \phi(T)_*\Phi(b) = \Phi T_*(b) = \Phi\gamma_P(T \otimes b)$. We have proved that the first diagram commutes.

Recall that $H^0_P(P/Q;m') = C^0_P(P/Q;m')$, and that if $c' \in C^0_P(P/Q;m') = \mathrm{Hom}_t(\hat{C}^P_0(P/Q),M')$ then $\xi_P(c') = c'(\mathrm{id}_{P/Q})$. Also observe that $f'_\phi : P/Q \to (G/\phi(Q))'$ can be considered as a P-equivariant singular 0-simplex in $(G/\phi(Q))'$, and that $(f'_\phi)_G$, the corresponding G-equivariant singular 0-simplex in $G/\phi(Q)$, equals $\mathrm{id}_{G/\phi(Q)}$. It follows that if $c \in C^0_G(G/\phi(Q);m)$ then $\xi_P((f_\phi,\phi,\Psi)^*(c)) = ((f'_\phi)^*(\phi,\Phi)^*(c))(\mathrm{id}_{P/Q}) = ((\phi,\Psi)^*(c))(f'_\phi) = \Psi(c((f'_\phi)_G)) = \Psi(c(\mathrm{id}_{G/\phi(Q)})) = \Psi\xi_G(c)$. This shows that the second diagram commutes. q.e.d.

2. TRANSFER HOMOMORPHISMS

In this section P denotes a fixed closed subgroup of G such that the space of right cosets $P\backslash G$ consists of s elements, that is,

$$P\backslash G = \{Pg_1,\ldots,Pg_s\}.$$

Since P is assumed to be closed in G it follows that each point in $P\backslash G$ ($P\backslash G$ has the quotient topology from the projection $\pi\colon G \to P\backslash G$) is closed in $P\backslash G$. It follows that $P\backslash G$ has the discrete topology.

We say that a G-map

$$\beta : G/H \to G/H'$$

(H and H' denote arbitrary subgroups of G) is of "type P" if $\beta(eH) = p_0H$, where $p_0 \in P$. In this case we have $H \subset p_0H'p_0^{-1}$, and hence $P \cap H \subset p_0(P \cap H)p_0^{-1}$. Thus we can define a P-map

$$\beta^{!} : P/P \cap H \to P/P \cap H'$$

by the condition $\beta^{!}(e(P \cap H)) = p_0(P \cap H')$. We have $\beta^{!}(p(P \cap H) = pp_0(P \cap H')$, $p \in P$. Moreover, the P-map $\beta^{!}$ depends only on the G-map β of "type P", and not on the specific choice of the element $p_0 \in P$. For if $\beta(eH) = p_1H'$, where $p_1 \in P$, then $p_1^{-1}p_0 \in P \cap H'$ and hence $p_0(P \cap H') = p_1(P \cap H')$. We have shown how any G-map $\beta : G/H \to G/H'$ of "type P" determines a P-map $\beta^{!} : P/P \cap H \to P/P \cap H'$.

Let as before $\mathfrak{F}$ be an orbit type family for G, and let $\mathfrak{F}'$ be an orbit type family for P, such that, if $H \in \mathfrak{F}$ then $P \cap H \in \mathfrak{F}'$. Now let k' and k be convariant coefficient systems for $\mathfrak{F}'$ and $\mathfrak{F}$, respectively, over the ring R. Let

$$\Lambda : k \to k'$$

be a natural transformation of transfer type with respect to the inclusion $P \hookrightarrow G$. By this we mean that for every $H \in \mathcal{F}$ we have a homomorphism of left R-modules

$$\Lambda : k(G/H) \to k'(P/P \cap H)$$

such that if $\beta : G/H \to G/K$, where also $K \in \mathcal{F}$, is a G-map of "type P", then the diagram

$$\begin{array}{ccc} k(G/H) & \xrightarrow{\Lambda} & k'(P/P \cap H) \\ \beta_* \downarrow & & \downarrow (\beta^!)_* \\ k(G/K) & \xrightarrow{\Lambda} & k'(P/P \cap K) \end{array}$$

commutes.

Let (Y,B) be a G-pair. We denote by (Y',B') the P-pair obtained by restricting the G-action to the subgroup P. We shall construct transfer homomorphisms

$$(\tau^!,\Lambda) : H_n^G(Y,B;k) \to H_n^P(Y',B';k')$$

for every n.

We begin by defining for each element $Pg \in P\backslash G$ an induced chain map

$$(Pg)_{\#} : \hat{S}^G(Y,B;k) \to S^P(Y',B';k').$$

(Observe that $(Pg)_{\#}$ has as domain the "roof" chain complex $\hat{S}^G(Y,B;k)$ and as range the chain complex $S^P(Y',B';k')$ which gives the equivariant singular homology modules of the P-pair (Y',B')). Let for the moment $g \in G$ be a fixed element of G. We then define

$$(g)_{\#} \; : \hat{C}^G_n(Y,B;k) \to \hat{C}^P_n(Y',B';k)$$

as follows. Let $T : \Delta_n \times G/K \to Y$ be an equivariant singular n-simplex belonging to $\mathfrak{F}$ in Y. Consider the composite map

$$\Delta_n \times P/P \cap gKg^{-1} \xrightarrow{\eta} \Delta_n \times G/gKg^{-1} \xrightarrow{[g]} \Delta_n \times G/K \xrightarrow{T} Y$$

where $\eta(x,p(P \cap gKg^{-1})) = (x,p(gKg^{-1}))$, $\eta \in P \subset G$, and $[g]$ is the G-map, in fact G-homeomorphism, determined by the condition $[g](x,e(gKg^{-1})) = (x,gK)$. The map $T[g]\eta : \Delta_n \times P/P \cap gKg^{-1} \to Y'$, when considered as a map into the P-space Y', is a P-equivariant singular n-simplex belonging to $\mathfrak{F}'$ in Y'.
Now set

$$(g)_{\#}(T \otimes a) = T[g]\eta \otimes \Lambda[g]_*^{-1}(a),$$

where $a \in k(G/K)$ and $[g]_* : k(G/gKg^{-1}) \to k(G/K)$ is the isomorphism determined by the G-homeomorphism $[g]$, and $\Lambda : k(G/gKg^{-1}) \to k'(P/P \cap gKg^{-1})$. This defines the homomorphism $(g)_{\#} \; : \hat{C}^G_n(Y,B;k) \to \hat{C}^P_n(Y',B';k')$. Clearly $(g)_{\#}$ commutes with the boundary homomorphism.

Now let $p \in P$. We shall show that $(g)_{\#}(T \otimes a) - (pg)_{\#}(T \otimes a) \in \bar{C}^P_n(Y',B';k')$. Consider the diagram

$$\begin{array}{ccccc}
\Delta_n \times P/P \cap gKg^{-1} & \xrightarrow{\eta} & \Delta_n \times G/gKg^{-1} & & \\
\Big\downarrow \{p^{-1}\} & & \Big\downarrow [p^{-1}] & \searrow^{[p]} & \\
 & & & & \Delta_n \times G/K \\
 & & & \nearrow_{[pg]} & \\
\Delta_n \times P/P \cap (pg)K(pg)^{-1} & \xrightarrow{\eta} & \Delta_n \times G/(pg)K(pg)^{-1} & &
\end{array}$$

where $[p^{-1}]$ and $[pg]$ denote the G-homeomorphisms determined by the conditions $[p^{-1}](x,gKg^{-1}) = (x,p^{-1}(pg)K(pg)^{-1}))$ and $[pg](x,e((pg)K(pg)^{-1})) = (x,pgK)$, and $\{p^{-1}\}$ is the P-homeomorphism determined by the condition $\{p^{-1}\}(x,e(P\ gKg^{-1})) = (x,p^{-1}(P \cap (pg)K(pg)^{-1}))$. The diagram commutes. We have $(g)_{\#}(T \otimes a) = T[g]\eta \otimes \Lambda[g]_*^{-1}(a)$, and $(pg)_{\#}(T \otimes a) = T[pg]\eta \otimes \Lambda[pg]_*^{-1}(a)$. We claim that $T[g]\eta \otimes \Lambda[g]_*^{-1}(a) \sim T[pg]\eta \otimes \Lambda[pg]_*^{-1}(a)$. Since $\{p^{-1}\}$ is a P-map, which covers $id : \Delta_n \to \Delta_n$, and $T[g]\eta = T[pg]\eta\{p^{-1}\}$, it only remains to show that $\{p^{-1}\}_*(\Lambda[g]_*^{-1}(a)) = \Lambda[pg]_*^{-1}(a)$. This is seen as follows. Let $\{p^{-1}\}_0 : P/P \cap gKg^{-1} \to P/P \cap (pg)K(pg)^{-1}$ and $[p^{-1}]_0 : G/gKg^{-1} \to G/(pg)K(pg)^{-1}$ be the maps obtained by restricting $\{p^{-1}\}$ and $[p^{-1}]$ to the orbit $d^0 \in \Delta_n$. Then the G-map $[p^{-1}]_0$ is of "type P" and the corresponding P-map is $\{p^{-1}\}_0$, that is, $([p^{-1}]_0)^! = \{p^{-1}\}$. It follows that $\{p^{-1}\}_*\Lambda = \Lambda[p^{-1}]_*$, and hence $\{p^{-1}\}_*(\Lambda[g]_*^{-1}(a)) = \Lambda[p^{-1}]_*[g]_*^{-1}(a) = \Lambda[pg]_*^{-1}(a)$.

We have shown that if $p \in P$ then $\pi(pg)_{\#} = \pi(g)_{\#} : \hat{C}_n^G(Y,B;k) \to C^P(Y',B';k')$, where π denotes the natural projection from $\hat{C}_n^P(Y',B';k')$ onto $C_n^P(Y',B';k')$. Thus we can for each element $Pg \in P\backslash G$ define an induced homomorphism $(Pg)_{\#}: \hat{C}_n^G(Y,B;k) \to C^P(Y',B';k')$ by defining $(Pg)_{\#} = \pi(\bar{g})_{\#}$, where $\bar{g}$ is any representative for the right conset Pg, that is, $\bar{g} \in Pg$. Since $(g)_{\#}$ commutes with the boundary it follows that the homomorphisms $(Pg)_{\#}$ form a chain map $(Pg)_{\#} : \hat{S}^G(Y,B;k) \to S^P(Y',B';k')$.

We now define

$$\hat{\tau}_{\#} : S^G(Y,B;k) \to S^P(Y',B';k')$$

to be the chain map

$$\hat{\tau}_{\#} = \sum_{i=1}^{s} (Pg_i)_{\#}.$$

Thus $\hat{\tau}_{\#}(T \otimes a) = \pi\left(\sum_{i=1}^{s} (g_i)_{\#}(T \otimes a)\right)$, where $g_1,\ldots,g_s \in G$ form any complete set of representatives for the set of right cosets $P\backslash G$. We shall now show that $\hat{\tau}_{\#}$ induces a chain map

$$\tau_{\#} : S^G(Y,B;k) \to S^P(Y',B';k').$$

Assume that $T \otimes a \sim T' \otimes a'$, where $T : \Delta_n \times G/K \to Y$, $T' : \Delta_n \times G/K' \to Y'$, and $a \in k(G/K)$, $a' \in k(G/K')$. Let $h : \Delta_n \times G/K \to \Delta_n \times G/K'$ be a G-map which covers $\mathrm{id} : \Delta_n \to \Delta_n$, such that $T = T'h$ and $h_*(a) = a'$. Let $g_0 \in G$ be such that $h(d^0,eK) = (d^0,g_0K')$. Now let $g \in G$ be any element of G and consider the diagram

$$\begin{array}{ccccc}
\Delta_n \times P/P \cap gKg^{-1} & \xrightarrow{\eta} & \Delta_n \times G/gKg^{-1} & \xrightarrow{[g]} & \Delta_n \times G/K \\
\downarrow r & & \downarrow \bar{h} & & \downarrow \\
\Delta_n \times P/P \cap (gg_0)K'(gg_0)^{-1} & \xrightarrow{\eta} & \Delta_n \times G/(gg_0)K'(gg_0)^{-1} & \xrightarrow{[gg_0]} & \Delta_n \times G/K'
\end{array}$$

We claim that the image of the map $h[g]\eta$ is the P-subset $P(\Delta_n \times \{gg_0K'\}) = \{(x,p(gg_0)K') \in \Delta_n \times G/K' \mid p \in P\}$ of $\Delta_n \times G/K'$. This is seen as follows. Let $\bar{\pi} : \Delta_n \times G/K' \to \Delta_n \times P\backslash G/K'$ be the natural projection. The set of double cosets $P\backslash G/K'$ is discrete since $P\backslash G$ is discrete. Denote for convenience $Q = P \cap gKg^{-1}$. Since the subset $\bar{\pi}h[g]\eta(\Delta_n \times \{eQ\}) \subset \Delta_n \times P\backslash G/K'$ is connected and the map $\bar{\pi}h[g]\eta$ covers $\mathrm{id} : \Delta_n \to \Delta_n$, and since moreover $(d^0,P(gg_0)K') \in \bar{\pi}h[g]\eta(\Delta_n \times \{eQ\})$, it follows that $\bar{\pi}h[g]\eta(\Delta_n \times \{eQ\}) = \Delta_n \times \{P(gg_0)K'\} \subset \Delta_n \times P\backslash G/K'$. From this and the fact that $\mathrm{Im}(h[g]\eta)$ is a P-subset it follows that $\mathrm{Im}(h[g]\eta) = P(\Delta_n \times \{(gg_0)K'\})$. Now observe that the map $[gg_0]\eta$ is a P-homeomorphism onto the P-subset $P(\Delta_n \times \{gg_0K\})$. Thus we define the map r in the diagram by $r = ([gg_0]\eta\ )^{-1}h[g]\eta$. Then r is a P-map, which covers $\mathrm{id} : \Delta_n \to \Delta_n$. We have $T[g]\eta = T'[gg_0]\eta r$.

We now claim that $(r)_*\Lambda[g]_*^{-1}(a) = \Lambda[gg_0]_*^{-1}(a')$. This is seen as follows. Define the G-map $\bar{h}$ in the diagram by $\bar{h} = [gg_0]^{-1}h[g]$. The whole diagram now commutes. Restricting the maps $\bar{h}$ and r to the orbit over $d^0 \in \Delta_n$ we get the G-map $\bar{h}_0 : G/gKg^{-1} \to G/(gg_0)^{-1}K'(gg_0)^{-1}$ and the P-map $r_0 : P/P \cap gKg^{-1} \to P/P \cap (gg_0)K'(gg_0)^{-1}$. Observe that $gKg^{-1} \subset (gg_0)K'(gg_0)^{-1}$ and that $\bar{h}_0$ and r_0 in fact are the natural projections. Thus in particular the G-map $\bar{h}$ is of "type P" and its corresponding P-map is r_0. It follows that $r_*\Lambda = \Lambda\bar{h}_*$. Thus $r_*\Lambda[g]_*^{-1}(a) = \Lambda\bar{h}_*[g]_*^{-1}(a) = \Lambda[gg_0]_*^{-1}h_*(a) = \Lambda[gg_0]_*^{-1}(a')$. We have now shown that $\pi(g)_\#(T \otimes a) = \pi(gg_0)_\#(T' \otimes a')$.

If $g_1,\ldots,g_s \in G$ is any complete set of representatives for the set of right cosets $P\backslash G$ then also $g_1g_0,\ldots,g_sg_0 \in G$ is a complete set of representatives. Thus $\hat{\tau}_\#(T \otimes a) = \pi(\sum_{i=1}^{s} (g_i)_\#(T \otimes a)) = $
$= \pi(\sum_{i=1}^{s} (g_ig_0)_\#(T' \otimes a')) = \hat{\tau}_\#(T' \otimes a')$. We have proved that $\hat{\tau}_\#$ induces a chain map $\tau_\#: S^G(Y,B;k) \to S^P(Y',B';k')$. Moreover it is clear from the way the chain map $\tau_\#$ is defined that $\tau_\#$ commutes with the chain maps induced by a G-map $f : (X,A) \to (Y,B)$ and its corresponding P-map f' : $(X',Y') \to (Y',B')$. We use the notation $(\tau^!,\Lambda) : H_n^G(Y,B;k) \to H_n^P(Y',B';k')$ for the homomorphism induced by the chain map $\tau_\#$. We have proved.

THEOREM 2.1. Assume that P is a closed subgroup of G such that $P\backslash G$ is a finite set. Let the covariant coefficient systems k' and k for P and G, respectively, be as above, and let $\Lambda : k \to k'$ be a natural transformation of transfer type with respect to $P \hookrightarrow G$. For any G-pair (Y,B) we have transfer homomorphisms

$$(\tau^!,\Lambda) : H_n^G(Y,B;k) \to H_n^P(Y',B';k')$$

for every n. The homomorphisms $(\tau^!,\Lambda)$ commute with the boundary homomorphism and with homomorphisms induced by G-maps. q.e.d.

We shall now study the composite of the transfer homomorphism $(\tau^!,\Lambda)$ followed by the homomorphism $(i,\Phi)_*$ induced by the inclusion $i : P \hookrightarrow G$. Let $\mathfrak{F}'$ and $\mathfrak{F}$ be orbit type families for P and G, respectively, and let k' and k be covariant coefficient system for $\mathfrak{F}'$ and $\mathfrak{F}$, respectively, over the ring R. We assume, that if $H \varepsilon \mathfrak{F}$ then $P \cap H \varepsilon \mathfrak{F}'$, and if $Q \varepsilon \mathfrak{F}'$, then also $Q \varepsilon \mathfrak{F}$. Let $\Lambda : k \to k'$ be a natural transformation of transfer type with respect to $P \hookrightarrow G$, and let $\Phi : k' \to k$ be a natural transformation with respect to $i : P \hookrightarrow G$. Moreover let $\Theta : k \to k$ be a homomorphism from k to itself, that is, a natural transformation from k to itself with respect to the identity homomorphism $id : G \to G$. We assume that the following condition is satisfied. For every $H \varepsilon \mathfrak{F}$ the diagram

$$\begin{array}{ccccc} k(G/H) & \xrightarrow{\ \Lambda\ } & k'(P/P \cap H) & \xrightarrow{\ \Phi\ } & k(G/P \cap H) \\ & \searrow^{\Theta} & & & \downarrow \rho_* \\ & & & & k(G/H) \end{array}$$

commutes. Here $\rho : G/P \cap H \to G/H$ denotes the natural projection.

THEOREM 2.2. Assume that P is a closed subgroup of G such that $P\backslash G$ consists of s elements. Let $\Lambda : k \to k'$, $\Phi : k' \to k$, and $\Theta : k \to k$ be as above, and assume that the above condition is satisfied. Then, for any G-pair (Y,B) and every integer n, the composite homomorphism

$$H_n^G(Y,B;k) \xrightarrow{(\tau^!,\Lambda)} H_n^P(Y',B';k') \xrightarrow{(i,\Phi)_*} H_n^G(Y,B;k)$$

equals $s\Theta_*$. In particular if $\Theta = id$ this composite equals multiplication by s.

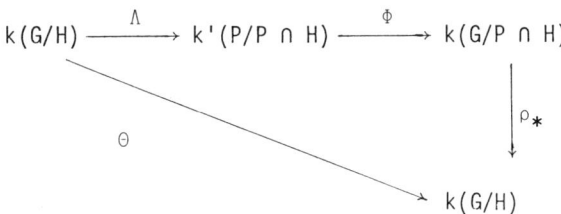

PROOF. Let $T : \Delta_n \times G/K \to Y$ be a G-equivariant singular n-simplex belonging to $\mathfrak{F}$ in Y, and let $a \varepsilon k(G/K)$. Let $Pg \varepsilon P\backslash G$ and consider the composite homomorphism $(i,\Phi)_\#(Pg)_\# : \hat{C}^P_n(Y,B;k) \to C^P_n(Y',B';k') \to C^G_n(Y,B;k)$. We have

$$(i,\Phi)_\#(Pg)_\#(T \otimes a) = \pi((T[g]\eta)_G \otimes \Phi\Lambda[g]_*^{-1}(a)).$$

Here $T[g]\eta$ is the P-equivariant singular n-simplex belonging to $'$ in Y' as in (2.1), and $(T[g]\eta)_G$ is its corresponding G-equivariant singular n-simplex belonging to $\mathfrak{F}$ in Y, as defined in section 1. of this chapter. It is immediately seen that $(T[g]\eta)_G = T[g]\rho$, where $\rho : \Delta_n \times G/P \cap gKg^{-1} \to \Delta_n \times G/gKg^{-1}$ denotes the natural projection. We have

$$T[g]\rho \otimes \Phi\Lambda[g]_*^{-1}(a) \sim T \otimes [g]_*\rho_*\Phi\Lambda[g]_*^{-1}(a) = T \otimes [g]_*\Theta[g]_*^{-1}(a) = T \otimes \Theta(a).$$

Thus

$$(i,\Phi)_\#(Pg)_\#(T \otimes a) = \pi\hat{\Theta}_\#(T \otimes a).$$

It follows that already on the chain level we have $(i,\Phi)_\#\tau_\# = s\Theta_\# : C^G_n(Y,B;k) \to C^G_n(Y,B;k)$. q.e.d.

We shall show that the transfer homomorphisms compose in a natural way. Assume, in addition to the assumptions made in establishing Theorem 2.1., that N is a closed subgroup of P such that $N\backslash P$ is a finite set, and that $\mathfrak{F}_1$ is an orbit type family for N, such that, if $Q \in \mathfrak{F}'$ then $N \cap Q \in \mathfrak{F}_1$. Moreover let k_1 be a covariant coefficient system for $\mathfrak{F}_1$ and $\Lambda_1 : k' \to k_1$ a natural transformation of transfer type with respect to $N \hookrightarrow P$. Observe that $\Lambda_1\Lambda : k \to k_1$ is then a natural transformation of transfer type with respect to $N \hookrightarrow G$.

PROPOSITION 2.3. Let the assumptions and notation be as above. Then

$$(\tau^{!},\Lambda_1)(\tau^{!},\Lambda) = (\tau^{!},\Lambda_1\Lambda).$$

PROOF. Let $g \in G$, $p \in P$ and consider the homomorphisms (g) : $\hat{C}_n^G(Y,B;k) \to \hat{C}_n^P(Y',B';k')$ and $(p)_\# : \hat{C}_n^P(Y',B';k') \to \hat{C}_n^N(Y_1,B_1;k_1)$ (here (Y_1,B_1) denotes the N-pair obtained from the G-pair (Y,B) by restricting the action to the subgroup N). It is easy to see that $(p)_\#(g)_\# = (pg)_\#$. The proposition now follows from this fact for if $g_1,\dots,g_s \in G$ is a complete set of representatives for $P\backslash G$ and $p_1,\dots,p_r \in P$ is a complete set of representatives for $N\backslash P$ then the elements $p_i g_j \in G$, $1 \leq i \leq s$, form a complete set of representatives for $N\backslash G$. q.e.d.

The construction of the transfer homomorphism in cohomology is dual to the construction in homology. We give the necessary details below, using the same diagrams and constructions we already used in constructing the transfer homomorphism in homology.

Assume that $\mathfrak{F}'$ and $\mathfrak{F}$ are orbit type families for P and G, respectively, such that if $H \in \mathfrak{F}$ the $P \cap H \in \mathfrak{F}'$. Let m' and m be contravariant coefficient systems for $\mathfrak{F}'$ and $\mathfrak{F}$, respectively, over the ring R. Let

$$\Omega : m' \to m$$

be a natural transformation of transfer type with respect to the inclusion $P \hookrightarrow G$. By this we mean that for any $H \in \mathfrak{F}$ we have a homomorphism of right R-modules

$$\Omega : m'(P/P \cap H) \to m(G/H)$$

such that if $\beta : G/H \to G/K$, where also $K \in \mathfrak{F}$, is a G-map of "type P", then the diagram

$$\begin{array}{ccc} m'(P/P\cap K) & \xrightarrow{\Omega} & m(G/K) \\ {\scriptstyle (\beta^{!})^*}\downarrow & & \downarrow{\scriptstyle \beta^*} \\ m'(P/P\cap H) & \xrightarrow{\Omega} & m(G/H) \end{array}$$

commutes.

THEOREM 2.4. Assume that P is a closed subgroup of G such that $P\backslash G$ is a finite set. Let the contravariant coefficient systems m' and m for P and G, respectively, be as above, and let $\Omega : m' \to m$ be a natural transformation of transfer type with respect to $P \hookrightarrow G$. For any G-pair (Y,B) we have transfer homomorphisms

$$(\tau_{!},\Omega) : H^n_P(Y',B';m') \to H^n_G(Y,B;m)$$

for every n. The homomorphisms $(\tau_{!},\Omega)$ commute with the boundary homomorphism and with homomorphisms induced by G-maps.

PROOF. Let $g \in G$ and define

$$(g)^{\#} : C^n_P(Y',B';m') \to \hat{C}^n_G(Y,B;m)$$

as follows. Let $c \in C^n_P(Y',B';k')$ and define $(g)^{\#}(c)$ by the following. If $T : \Delta_n \times G/K \to Y$ is a G-equivariant singular n-simplex belonging to $\mathcal{F}$ in Y we define the value of $(g)^{\#}(c)$ on T by (see (2.1))

$$((g)^{\#}(c))(T) = ([g]^*)^{-1}\Omega c(T[g]_{\eta}).$$

Here $c(T[g]_{\eta}) \in m'(P/P\cap gKg^{-1})$ and $\Omega : m'(P/P\cap gKg^{-1}) \to m(G/gKg^{-1})$, and $[g]^*$ is an isomorphism $[g]^* : m(G/K) \to m(G/gKg^{-1})$. This defines the homomorphism $(g)^{\#}$ and it is clear that is commutes with the coboundary.

Now let $p \in P$. We claim that $(pg)^{\#} = (g)^{\#}$. Consider the diagram (2.2.). Since $c \in C^n_P(Y',B';k')$ (no roof!) it follows that $c(T[g]\eta) = c(T[pg]\eta\{p^{-1}\}) = \{p^{-1}\}^* c(T[pg]\eta)$. Thus we have

$$((g)^{\#}(c)(T) = ([g]^*)^{-1}\Omega c(T[g]\eta) = ([g]^*)^{-1}\Omega\{p^{-1}\}^* c(T[pg]\eta) =$$

$$([g]^*)^{-1}[p^{-1}]^*\Omega c(T[pg]\eta) = ((pg)^{\#}(c))(T).$$

It follows that each element $Pg \in P\backslash G$ gives rise to homomorphisms

$$(Pg)^{\#} : C^n_P(Y',B';m') \to \hat{C}^n_G(Y,B;m)$$

for all n, defined by $(Pg)^{\#} = (\bar{g})^{\#}$, where $\bar{g}$ is any representative for the right coset Pg.

We now define

$$\hat{\tau}^{\#} : C^n_P(Y',B';m') \to \hat{C}^n_G(Y,B;m)$$

to be the homomorphism

$$\hat{\tau}^{\#} = \sum_{i=1}^{s} (Pg_i)^{\#}.$$

Thus $\tau^{\#}(c) = \sum_{i=1}^{s} (g_i)^{\#}(c)$, where $g_1,\ldots,g_s \in G$ form any complete set of representatives for the set of right cosets $P\backslash G$. We claim that the image of $\hat{\tau}^{\#}$ lies $C^n_G(Y,B;m)$ and $\hat{\tau}^{\#}$ thus induces

$$\tau^{\#} : C^n_P(Y',B';m') \to C^n_G(Y,B;m).$$

This is seen as follows. Let $T : \Delta_n \times G/K \to Y$ and $T' : \Delta_n \times G/K \to Y$ be equivariant singular n-simplexes belonging to in Y, and assume that $h : \Delta_n \times G/K \to \Delta_n \times G/K'$ is a G-map, which covers $id : \Delta_n \to \Delta_n$, such that $T = T'h$. We must show that $(\hat{\tau}^{\#}(c))(T) = h^*(\hat{\tau}^{\#}(c))(T')$. Let $g_0 \in G$ be such that $h(d^0,eK) = (d^0,g_0K)$. Now let $g \in G$ and consider the diagram (2.3.).

We then have

$$((g)^{\#}(c))(T) = ([g]^*)^{-1}\Omega c(T([g]n) = ([g]^*)^{-1}\Omega c(T'[gg_0]nr) =$$

$$([g]^*)^{-1}\Omega r^*c(T'[gg_0]n) = ([g]^*)^{-1}\bar{h}^*\Omega c(T'[gg_0]n) =$$

$$h^*([gg_0]^*)^{-1}\Omega c(T'[gg_0]n) = h^*(gg_0)^{\#}(c)(T').$$

Now if $g_1,\ldots,g_s \in G$ is any complete set representatives for $P\backslash G$ the same is true for $g_1g_0,\ldots,g_sg_0 \in G$. Thus by what we just showed it follows that $(\hat{\tau}^{\#}(c))(T) = \sum_{i=1}^{s}((g_i)^{\#}(c))(T) = \sum_{i=1}^{s} h^*((g_ig_0)^{\#}(c))(T') = h^*(\hat{\tau}^{\#}(c)(T'))$. This proves our claim and thus $\hat{\tau}^{\#}$ induces $\tau^{\#}$. The homomorphisms $\tau^{\#}$ form a cochain map $\tau^{\#} : S_P(Y',B';m') \to S_G(Y,B;m)$. We denote the induced homomorphism on cohomology by $(\tau_!,\Omega) : H_P^n(Y',B';m') \to H_G^n(Y,B;m)$ and call it the transfer homomorphism. q.e.d.

We shall now give the cohomology version ot Theorem 2.2. For this let $\mathfrak{F}'$ and $\mathfrak{F}$ be orbit type families for P and G, respectively, such that if $H \in \mathfrak{F}$ then $P \cap H \in \mathfrak{F}'$ and if $Q \in \mathfrak{F}$ then also $Q \in \mathfrak{F}$. Let m' and m be contravariant coefficient systems for $\mathfrak{F}'$ and $\mathfrak{F}$, respectively. Let $\Omega : m' \to m$ be a natural transformation of transfer type with respect to $P \hookrightarrow G$, and let $\Psi : m \to m'$ be a natural transformation with respect to $i : P \hookrightarrow G$. Moreover let $\Theta : m \to m$ be a homomorphism from the contravariant coefficient system m to itself. (The homomorphism induced by Θ on equivariant singular cohomology is denoted by Θ_*.) We assume that the following condition is satisfied. For every $H \in \mathfrak{F}$ the diagram

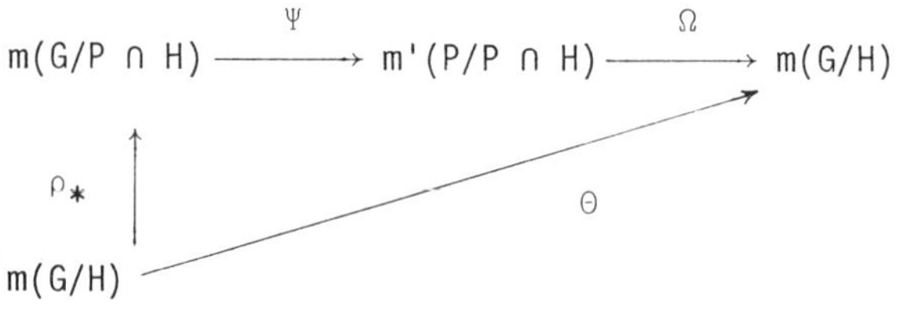

commutes. Here $\rho : G/P \cap H \to G/H$ denotes the natural projection.

THEOREM 2.5. Assume that P is a closed subgroup of G such that $P\backslash G$ consists of s elements. Let $\Omega : m' \to m$, $\Psi : m \to m'$, and $\Theta : m \to m$ be as above, and assume that the above condition is satisfied. Then for any G-pair (Y,B) and every integer n, the composite homomorphism

$$H^n_G(Y,B;m) \xrightarrow{(i,\Psi)^*} H^n_P(Y',B';m) \xrightarrow{(\tau_!,\Omega)} H^n_G(Y,B;m)$$

equals $s\Theta_*$. In particular if $\Theta = id$ this composite equals multiplication by s.

PROOF. Let $Pg \in P\backslash G$, and consider the composite homomorphism $(Pg)^{\#}(i,\Psi)^{\#} : C^n_G(Y,B;m) \to C^n_P(Y',B';m') \to \hat{C}^n_G(Y,B;m)$. Let $c \in C^n_G(Y,B;m)$. The value of $(Pg)^{\#}(i,\Psi)^{\#}(c)$ on an equivariant singular n-simplex $T : \Delta_n \times G/K \to Y$ belonging to $\mathfrak{F}$ equals $((Pg)^{\#}(i,\Psi)^{\#}(c))(T) = ([g]^*)^{-1}\Omega(i,\Psi)^{\#}(c))(T[g]) = ([g]^*)^{-1}\Omega\Psi c((T[g]\eta)_G)$. (The notation is the same as in (2.1) and the proof of Theorem 2.2.). We have $(T[g]\eta)_G = T[g]\rho$, where $\rho : \Delta_n \times G/P \cap gKg^{-1} \to \Delta_n \times G/gKg^{-1}$ denotes the natural projection, and $c(T[g]\eta) = \rho^*[g]^*c(T)$. Hence

$$((Pg)^{\#}(i,\Psi)^{\#}(c))(T) = ([g]^*)^{-1}\Omega\Psi\rho^*[g]^*c(T) = \Theta(c(T)) = \Theta_{\#}(c))(T).$$

The result follows. q.e.d.

REMARK. The transfer homomorphisms in cohomology also compose in a natural way. The cohomology version of Proposition 2.3. reads

$$(\tau_!,\Omega)(\tau_!,\Omega_1) = (\tau_!\Omega\Omega_1)$$

where $\Omega_1 : m_1 \to m'$ is a natural transformation of transfer type with respect to $N \hookrightarrow P$.

3. THE KRONECKER INDEX AND THE CUP-PRODUCT

In this section we assume that R is a commutative ring. By $\mathcal{F}$ we denote an orbit type family for G.

DEFINITION 3.1. Let k and m be a covariant and a contravariant, respectively, coefficient system for $\mathcal{F}$ over R. A pairing ω of k and m consists of the following. For each $H \varepsilon \mathcal{F}$ we have a homomorphism of R-modules

$$\omega : m(G/H) \otimes_R k(G/H) \to R,$$

such that if $\alpha : G/H \to G/K$, where also $K \varepsilon \mathcal{F}$, is a G-map and $b \varepsilon m(G/K)$, $a \varepsilon k(G/K)$, then

$$\omega(b \otimes_R \alpha_*(a)) = \omega(\alpha^*(b) \otimes_R a).$$

Let X be a G-space, and let $\hat{c} \varepsilon \hat{C}^n_G(X;m)$ and $\hat{\sigma} \in \hat{C}^G_n(X;k)$. Assume that we are given a pairing ω of k and m. We then define the Kronecker index of $\hat{c}$ and $\hat{\sigma}$, denoted $\langle \hat{c},\hat{\sigma}\rangle \varepsilon R$, as follows. If $\hat{\sigma} = \sum_{i=1}^{q} T_i \otimes_R a_i$, we set

$$\langle \hat{c},\hat{\sigma}\rangle = \omega(\sum_{i=1}^{q} \hat{c}(T_i) \otimes_R a_i).$$

It is immediately seen that this gives us a well-defined homomorphism of R-modules

$$\langle\ ,\ \rangle : \hat{C}^n_G(X;m) \otimes_R \hat{C}^G_n(X;k) \to R.$$

Let T be an equivariant singular $(n+1)$-simplex belonging to $\mathcal{F}$ in X and $a \varepsilon k(G/t(T))$. We then have

$$\langle \hat{c},\hat{\partial}_{n+1}(T \otimes a)\rangle = \langle \hat{c}, \sum_{i=1}^{n+1} (-1)^i T^{(i)} \otimes a\rangle =$$

$$\omega\left(\sum_{i=0}^{n+1} (-1)^i \hat{c}(T^{(i)}) \otimes_R a\right) = \omega(\hat{\delta}_n \hat{c}(T) \otimes_R a) = \left\langle \hat{\delta}_n \hat{c}, T \otimes a \right\rangle .$$

It follows that

$$\left\langle \hat{c}, \hat{\partial}_{n+1}(\hat{\sigma}) \right\rangle = \left\langle \hat{\delta}_n \hat{c}, \hat{\sigma} \right\rangle$$

for any $\hat{c} \in \hat{C}^n_G(X;m)$ and $\hat{\sigma} \in \hat{C}^G_{n+1}(X;k)$.

Now assume that $c \in C^n_G(X;m)$ and $\sigma \in C^G_n(X;k)$. We claim that the definition

$$\langle c, \sigma \rangle = \langle c, \hat{\sigma} \rangle ,$$

where $\hat{\sigma} \in \hat{C}^G_n(X;k)$ is any representative for σ, gives a well-defined homomorphism

$$\langle \, , \, \rangle : C^n_G(X;m) \otimes_R C^G_n(X;k) \to R.$$

This is seen as follows. Assume that $T \otimes a \sim T' \otimes a'$, and let $h : \Delta_n \times G/t(T) \to \Delta_n \times G/t(T')$ be a G-map, which covers $id : \Delta_n \to \Delta_n$, such that $T = T'h$ and $h_*(a) = a'$. Since $c \in C^n_G(X,m)$ it follows that $c(T) = h^*c(T')$, and hence we have

$$\langle c, T \otimes a \rangle = \omega(c(T) \otimes_R a) = \omega(h^*c(T') \otimes_R a) =$$

$$\omega(c(T') \otimes_R h_*(a)) = \omega(c(T') \otimes_R a') = \langle c, T' \otimes a' \rangle .$$

This proves our claim.

Let (X,A) be a G-pair. It follows directly from the definitions that the already established pairing $\langle \, , \, \rangle$ for the absolute case induces a pairing

$$\langle \, , \, \rangle : C^n_G(X,A;m) \otimes_R C^G_n(X,A;k) \to R.$$

Since now

$$\langle c, \partial_{n+1}\sigma \rangle = \langle \delta_n c, \sigma \rangle ,$$

where $c \in C_G^n(X,A;m)$ and $\sigma \in C_{n+1}^G(X,A;k)$, it follows that we have an induced pairing

$$\langle \, , \, \rangle : H_G^n(X,A;m) \otimes_R H_n^G(X,A;k) \to R.$$

This map, $\langle \, , \, \rangle$ is a homomorphism of R-modules, and we call it the Kronecker index. The Kronecker index gives rise to the homomorphism of R-modules

$$\nu : H_G^n(X,A;m) \to \mathrm{Hom}_R(H_n^G(X,A),R)$$

defined by $\nu(\eta)(\xi) = \langle \eta, \xi \rangle$, where $\eta \in H_G^n(X,A;m)$ and $\xi \in H_n^G(X,A;k)$.

For a G-space of the form G/H, where $H \in \mathcal{F}$, the Kronecker index agrees with the given pairing ω. To be more precise we have the following proposition.

PROPOSITION 3.2. Let $H \in \mathcal{F}$. Then the diagram

$$\begin{array}{ccc} H_G^0(G/H;m) \otimes_R H_0^G(G/H;k) & \xrightarrow{\langle \, , \, \rangle} & R \\ \xi \otimes \gamma \cong \downarrow & & \downarrow \mathrm{id} \\ m(G/H) \otimes_R k(G/H) & \xrightarrow{\omega} & R \end{array}$$

commutes. Here γ and ξ are the isomorphisms given by the dimension axiom.

PROOF. Let $c \in C_G^0(G/H;m) = H_G^0(G/H;m)$ and $T \otimes a \in C_0^G(G/H;k) = H_0^G(G/H;k)$ where $T : G/K \to G/H$ is an equivariant singular 0-simplex in G/H and $a \in k(G/H)$. We have $c(T) = T^*c(\mathrm{id}_{G/H}) \in m(G/K)$. Thus

$$\omega(\xi \otimes \gamma)(c \otimes_R (T \otimes a)) = \omega(\xi(c) \otimes_R \gamma(T \otimes a)) =$$

$$\omega(c(\mathrm{id}_{G/H}) \otimes_R T_*(a)) = \omega(T^*c(\mathrm{id}_{G/H}) \otimes_R a) =$$

$$\omega(c(T) \otimes_R a) = \langle c, T \otimes a \rangle \quad . \qquad \text{q.e.d.}$$

We shall now construct a cup-product in equivariant singular cohomology. We assume in the following that the orbit type family $\mathcal{F}$ is such that $G \varepsilon \mathcal{F}$.

DEFINITION 3.3. A contravariant coefficient system m for $\mathcal{F}$, over the ring R, is called a commutative ring coefficient sysstem if the following condition is satisfied. Each $m(G/H)$, $H \varepsilon \mathcal{F}$, is a commutative ring and all induced homomorphisms are ring homomorphisms, and moreover $m(G/G) = R$ and the R-module structure on each $m(G/H)$ is the same as the one induced by the ring homomorphism $\pi^* : R = m(G/G) \to m(G/H)$.

Assume from now that m is a commutative ring coefficient system. Let $\hat{c} \varepsilon \hat{C}^n_G(X;m)$ and $\hat{c}_1 \varepsilon \hat{C}^p_G(X;m)$. We define the cup-product $\hat{c} \cup \hat{c}_1 \varepsilon \hat{C}^{n+p}_G(X;m)$ by the following. Let $T : \Delta_{n+p} \times G/K \to X$ be an equivariant singular (n+p)-simplex belonging to $\mathcal{F}$ in X. We use the notation

$$\alpha_n : \Delta_n \times G/K \to \Delta_{n+p} \times G/K, \quad \text{and}$$

$$\beta_p : \Delta_p \times G/K \to \Delta_{n+p} \times G/K$$

for the front n-face and back p-face, respectively, of $\Delta_{n+p} \times G/K$, that is, $\alpha_n((x_0,\ldots,x_n),gK) = ((x_0,\ldots,x_n,0,\ldots,0),gK)$ and $\beta_p((x_0,\ldots,x_p),gK) = ((0,\ldots,0,x_0,\ldots,x_p),gK)$. We now define the value of $\hat{c} \cup \hat{c}_1$ on T to be

$$(\hat{c} \cup \hat{c}_1) = (\hat{c}(T\alpha_n))(\hat{c}_1(T\beta_n)) \varepsilon m(G/K).$$

This defines a homomorphism of R-modules

$$\cup : \hat{C}^n_G(X;m) \otimes_R \hat{C}^p_G(X;m) \to \hat{C}^{n+p}_G(X;m).$$

The formula

$$\hat{\delta}(\hat{c} \cup \hat{c}_1) = (\hat{\delta}\hat{c}) \cup \hat{c}_1 + (-1)^n \hat{c} \cup (\hat{\delta}\hat{c}_1)$$

is established by the standard calculation.

We now claim that if $c \varepsilon C_G^n(X;m)$ and $c_1 \varepsilon C_G^p(X;m)$ then also $c \cup c_1 \varepsilon C_G^{n+p}(X;m)$. This is seen as follows. Let $T : \Delta_{n+p} \times G/K \to X$ and $T' : \Delta_{n+p} \times G/K' \to X$ and assume that $h : \Delta_{n+p} \times G/K \to \Delta_{n+p} \times G/K'$ is a G-map, which covers $id : \Delta_{n+p} \to \Delta_{n+p}$, such that $T = T'h$. We have to show that

$$(c \cup c_1)(T) = h^*(c \cup c_1)(T').$$

The G-map h determines G-maps $h_\alpha : \Delta_n \times G/K \to \Delta_n \times G/K'$ and $h_\beta : \Delta_p \times G/K \to \Delta_p \times G/K'$, which cover the identity, such that $h\alpha_n = \alpha_n h_\alpha$ and $h\beta_p = \beta_p h_\beta$. Moreover $h_\beta^* = h_\beta^* = h^* : m(G/K') \to m(G/K)$. Since now $T\alpha_n = T'\alpha h_\alpha$ and $T\beta_p = T'\beta_p h_\beta$ it follows that

$$(c \cup c_1)(T) = (c(T'\alpha_n h_\alpha))(c_1(T'\beta_p h)) =$$

$$(h^*c(T'\alpha_n))(h^*c_1(T'\beta_p)) = h^*((c(T'\alpha_n))(c_1(T'\beta_p))) =$$

$$h^*(c \cup c_1)(T'),$$

where we used the fact that h^* is a ring homomorphism. This proves our claim.

If (X,A) is a G-pair and, for example, $c \in C_G^n(X,A;m)$ then also $c \cup c_1 = C_G^{n+p}(X,A;m)$. In particular we have

$$\cup : C_G^n(X,A;m) \otimes_R C_G^p(X,A;m) \to C_G^{n+p}(X,A;m).$$

Since now $\delta(c \cup c_1) = (\delta c) \cup c_1 + (-1)^n c \cup (\delta c_1)$, that is, the homomorphisms $\cup$ form a cochain map, it follows that we get a cup-product on the cohomology level

$$\cup : H^n_G(X,A;m) \otimes_R C^p_G(X,A;m) \to C^{n+p}_G(X,A;m).$$

We shall conclude by showing that the cup-product is commutative. The proof in Artin-Braun [1] for the commutativity (also sometimes called anticommutativity) of the cup-product in ordinary singular cohomology carries over to our situation without any difficulties. (See section 22 in Artin-Braun [1] for more details than we give below.)

The reader should recall the notion of an equivariant linear q-simplex $v^0 \ldots v^q \times \mathrm{id} : \Delta_q \times G/K \to \Delta_n \times G/K$ in $\Delta_n \times G/K$ and the definition of the linear chain groups $\hat{C}^G_q Q(\Delta_n \times G/K)$ and the corresponding chain complex $\hat{S}^G Q(\Delta_n \times G/K)$, as defined in Section 6. of Chapter I. We shall also use the join homomorphism $v\cdot : C^G_{q+1} Q(\Delta_n \times G/K) \to C^G_{q+1} Q(\Delta_n \times G/K)$ and its properties with respect to the boundary homomorphism. For this we again refer to the beginning of Section 6. of Chapter I.

Define the homomorphism

$$\hat{\rho}_q : \hat{C}^G_q Q(\Delta_n \times G/K) \to \hat{C}^G_q Q(\Delta_n \times G/K)$$

by

$$\hat{\rho}_q(v^0 \ldots v^q_q \times \mathrm{id}) = (-1)^{q(q+1)/2}(v^q \ldots v^0 \times \mathrm{id}).$$

It is easily seen that the homomorphisms $\hat{\rho}_q$ commute with the boundary and thus form a chain map $\hat{\rho}_{\#}$. We now inductively define homomorphisms

$$\hat{D}_q : \hat{C}^G_q Q(\Delta_n \times G/K) \to \hat{C}^G_{q+1} Q(\Delta_n \times G/K)$$

by setting $\hat{D}_0 = 0$, and

$$\hat{D}_q(\sigma) = v^0\cdot(\sigma - \hat{\rho}_q(\sigma) - \hat{D}_{q-1}(\hat{\partial}_q(\sigma))), \quad q \geq 1$$

where $\sigma = v^0 \ldots v^q \times \mathrm{id}$. A formal computation and induction shows that the homomorphisms $\hat{D}_q$ form a chain homotopy from the identity map to $\hat{\rho}_{\#}$.

Let X be a G-space. Define the homomorphisms

$$\hat{\rho}_n : \hat{C}^G_n(X) \to \hat{C}^G_n(X)$$

$$\hat{D}_n : \hat{C}^G_n(X) \to \hat{C}^G_{n+1}(X)$$

as follows. If $T : \Delta_n \times G/K \to X$ is an equivariant singular n-simplex belonging to $\mathfrak{F}$ in X we set

$$\hat{\rho}_n(T) = \hat{T}_{\#}\hat{\rho}_n(d^0 \ldots d^n \times \mathrm{id}),$$

$$\hat{D}_n(T) = \hat{T}_{\#}\hat{D}_n(d^0 \ldots d^n \times \mathrm{id}).$$

(Recall that $d^0 \ldots d^n \times \mathrm{id} : \Delta_n \times G/K \to \Delta_n \times G/K$ is the identity map.) It is easy to see that these "new" homomorphisms $\hat{\rho}_n$ form a chain map $\hat{\rho} : S^G(X) \to S^G(X)$, and that the new homomorphisms $\hat{D}_n$ form a chain homotopy from the identity map to $\hat{\rho}_{\#}$.

We now claim that the homomorphisms $\hat{\rho}_n : \hat{C}^G_n(X) \to \hat{C}^G_n(X)$ and $\hat{D}_n : \hat{C}^G_n(X) \to \hat{C}^G_{n+1}(X)$ both "preserve the relation $\sim$" (see Def. 4.4. in Chapter I). This is easily proved in exactly the same way as Proposition 6.1. in Chapter I. It follows that $\hat{\rho}_n$ and $\hat{D}_n$ have duals $\hat{\rho}^n : \hat{C}^n_G(X,A;m) \to \hat{C}^n_G(X,A;m)$ and $\hat{D}^{n+1} : \hat{C}^{n+1}_G(X,A;m) \to \hat{C}^n_G(X,A;m)$ which restrict to give

$$\rho^n : C^n_G(X,A;m) \to C^n_G(X,A;m)$$

$$D^{n+1} : C^{n+1}_G(X,A;m) \to C^n_G(X,A;m).$$

The homomorphisms ρ^n form a cochain map $\rho^{\#}$ and the homomorphisms D^n form a cochain homotopy from the identity map to $\rho^{\#}$.

We can now show that the cup-product is commutative. Let $y \varepsilon H_G^n(X,A;m)$, $y_1 \varepsilon H_G^p(X,A;m)$ and let $c \varepsilon C_G^n(X,A;m)$ and let $c_1 \varepsilon C_G^p(X,A;m)$ be cocycles representing y and y_1, respectively. The cohomology class $y \cup y_1$ is represented by the cocycle $c \cup c_1$. It now follows from what we showed above that the cocycle $\rho^{n+p}((\rho^n c) \cup (\rho^p c_1))$ also represents $y \cup y_1$. Let $T : \Delta_{n+p} \times G/K \to X$ be an equivariant singular (n+p)-simplex belonging to $\mathcal{F}$ in X. Since we have $T(d^{n+p}\ldots d^0)\alpha_n(d^n\ldots d^0) = T(d^p\ldots d^{n+p}) = T\beta_n$ and $T(d^{n+p}\ldots d^0)\beta_p(d^p\ldots d^0) = T(d^0\ldots d^p) = T\alpha_p$ it follows that

$$\rho^{n+p}((\rho^n c) \cup (\rho^p c_1))(T) = (-1)^{np}(c(T\beta_n))(c_1(T\alpha_n)) =$$

$$(-1)^{np}(c_1(T\alpha_n))(c(T\beta_n)) = (-1)^{np}(c_1 \cup c)(T).$$

(The sign is as stated since $((n+p)(n+p+1) + n(n+1) + p(p+1))/2 \equiv np(\text{mod } 2)$.) This proves that

$$y \cup y_1 = (-1)^{np}(y_1 \cup y),$$

for $y \varepsilon H_G^n(X,A;m)$ and $y_1 \varepsilon H_P^p(X,A;m)$.

BIBLIOGRAPHY

[1] E. Artin and H. Braun, Introduction to Algebraic Topology, Merrill Publishing Co., Columbus, Ohio, 1969.

[2] G. Bredon, Equivariant cohomology theories, Bull. Amer. Math. Soc. 73 (1967), 266-268.

[3] - , Equivariant cohomology theories, Lecture Notes in Math., volume 34, Springer-Verlag, Berlin and New York, 1967.

[4] Th. Bröker, Singuläre Definition der äquivarianten Bredon Homologie, Manuscripta Math. 5 (1971), 91-102.

[5] S. Eilenberg, Singular homology theory, Ann. of Math. 45 (1944), 407-447.

[6] S. Eilenberg and N.E. Steenrod, Foundations of Algebraic Topology, Princeton Univ. Press, Princeton, N.J., 1952.

[7] S. Illman, Equivariant singular homology and cohomology for actions of compact Lie groups, Proc. Conference on Transformation Groups (University of Massachusetts, Amherst, 1971) Lecture Notes in Math. Springer-Verlag volume 298, Berlin and New York, 1972.

[8] - , Equivariant Algebraic Topology, Thesis, Princeton University, Princeton, N.J., 1972.

[9] - , Equivariant singular homology and cohomology, Bull. Amer. Math. Soc. 79 (1973), 188-192.

University of Helsinki

Memoirs of the American Mathematical Society
Number 157

Jonathan Leech

Two Papers:
$\mathcal{H}$-coextensions of monoids
and
The structure of a band of groups

Published by the
AMERICAN MATHEMATICAL SOCIETY
Providence, Rhode Island

VOLUME 1 · ISSUE 2 · NUMBER 157 (second of 3 numbers) · MARCH 1975

ABSTRACT

In the first paper, the H-coextension problem for monoids is studied in full generality. This is done by means of a factor system technique which reduces to classical group extension theory when all monoids under consideration are groups. This technique is made possible by a correspondence between the sub-H congruences on a monoid and the subfunctors of a certain group-valued functor Γ. This correspondence is the precise generalization to semigroups of the basic congruence-normal subgroup correspondence for group theory. The paper concludes with the relationships between H-coextension theory and an appropriate cohomology theory for monoids.

In the second paper, the structure of a band of groups is studied by means of the techniques of the first paper.

AMS (MOS) subject classifications (1970). Primary 20M10; Secondary 18H10.

ISBN 0-8218-1857-0.

TABLE OF CONTENTS

PREFACE

This memoir consists of two papers, both of which deal with the heuristic border between group theory and semigroup theory, i.e., the H-coextension problem. By an H-coextension of a semigroup S_0 is meant an epimorphism of semigroups, $f: S \to S_0$, such that the f-induced congruence on S is contained in the Green's relation, H_S. The H-coextension problem consists of determining how S is built from S_0 and from groups.

The importance of the H-coextension problem in semigroup theory arises from the following considerations. For every semigroup S there exists a congruence G which is maximum among those congruences which are contained in H_S. The canonical epimorphism $S \to S/G$ is clearly an H-coextension. Moreover, S/G is reduced, i.e., $G_{S/G}$ is the identity relation. Studies by a variety of researchers have indicated that a good way to approach the semigroup classification problem is first to classify reduced semigroups and then to study H-coextensions of these reduced semigroups. Although the first problem is intrinsically much harder than the second, the second is by no means trivial.

In the first paper, the H-coextension problem is studied for arbitrary (i.e., not necessarily reduced) monoids. The condition of an identity is added only for convenience, since, if one is careful, this leads to no real loss in generality.

In the second paper of this memoir the H-coextension problem is studied for the case where the underlying semigroup is a band, i.e., a semigroup of idempotents. Such H-coextensions are called bands of groups. Their structure is of interest for two reasons. In the first place a band of groups is a special case of a union of groups, the structure of which has been of interest for some time. In the second place H-coextension theory for bands is sufficiently simpler than the general case to make it a good source of both examples and counterexamples. In order to make this second paper self-contained, there is a small amount of overlap with the first paper.

ACKNOWLEDGEMENT

The author expresses his appreciation to A. H. Clifford, whose helpful remarks have been the source of numerous improvements in the exposition of both papers.

$\mathcal{H}$-COEXTENSIONS OF MONOIDS

In this paper we give a generalization of group extension theory to algebraic semigroups. We recall that a group extension is a short exact sequence of groups $1 \to K \to \overline{G} \to G \to 1$. Given such a sequence, $\overline{G}$ is called an extension of K by G, or a coextension of G by K. More generally, a coextension of G is an epimorphism of groups, $p : \overline{G} \to G$. We can form a category of coextensions of G, $\underline{\underline{Gr}}_G$, by requiring that a morphism $f : (\overline{G},p) \to (\overline{G}',p')$ be a group morphism $f : \overline{G} \to \overline{G}'$ such that $p = p' \circ f$.

The situation we have described has an obvious generalization to monoids, i.e., to semigroups with identity. Hence we can talk about the category of homomorphisms onto a monoid S to be denoted $\underline{\underline{Mon}}_S$. Unlike $\underline{\underline{Gr}}_G$ it is impossible to obtain a Schreier Theory to describe $\underline{\underline{Mon}}_S$. However, there is an important subcategory of $\underline{\underline{Mon}}_S$ which can be described by a generalized version of Schreier Theory. We call a coextension of S, $(\overline{S},p)$, an $\mathcal{H}$-coextension if the congruence on $\overline{S}$ induced by p is contained in the Green's relation $\mathcal{H}$. The full subcategory of $\underline{\underline{Mon}}_S$ determined by the $\mathcal{H}$-coextensions is denoted $\mathcal{H}(\underline{\underline{Mon}}_S)$. It is the purpose of this paper to describe this category.

This paper has been divided into five chapters. In the first chapter we discuss the categories which will be the domains of the group valued functors which are of interest. Besides the quasiorders $\mathcal{L}(S)$ and $\mathcal{R}(S)$, where S is a monoid, we define two new categories $\mathbb{D}(S)$ and $\mathcal{D}(S)$ and present some of their basic properties. A more detailed account of the $\mathcal{D}$-category will appear in another paper [2]. In the second chapter we study the lattice of congruences on a monoid which are contained in its Green's relation $\mathcal{H}$. In the third chapter we give our description of the category

---

Received by the editors. April 27, 1970.

$\mathcal{H}(\underline{\underline{\text{Mon}}}_S)$. The fourth chapter is devoted to discussing split coextensions of S . In the final chapter we discuss abelian coextensions of S , i.e., $\mathcal{H}$-coextensions of S where the Schützenberger groups of S are coextended by abelian groups, and we also show the relationships between abelian coextensions and the lower $\mathbb{D}(S)$-cohomology functors.

Many of the basic concepts and constructions of this paper were first introduced by the author in his 1969 dissertation at UCLA. In the second half of this dissertation, he was interested in the problem of extending groups by monoids, i.e., the problem of constructing exact sequences $K \to \overline{S} \to S$ where K is a group, $\overline{S}$ and S are monoids, and $K \to \overline{S}$ is a monoid embedding such that im(K) is a normal subgroup, i.e., $xK = Kx$ for all $x \in \overline{S}$, and $\overline{S} \to S$ is the quotient map essentially obtained by shrinking each coset of im(K) to a point. Sometime later he was able to use this technique to give a description of bands of groups, i.e., $\mathcal{H}$-coextensions of bands where a band is semigroup all of whose elements are idempotents. A paper on the structure of bands of groups forms the second part of this memoir.

As any semigrouper knows, the $\mathcal{H}$-coextension problem has been explicitly or implicitly connected with much that has gone on in the study of algebraic semigroups. Instead of giving a detailed account of various results bearing on the $\mathcal{H}$-coextension problem, we refer the reader to the introduction of Grillet's paper [1].

Finally, it should be mentioned that although we have restricted our attention in this paper to monoids, all of our main results can easily be extended to arbitrary semigroups. The reason for looking only at monoids is merely convenience.

Chapter One

The Categories, $\mathbb{D}(S)$ and $\mathcal{D}(S)$

Section 1a. The category, $\mathbb{D}(S)$. If S is a monoid, then $\mathbb{D}(S)$ is the small category whose set of objects is S, such that for every $x, y \in S$, $\mathrm{Hom}(x,y) = \{\langle u,x,v\rangle \mid uxv = y\}$, and the morphism composition is defined by $\langle u',uxv,v'\rangle \circ \langle u,x,v\rangle = \langle u'u,x,vv'\rangle$. Since S has associative multiplication, the composition of $\mathbb{D}(S)$ is associative. Since S is a monoid, $\langle 1,x,1\rangle = 1_x$ for all $x \in S$. If T is also a monoid and $f : S \to T$ is a morphism of monoids, then f induces a functor, $\mathbb{D}(f) : \mathbb{D}(S) \to \mathbb{D}(T)$, defined by $\mathbb{D}(f)x = f(x)$ and $\mathbb{D}(f)\langle u,x,v\rangle = \langle f(u),f(x), f(v)\rangle$. Clearly, if Mon is the category of monoids and Cat is the category of small categories, then $\mathbb{D} : \underline{\underline{\mathrm{Mon}}} \to \underline{\underline{\mathrm{Cat}}}$ is a well defined functor.

1.1. $\mathbb{D}(S)$ contains two important subcategories, $\mathbb{L}(S)$ and $\mathbb{R}(S)$. Both $\mathbb{L}(S)$ and $\mathbb{R}(S)$ have S as their set of objects. $\mathbb{L}(S)$ is characterized by the fact that all of its morphisms are of the form $\langle u,x,1\rangle$, while $\mathbb{R}(S)$ is characterized by the fact that all of its morphisms are of the form $\langle 1,x,v\rangle$. $\mathbb{R}(S)$ and $\mathbb{L}(S)$ together generate $\mathbb{D}(S)$. If $\langle u,x,v\rangle$ is a $\mathbb{D}(S)$-morphism, then $\langle u,x,v\rangle = \langle u,xv,1\rangle \circ \langle 1,x,v\rangle = \langle 1,ux,v\rangle \circ \langle u,x,1\rangle$. $\langle u,xv,1\rangle \circ \langle 1,x,v\rangle$ is called the $\mathbb{L} \circ \mathbb{R}$ decomposition of $\langle u,x,v\rangle$, while $\langle 1,ux,v\rangle \circ \langle u,x,1\rangle$ is called the $\mathbb{R} \circ \mathbb{L}$ decomposition of $\langle u,x,v\rangle$.

1.2. Lemma. The $\mathbb{L} \circ \mathbb{R}$ and $\mathbb{R} \circ \mathbb{L}$ decompositions of $\langle u,x,v\rangle$ are unique.

Proof. Given $\langle u,x,v\rangle = \alpha\circ\beta$ with $\alpha \in \mathrm{Mor}(\mathbb{L}(S))$ and $\beta \in \mathrm{Mor}(\mathbb{R}(S))$, then β must have the form $\langle 1,x,v'\rangle$ and α must have the form $\langle u',xv',1\rangle$. Hence $\langle u,x,v\rangle = \langle u',xv',1\rangle \circ \langle 1,x,v'\rangle = \langle u',x,v'\rangle$, so that $u' = u$ and $v' = v$.

The $\mathbb{R} \circ \mathbb{L}$ part follows by the dual argument.

1.3. Let K be a category. By $[\mathbb{D}(S),K]$ we denote the category of functors and natural transformations from $\mathbb{D}(S)$ to K. If $F : \mathbb{D}(S) \to K$ is such a functor, $F_L = F|\mathbb{L}(S) : \mathbb{L}(S) \to K$ and $F_R = F|\mathbb{R}(S) : \mathbb{R}(S) \to K$ are also functors. If a pair of functors, $F_L : \mathbb{L}(S) \to K$ and $F_R : \mathbb{R}(S) \to K$, are such that they are induced by restriction of a functor $F : \mathbb{D}(S) \to K$, then we say that F extends the pair (F_L,F_R) to $\mathbb{D}(S)$. Since $\mathbb{L}(S)$ and $\mathbb{R}(S)$ together generate $\mathbb{D}(S)$, such an extension must be unique.

1.4. Theorem. *Let* $F_L,F_R : \mathbb{L}(S), \mathbb{R}(S) \to K$. *Then there exists a functor* $F : \mathbb{D}(S) \to K$ *which extends* F_L *and* F_R *to* $\mathbb{D}(S)$ *if and only if for all* $u, x, v \in S$:

(i) $F_L(x) = F_R(x)$

(ii) $F_R(1,ux,v) \circ F_L(u,x,1) = F_L(u,xv,1) \circ F_R(1,x,v)$.

If the extension exists, then it is uniquely determined by $F(x) = F_R(x)$ *and* $F(u,x,v)$ *being either of the equal morphism compositions in* (ii). *If* $F, G : \mathbb{D}(S) \to K$ *are functors and* $\sigma = \{\sigma(x) : F(x) \to G(x)\}$ *is a collection of morphisms in* K, *then* $\sigma : F \to G$ *is a natural transformation if and only if both* $\sigma : F_L \to G_L$ *and* $\sigma : F_R \to G_R$ *are natural transformations.*

Proof. Clearly, if an extension F exists, it must satisfy the above conditions. Conversely, if the conditions are satisfied, let $F : \mathbb{D}(S) \to K$ be defined by $F(x) = F_R(x)$ and $F(u,x,v)$ being the above composition of morphisms. If $(u,x,v), (u',uxv,v') \in Mor(\mathbb{D}(S))$, then

$$F(u',uxv,v') \circ F(u,x,v)$$

$$= F_R(1,u'uxv,v') \circ F_L(u',uxv,1) \circ F_R(1,ux,v) \circ F_L(u,x,1)$$

$$= F_R(1,u'uxv,v') \circ F_R(1,u'ux,v) \circ F_L(u',ux,1) \circ F_L(u,x,1)$$

$$= F_R(1,u'ux,vv') \circ F_L(u'u,x,1) = F(u'u,x,vv') \ .$$

Hence F respects morphism composition. $F(1_x) = 1_{F(x)}$ comes from $F_R(x) = F_L(x)$ and $1_x = 1_x \circ 1_x \in \mathbb{L}(S) \cap \mathbb{R}(S)$. If $\sigma : F \to G$ is a natural transformation, then surely both $\sigma : F_R(x) \to G_R(x)$ and $\sigma : F_L(x) \to G_L(x)$ are natural transformations. The converse is seen from the fact that commutativity of the inner squares implies commutativity of the outer rectangle in the following diagram.

$$\begin{array}{ccccc}
F(x) & \xrightarrow{F(u,x,1)} & F(ux) & \xrightarrow{F(1,ux,v)} & F(uxv) \\
\downarrow{\scriptstyle \sigma(x)} & & \downarrow{\scriptstyle \sigma(ux)} & & \downarrow{\scriptstyle \sigma(uxv)} \\
G(x) & \xrightarrow{G(u,x,1)} & G(ux) & \xrightarrow{G(1,ux,v)} & G(uxv) \ .
\end{array}$$

1.5. We now proceed to give several simple examples of functors whose domain is $\mathbb{D}(S)$.

(i) Let $X \in K$ and let $f : S \to \text{End}\ (X)$ be a monoid homomorphism. Define $F : \mathbb{D}(S) \to K$ by $F(x) = X$, and $F(u,x,v) = f(u)$.

(ii) Let $X \in K$ and let $g : S \to \text{End}\ (X)$ be a monoid antihomomorphism. Define $G : \mathbb{D}(S) \to K$ by $G(x) = X$ and $G(u,x,v) = G(v)$.

(iii) Let $X \in K$, $f : S \to \text{End}\ (X)$ be a monoid homomorphism, and $g : S \to \text{End}\ (X)$ be a monoid antihomomorphism such that f and g commute, i.e., for all $u, v \in S$, $g(v) \circ f(u) = f(u) \circ g(v)$. Define $F : \mathbb{D}(S) \to K$ by $F(x) = X$, and $F(u,x,v) = f(u) \circ g(v)$.

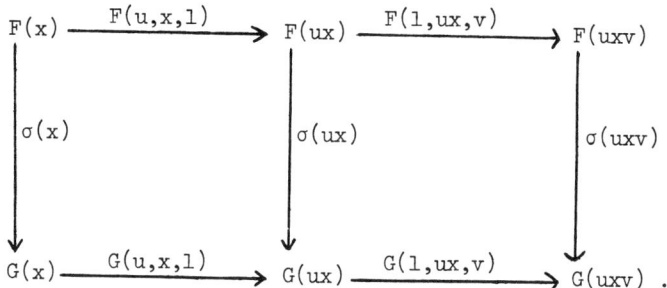

(iv) Let C be a congruence on S. Define $C : \mathbb{D}(S) \to \underline{\underline{\text{Set}}}$, the category of sets, by $C(x) =$ the C-equivalence class of x, and if $y \in C(x)$, then $C(u,x,v)[y] = uyv$.

(v) Throughout much of this paper we shall be interested in functors from $\mathbb{D}(S)$ to $\underline{\underline{Gr}}$ and $\underline{\underline{AG}}$ where $\underline{\underline{Gr}}$ is the category of groups and $\underline{\underline{AG}}$ is the category of abelian groups. If $F, F': \mathbb{D}(S) \to \underline{\underline{Gr}}$ are functors, then we say that F' is a subfunctor of F , denoted $F' \leq F$, if for all $x \in S$, $F'(x)$ is a subgroup of $F(x)$, and for all morphisms, $\langle u,x,v\rangle$, we have $F'(u,x,v) = F(u,x,v)|F'(x)$. Furthermore, if $F'(x)$ is always a normal subgroup of $F(x)$, we say that F' is a normal subfunctor of F , denoted $F' \trianglelefteq F$. In this case we can form a quotient functor, F/F' , in the obvious manner, so that $F' \to F \to F/F'$ is an exact sequence of functors.

1.6. Lemma. *$\langle u,x,v\rangle$ is an isomorphism in $\mathbb{D}(S)$ if and only if $u, v \in G(S)$, the group of units of S .*

Proof. Let $\langle u,x,v\rangle: x \to y$. If $u, v \in G(S)$, then $\langle u^{-1},y,v^{-1}\rangle = \langle u,x,v\rangle^{-1}$ is clear. If $\langle u,x,v\rangle$ is an isomorphism, then let $\langle \bar{u},y,\bar{v}\rangle = \langle u,x,v\rangle^{-1}$. Then $\langle 1,x,1\rangle = \langle \bar{u},y,\bar{v}\rangle \circ \langle u,x,v\rangle = \langle \bar{u}u,x,v\bar{v}\rangle$, while $\langle 1,y,1\rangle = \langle u,x,v\rangle \circ \langle \bar{u},y,\bar{v}\rangle = \langle u\bar{u},y,\bar{v}v\rangle$, so that $\bar{u}u = 1 = u\bar{u}$ and $\bar{v}v = 1 = v\bar{v}$ follow.

1.7. If K is any small category, then $P(K)$ is the underlying preorder obtained by identifying all morphisms in any hom set. Clearly $L(S) = P(\mathbb{L}(S))$, $R(S) = P(\mathbb{R}(S))$, while $J(S) = P(\mathbb{D}(S))$.

Section 1.b. The category, $\mathcal{D}(S)$. Let $\mathcal{D}(S)$ be the quotient category of $\mathbb{D}(S)$ such that if $\delta : \mathbb{D}(S) \to \mathcal{D}(S)$ is the quotient functor, then δ induces functors $\delta_L : L(S) \to \mathcal{D}(S)$ and $\delta_R : R(S) \to \mathcal{D}(S)$ such that

$$
\begin{array}{ccccc}
\mathbb{L}(S) & \longrightarrow & \mathbb{D}(S) & \longleftarrow & \mathbb{R}(S) \\
\downarrow & & \downarrow{\scriptstyle \delta} & & \downarrow \\
L(S) & \xrightarrow{\delta_L} & \mathcal{D}(S) & \xleftarrow{\delta_R} & R(S)
\end{array}
$$

commutes, and $(\delta_L, \delta, \delta_R)$ is universal. Hence if $F : \mathbb{D}(S) \to K$ is also a functor such that F_L factors through $L(S)$ and F_R factors through $R(S)$, then there is a (necessarily) unique functor, $F : \mathcal{D}(S) \to K$ such that $F = F \circ \delta$.

1.8. The pair $(\mathcal{D}(S), \delta)$ must exist. For if P_L is the congruence on $\mathbb{L}(S)$ which gives rise to $L(S)$ and P_R is the congruence on $\mathbb{R}(S)$ which gives rise to $R(S)$, then let $\approx$ be the smallest congruence on $\mathbb{D}(S)$ such that $P_L \cup P_R \subseteq \approx$. We now set $\mathcal{D}(S) = \mathbb{D}(S)/\approx$ and let $\delta : \mathbb{D}(S) \to \mathcal{D}(S)$ be the induced functor. Clearly $(\mathcal{D}(S), \delta)$ is the desired pair.

1.9. Definition. $\sim$ is the equivalence relation on $\mathbb{D}(S)$ defined by: $\langle u,x,v\rangle \sim \langle u',x,v'\rangle$ if and only if $ux = u'x$ and $xv = xv'$. Clearly $\langle u,x,v\rangle \sim \langle u',x,v'\rangle$ implies $uxv = u'xv'$.

1.10. Lemma. *$\approx$ is the congruence on $\mathbb{D}(S)$ which is generated by $\sim$. Hence a functor, $F : \mathbb{D}(S) \to K$, factors through $\mathcal{D}(S)$ if and only if $\langle u,x,v\rangle \sim \langle u',x,v'\rangle$ implies $F(u,x,v) = F(u',x,v')$.*

Proof. $P_L \cup P_R \subseteq \sim \subseteq \approx$.

1.11. If $u,x,v \in S$, then $[u,x,v] = \delta(\langle u,x,v\rangle)$.

1.12. Lemma. *If $uxLxRxv$, then $[u,x,1]$ and $[1,x,v]$ are isomorphisms in $\mathcal{D}(S)$.*

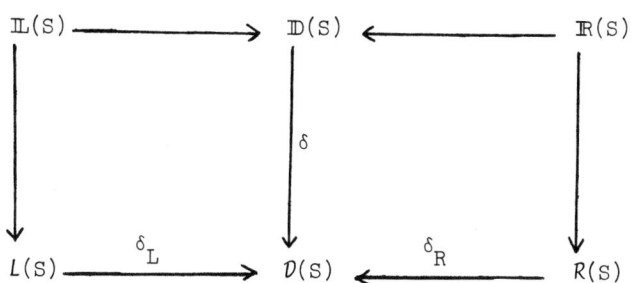

Proof. $[u,x,1]$ can be considered to be a morphism in $\mathcal{L}(S)$. But in $\mathcal{L}(S)$, $[u,x,1]$ is an isomorphism. Hence $[u,x,1]$ is an isomorphism in $\mathcal{D}(S)$. Dually, $[1,x,v]$ is an isomorphism in $\mathcal{D}(S)$.

1.13. The simple facts about $\mathcal{D}(S)$ mentioned in 1.10 and 1.12 above are all that we shall need to know about the structure of $\mathcal{D}(S)$ in this paper. We now mention other properties of $\mathcal{D}(S)$ which are not used in this paper and hence not proved.

(i) For all $x \in S$, $\mathrm{Aut}_{\mathcal{D}(S)}(x) \cong H_L(x)$, the left Schützenberger group of S .

(ii) $[u,x,v]$ is an isomorphism in $\mathcal{D}(S)$ if and only if $ux\mathcal{L}x\mathcal{R}xv$.

(iii) $x \cong y$ in $\mathcal{D}(S)$ if and only if $x\mathcal{D}y$.

(iv) x is the unique terminal object of $\mathcal{D}(S)$ if and only if $x = 0$.

1.14. It should be clear that if. $f : S \to T$ is a monoid morphism, then f induces a unique factor $\mathcal{D}(f) : \mathcal{D}(S) \to \mathcal{D}(T)$, defined by $\mathcal{D}(f)(x) = f(x)$ and $\mathcal{D}(f)[u,x,v] = [f(u),f(x),f(v)]$. We therefore have a functor $\mathcal{D} : \underline{\underline{\mathrm{Mon}}} \to \underline{\underline{\mathrm{Cat}}}$, with $\delta : \mathbb{D} \to \mathcal{D}$ being a natural transformation.

Chapter Two

Congruences Under H

We say that a congruence, C, on a monoid is under H, is a sub-H congruence, or is contained in H, if $C \subseteq H$. Since H is an equivalence relation on the monoid, it is well known that there must exist a unique maximal sub-H congruence, G, so that $C \subseteq H$ if and only if $C \subseteq G$. Throughout this chapter, unless otherwise clear from the context, C will always denote a sub-H congruence (although "$C \subseteq H$" will sometimes be stated for emphasis).

Section 2a. *The structure functors of sub-H congruences.* Corresponding to each sub-H congruence, C, we shall construct a group-valued functor, $\Sigma_C : D(S) \to \underline{\underline{Gr}}$. The function, $C \to \Sigma_C$, turns out to be an order isomorphism between the lattice of sub-H congruences, $\{C \subseteq H\}$, and the lattice of subfunctors of $\Gamma = \Sigma_G$, $\{F \leq \Gamma\}$. These functors enable us to study sub-H congruences not only locally (the way H is often studied), but also globally.

2.1. Lemma. *Let C be a congruence on S and let $\pi : S \to S/C$ be the canonical morphism. Then the following are equivalent:*

(i) $C \subseteq H$.

(ii) *for all* $x, y \in S$, xHy *if and only if* $\pi(x) H \pi(y)$.

(iii) *the functor* $H(\pi) : H(S) \to H(S/C)$ *is an equivalence.*

(iv) *both functors,* $L(\pi) : L(S) \to L(S/C)$ *and* $R(\pi) : R(S) \to R(S/C)$, *are equivalences.*

When these conditions hold, so do the following:

(v) $L(x) = \pi^{-1}(L(\pi(x)))$ *and* $L(\pi(x)) = \pi(L(x))$, *with similar equalities holding for* R, H, D, *and* J.

(vi) *If* $e \in E(S)$, *the set of idempotents of* S, *then*

$\pi_e = \pi|H(e) : H(e) \to H(\pi(e))$ <u>is a surmorphism of groups</u>.

(vii) <u>The induced functor</u>, $C : \mathbb{D}(S) \to$ <u>Sets</u>, <u>factors through</u> $\mathcal{D}(S)$.

<u>Proof</u>. The equivalence of conditions (i) through (iv) should be clear, as well as parts (v) and (vi). For part (vii), let $\langle u,x,v\rangle \sim \langle u',x,v'\rangle$. Let $y \in C(x)$. Since $C(x) \subseteq H(x)$, there exist $a, b \in S$ such that $y = ax = xb$. Therefore:

$$uyv = u(xb)v = u'(xb)v = u'(ax)v = u'(ax)v' = u'yv'$$

so that $C(u,x,v) = C(u',x,v')$. Hence C factors through $\mathcal{D}(S)$.

2.2. We shall denote the induced functor from $\mathcal{D}(S)$ to <u>Sets</u>, mentioned in part (vii) above, by C also.

2.3. Of course, we expect the reader to be familiar with the left and right Schützenberger groups of an H-class. We nonetheless outline some of their main properties. The left Schützenberger group of $H(x)$ is the quotient group of the submonoid, $\{u \in S | uH(x) \subseteq H(x)\} = L(H(x))$, where two elements of $L(H(x))$, a and b, are identified if and only if $ax = bx$. Denoting this group by $H_L(x)$, there arises a simply transitive left action of $H_L(x)$ upon $H(x)$ which is well defined by $[u]\cdot y = uy$. Moreover, this action is R-invariant, i.e., if xRy, then $H_L(x) = H_L(y)$, and if $y = xv$, then for all $g \in H_L(x)$, $g\cdot(xv) = (g\cdot x)v$. Dually, if $R(H(x)) = \{v \in S | H(x)v \subseteq H(x)\}$, then the right Schützenberger group of $H(x)$, $H_R(x)$, is the quotient group of $R(H(x))$, obtained by identifying two elements, a and b, of $R(H(x))$ if and only if $xa = xb$. Here there arises an L-invariant right action, $H(x) \times H_R(x) \to H(x)$ defined by $y\cdot[v] = yv$. These two actions together form a biaction, $H_L(x) \times H(x) \times H_R(x) \to H(x)$, and for each $y \in H(x)$ there is induced an isomorphism, $(\ )_y^R : H_L(x) \to H_R(x)$, defined implicitly by $g \cdot y = y \cdot (g)_y^R$. We denote the inverse of $(\ )_y^R$ by $(\ )_y^L$. Of course $(\ )_y^L$ is implicitly defined by $y \cdot g = (g)_y^L \cdot y$.

2.4. Lemma. If $C \subseteq H$ and $H_C(x)$ is the set of C-equivalence classes in $H(x)$, then the biaction $H_L(x) \times H(x) \times H_R(x) \to H(x)$ induces a biaction $*$: $H_L(x) \times H_C(x) \times H_R(x) \to H_C(x)$ defined by $g*C(y)*h \to C(g \cdot y \cdot h)$. Both one-sided actions on $H_C(x)$ are transitive. If $\Sigma(y) = \{g \in H_R(x) | C(y)*g = C(y)\}$ and $\Sigma'(y) = \{g \in H_L(x) | g*C(y) = C(y)\}$, where $y \in H(x)$, then $\Sigma(y) = \Sigma(x)$ and $\Sigma'(y) = \Sigma'(x)$, for all $y \in H(x)$, and $\Sigma'(x) \cong \Sigma(x)$ under the restriction of $(\ )_x^R$ to $\Sigma'(x)$. Hence $\Sigma'(x)$ and $\Sigma(x)$ are the kernels of the actions, $H_L(x) \times H_C(x) \to H_C(x)$ and $H_C(x) \times H_R(x) \to H_C(x)$, and both induced actions, $H_L(x)/\Sigma'(x) \times H_C(x) \to H_C(x)$ and $H_C(x) \times H_R(x)/\Sigma(x) \to H_C$, are simply transitive. Finally, both restricted actions, $\Sigma'(x) \times C(x) \to C(x)$ and $C(x) \times \Sigma(x) \to C(x)$, are simply transitive, and $\Sigma'(x) \times C(x) \times \Sigma(x) \to C(x)$ is a biaction.

Proof. If $g \in H_R(x)$, then let $\bar{g} \in S$ be such that $y \cdot g = y\bar{g}$ for all $y \in H(x)$. Since C is a congruence, $C(y) \cdot g = C(y)\bar{g} \subseteq C(y\bar{g}) = C(y \cdot g)$ so that $(\ ) \cdot g$ is a function from $C(y)$ to $C(y \cdot g)$. But $(\ ) \cdot g^{-1}: C(y \cdot g) \to C(y)$ must clearly be the inverse of this function. Hence $(\ ) \cdot g : C(y) \cong C(y \cdot g)$ (in Sets). It should now be clear that $\langle C(y), g \rangle \to C(y \cdot g)$ indeed defines a transitive action, $H_C(x) \times H_R(x) \to H_C(x)$. Dually, there exists a transitive action, $H_L(x) \times H_C(x) \to H_C(x)$, such that together both actions form the biaction stated in the theorem. Let $h \in \Sigma(x)$, and let $y \in H(x)$. Then $y = g \cdot x$ for some $g \in H_L(x)$, and since $x \cdot h\, C\, x$, $g \cdot x \cdot h\, C\, g \cdot x$ (by the above biaction), i.e., $y \cdot h\, C\, y$ and $C(y) \cdot h = C(y)$ follows. Hence $\Sigma(x) \subseteq \Sigma(y)$, and by the reverse argument, $\Sigma(x) = \Sigma(y)$ follows. Hence $\Sigma(x)$ is the kernel of the action of $H_R(x)$ upon $\mathcal{H}_C(x)$, and the induced action, $\mathcal{H}_C(x) \times H_R(x)/\Sigma(x) \to \mathcal{H}_C(x)$, must be simply transitive. Clearly, $\mathcal{C}(x) \times \Sigma(x) \to \mathcal{C}(x)$ is a well defined action, obtained by restriction of the action, $\mathcal{H}(x) \times H_R(x) \to \mathcal{H}(x)$. To show that the action is transitive, let $y = y \cdot g \in \mathcal{C}(x)$ (where $g \in H_R(x)$). Then $x \,\mathcal{C}\, x \cdot g$ implies that $\mathcal{C}(x) \cdot g = \mathcal{C}(x)$ and $g \in \Sigma(x)$. The simple transitivity of the action follows from the fact that it is the restriction of a simply transitive action. The facts about $\Sigma'(x)$ follow in similar fashion. Clearly, $(\ )^R_x : \Sigma'(x) \to \Sigma(x)$ and $(\ )^L_x : \Sigma(x) \to \Sigma'(x)$, from which $(\ )^R_x : \Sigma'(x) \cong \Sigma(x)$ follows. We are done.

2.5. Corollary. Let $\mathcal{C} \subseteq \mathcal{H}$. Then for each $x \in S$, the following subsets of $H_R(x)$ are equal to $\Sigma(x)$:

(a) $\{g \mid \mathcal{C}(x) \cdot g = \mathcal{C}(x)\}$. (Here $\mathcal{C}(x) \cdot g = \{y \cdot g \mid y \in \mathcal{C}(x)\}$.)

(b) $\{g \mid \mathcal{C}(x) \cdot g \subseteq \mathcal{C}(x)\}$.

(c) $\{g \mid x \cdot g \in \mathcal{C}(x)\}$.

2.6. Corollary. For all $y \in \mathcal{H}(x)$, $y \cdot \Sigma(x) = \mathcal{C}(y)$.

2.7. The above assignment, $x \to \Sigma(x)$, is the first step in constructing a functor, $\Sigma : \mathcal{D}(S) \to \underline{\underline{Gr}}$, where $\underline{\underline{Gr}}$ is the category of groups. We first need a lemma.

2.8. Lemma. Let $\langle u,x,v\rangle \sim \langle u',x,v'\rangle$ and let $x \,\mathcal{H}\, \bar{x}$. Then $\langle u,\bar{x},v\rangle \sim \langle u',\bar{x},v'\rangle$.

Proof. If $\bar{x} = ax = xb$, then $u\bar{x} = uxb = u'xb = u'\bar{x}$ and $\bar{x}v = axv = axv' = \bar{x}v'$.

2.9. Theorem. Let $\mathcal{C} \subseteq \mathcal{H}$. Then $\mathcal{C}$ induces a pair of functors, Σ, $\Sigma' : \mathcal{D}(S) \to \underline{\underline{Gr}}$, defined by $\Sigma(x) = \{g \in H_R(x) \mid x \cdot g \in \mathcal{C}(x)\}$, $\Sigma'(x) = \{g \in H_L(x) \mid g \cdot x \in \mathcal{C}(x)\}$, and if $[u,x,v] : x \to uxv$, then

$\Sigma(u,x,v) : \Sigma(x) \to \Sigma(uxv)$ _and_ $\Sigma'(u,x,v) : \Sigma'(x) \to \Sigma'(uxv)$ _are implicitly defined by the equations:_ $u(x \cdot g)v = uxv \cdot \Sigma(u,x,v)[g]$ _and_ $u(g' \cdot x)v = \Sigma'(u,x,v)[g'] \cdot uxv$, _where_ $g \in \Sigma(x)$ _and_ $g' \in \Sigma'(x)$. $(\ )^R = \{(\ )^R_x : \Sigma'(x) \to \Sigma(x) | x \in S\}$ _is a natural equivalence of these two functors._

Proof. We have already seen that $\Sigma(x)$ and $\Sigma'(x)$ are groups. Since C is a congruence, if $g \in \Sigma(x)$, then $x \cdot g \, C \, x$ and hence $u(x \cdot g)v \, C \, uxv$. Hence by simply transitivity of $C(uxv) \times \Sigma(uxv) \to C(uxv)$, there exists a unique element in $\Sigma(uxv)$, $\Sigma(u,x,v)[g]$, such that $u(x \cdot g)v = uxv \cdot \Sigma(u,x,v)[g]$. To show that $\Sigma(u,x,v) : \Sigma(x) \to \Sigma(uxv)$ is a homomorphism, let g , $h \in \Sigma(x)$. There must exist $\bar{g}_{-1}, \bar{g}_1, \bar{h}_1, \bar{h}_2 \in S$ such that $\bar{g}_1 ux = uxg_1 = u(x \cdot g)$ and $u(x\bar{h}_1)v = uxv\bar{h}_2 = u(x \cdot h)v = uxv \cdot \Sigma(u,x,v)[h]$. Thus we have $uxv \cdot \Sigma(u,x,v)[gh] = u(x \cdot gh)v = u(x\bar{g}_1\bar{h}_1)v = \bar{g}_{-1}(ux\bar{h}_1 v) = \bar{g}_{-1}uxv\bar{h}_2 = u(x \cdot g)v\bar{h}_2 = (uxv \cdot \Sigma(u,x,v)[g])\bar{h}_2 = uxv \cdot \Sigma(u,x,v)[g \, \Sigma(u,x,v)[h]$. Hence for each $\langle u,x,v \rangle \in \mathbb{D}(S)$, $\Sigma(u.x.v)$ is a well defined homomorphism. That $\Sigma(u,uxv,\bar{v}) \circ \Sigma(u,x,v) = \Sigma(u\bar{u},x,v\bar{v})$ is essentially a statement of the fact that $\bar{u}(u(x \cdot g)v)\bar{v} = (\bar{u}u)(x \cdot g)(v\bar{v})$. Clearly $\Sigma(1,x,1) = 1_{\Sigma(x)}$ so that $\Sigma : \mathbb{D}(S) \to \underline{\underline{Gr}}$ is a well defined functor. A similar argument shows that $\Sigma' : \mathbb{D}(S) \to \underline{\underline{Gr}}$ is well defined. If $g' \in \Sigma'(x)$, then $uxv \cdot (\Sigma'(u,x,v)[g'])^R_{uxv} = \Sigma'(u,x,v)[g'] \cdot uxv = u(g' \cdot x)v = u(x \cdot (g')^R_x)v = uxv \cdot \Sigma(u,x,v)[(g')^R_x]$, and $(\ )^R_{uxv} \circ \Sigma'(u,x,v) = \Sigma(u,x,v) \circ (\ )^R_x$ follows. Hence $(\ )^R : \Sigma' \cong \Sigma$. Finally, let $\langle u,x,v \rangle \sim \langle u',x,v' \rangle$. If $g \in \Sigma(x)$, then by Lemma 2.8, $uxv \cdot \Sigma(u,x,v)[g] = u(x \cdot g)v = u'(x \cdot g)v' = u'xv' \cdot \Sigma(u',x,v')[g]$. Hence $\Sigma(u,x,v) = \Sigma(u',x,v')$ and Σ factors through $\mathcal{D}(S)$. We denote the induced functor from $\mathcal{D}(S)$ to $\underline{\underline{Gr}}$ by Σ also.

2.10. _Definition._ The functors, Σ and Σ' , constructed above will be referred to as the right and left structure functors of C. When various sub-H congruences are under consideration, they will often be denoted by Σ_C and Σ'_C. If $C = G$, the maximal congruence under H, then Σ_G and Σ'_G will be denoted by Γ and Γ'.

2.11. *Definition.* **LGA** is the category of left group actions. Its objects are actions, $G \times X \to X$, where G is a group and X is a set. A morphism of actions is a pair, $\langle \gamma, f \rangle : \langle G, X \rangle \to \langle H, Y \rangle$, where $\gamma : G \to H$ is a group morphism, $f : X \to Y$ is a function, and the following diagram commutes:

$$\begin{array}{ccc} G \times X & \longrightarrow & X \\ \downarrow{\scriptstyle \gamma \times f} & & \downarrow{\scriptstyle f} \\ H \times Y & \longrightarrow & Y \end{array}$$

Dually one defines the category of right group actions, **RGA**. Both categories can be considered to be subcategories of **TSGA**, the category of two-sided group actions.

2.12. *Theorem. Let* $C \subseteq H$. *Then* $C : \mathcal{D}(S) \to$ **Sets** *and* $\Sigma', \Sigma : \mathcal{D}(S) \to$ **Gr** *naturally form functors:*

(i) $(\Sigma', C) : \mathcal{D}(S) \to$ **LGA**.

(ii) $(C, \Sigma) : \mathcal{D}(S) \to$ **RGA**.

(iii) $(\Sigma', C, \Sigma) : \mathcal{D}(S) \to$ **TSGA**.

$(\Sigma', C, \Sigma)(x)$ *is the biaction,* $\Sigma'(x) \times C(x) \times \Sigma(x) \to C(x)$, *and* $(\Sigma', C, \Sigma)(u,x,v)$ *is the triple,* $\langle \Sigma'(u,x,v), C(u,x,v), \Sigma(u,x,v) \rangle$, *while*

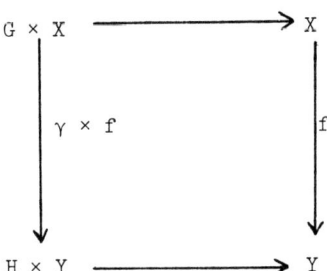

(Σ', C) and (C, Σ) are defined similarly.

Proof. We need only consider the functor, (Σ', C, Σ), since (Σ', C) and (C, Σ) are both subfunctors of (Σ', C, Σ) in the obvious manner. So let $g \in \Sigma'(x)$, $y \in C(x)$, and $h \in \Sigma(x)$. Also let $k \in \Sigma(x)$ be such that $x \cdot k = y$, and let $\tilde{g} = (g)_x^R$. We must have:

$$\begin{aligned} u(g \cdot y \cdot h)v &= u(g \cdot x \cdot kh)v = u(x \cdot \tilde{g}kh)v \\ &= uxv \cdot \Sigma(u,x,v)[\tilde{g}kh] \\ &= \Sigma'(u,x,v)[g] \cdot uxv \cdot \Sigma(u,x,v)[kh] \\ &= \Sigma'(u,x,v)[g] \cdot (u(x \cdot k)v) \cdot \Sigma(u,x,v)[h] \\ &= \Sigma'(u,x,v)[g] \cdot uyv \cdot \Sigma(u,x,v)[h] . \end{aligned}$$

Hence the diagram:

$$\begin{array}{ccc} \Sigma'(x) \times C(x) \times \Sigma(x) & \longrightarrow & C(x) \\ \downarrow & & \downarrow \\ \Sigma'(uxv) \times C(uxv) \times \Sigma(uxv) & \longrightarrow & C(uxv) \end{array}$$

commutes, showing that $\langle \Sigma'(u,x,v), C(u,x,v), \Sigma(u,x,v) \rangle$ is a morphism in TSGA. The functorial property is immediate from those of Σ', C, and Σ.

2.13. Lemma. Let $C \subseteq H$. If $ux \, L \, x \, R \, xv$, then:

(i) $uC(x) = C(ux)$ and $C(x)v = C(xv)$.

(ii) $\Sigma(ux) = \Sigma(x)$ and $\Sigma'(xv) = \Sigma'(x)$.

(iii) $\Sigma(u,x,1) = 1_{\Sigma(x)}$ and $\Sigma'(1,x,v) = 1_{\Sigma'(x)}$.

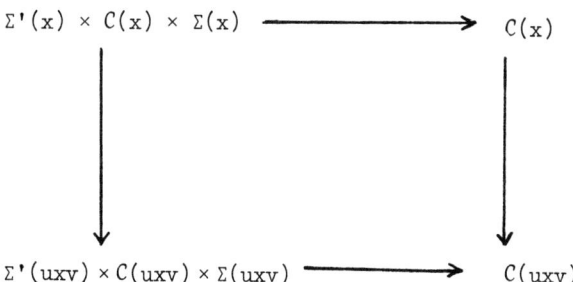

Proof. Since $[u,x,1]$ is an isomorphism in $\mathcal{D}(S)$, $uC(x) = C(u,x,1)[C(x)] = C(ux)$, and $\Sigma(u,x,1) : \Sigma(x) \cong \Sigma(ux)$. But since $H_R(ux) = H_R(x)$, with $u(x\cdot g) = (ux)\cdot g$ for all $g \in H_R(x)$, we must have $\Sigma(u,x,1) = 1_{H_R(x)}|\Sigma(x)$. Hence $\Sigma(x) = \Sigma(ux)$ and $\Sigma(u,x,1) = 1$.

2.14. *Corollary. Let* $u, \bar{u}, x, \bar{x}, v, \bar{v} \in S$, *with* $xC\bar{x}$, $\bar{u}\bar{x}Cux$, *and* $\bar{x}\bar{v}Cxv$. *Then* $\Sigma(u,x,v) = \Sigma(\bar{u},\bar{x},v)$, *while* $\Sigma'(u,x,v) = \Sigma'(u,\bar{x},\bar{v})$. *These conditions are satisfied if* $\bar{u}Cu$ *and* $vC\bar{v}$.

Proof. Let $a, b \in S$ be such that $\bar{x} = ax$ and $\bar{u}\bar{x} = bux$ (a, b exist since $\bar{x}Cx$ and $\bar{u}\bar{x}Cux$). Then $\Sigma(a,x,1) = 1_{\Sigma(x)}$ and $\Sigma(b,uxv,1) = 1_{\Sigma(uxv)}$. Since $\langle\bar{u}a,x,v\rangle \sim \langle bu,x,v\rangle$, we also have $\Sigma(\bar{u}a,x,v) = \Sigma(bu,x,v)$. Hence $\Sigma(\bar{u},\bar{x},v) = \Sigma(\bar{u},\bar{x},v) \circ \Sigma(a,x,1) = \Sigma(\bar{u}a,x,v) = \Sigma(bu,x,v) = \Sigma(b,uxv,1) \circ \Sigma(u,x,v) = \Sigma(u,x,v)$. The proof of the Σ' part is dual.

2.15. *Lemma. Let* $g \in \Sigma(x)$, $h \in H_R(x)$, *and* $\bar{h} \in S$ *with* $x \cdot h = x\bar{h}$. *Then* $g^h = h^{-1}gh = \Sigma(1,x,\bar{h})[g] = ((g)^L_x)^R_{x\cdot h}$. *Dually, if* $g \in \Sigma'(x)$, $h \in H_L(x)$, *and* $\bar{h} \in S$ *with* $h \cdot x = \bar{h}x$, *then* ${}^h g = hgh^{-1} = \Sigma'(\bar{h},x,1)[g] = ((g)^R_x)^L_{h\cdot x}$.

Proof. $(x\bar{h}) \cdot g^h = x \cdot (hg^h) = x \cdot gh = (x \cdot g)\bar{h} = x\bar{h} \cdot \Sigma(1,x,\bar{h})[g]$, and $g^h = \Sigma(1,x,\bar{h})[g]$ follows. Also $(x\bar{h}) \cdot ((g)^L_x)^R_{x\cdot h} = (g)^L_x \cdot (x\bar{h}) = (x \cdot g)\bar{h} = x\bar{h} \cdot \Sigma(1,x,\bar{h})[g] = x\bar{h} \cdot g^h$, and $g^h = ((g)^L_x)^R_{x\cdot h}$ follows.

2.16. *Corollary. If* $\Sigma : \mathcal{D}(S) \to \underline{GR}$ *is the right structure functor of* C, *and* $F : \mathcal{D}(S) \to \underline{GR}$ *is a subfunctor of* Σ, *then it is a normal subfunctor of* Σ. *If* Σ *is considered to have* $\mathbb{D}(S)$ *as its domain and* $F : \mathbb{D}(S) \to \underline{GR}$ *is a subfunctor of* Σ, *then* F *is a normal subfunctor of* Σ *which factors through* $\mathcal{D}(S)$. *Hence the* $\mathbb{D}(S)$*-domained subfunctors of* Σ *are order isomorphic with*

<u>the</u> $\mathcal{D}(S)$-<u>domained subfunctors of</u> Σ <u>in the canonical manner. In both situations,</u> $F(x) \trianglelefteq H_R(x)$ <u>for all</u> $x \in S$, <u>and</u> $y \in \mathcal{H}(x)$ <u>implies</u> $F(y) = F(x)$.

<u>Proof</u>. Let $F : \mathcal{D}(S) \to \underline{\underline{GR}}$ be a subfunctor of Σ . Then for all $x \in S$, $F(x) \leq H_R(x)$. If $y \in \mathcal{H}(x)$, let $a, b \in S$ be such that $y = ax = xb$. Since $[a,x,1], [1,x,b] : x \to y$ are isomorphisms, $F(y) = F(a,x,1)[F(x)] = F(1,x,b)[F(x)]$. But $F(a,x,1) = \Sigma(a,x,1)|F(x) = 1_{\Sigma(x)}|F(x)$ by Lemma 2.13, so that $F(y) = F(x)$. $F(1,x,b) = \Sigma(1,x,b)|F(x) = (\ )^h|F(x)$ by Lemma 2.15 where $x \cdot h = y$. Hence $[F(x)]^h = F(1,x,b)[F(x)] = F(y) = F(x)$. Hence $F \trianglelefteq \Sigma$, and we are done with the $\mathcal{D}(S)$-domained part of the proof. If $F \leq \Sigma$, where F is $\mathbb{D}(S)$-domained, then if $\langle u,x,v\rangle \sim \langle u',x,v'\rangle$, we must have $F(u,x,v) = \Sigma(u,x,v)|F(x) = \Sigma(u',x,v')|F(x) = F(u',x,v')$, so that F factors through $\mathcal{D}(S)$. The remainder of the corollary is now clear.

2.17. <u>Lemma</u>. <u>Let</u> $C_1 \subseteq C_2 \subseteq \mathcal{H}$. <u>If</u> Σ_1 <u>and</u> Σ_2 <u>are the right structure functors of</u> C_1 <u>and</u> C_2 <u>respectively, then</u> $\Sigma_1 \leq \Sigma_2$, <u>with equality holding if and only if</u> $C_1 = C_2$. <u>In particular, if</u> Γ <u>is the right structure functor of</u> $\mathcal{G}$, <u>the maximum sub-</u>$\mathcal{H}$ <u>congruence, then</u> $\{C \to \Sigma_C\}$ <u>is an order embedding of</u> $\{C \subseteq \mathcal{H}\}$ <u>into</u> $\{F \leq \Gamma\}$

<u>Proof</u>. Since $C_1(x) = x \cdot \Sigma_1(x)$ and $C_2(x) = x \cdot \Sigma_2(x)$, it is clear that $C_1 \subseteq C_2$ if and only if $\Sigma_1(x) \leq \Sigma_2(x)$ for all $x \in S$. If indeed $C_1 \subseteq C_2$, then for all $g \in \Sigma_1(x)$ we have $uxv \cdot \Sigma_2(u,x,v)[g] = u(x \cdot g)v = uxv \cdot \Sigma_1(u,x,v)[g]$ so that $\Sigma_1(u,x,v) = \Sigma_2(u,x,v)|\Sigma_1(x)$. The lemma follows.

2.18. <u>Lemma</u>. Let $g \in \Sigma(x)$, and $h \in \Sigma(y)$. Then $(x \cdot g)(y \cdot h) = xy \cdot (\Sigma(1,x,y)[g]\Sigma(x,y,1)[h])$.

<u>Proof</u>. Let $g' = (g)^L_x$. Then using 2.12 and 2.14 we obtain:

$$(x \cdot g)(y \cdot h) = (g' \cdot x)(y \cdot h)$$
$$= [(g' \cdot x)y] \cdot \Sigma(g' \cdot x,y,1)[h]$$
$$= [(g' \cdot x)y] \cdot \Sigma(x,y,1)[h]$$
$$= [\Sigma'(1,x,y)[g'] \cdot xy] \cdot \Sigma(x,y,1)[h]$$
$$= [xy \cdot \Sigma(1,x,y)[g]] \cdot \Sigma(x,y,1)[h]$$
$$= xy \cdot (\Sigma(1,x,y)[g]\ (x,y,1[h])\ .$$

2.19. Theorem. *Let G be the maximum congruence on S which is contained in H, and let Γ be its right structure functor. Then the function, $C \to \Sigma_C$, is an order isomorphism of the lattice of congruences under H, $\{C \subseteq H\}$, with the lattice of subfunctors of Γ, $\{F \leq \Gamma\}$. In particular, if $F \leq \Gamma$, then F corresponds to the congruence whose induced partition of S is* $\{x \cdot F(x) | x \in S\}$.

Proof. We need to show that $C \to \Sigma_C$ is surjective. Hence let $F \leq \Gamma$. We first need to show that $\{x \cdot F(x)\}$ is a congruence partition. Let $z \in x \cdot F(x) \cap y \cdot F(y)$. Then $H(z) = H(x) = H(y)$, and by the corollary to Lemma 13, $F(x) = F(y) = F(z)$. If $z = x \cdot g$ where $g \in F(x) = F(z)$, we obtain $x \cdot F(x) = (z \cdot g^{-1}) \cdot F(z) = z \cdot F(z)$. Similarly, $y \cdot F(y) = z \cdot F(z)$ so that $x \cdot F(x) = y \cdot F(y)$. Hence $\{x \cdot F(x)\}$ is a partition. If $x, y \in S$, then $(x \cdot F(x))(y \cdot F(y)) = xy \cdot (\Gamma(1,x,y)F(x))(\Gamma(x,y,1)F(y)) \subseteq xy \cdot F(xy)F(xy) = xy \cdot F(xy)$ since $F \leq \Gamma$. Hence $\{x \cdot F(x)\}$ is a congruence partition. Let C be the corresponding congruence. Since $C(x) = x \cdot F(x) \subseteq x \cdot H_R(x) = H(x)$ for all $x \in S$, $C \subseteq H$. Let $\Sigma = \Sigma_C$. Then for all $x \in S$, $x \cdot F(x) = C(x) = x \cdot \Sigma(x)$, and $F(x) = \Sigma(x)$ follows by simple transitivity of the action. Hence $F = \Sigma$ and surjectivity has been shown.

2.20. Notation. If C_1 and C_2 are congruences on S, then $C_1 \vee C_2$ is the congruence on S generated by C_1 and C_2, i.e., the smallest congruence on S containing both C_1 and C_2. $C_1 \wedge C_2 = C_1 \cap C_2$ is the smallest congruence on S contained in C_1 and C_2. If $F : \mathcal{D}(S) \to \underline{\underline{Gr}}$ and F_1, F_2 are normal subfunctors of F, then F_1F_2 and $F_1 \cap F_2$ are the normal subfunctors of F defined by $(F_1F_2)(x) = F_1(x)F_2(x)$ and $(F_1 \cap F_2)(x) = F_1(x) \cap F_2(x)$, with $(F_1F_2)(u,x,v)$ and $(F_1 \cap F_2)(u,x,v)$ given by restriction. F_1F_2 is the smallest subfunctor of F containing both F_1 and F_2, while $F_1 \cap F_2$ is the largest subfunctor of F contained in both F_1 and F_2.

2.21. Corollary. *Let* $C_1 \leftrightarrow \Sigma_1$ *and* $C_2 \leftrightarrow \Sigma_2$. *Then* $C_1 \wedge C_2 \leftrightarrow \Sigma_1 \cap \Sigma_2$, *while* $C_1 \vee C_2 \leftrightarrow \Sigma_1\Sigma_2$.

2.22. Corollary. *The lattice* $\{C \subseteq \mathcal{H}\}$ *is modular.*

2.23. Corollary. *The results of Theorem* 2.19 *and its first corollary remain the same if we consider the functors to be* $\mathbb{D}(S)$-*domained.*

Proof. This is an immediate consequence of 2.16.

2.24. We have seen that much of the study of sub-$\mathcal{H}$ congruences reduces to the study of subfunctors of Γ. The question remains as to what Γ looks like. In particular, what is $\Gamma(x)$ for all $x \in S$? The following trivial result yields an answer.

2.25. Theorem. *Let* $g \in H_R(x)$. *Then* $g \in \Gamma(x)$ *if and only if for all* $u, v \in S$, $u(x \cdot g)v \,\mathcal{H}\, uxv$.

Proof. Let $g \in H_R(x)$ be such that $u(x{\cdot}g)v \,\mathcal{H}\, x \cdot g$ for all $u, v \in S$. Let R_g be the relation:

$$\{\langle y,y\rangle \mid y \in S\} \cup \{\langle uxv,\ u(x\cdot g)v\rangle \mid u,v \in S\} \cup \{\langle u(x\cdot g)v,\ uxv\rangle \mid u,\ v \in S\}\ .$$

R_g is clearly reflexive, symmetric, compatible with multiplication, and contained in H. Hence R_g^t, the transitive closure of R_g, is under H. But R_g^t is a congruence. Hence $R_g^t \subseteq G$ and $\langle x,\ x\cdot g\rangle \in G$ follows.

Before stating the correspondence theorem, we shall need the following collection of facts.

2.26. Lemma. Let $f : S \to T$ be a homomorphism. For each $x \in S$, $f[H(x)] \subseteq H(f(x))$ and f induces a pair of homomorphisms, $f_x : H_L(x) \to H_L(f(x))$ and $f^x : H_R(x) \to H_R(f(x))$, defined implicitly by the equations, $f(g\cdot x) = f_x(g)\cdot f(x)$ and $f(x\cdot g) = f(x)\cdot f^x(g)$. Moreover, $(\ )^R_{f(x)} \circ f_x = f^x \circ (\ )^R_x$, and $(\ )^L_{f(x)} \circ f^x = f_x \circ (\ )^L_x$. If $C_1 \subseteq H_S$, $C_2 \subseteq H_T$, $(f\times f)C_1 \subseteq C_2$, $C_1 \leftrightarrow \Sigma_1$, Σ_1', and $C_2 \leftrightarrow \Sigma_2$, Σ_2', then $\{f_x\} : \Sigma_1' \to \Sigma_2' \circ \mathcal{D}(f)$ and $\{f^x\} : \Sigma_1 \to \Sigma_2 \circ \mathcal{D}(f)$ are well defined natural transformations such that the following diagram commutes:

$$\begin{array}{ccc}
\Sigma_1' & \xrightarrow[(\ )^R_S]{\cong} & \Sigma_1 \\
\Big\downarrow \{f_x\} & & \Big\downarrow \{f^x\} \\
\Sigma_2' \circ \mathcal{D}(f) & \xrightarrow[(\ )^R_T]{\cong} & \Sigma_2 \circ \mathcal{D}(f)
\end{array}$$

Proof. The first half of the lemma is well known. If $g \in \Sigma_1(x)$, then $f(x)\cdot f^x(g) = f(x\cdot g)\ C_2 f(x)$, since $x\cdot g\ C_1 x$. Hence $f^x(g) \in \Sigma_2(f(x))$ and $f^x : \Sigma_1(x) \to \Sigma_2(f(x))$. Moreover,

$$f(uxv)\cdot\Sigma_2(f(u),f(x),f(v))[f^x(g)] = f(u)(f(x)\cdot f^x(g))f(v)$$

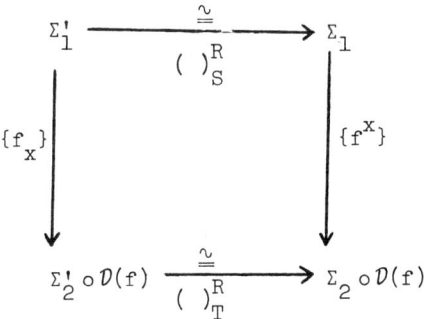

$$= f(u(x \cdot g)v)$$
$$= f(uxv \cdot \Sigma_1(u,x,v)[g])$$
$$= f(uxv) \cdot f^{uxv}(\Sigma_1(u,x,v)[g])$$

and $\{f^x\} : \Sigma_1 \to \Sigma_2 \circ \mathcal{D}(f)$ follows. The final statement is just the appropriate restriction of $(\ )^R_{f(x)} \circ f_x = f^x \circ (\ )^R_x$.

2.27. Theorem. *Let $C_0 \subseteq H$, and let $\pi : S \to S/C_0$ be the canonical morphism. Then the map $C \to C^\pi = (\pi \times \pi)[C]$ induces by restriction an order isomorphism between $\{C_0 \subseteq C \subseteq H_S\}$ and $\{C \subseteq H_{S/C_0}\}$, whose inverse is the map $C \to (\pi \times \pi)^{-1}[C]$. Moreover, for all $x \in S$, both $\pi_x : H_L(x) \to H_L(\pi(x))$ and $\pi^x : H_R(x) \to H_R(\pi(x))$ are surmorphisms whose kernels are $\Sigma'_0(x)$ and $\Sigma_0(x)$ respectively, where Σ'_0 and Σ_0 are the structure functors of C_0. If $C \leftrightarrow \Sigma$ where $C_0 \subseteq C \subseteq H_S$ and $\Sigma^\pi \leftrightarrow C^\pi$, then $\pi^x[\Sigma(x)] = \Sigma^\pi(x)$ for all $x \in S$, and the $\{\pi^x\}$ induce a natural isomorphism, $\Sigma/\Sigma_0 \cong \Sigma^\pi \circ \mathcal{D}(\pi)$. In particular, $\Gamma_{S/\Sigma_0} \cong \Gamma_{S/C_0} \circ \mathcal{D}(\pi)$.*

Proof. The first part is universal algebraic trivia. Since $\pi[H(x)] = H(\pi(x))$, both π_x and π^x must be surjective. $g \in \mathrm{Ker}(\pi^x)$ means $\pi(x)\cdot\pi^x(g) = \pi(x\cdot g) = \pi(x)$, i.e., $g \in \mathrm{Ker}(\pi^x)$ if and only if $x \cdot g\, C_0\, x$. Hence $\mathrm{Ker}(\pi^x) = \Sigma_0(x)$. From the previous lemma it is clear that $\{\pi^x\} : \Sigma \to \Sigma^\pi \circ \mathcal{D}(f)$. If $g \in H_R(x)$ and $\pi^x(g) \in \Sigma^\pi(\pi(x))$, then $\pi(x) \cdot \pi^x(g) = \pi(x \cdot g) C^\pi \pi(x)$ so that $x \cdot g\, C\, x$, and hence $g \in \Sigma(x)$. Since π^x is surjective, it follows that $\{\pi^x\} : \Sigma \to \Sigma^\pi \circ \mathcal{D}(x)$ is surjective. Since Σ_0 is the kernel of this natural transformation, $\Sigma/\Sigma_0 \cong \Sigma^\pi \circ \mathcal{D}(f)$ follows.

Section 2b. *Abelian congruences, central congruences, and normal subgroups.* If $A \xrightarrow{i} G \xrightarrow{\pi} Q$ is an extension of a group A by a group Q, where A is abelian, then it is well known that there is an induced action, $A \times Q \to A$, defined by $\langle a, q\rangle \to a^{\bar{q}}$, where $\pi(\bar{q}) = q$, and moreover this action is independent of the choice of $\bar{q} \in \pi^{-1}(q)$. We now generalize this fact to sub-$\mathcal{H}$ congruences.

2.28. *Lemma. If* $\mathcal{C} \subseteq \mathcal{H}$, $\pi : S \to S/\mathcal{C}$ *is the induced morphism, and* Σ *is the right structure functor of* $\mathcal{C}$, *then the following are equivalent:*

(i) *For all* $x \in S$, $\Sigma(x)$ *is abelian.*

(ii) *If* $u, x, v, \bar{u}, \bar{x}, \bar{v} \in S$ *are such* $x\mathcal{C}\bar{x}$, $ux\,\mathcal{C}\,\bar{u}\bar{x}$, *and* $xv\,\mathcal{C}\,\bar{x}\bar{v}$, *then* $\Sigma(u,x,v) = \Sigma(\bar{u},\bar{x},\bar{v})$.

(iii) *There exists a functor,* $F : \mathcal{D}(S/\mathcal{C}) \to \underline{\underline{Gr}}$, *such that*

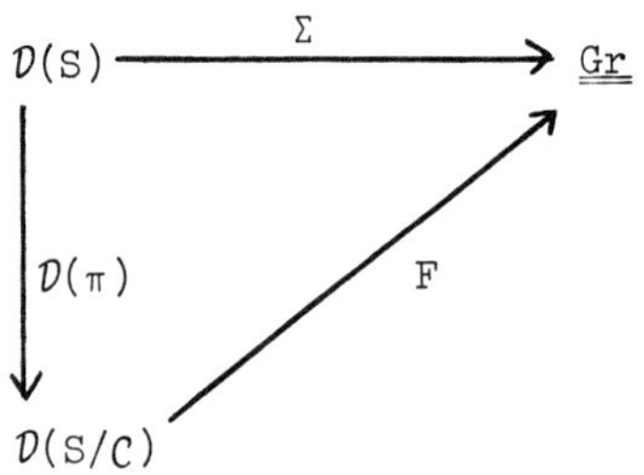

commutes. When these conditions hold, then F *is uniquely determined.*

Proof. We clearly have (ii) $\leftrightarrow$ (iii). By 2.15, (ii) $\to$ (i), since if (ii) holds, then all conjugations of $\Sigma(x)$ are trivial. Conversely, if (i) holds and the hypotheses of (ii) are satisfied, then we may find $a, b \in S$ such that $xa = \bar{x}$ and $\overline{xv}b = xv$. Since (i) holds, we must have $\Sigma(1,x,a) = 1$ and $\Sigma(1,\overline{u}\,\overline{x}\,\overline{v},b) = 1$ by 2.15 again. Hence:

$$\begin{aligned}\Sigma(\bar{u},\bar{x},\bar{v}) &= \Sigma(1,\overline{uxv},b) \circ \Sigma(\bar{u},\bar{x},\bar{v})\\ &= \Sigma(\bar{u},\bar{x},\bar{v}b)\\ &= \Sigma(\bar{u},\bar{x},\bar{v}b) \circ \Sigma(1,x,a)\end{aligned}$$

$= \Sigma(\bar{u},x,a\bar{v}b)$

$= \Sigma(\bar{u},x,v)$

$= \Sigma(u,x,v)$ by 2.14.

2.29. Lemma. If S is regular, then conditions (i) to (iii) above are equivalent with:

(iv) If $u\,C\,\bar{u}$, $x\,C\,\bar{x}$, $v\,C\,\bar{v}$, then $\Sigma(u,x,v) = (\bar{u},\bar{x},\bar{v})$.

Proof. Clearly (ii) → (iv). If $e \in E(S)$, then $C(e) \cong \Sigma(e)$ under the isomorphism defined implicitly by $e\bar{g} = e \cdot g$. Moreover, we have $\Sigma(1,e,\bar{g}) = (\ )^g$. Hence, if (iv) holds, we must have $(\ )^g = \Sigma(1,e,\bar{g}) = \Sigma(1,e,1) = 1_{\Sigma(e)}$, for all $g \in \Sigma(e)$, i.e., $\Sigma(x)$ is abelian $e \in E(S)$. But since S is regular, for all $x \in S$, there exists $e \in L(x) \cap E(S)$. Hence $\Sigma(x,e,1) : \Sigma(e) \cong \Sigma(x)$, so that $\Sigma(x)$ must be abelian.

2.30. The result of 2.29 no longer holds if we do not assume that S is regular. We construct a counterexample. Let G be a nonabelian group and let $\mu : F_X \to G$ be a surjective morphism, where F_X is a free monoid. Let $S = F_X \cup G \cup \{0\}$. We define a multiplication, $*$, on S by:

(i) $0*x = x*0 = 0$ for all $x \in S$.

(ii) $g*h = 0$ for all $g, h \in G$.

(iii) $w*g = \mu(w)g$ for all $w \in F_X$, $g \in G$.

(iv) $g*w = g\mu(w)$ for all $g \in G$, $w \in F_X$.

(v) $w_1*w_2 = w_1w_2$ for all $w_1, w_2 \in F_X$.

It is easily seen that H is a congruence whose only nontrivial class is G, whose Schützenberger group is isomorphic with the original group, G. Moreover, it should be clear that Γ satisfies condition (iv).

2.31. Definition. We call a congruence, $C \subseteq H$, satisfying conditions (i) → (iii) of 2.28, an abelian congruence.

2.32. We now turn our attention to a special type of congruence which behaves even more like a group congruence than do sub-H congruences in general. We recall that a subgroup, $K \subseteq G(S)$ -- the group of units of S -- is said to be normal if for all $x \in S$ we have $xK = Kx$. If K is normal, this is of course, denoted $K \triangleleft S$. When $K \triangleleft S$, then the set of cosets, $P = \{Kx \mid x \in S\}$, is a partition of S. We denote the corresponding equivalence, $\tilde{K}$.

2.33. Lemma. If $K \triangleleft S$, then the equivalence, $\tilde{K}$, is a congruence which is contained in H. Moreover, for all $x, y \in S$, $\tilde{K}(x)y = x\tilde{K}(y) = \tilde{K}(xy)$, i.e., $\tilde{K}$ is a surjective congruence.

Proof. Since $\tilde{K}(x) = xK = Kx$ for all $x \in S$, it follows that $\tilde{K}(x)y = (Kx)y = K(xy) = \tilde{K}(xy)$ and similarly that $x\tilde{K}(y) = x(yK) = \tilde{K}(xy)$. Hence $\tilde{K}$ is a surjective congruence. Since $xK \subseteq R(x)$ and $Kx \subseteq L(x)$ for all subgroups, $K \leq G(S)$, if $K \triangleleft S$, $xK = Kx \subseteq H(x)$ for all $x \in S$. Hence $K \subseteq H$.

2.34. In the following sequence of results, we denote $\Sigma_{\tilde{K}}$ by Σ_K.

2.35. Theorem. If K is a normal subgroup of S, then Σ_K is a surjective subfunctor of Γ, i.e., $\Sigma_K[u,x,v]$ is surjective for all $u, x, v \in S$. Conversely, if Σ is a surjective subfunctor of Γ, then $\Sigma(1)$ is normal and $\Sigma = \Sigma_{\Sigma(1)}$. Hence there is an order isomorphism between the normal subgroups of S and the surjective subfunctors of Γ. Finally K is a normal subgroup of S if and only if K is a subgroup of $\Gamma(1)$, and $\Gamma(x,1,1)K = \Gamma(1,1,x)K$ for all $x \in S$.

Proof. Since $\tilde{K}$ is surjective, for all $u, x, v \in S$ we have $u\tilde{K}(x)v = \tilde{K}(uxv)$. But since the functorial action $\tilde{K} \times \Sigma_K \to \tilde{K}$ is simply transitive we must have $uxv \cdot \Sigma_K(uxv) = \tilde{K}(uxv) = u\tilde{K}(x)v = u(x \cdot \Sigma_K(x))v = uxv \cdot \Sigma_K[u,x,v](\Sigma_K(x))$, and hence $\Sigma_K(uxv) = \Sigma_K[u,x,v](\Sigma_K(x))$. Conversely, if Σ is surjective and $K = \Sigma(1)$, then for all $x \in S$ we have $xK = x(1 \cdot K) = x \cdot \Sigma[x,1,1]K = x \cdot \Sigma(x) = x \cdot \Sigma[1,1,x]K = (K \cdot 1)x = Kx$, and K is normal. Moreover for all $x \in S$ we have $x \cdot \Sigma_K(x) = xK = x \cdot \Sigma[1,1,x]K = x \cdot \Sigma[1,1,x](\Sigma(1)) = x \cdot \Sigma(x)$ so that $\Sigma_K(x) = \Sigma(x)$ by simple transitivity, and $\Sigma_K = \Sigma$ follows. Finally, if K is normal, then clearly $K = \Sigma_K(1) \subseteq \Gamma_S(1)$ and $\Gamma_S[x,1,1]K = \Sigma_K[x,1,1]K = \Sigma_K(x) = \Sigma_K[1,1,x]K = \Gamma_S[1,1,x]K$. Conversely, if $K \subseteq \Gamma_S(1)$ and $\Gamma_S[x,1,1]K = \Gamma_S[1,1,x]K$ for all $x \in S$, then clearly $xK = x \cdot \Gamma_S[x,1,1]K = x \cdot \Gamma_S[1,1,x]K = Kx$.

2.36. It is clear that if $\{K_\delta \mid \delta \in \Delta\}$ is a family of normal subgroups of S then so is $K = \bigvee_{\delta\in\Delta} K_\delta$, the subgroup generated by the K_δ. Moreover, we have $\Sigma_K = \bigvee_{\delta\in\Delta} \Sigma_{K_\delta}$. Hence, S contains a maximum normal subgroup, N_S, and since 1 is a minimum normal subgroup, the collection of normal subgroups is a complete lattice such that the embedding $K \to \tilde{K}$ of normal subgroups into sub-$\mathcal{H}$ congruences is an upper semilattice embedding. However it is not a lattice embedding since the intersection of normal subgroups need not be normal, as the following example shows.

2.37. Example. Let $S = K^3 \cup K \cup 0$. We define multiplication by: $0x = x0 = 0 = 0_S$ for all $x \in S$; $(a,b,c)(a',b',c') = (aa',bb',cc')$ for $(a,b,c), (a',b',c') \in K^3$; $(a)(b) = 0$ for $a, b \in K$; $(a,b,c)(e) = (abe)$ and $(e)(a,b,c) = (ebc)$. It is an easy check to see that S is a normal monoid with $G(S) = K \times K \times K$. Also, $K \times K \times \{1\}$ and $K \times \{1\} \times K$ are normal subgroups. However, their intersection, $K \times \{1\} \times \{1\}$, is not normal.

2.38. A special type of normal subgroup is a group, K, which is contained in $G(S) \cap Z(S)$, where $Z(S)$ is the center of S. Such a normal subgroup is called a central subgroup. Clearly if $\{K_\delta \mid \delta \in \Delta\}$ is a set of central subgroups of S, then so are $\bigvee_{\delta\in\Delta} K_\delta$ and $\bigcap_{\delta\in\Delta} K_\delta$. In particular there exists a maximum central subgroup and a minimum central subgroup, $\{1_S\}$. Hence the map $K \to \tilde{K}$ is a lattice embedding of the lattice of central subgroups of S into the lattice of sub-$\mathcal{H}$ congruences of S. Moreover, the following theorem holds.

2.39. Theorem. A normal subgroup, K, is central if and only if Σ_K factors through $\mathcal{J}(S)$.

Proof. If K is a central subgroup, then if $k \in K$, $u(xk)v = uxvk$ must hold for all $u, x, v \in S$. Hence if $g \in \Sigma_K(x)$ is such that $x \cdot g = xk$, then $uxv \cdot \Sigma[u,x,v](g) = uxvk$. But if $\bar{u}x\bar{v} = uxv$, then again $\bar{u}x\bar{v} \cdot \Sigma[\bar{u},x,\bar{v}](g) = \bar{u}x\bar{v}k = uxvk$ so that $\Sigma[u,x,v](g) = \Sigma[\bar{u},x,\bar{v}](g)$, and $\Sigma_K[u,x,v] = \Sigma_K[\bar{u},x,\bar{v}]$ follows. Hence Σ_K factors through $\mathcal{J}(S)$. Conversely, if Σ_K factors through $\mathcal{J}(S)$, then for all $k \in K$, $x \in S$ we have $kx = x \cdot \Sigma_K[1,1,x](k) = x \cdot \Sigma_K[x,1,1](k) = xk$, and K is a central subgroup of S.

2.40. In general we say that a sub-$\mathcal{H}$ congruence, is a central congruence if its structure functors factor through $\mathcal{J}(S)$.

Chapter Three

$\mathcal{H}$-Coextensions of Monoids

Let S be a monoid, and let $\underline{\underline{Mon}}_S$ denote the category of monoids over S , whose objects are pairs, $(\overline{S},p)$, where $\overline{S}$ is a monoid and $p : \overline{S} \to S$ is a surmorphism (i.e., a homomorphism onto). A morphism $f : (\overline{S}_1, p_1) \to (\overline{S}_2,p_2)$ is a monoid morphism, $f : \overline{S}_1 \to \overline{S}_2$, such that $p_2 \circ f = p_1$. $\mathcal{H}(\underline{\underline{Mon}}_S)$ is the full subcategory of $\underline{\underline{Mon}}_S$ whose objects are those pairs, $(\overline{S},p)$, for which $\mathcal{C}_p \subseteq \mathcal{H}_{\overline{S}}$, where $\mathcal{C}_p$ is the congruence on $\overline{S}$ induced by p . We call the objects of $\mathcal{H}(\underline{\underline{Mon}}_S)$, $\mathcal{H}$-coextensions of S . We call the morphisms of $\mathcal{H}(\underline{\underline{Mon}}_S)$, $\mathcal{H}$-morphisms.

Section 3a. Constructing $\mathcal{H}(\underline{\underline{Mon}}_S)$. This section has two main results: the First Main Theorem and the Second Main Theorem, to be referred to as FMT and SMT, respectively. The FMT deals with the objects of $\mathcal{H}(\underline{\underline{Mon}}_S)$, while the SMT deals with the morphisms of $\mathcal{H}(\underline{\underline{Mon}}_S)$. Before stating the FMT we present a collection of definitions and special notations.

3.1. Definition. A pair of functors, $F_L : \mathcal{L}(S) \to \underline{\underline{Gr}}$ and $F_R : \mathcal{R}(S) \to \underline{\underline{Gr}}$, is said to be compatible when for all $x \in S$, $F_L(x) = F_R(x)$. When F_L and F_R are compatible functors, the situation will be denoted by: $(F_L,F_R) : (\mathcal{L},\mathcal{R})(S) \to \underline{\underline{Gr}}$.

3.2. Let $(F_L,F_R) : (\mathcal{L},\mathcal{R})(S) \to \underline{\underline{Gr}}$ be a compatible pair of functors. For all $\langle u,x,v\rangle \in \mathrm{Mor}(\mathbb{D}(S))$ we let $F_{LR}(u,x,v) = F_L(u,xv,1) \circ F_R(1,x,v)$ and $F_{RL}(u,x,v) = F_R(1,ux,v) \circ F_L(u,x,1)$. Clearly (F_L,F_R) can be extended to $\mathcal{D}(S)$ if and only if we have $F_{LR} = F_{RL}$.

3.3. Let $A \subseteq S$. Then:

3.3i. $r(A) = \{x \in S | ax = a$ for all $a \in A\}$.

3.3ii. $\ell(A) = \{x \in S | xa = a$ for all $a \in A\}$.

3.3iii. $R(A) = \{x \in S | Ax \subseteq A\}$.

3.3iv. $L(A) = \{x \in S | xA \subseteq A\}$.

When $A = \{a\}$, $r(\{a\}) = r(a)$ and $\ell(\{a\}) = \ell(a)$.

3.4. If G is a group and $g \in G$, then $(\ )^g$ denotes the conjugation-by-g function, i.e., $(x)^g = x^g = g^{-1}xg$. Moreover, $(\ )^{-g}$ denotes $(\ )^{g^{-1}} = ((\ )^g)^{-1}$.

3.5. Often we will use exponential notation:

3.5i. $F_L(u,x,1)[\ ] = [\ ]^{(u,x,1)}$.

3.5ii. $F_R(1,x,v)[\ ] = [\ ]^{(1,x,v)}$.

3.5iii. $F(u,x,v)[\ ] = [\ ]^{(u,x,v)}$.

3.6. If $A \subseteq S$, and $F_L : L(S) \to \underline{\underline{Gr}}$, then $A \times F_L$ is the set, $\{<x,g> | x \in A$ and $g \in F_L(x)\}$.

3.7. <u>Definition</u>. Let $F_L : L(S) \to \underline{\underline{Gr}}$. A <u>factor system</u> for F_L (or just a factor system when there is only one F_L around) is a function, $\alpha : S \times S \to \bigcup_{x \in S} F_L(x)$, such that $\alpha(x,y) \in F_L(xy)$ for all $x, y \in S$. If $\alpha(1,1) = 1 \in F_L(1)$, then α is said to be a <u>unitary</u> factor system.

3.8. <u>Definition</u>. Let $(\bar{S},p) \in H(\underline{\underline{Mon}}_S)$. Then a <u>lifting</u> from S to $(\bar{S},p)$ is a function $x \to \bar{x}$, from S to $\bar{S}$, such that for all $x \in S$, $p(\bar{x}) = x$. In general, $x \to \bar{x}$ will not be a homomorphism, nor can one find a lifting which will be a homomorphism. $x \to \bar{x}$ will be called a unitary lifting if $\bar{1} = 1$.

We are now ready for a statement of FMT. Its proof will follow as a consequence of several lemmas.

3.9. *The First Main Theorem. Let* $(F_L, F_R) : (L,R)(S) \to \underline{\underline{Gr}}$ *, and let* α *be a factor system for* F_L *, such that* F_L *,* F_R *, and* α *together satisfy:*

HI. *For all* $u, x, v \in S$ *and* $g \in F_L(x)$,

$$\alpha(u,xv)F_{LR}(u,x,v)[g]\alpha(x,v)^{(u,xv,1)} = \alpha(u,x)^{(1,ux,v)}F_{RL}(u,x,v)[g]\alpha(ux,v).$$

HII. *For all* $x \in S$,

$$\bigcup_{u\in\ell(x)}(F_{L(u)}{}^{(1,u,x)}\alpha(u,x)) = F_L(x) = \bigcup_{v\in r(x)}(\alpha(x,v)F_L(v)^{(x,v,1)}) .$$

Then defining a multiplication on $S \times F_L$ *by*

HIII $\quad \langle x,a\rangle\langle y,b\rangle = \langle xy, a^{(1,x,y)}\alpha(x,y)b^{(x,y,1)}\rangle$,

we obtain a monoid denoted by $S \times (F_L, F_R, \alpha)$. *If* $\pi : S \times (F_L,F_R,\alpha) \to S$ *is the first coordinate projection,* $\langle x,a\rangle \to x$, *then* $(S \times (F_L,F_R,\alpha),\pi) \in H(\underline{\underline{Mon}}_S)$.

Conversely, if $(\overline{S},p) \to H(\underline{\underline{Mon}}_S)$ *and* $x \to \overline{x}$ *is a lifting from* S *to* $\overline{S}$, *then* $x \to \overline{x}$ *induces a triple* (F_L,F_R,α) *defined by:*

HIV. *For all* $x \in S$, $F_L(x) = F_R(x) = \Sigma(\overline{x})$, *where* $\Sigma = \Sigma_{C_p}$.

HV. *For all* $x, y \in S$, $\alpha(x,y)$ *is implicitly defined by:*

$$\overline{xy} \cdot \alpha(x,y) = \overline{x} \cdot \overline{y} .$$

HVI. *For all* $u, x, v \in S$,

$$F_L(u,x,1) = \Sigma(\overline{u},\overline{x},1) .$$

$$F_R(1,x,v) = (\ )^{R}_{\overline{xv}} \circ \Sigma'(1,\overline{x},\overline{v}) \circ (\ )^{L}_{\overline{x}}$$
$$= (\ )^{-\alpha(x,v)} \circ \Sigma(1,\overline{x},\overline{y}) .$$

Then $(F_L,F_R) : (L,R)(S) \to \underline{\underline{Gr}}$, *and* (F_L,F_R,α) *satisfies conditions* HI *and* HII *above. Moreover,* $(S \times (F_L,F_R,\alpha),\pi) \cong (\overline{S},P)$ *under the isomorphism,* $\langle x,g\rangle \to \overline{x} \cdot g$. α *will be a unitary factor system if and only if the lifting is unitary.*

3.10. *Remark.* Condition HI above is equivalent to the following pair of conditions:

HIa. *For all* $u, x, v \in S$, $\alpha(u,x,v)\alpha(x,v)^{(u,xv,1)} = \alpha(u,x)^{(1,ux,v)}\alpha(ux,v)$.

$\mathcal{H}$IIb. For all $u, x, v \in S$, $F_{LR} = (\ )^{\tilde{\alpha}(u,x,v)} \circ F_{RL}$, where $\tilde{\alpha}(u,x,v) = \alpha(ux,v)[\alpha(x,v)^{-1}]^{(u,xv,1)} = [\alpha(u,x)^{-1}]^{(1,ux,v)}\alpha(u,xv)$.

3.11. Let $(F_L,F_R) : (\mathcal{L},\mathcal{R})(S) \to \underline{\underline{Gr}}$, and let α be a factor system for (F_L,F_R). We denote by $S \times (F_L,F_R,\alpha)$, the set $S \times F_L$ together with the multiplication given by $\mathcal{H}$III above. This multiplication will not, in general, be associative.

3.12. Lemma. Let $(F_L,F_R) : (\mathcal{L},\mathcal{R})(S) \to \underline{\underline{Gr}}$, and let α be a factor system for (F_L,F_R). Then $S \times (F_L,F_R,\alpha)$ has an associative multiplication if and only if (F_L,F_R,α) satisfies condition $\mathcal{H}$I above. When $\mathcal{H}$I is satisfied, $S \times (F_L,F_R,\alpha)$ is a monoid whose identity element is $\langle 1,\alpha(1,1)^{-1}\rangle$, and whose group of units is the set $G(S) \times F_L$.

Proof. Let $A(u,x,v,g)$ denote the product on the left side of $\mathcal{H}$I, and let $B(u,x,v,g)$ denote the product on the right of $\mathcal{H}$I. Let $\langle u,a\rangle$, $\langle x,g\rangle$, and $\langle v,c\rangle$ be elements of $S \times F_L$. Multiplying out $\langle u,a\rangle \cdot (\langle x,g\rangle \cdot \langle v,c\rangle)$ we get $\langle uxv, F_R(1,u,xv)[a]\cdot A(u,x,v,g)\cdot F_L(ux,v,1)[c]\rangle$. By multiplying out $(\langle u,a\rangle \cdot \langle x,g\rangle) \cdot \langle v,c\rangle$ we get $\langle uxv, F_L(1,u,xv)[a]\cdot B(u,x,v,g)\cdot F_R(ux,v,1)[c]\rangle$. By cancelling $F_R(1,u,xv)[a]$ on the left and $F_R(ux,v,1)[c]$ on the right we see that $S \times (F_L,F_R,\alpha)$ is associative if and only if for all $u,x,v \in S$ and $g \in F_L(x)$ we have $A(u,x,v,g) = B(u,x,v,g)$. Hence assuming associativity, $A(1,1,x,1) = B(1,1,x,1)$ tells us that $\alpha(1,x)\alpha(1,x) = F_R(1,1,x)[\alpha(1,1)]\alpha(1,x)$, and hence $\alpha(1,x) = F_R(1,1,x)[\alpha(1,1)]$ for all $x \in S$. Similarly $\alpha(x,1) = F_L(x,1,1)[\alpha(1,1)]$. Hence for all $\langle x,g\rangle \in S \times F_L$ we have $\langle 1,\alpha(1,1)^{-1}\rangle \cdot \langle x,g\rangle = \langle x,g\rangle = \langle x,g\rangle \cdot \langle 1,\alpha(1,1)^{-1}\rangle$. Since for all $g, h \in G(S)$, $[g,h,1]$ and $[1,g,h]$ are isomorphisms in $\mathcal{L}(S)$ and $\mathcal{R}(S)$, $F_L(g,h,1)$ and $F_R(1,g,h)$ are isomorphisms in $\underline{\underline{Gr}}$. So let $\langle g,a\rangle$ and $\langle g^{-1},b\rangle$ be elements of $G(S) \times F_L$. Then

$\langle g,a\rangle \cdot \langle g^{-1},b\rangle = \langle 1, a^{(1,g,g^{-1})}\alpha(g,g^{-1})b^{(g,g^{-1},1)}\rangle$. As $F_L(g,g^{-1},1)$ is an isomorphism of $F_R(g^{-1})$ with $F_R(1)$, the equation $a^{(1,g,g^{-1})}\alpha(g,g^{-1})b^{(g,g^{-1},1)} = \alpha(1,1)^{-1}$ can be solved for b. For this value of b we thus have $\langle g,a\rangle \cdot \langle g^{-1},b\rangle = \langle 1,\alpha(1,1)^{-1}\rangle$. Hence $\langle g,a\rangle$ has a right inverse. By a symmetric argument $\langle g,a\rangle$ also has a left inverse. Hence $G(S) \times F_L \subseteq G(S \times (F_L,F_R,\alpha))$. If $\pi : S \times (F_L,F_R,\alpha) \to S$ is the first coordinate projection, then π is an epimorphism. As π takes units to units, $G(S \times (F_L,F_R,\alpha)) \subseteq G(S) \times F_L$. Hence equality holds.

3.13. <u>Corollary to the proof.</u> <u>If</u> $S \times (F_L,F_R,\alpha)$ <u>is a monoid, then for all</u> $x \in S$, $\alpha(x,1) = F_L(x,1,1)[\alpha(1,1)]$ <u>and</u> $\alpha(1,x) = F_R(1,1,x)[\alpha(1,1)]$. <u>In particular, if</u> α <u>is unitary, then for all</u> $x \in S$, $\alpha(x,1) = 1 = \alpha(1,x)$.

3.14. One might be tempted to conjecture that for all $\langle x,g\rangle \in S \times F_L$ we have $\mathcal{H}(\langle x,g\rangle) = \mathcal{H}_S(x) \times F_L$. This however is not the case. In general we only have $\mathcal{H}(\langle x,g\rangle) \subseteq \mathcal{H}_S(x) \times F_L$ resulting from $\pi[\mathcal{H}(x,g)] \subseteq \mathcal{H}_S(x)$.

3.15. <u>Lemma.</u> <u>Let</u> $S \times (F_L,F_R,\alpha)$ <u>be a monoid.</u> <u>Then</u> $(S \times (F_L,F_R,\alpha),\pi) \in \mathcal{H}(\underline{\underline{\mathrm{Mon}}}_S)$ <u>if and only if condition</u> $\mathcal{H}$II <u>is satisfied.</u>

<u>Proof.</u> Since $(F_L,F_R) : (\mathcal{L},\mathcal{R})(S) \to \underline{\underline{\mathrm{Gr}}}$, for all $u \in \ell(x)$ and $v \in r(x)$ we must have $F_L(u,x,1) = F_R(1,x,v) = 1_{F_L(x)}$. Hence, since $gF_L(x) = F_L(x)g = F_L(x)$ for all $g \in F_L(x)$, condition $\mathcal{H}$II is equivalent with:

$\mathcal{H}$II'. $(\ell(x) \times F_L)\langle x,g\rangle = \{x\} \times F_L(x) = \langle x,g\rangle(r(x) \times F_L$.

Hence $\mathcal{H}$II implies that $\{x\} \times F_L(x) \subseteq \mathcal{H}(x,1)$ for all $x \in S$, i.e., $(S \times (F_L,F_R,\alpha),\pi) \in \mathcal{H}(\underline{\underline{\mathrm{Mon}}}_S)$. Conversely, if $\{x\} \times F_L(x) \subseteq \mathcal{H}(x,1)$ for all $x \in S$, then for each $\langle x,g\rangle \in S \in F_L$ there exist sets A, $B \subseteq S \times F_L$ such that $A\langle x,g\rangle = \langle x,g\rangle B = \{x\} \times F_L(x)$. But since $\langle u,a\rangle\langle x,g\rangle \in \{x\} \times F_L(x)$ if and only if $u \in \ell(x)$, we obtain $A \subseteq \ell(x) \times F_L$. Since $(\ell(x) \times F_L)\langle x,g\rangle \subseteq \{x\} \times F_L(x)$ always holds, for all $g \in F_L(x)$ we obtain $(\ell(x) \times F_L)\langle x,g\rangle = \{x\} \times F_L(x)$. Similarly, $\langle x,g\rangle(r(x) \times F_L) = \{x\} \times F_L(x)$ and $\mathcal{H}$II' follows.

We have now proved the first half of the FMT.

3.16. Lemma. *Let $(\overline{S},p) \in H(\underline{\mathrm{Mon}}_S)$, and let $x \to \overline{x}$ be a lifting. If $\Sigma = \Sigma_{C_p}$, and F_L is defined by $F_L(x) = \Sigma(\overline{x})$ and $F_L(u,x,1) = \Sigma(\overline{u},\overline{x},1)$, then $F_L : L(S) \to \underline{\underline{\mathrm{Gr}}}$ is a well defined functor. Similarly, $\tilde{F}_R : R(S) \to \underline{\underline{\mathrm{Gr}}}$ is a well defined functor, where $\tilde{F}_R(x) = \Sigma'(\overline{x})$ and $\tilde{F}_R(1,x,v) = \Sigma'(1,\overline{x},\overline{v})$.*

Proof. Since $\overline{ux}\, C_p\, \overline{u}\,\overline{x}$, $\Sigma(\overline{ux}) = \Sigma(\overline{u}\,\overline{x})$, and $F_L(u,x,1)$ is a well defined morphism. If $\langle u_1,ux,1\rangle : ux \to u_1ux$, then $F_L(u_1,ux,1) \circ F_L(u,x,1) = \Sigma(\overline{u_1},\overline{ux},1) \circ \Sigma(\overline{u},\overline{x},1)$. If $a \in \overline{S}$ is such that $a\overline{x} = \overline{ux}$, then by 2.14, $\Sigma(\overline{u},\overline{x},1) = \Sigma(a,\overline{x},1)$, so that $F_L(u,ux,1) \circ F_L(u,x,1) = \Sigma(\overline{u_1}a, \overline{x},1)$. Now $F_L(u_1u,x,1) = \Sigma(\overline{u_1u},\overline{x},1)$. But $\overline{u_1}a\overline{x}\, C\, \overline{u_1}\,\overline{ux}\, C\, \overline{u_1}\,\overline{u}\,\overline{x}$, so that $\Sigma(\overline{u_1u},\overline{x},1) = \Sigma(\overline{u_1}a,\overline{x},1)$ by 2.14. Hence $F_L : \mathbb{L}(S) \to \underline{\underline{\mathrm{Gr}}}$. Finally, if $\langle u,x,1\rangle \sim \langle u',x,1\rangle$, then $\overline{u}\,\overline{x}\, C_p\, \overline{u'}\,\overline{x}$ so that by 2.14 again, $\Sigma(\overline{u},\overline{x},1) = \Sigma(\overline{u'},\overline{x},1)$. Hence $F_L : L(S) \to \underline{\underline{\mathrm{Gr}}}$ is a well defined functor.

3.17. Corollary. *$F_R : R(S) \to \underline{\underline{\mathrm{Gr}}}$ is a well defined functor where $F_R(x) = \Sigma(\overline{x})$ and $F_R(1,x,v) = (\ )^R_{\overline{xv}} \circ \Sigma'(1,\overline{x},\overline{v}) \circ (\ )^L_{\overline{x}}$. Hence $(F_L,F_R) : (L,R)(S) \to \underline{\underline{\mathrm{Gr}}}$ is a compatible pair of functors.*

3.18. Lemma. *Under the conditions of 3.16, let α be the factor system for (F_L,F_R) defined implicitly by $\overline{x}\,\overline{y} = \overline{xy}\cdot\alpha(x,y)$. Then:*

i. *For all* $x,\ v \in S$, $F_R(1,x,v) = (\ )^{-\alpha(x,v)} \circ \Sigma(1,\overline{x},\overline{v})$.

ii. *For all* $\langle x,g\rangle, \langle y,h\rangle \in S \times F_L$,

$$(\overline{x}\cdot g)(\overline{y}\cdot h) = \overline{xy}\cdot(F_R(1,x,y)[g]\alpha(x,y)F_L(x,y,1)[h]).$$

Proof. $(\ )^R_{\overline{x}\,\overline{v}} = (\ )^R_{\overline{xv}\cdot\alpha(x,v)} = (\ )^{\alpha(x,v)} \circ (\ )^R_{\overline{xv}}$, so that $(\ )^R_{\overline{xv}} = (\ )^{-\alpha(x,v)} \circ (\ )^R_{\overline{x}\,\overline{v}}$. Hence $F_R(1,x,v) = (\ )^R_{\overline{xv}} \circ \Sigma'(1,\overline{x},\overline{v}) \circ (\ )^L_{\overline{x}} = (\ )^{-\alpha(x,v)} \circ (\ )^R_{\overline{x}\,\overline{v}} \circ \Sigma'(1,\overline{x},\overline{v}) \circ (\ )^L_{\overline{x}} = (\ )^{-\alpha(x,v)} \circ \Sigma(1,\overline{x},\overline{v})$. If $\langle x,g\rangle$, $\langle y,h\rangle \in S \times F_L$, then setting $\tilde{g} = (g)^L_{\overline{x}}$,

$$(\overline{x}\cdot g)(\overline{y}\cdot h) = (\tilde{g}\cdot\overline{x})(\overline{y}\cdot h)$$
$$= \Sigma'(1,\overline{x},\overline{y})[\tilde{g}]\cdot\overline{x}\overline{y}\cdot\Sigma(\overline{x},\overline{y},1)\ ,\ \text{by 2.18},$$
$$= \Sigma'(1,\overline{x},\overline{y})[\tilde{g}]\cdot\overline{xy}\cdot\alpha(x,y)\Sigma(\overline{x},\overline{y},1)$$
$$= \overline{xy}\cdot(F_R(1,x,y)[g]\alpha(x,y)F_L(x,y,1)[h])\quad \text{by 3.15.}$$

3.19. <u>Corollary</u>. <u>Under the conditions of</u> (3.16), <u>then defining</u> (F_L,F_R,α) <u>as in</u> 3.16, 17, <u>and</u> 18, $(S\times(F_L,F_R,\alpha),\pi)\cong(\overline{S},p)$ <u>under the morphism</u>, $\langle x,g\rangle\to\overline{x}\cdot g$.

<u>Proof</u>. Since $\overline{S} = \dot{\cup}\overline{x}\cdot\Sigma(\overline{x}) = \dot{\cup}\overline{x}\cdot F_L(x)$ (where $\dot{\cup}$ denotes disjoint union), the above function is a bijection. By the previous lemma, this function is also an isomorphism.

3.20. <u>Proof of 3.9</u>. The first part of this theorem is an immediate consequence of 3.11, 3.12, and 3.15. For the second part of the theorem, as demonstrated in 3.16 to 3.19, $(\overline{S},P)\cong(S\times(F_L,F_R,\alpha),\pi)$ where (F_L,F_R,α) are defined as in $\mathcal{H}$IV, $\mathcal{H}$V, and $\mathcal{H}$VI. That (F_L,F_R,α) satisfies $\mathcal{H}$I and $\mathcal{H}$II is a consequence of 3.12 and 3.15. We are done.

We now turn our attention to the morphisms of $\mathcal{H}(\underline{\underline{\mathrm{Mon}}}_S)$. We start by studying a special type of isomorphism, an equivalence morphism. We first present some notation.

3.21. Let $(F_L,F_R) : (\mathcal{L},\mathcal{R})(S)\to\underline{\mathrm{Gr}}$. $C^2(F_L)$ denotes the set of all factor systems of F_L. $Z^2(F_L,F_R)$ denotes the set of all associative factor systems for (F_L,F_R), i.e., $\alpha\in Z^2(F_L,F_R)$ if and only if $S\times(F_L,F_R,\alpha)$ is a semigroup (and hence a monoid). $C^1(F_L)$ denotes the set of all functions, $\phi : S\to\bigcup_{x\in S}F_L(x)$ such that for all $x\in S$, $\phi(x)\in F_L(x)$. If (F_L,F_R,α) is a triple and $\phi\in C^1(F_L)$, then $(F_L^\phi,F_R^\phi,\alpha^\phi) = (F_L,F_R,\alpha)^\phi$ is the triple defined by:

i. $F_L^\phi(u,x,1) = F_L(u,x,1)$.

ii. $F_R^\phi(1,x,v) = (\ )^{\phi(xv)} \circ F_R(1,x,v) \circ (\ )^{-\phi(x)}$.

iii. $\alpha^\phi(x,y) = \phi(xy)^{-1}\phi(x)^{(1,x,y)}\alpha(x,y)\phi(y)^{(x,y,1)}$.

3.22. Lemma. If $(F_L,F_R) : (L,R)(S) \to \underline{\underline{Gr}}$ and $\phi \in C^1(F_L)$, then $(F_L^\phi,F_R^\phi) : (L,R)(S) \to \underline{\underline{Gr}}$. If $\alpha \in C^2(F_L)$, then $S \times (F_L,F_R,\alpha) \cong S \times (F_L^\phi,F_R^\phi,\alpha^\phi)$ under the function $\langle x,g\rangle \to \langle x,\phi(x)^{-1}g\rangle$.

Proof. Clearly $F_L^\phi : L(S) \to \underline{\underline{Gr}}$. It is also clear F_R must be a functor from $\mathbb{R}(S)$ to $\underline{\underline{Gr}}$. If $\langle 1,x,v\rangle \sim \langle 1,x,v'\rangle$, then $\phi(xv) = \phi(xv')$, $\phi(x) = \phi(x)$, and $F_R(1,x,v) = F_R(1,x,v')$ so that $F_R^\phi(1,x,v) = F_R^\phi(1,x,v')$. Hence $F_R : R(S) \to \underline{\underline{Gr}}$. Clearly $\alpha^\phi \in C^2(F_L)$, and $\langle x,g\rangle \to \langle x,\phi(x)^{-1}g\rangle$ is a bijection from $S \times F_L$ to $S \times F_L^\phi$. So let $\langle x,g\rangle, \langle y,h\rangle \in S \times F_L$. Then if $\circ$ and $*$ denote the compositions of $S \times (F_L,F_R,\alpha)$ and $S \times (F_L^\phi,F_R^\phi,\alpha^\phi)$ respectively,

$$\langle x,\phi(x)^{-1}g\rangle * \langle y,\phi(y)^{-1}h\rangle = \langle xy,(\phi(x)^{-1}g)^{F_R^\phi(1,x,y)}\alpha^\phi(x,y)(\phi(y)^{-1}h)^{F_L^\phi(x,y,1)}\rangle$$

$$= \langle xy,(\phi(x)^{-1}g)^{F_R^\phi(1,x,y)}\phi(xy)^{-1}\phi(x)^{(1,x,y)}\alpha(x,y)\phi(y)^{(x,y,1)}(\phi(y)^{-1}h)^{(x,y,1)}\rangle$$

$$= \langle xy,(\phi(x)^{-1}g)^{F_R^\phi(1,x,y)}\phi(xy)^{-1}\phi(x)^{(1,x,y)}\alpha(x,y)h^{(x,y,1)}\rangle$$

$$= \langle xy,(\ )^{\phi(xy)} \circ F_R(1,x,y) \circ (\phi(x)^{-1}g)^{-\phi(x)}\phi(xy)^{-1}\phi(x)^{(1,x,y)}\phi(x,y)h^{(x,y,1)}\rangle$$

$$= \langle xy,\phi(xy)^{-1}\, g^{(1,x,y)}\alpha(x,y)h^{(x,y,1)}\rangle$$

$$= \text{Image } (\langle xy,g^{(1,x,y)}\alpha(x,y)h^{(x,y,1)}\rangle$$

$$= \text{Image } (\langle x\,,\,g\rangle \circ \langle y,h\rangle) .$$

Hence $\langle x,g\rangle \to \langle x,\phi(x)^{-1}g\rangle$ is an isomorphism of $S \times (F_L,F_R,\alpha)$ with $S \times (F_L^\phi,F_R^\phi,\alpha^\phi)$.

3.23. The isomorphism, $\phi^* = (\langle x,g\rangle \to \langle x,\phi(x)^{-1}g\rangle)$, will be called an equivalence morphism.

3.24. Corollary. $S \times (F_L,F_R,\alpha)$ *is a monoid if and only if* $S \times (F_L,F_R,\alpha)^\phi$ *is a monoid. If* $S \times (F_L,F_R,\alpha)$ *is a monoid, then the map* $\phi^* = (\langle x,g\rangle \to \langle x,\phi(x)^{-1}g\rangle)$ *is an isomorphism in* $\underline{\underline{\mathrm{Mon}}}_S$ *whose inverse is* $(\phi^*)^{-1} = (-\phi)^* = (\langle x,g\rangle \to \langle x,\phi(x)g\rangle)$. *Finally* $(S \times (F_L,F_R,\alpha),\pi) \in H(\underline{\underline{\mathrm{Mon}}}_S)$ *if and only if* $(S \times (F_L,F_R,\alpha)^\phi,\pi) \in H(\underline{\underline{\mathrm{Mon}}}_S)$.

3.25. Definition. We say that two H-coextensions, $(S \times (F_L,F_R,\alpha),\pi)$ and $(S \times (G_L,G_R,\beta),\pi)$, are equivalent if $F_L = G_L$ and there exists $\phi \in C^1(F_L)$ such that $(G_L,G_R,\beta) = (F_L,F_R,\alpha)^\phi$. This relation is indeed an equivalence relation. Reflexivity is clear. If $(G_L,G_R,\beta) = (F_L,F_R,\alpha)^\phi$, then $(F_L,F_R,\alpha) = (G_L,G_R,\beta)^{-\phi}$, where $-\phi \in C^1(F_L)$ is defined by $(-\phi)(x) = (\phi(x))^{-1}$ for all $x \in S$, and symmetry follows. If $(G_L,G_R,\beta) = (F_L,F_R,\alpha)^\phi$ and $(K_L,K_R,\gamma) = (G_L,G_R,\beta)^\psi$, then $(K_L,K_R,\gamma) = (F_L,F_R,\alpha)^{\phi*\psi}$ where $(\phi*\psi)(x) = \phi(x)\psi(x)$, and transitivity follows. The verification of these equalities involves a sequence of straightforward computations which we shall not give here. However, the fact that the above equivalence is really an equivalence will also follow from the following sequence of results, which show that equivalent coextensions essentially differ only by different choices of liftings, and in particular that a representation of $(\overline{S},p)$ by $(S \times (F_L,F_R,\alpha),\pi)$ is unique to within equivalence.

3.26. Lemma. *Let* $(\overline{S},P) \in H(\underline{\underline{\mathrm{Mon}}}_S)$, *and let* $x \to \overline{x}$ *and* $x \to \overline{\overline{x}}$ *be a pair of liftings. If* (F_L,F_R,α) *and* (G_L,G_R,β) *are the triples determined by* $x \to \overline{x}$ *and* $x \to \overline{\overline{x}}$ *respectively, then* $(G_L,G_R,\beta) = (F_L,F_R,\alpha)^\phi$ *where* $\phi \in C^1(F_L)$ *is defined implicitly by* $\overline{\overline{x}} = \overline{x} \cdot \phi(x)$. *Hence every triple of the form,* $(F_L,F_R,\alpha)^\phi$, *is obtained by computing the triple corresponding to the lifting* $x \to \overline{x} \cdot \phi(x)$.

Proof. Let $\phi \in C^1(F_L)$ be defined as above: $\bar{\bar{x}} = \bar{x} \cdot \phi(x)$. For all $x, y \in S$ we must have $\bar{\bar{x}}\bar{\bar{y}} = \overline{\overline{xy}} \cdot \beta(x,y) = \overline{xy} \cdot \phi(xy)\beta(x,y)$. We also have $\bar{\bar{x}}\bar{\bar{y}} = (\bar{x} \cdot \phi(x))(\bar{y} \cdot \phi(y)) = \overline{xy} \cdot (\phi(x)^{(1,x,y)}\alpha(x,y)\phi(y)^{(x,y,1)})$. Hence, $\phi(xy)\beta(x,y) = \phi(x)^{(1,x,y)}\alpha(x,y)\phi(y)^{(x,y,1)}$, or $\beta(x,y) = \alpha^{\phi}(x,y)$. That $F_L = G_L$ follows from the fact that $\Sigma(\bar{u},\bar{x},1) = \Sigma(\bar{\bar{u}},\bar{\bar{x}},1)$, which is a consequence of 2.14. Using 2.14 again, we have $\Sigma'(1,\bar{x},v) = \Sigma'(1,\bar{\bar{x}},\bar{\bar{v}})$ so that

$$G_R(1,x,v) = (\ )^R_{\overline{\overline{xv}}} \circ \Sigma'(1,\bar{\bar{x}},\bar{\bar{v}}) \circ (\ )^L_{\bar{\bar{x}}}$$

$$= (\ )^R_{\overline{xv}\cdot\phi(xv)} \circ \Sigma'(1,\bar{x},\bar{v}) \circ (\ )^L_{\bar{x}\cdot\phi(x)}$$

$$= (\ )^{\phi(xv)} \circ (\ )^R_{\overline{xv}} \circ \Sigma'(1,\bar{x},\bar{v}) \circ (\ )^L_{\bar{x}} \circ (\ )^{-\phi(x)}$$

$$= (\ )^{\phi(xv)} \circ F_R(1,x,v) \circ (\ )^{-\phi(x)} = F_R^{\phi}(1,x,v) .$$

3.27. Lemma. Let $(S \times (F_L,F_R,\alpha),\pi) \in H(\underline{\underline{\mathrm{Mon}}}_S)$. Then for all $\langle x,g\rangle \in S \times F_L$, $C_\pi(x,g) = \{x\} \times F_L(x)$ and $R(C_\pi(x,g)) = \mathcal{R}_S(x) \times F_L$. Setting $\Sigma_\pi = \Sigma_{C_\pi}$, then defining $R_x\colon \mathcal{R}_S(x) \times F_L \to F_L(x)$ by $R_x(v,a) = \alpha(x,v)a^{(x,v,1)}$, R_x is a surmorphism which induces an isomorphism, $R_x^* : \Sigma_\pi(x,g) \cong F_L(x)$, defined by $R_x^*([\langle v,a\rangle]) = R_x(v,a)$ where $[\langle v,a\rangle]$ is the member of $H_R(x,g)$ to which $\langle v,a\rangle$ belongs. Finally, under this identification of $\Sigma_\pi(x,g)$ with $F_L(x)$, for all $g, h \in F_L(x)$ we have $\langle x,g\rangle \cdot h = \langle x,gh\rangle$ and

$$\Sigma(\langle u,a\rangle,\langle x,g\rangle,\langle v,b\rangle) = (\ )^{(\alpha(x,v)^{(u,xv,1)}b^{(ux,v,1)})} \circ F_{LR}(u,x,v)$$

$$= (\ )^{(\alpha(ux,v)b^{(ux,v,1)})} \circ F_{RL}(u,x,v) .$$

Proof. The first assertion has already been proven. Since $\langle x,g\rangle\cdot\langle v,a\rangle = \langle x,gR_x(u,a)\rangle$, R_x must be a surmorphism which induces the isomorphism R_x^* of $\Sigma_\pi(x,g)$ with $F_L(x)$. This equality also gives us $\langle x,g\rangle \cdot h = \langle x,gh\rangle$ upon identifying $\Sigma_\pi(x,g)$ with $F_L(x)$. The final two equalities come from multiplying out $\langle u,a\rangle(\langle x,g\rangle \cdot h)\langle v,b\rangle$ two different ways and using associativity.

3.28. Lemma. *Let us identify* $\Sigma_\pi(x,g)$ *with* $F_L(x)$ *as above. If* (G_L,G_R,β) *is the triple induced by the lifting* $x \to \langle x,1\rangle$ *from* S *to* $(S \times (F_L,F_R,\alpha),\pi)$, *then* $(G_L,G_R,\beta) = (F_L,F_R,\alpha)$.

Proof. Clearly $F_L(x) = \Sigma_\pi(x,1) = G_L(x)$. Moreover, $\langle x,1\rangle\langle y,1\rangle = \langle xy,\alpha(x,y)\rangle = \langle xy,1\rangle \cdot \alpha(x,y)$ gives us $\alpha = \beta$. By the previous lemma again,

$$\begin{aligned} G_L(u,x,1) &= \Sigma_\pi(\langle u,1\rangle,\langle x,1\rangle,\langle 1,\alpha(1,1)^{-1}\rangle \\ &= (\ )^{(\alpha(ux,1)(\alpha(1,1)^{-1})^{(ux,1,1)})} \circ F_L(u,x,1) \\ &= (\ )^{(\alpha(1,1)^{(ux,1,1)}(\alpha(1,1)^{-1})^{(ux,1,1)})} \circ F_L(u,x,1) \text{ by 3.13} \\ &= F_L(u,x,1)\ , \\ G_R(1,x,v) &= (\ )^{-\alpha(x,v)} \circ \Sigma_\pi(\langle 1,\alpha(1,1)^{-1}\rangle,\langle x,1\rangle,\langle v,1\rangle) \\ &= (\ )^{-\alpha(x,v)} \circ (\ )^{\alpha(x,v)} \circ F_R(1,x,v) \\ &= F_R(1,x,v)\ . \end{aligned}$$

3.29. Theorem. *Let* $(S \times (F_L,F_R,\alpha),\pi) \in \mathcal{H}(\underline{\underline{\mathrm{Mon}}}_S)$, *and let* $\phi \in C^1(F)$. *If* $\Sigma_\pi(x,g)$ *is identified with* $F_L(x)$, *as in* (3.27), *then* $(F_L,F_R,\alpha)^\phi$ *can be found by computing the functors and factor systems corresponding to the lifting* $x \to \langle x,\phi(x)\rangle$.

Proof. This is a consequence of 3.26 and 3.28 above.

Equivalence morphisms between equivalent objects take us halfway towards representation of the morphisms of $\mathcal{H}(\underline{\underline{\mathrm{Mon}}}_S)$.

3.30. Definition. Let $(F_L,F_R),(G_L,G_R) : (\mathcal{L},\mathcal{R})(S) \to \underline{\underline{\mathrm{Gr}}}$. A *natural transformation* $\sigma : (F_L,F_R) \to (G_L,G_R)$ is a collection of homorphisms $\sigma(x):F_L(x) \to G_L(x)$

such that for all (u,x,v) the following diagram commutes:

$$\begin{array}{ccccc} F_L(xv) & \xleftarrow{F_R(1,x,v)} & F_L(x) & \xrightarrow{F_L(u,x,1)} & F_L(ux) \\ \downarrow{\scriptstyle \sigma(xv)} & & \downarrow{\scriptstyle \sigma(x)} & & \downarrow{\scriptstyle \sigma(ux)} \\ G_L(xv) & \xleftarrow{G_R(1,x,v)} & G_L(x) & \xrightarrow{G_L(u,x,1)} & G_L(ux) \end{array}$$

If $\alpha \in C^2(F_L)$, $\beta \in C^2(G_L)$, and for all $x, y \in S$, $\beta(x,y) = \sigma(xy)[\alpha(x,y)]$, we denote this by $\beta = C^2(\sigma)[\alpha]$. We let $\text{Hom}((F_L,F_R,\alpha),(G_L,G_R,\beta))$ denote the set of all natural transformations from (F_L,F_R) to (G_L,G_R) such that $\beta = C^2(\sigma)[\alpha]$.

3.31. <u>Lemma</u>. <u>Let</u> $S \times (F_L,F_R,\alpha)$ <u>and</u> $S \times (G_L,G_R,\beta)$ <u>be objects of</u> $\mathcal{H}(\underline{\text{Mon}}_S)$. <u>If</u> $\sigma : (F_L,F_R) \to (G_L,G_R)$, <u>then define</u> $\sigma^* : S \times F_L \to S \times G_L$ <u>by</u> $\sigma^*(x,g) = \langle x,\sigma(x)[g]\rangle$. <u>Then</u> σ^* <u>is a homomorphism (and hence a morphism of</u> $\mathcal{H}(\underline{\text{Mon}}_S)$) <u>if and only if</u> $\beta = C^2(\sigma)[\alpha]$. <u>Moreover</u>, $\sigma^*(x,1) = \langle x,1\rangle$. <u>Conversely, if</u> $f : S \times (F_L,F_R,\alpha) \to S \times (G_L,G_R,\beta)$ <u>is an</u> $\mathcal{H}$<u>-morphism such that</u> $f(x,1) = \langle x,1\rangle$ <u>for all</u> $x \in S$, <u>then there exists</u> $\sigma : (F_L,F_R) \to (G_L,G_R)$ <u>such that</u> $f = \sigma^*$. σ <u>is necessarily unique</u>.

<u>Proof</u>. The proof of the first part is a straightforward comparison of $\sigma^*(\langle x,g\rangle \cdot \langle y,h\rangle)$ with $\sigma^*(x,g) \cdot \sigma^*(y,h)$. For the converse, let $f : S \times (F_L,F_R,\alpha) \to S \times (G_L,G_B,\beta)$ be an $\mathcal{H}$-morphism for which $f(x,1) = \langle x,1\rangle$. For each $x \in S$ define $\sigma(x) : F_L(x) \to G_L(x)$ by $f(x,g) = \langle x,\sigma(x)[g]\rangle$. Looking at $f(\langle x,g\rangle \cdot \langle y,h\rangle)$ we get

$$(*)\quad \sigma(xy)[g^{(1,x,y)}\alpha(x,y)h^{(x,y,1)}] = (\sigma(x)[g])^{(1,x,y)}\beta(x,y)(\sigma(y)[h])^{(x,y,1)} .$$

We also have $\sigma(x)[1] = 1$. Hence setting $g = 1$ and $h = 1$ in $(*)$ we get

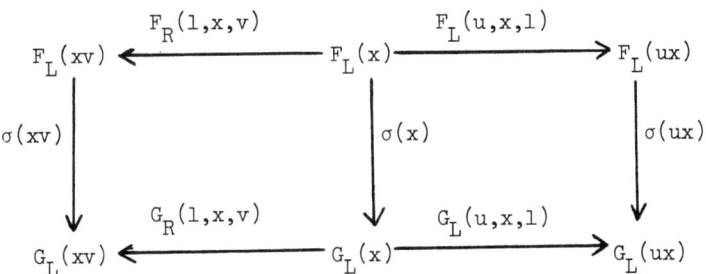

$C^2(\sigma)[\alpha] = \beta$. Let $g,h \in F_L(y)$. Now there exists $\langle x,a\rangle \in \mathcal{L}_S(y) \times F_L$ such that $\langle x,a\rangle\cdot\langle y,1\rangle = \langle y,g\rangle$, and hence $\langle x,a\rangle\cdot\langle y,h\rangle = \langle y,gh\rangle$. From (*) we get

$$\sigma(y)[gh] = \sigma(y)[a^{(1,x,y)}\alpha(x,y)h^{(x,y,1)}]$$
$$= (\sigma(x)[a])^{(1,x,y)}\beta(x,y)(\sigma(y)[h])^{(x,y,1)}$$
$$= (\sigma(x)[a])^{(1,x,y)}\beta(x,y)(\sigma(y)[h]) .$$

Since $h = 1$ gives $\sigma(y)[g] = (\sigma(x)[a])^{(1,x,y)}\beta(x,y)$, we have $\sigma(y)[gh] = \sigma(y)[g]\sigma(y)[h]$, so that $\sigma(y)$ is a homomorphism for all $x \in S$. Using this fact and $C^2(\sigma)[\alpha] = \beta$, (*) implies that $\sigma : (F_L,F_R) \to (G_L,G_R)$ is a functor by setting first $g = 1$ and letting h vary, and then letting g vary while fixing $h = 1$. That $f = \sigma^*$ is clear.

3.32. By the First Main Theorem of 3.9, every H-coextension of S, $(\overline{S},p)$, has a representation of the form, $(S \times (F_L,F_R,\alpha),\pi)$, and this representation is unique to within equivalence by 3.26. We now use these facts and the previous lemma to give a representation of the morphisms of $H(\underline{\underline{\mathrm{Mon}}}_S)$.

Let $f : (S \times (F_L,F_R,\alpha),\pi) \to (S \times (G_L,G_R,\beta),\pi)$ be a morphism. Define $\phi \in C^1(G_L)$ by $f(x,1) = \langle x,\phi(x)\rangle$. Then $\phi^* \circ f : (S \times (F_L,F_R,\alpha),\pi) \to (S \times (G_L,G_R,\beta)^\phi,\pi)$ is a morphism such that $\phi^* \circ f(x,1) = \langle x,1\rangle$. Hence by Lemma 3.31 above, there exists a natural transformation $\sigma : (F_L,F_R,\alpha) \to (G_L,G_R,\beta)^\phi$ such that $\phi^* \circ f = \sigma^*$. Hence, since ϕ^* is an isomorphism, $f = (\phi^*)^{-1} \circ \sigma^* = (-\phi)^* \circ \sigma^*$. Moreover, this representation is unique. For suppose that we have $(-\phi)^* \circ \sigma^* = (-\psi)^* \circ \tau^*$. Then we must have $\sigma^* = (-\phi)^{*-1} \circ (-\psi)^* \circ \tau^* = \phi^* \circ (-\psi)^* \circ \tau^*$. Since $\sigma^*(x,1) = \tau^*(x,1) = \langle x,1\rangle$, we must have $\phi^* \circ (-\psi)^*(x,1) = \langle x,1\rangle$. But $\phi^* \circ (-\psi)^*(x,1) = \phi^*(x,\psi(x)) = \langle x,\phi(x)^{-1}\psi(x)\rangle$. Hence for all $x \in S$, $\phi(x) = \psi(x)$ and $\phi^* = \psi^*$ follows. But this means $\sigma^* = \phi^* \circ (-\psi)^* \circ \tau^* = \tau^*$.

We have completed the proof of the Second Main Theorem.

3.33. <u>The Second Main Theorem</u>. <u>Let</u> $(S \times (F_L, F_R, \alpha), \pi), (S \times (G_F, G_R, \beta), \pi) \in H(\underline{\underline{Mon}}_S)$. <u>Then every morphism</u>, $f : (S \times (F_L, F_R, \alpha), \pi) \to (S \times (G_L, G_R, \beta), \pi)$, <u>has the unique form</u>, $f = (\phi^*)^{-1} \circ \sigma^*$, <u>where</u> $\phi \in C^1(G_L)$ <u>is defined implicitly by</u> $f(x,1) = \langle x, \phi(x) \rangle$ <u>and</u> $\sigma : (F_L, F_R, \alpha) \to (G_L, G_R, \beta)^{\phi}$. <u>Moreover, for all</u> $\langle x, g \rangle \in S \times F_L$, $f(x,g) = \langle x, \phi(x)\sigma(x)[g] \rangle$.

3.34. <u>Definition</u>. The representation in (3.33) of f as the composition, $(\phi^*)^{-1} \circ \sigma^*$, is called the standard representation of f. We will often use the notation, $f = (\phi_f^*)^{-1} \circ \sigma_f^*$.

3.35. <u>Corollary</u>. <u>Let</u> $(\phi^*)^{-1} \circ \sigma^* : (S \times (F_L, F_R, \alpha), \pi) \to (S \times (G_L, G_R, \beta), \pi)$ <u>and</u> $(\psi^*)^{-1} \circ \tau^* : (S \times (G_L, G_R, \beta), \pi) \to (S \times (K_L, K_R, \gamma), \pi)$ <u>be morphisms, given in their standard representation</u>. <u>Then defining</u> ${}^{\tau}\phi \in C^1(K_L)$ <u>by</u> ${}^{\tau}\phi(x) = \tau(x)[\phi(x)]$, <u>we obtain</u>:

$$(\psi^*)^{-1} \circ \tau^* \circ (\phi^*)^{-1} \circ \sigma^* = (\psi * {}^{\tau}\phi)^{*-1} \circ (\tau \circ \sigma)^* .$$

<u>Proof</u>. This is immediate from the fact that:

$$((\psi^*)^{-1} \circ \tau^* \circ (\phi^*)^{-1} \circ \sigma^*)\langle x,g \rangle = \langle x, \psi(x)\,{}^{\tau}\phi(x)(\tau \circ \sigma)(x)[g] \rangle .$$

3.36. <u>Corollary</u>. <u>If</u> $(S \times (F_L, F_R, \alpha), \pi) \xrightarrow{f} (S \times (G_L, G_R, \beta), \pi) \xrightarrow{g} (S \times (K_L, K_R, \gamma), \pi)$, <u>then</u> $\sigma_{g \circ f} = \sigma_g \circ \sigma_f$.

3.37. Before proceeding further, it should be remarked that everything which has been done above can be normalized in the sense that all α's can be considered to be unitary and all ϕ's can be considered to be unitary, i.e., $\phi(1) = 1$. This follows from the fact that we can always choose unitary liftings.

Section 3b. Semicentral coextensions and Connections with group theory. It should be clear by now that if S is a group, then $H(\underline{\underline{Mon}}_S) = \underline{\underline{Gr}}_S$, the category of groups over S . Here we might expect our representation of the objects of $H(\underline{\underline{Mon}}_S)$ to reduce to that given by Schreier extension theory for groups. However, the formulas we will obtain will not be those given by Schreier. Hence it would appear that our representation is not a generalization of Schreier theory. However, this is only apparent, since in this section we give an alternative description of our construction, which in the group case yields the Schreier formulas. In the latter part of this section we confront for the first time, a special class of objects, the semicentral coextensions.

3.38. As in (3.1) we say that a pair of functors, $F_L : \mathbb{L}(S) \to \underline{\underline{Gr}}$ and $F_R : \mathbb{R}(S) \to \underline{\underline{Gr}}$, is compatible if for all $x \in S$, $F_L(x) = F_R(x)$. This will be denoted, $(F_L,F_R) : (\mathbb{L},\mathbb{R})(S) \to \underline{\underline{Gr}}$. If (F_L,F_R) is a compatible pair, then as in (3.2) we can define F_{LR} and F_{RL} . The notions of (3.5), (3.6), and (3.7) also generalize. Finally, we can also talk about structures of the form, $S \times (F_L,F_R,\alpha)$, which will be monoids if and only if the condition HI of (3.9) holds. The proof is exactly that of Lemma 3.12, with the only exception being that we use the fact that $\langle g,h,1\rangle$ and $\langle 1,g,h\rangle$ are isomorphisms in $\mathbb{D}(S)$ for all $g, h \in G(S)$. Hence we can talk about the construction of Section 3a in a slightly more generalized form, and it is presently convenient to do so. We shall also assume that all factor systems under consideration are unitary, which can be done without any essential loss of generality.

3.39. Definition. A function, $F : \mathbb{D}(S) \to \underline{\underline{Gr}}$, is a left semi-functor if for all $x \in S$, $F(x)$ is a group, for all $\langle u,x,v\rangle \in \mathbb{D}(S)$, $F(u,x,v) : F(x) \to F(uxv)$ is a group morphism, and the following conditions hold:

(LSF - I). For all $u, x, v \in S$,

$$F(u,x,v) = F(u,xv,1) \circ F(1,x,v) = F(1,ux,v) \circ F(u,x,1) .$$

(LSF - II) $F_L = F|\mathbb{L}(S)$ is a functor.

3.40. Clearly, a left semifunctor, F, is a functor if and only if $F_R|\mathbb{R}(S)$ is a functor.

3.41. If $F : \mathbb{D}(S) \to \underline{\underline{Gr}}$ is a left semifunctor and α is a factor system for F_L, then we can define a composition on $S \times F_L$ by:

$$\langle x,a\rangle\langle y,b\rangle = \langle xy,\alpha(x,y)a^{(1,x,y)}b^{(x,y,1)}\rangle .$$

This structure will be denoted, $S \times (F,\alpha)$.

3.42. <u>Lemma</u>. $S \times (F,\alpha)$ <u>is a monoid if and only if for all</u> $u, x, v \in S$, $g \in F(u)$ <u>we have</u>:

$$\text{(i)} \quad \alpha(ux,v)\alpha(u,x)^{(1,ux,v)}(g^{(1,u,x)})^{(1,ux,v)} = \alpha(u,xv)(g)^{(1,u,xv)}\alpha(x,v)^{(u,xv,1)} .$$

<u>This condition is equivalent with the following pair of conditions</u>:

$$\text{(ii)} \quad \alpha(ux,v)\alpha(u,x)^{(1,ux,v)} = \alpha(u,xv)\alpha(x,v)^{(u,xv,1)} .$$

$$\text{(iii)} \quad (g^{(1,u,x)})^{(1,ux,v)} = (g^{(1,u,xv)})^{F(u,xv,1)[\alpha(x,v)]} .$$

<u>When</u> $S \times (F,\alpha)$ <u>is a monoid, then</u> $\alpha(x,1) = \alpha(1,x) = 1$ <u>for all</u> $x \in S$, <u>and</u> $G(S \times (F,\alpha)) = G(S) \times F_L$.

<u>Proof</u>. Let us denote the right side of (i) by $A(u,x,v,g)$ and the left side of (i) by $B(u,x,v,g)$. If $\langle u,g\rangle, \langle x,g\rangle, \langle v,k\rangle \in S \times F_L$, then, using the properties of LSF I and LSF II, we obtain:

$$(\langle u,g\rangle\langle x,h\rangle)\langle v,k\rangle = \langle uxv,A(u,x,v,g)h^{(u,x,v)}k^{(ux,v,1)}\rangle.$$

$$(\langle u,g\rangle(\langle x,h\rangle\langle v,k\rangle) = \langle uxv,B(u,x,v,g)h^{(u,x,v)}k^{(ux,v,1)}\rangle.$$

Hence, $S \times (F,\alpha)$ is associative if and only if (i) holds. That (i) is equivalent with (ii) and (iii) is an almost immediate observation. Now $\alpha(1,1) = 1$. Hence by letting $\langle u,x,v\rangle = \langle 1,1,v\rangle$ in (ii) we obtain $\alpha(1,v) = [\alpha(1,v)]^2$, so that $\alpha(1,v) = 1$ follows. Similarly $\alpha(u,1) = 1$. Hence for all $\langle x,g\rangle \in S \times (F,\alpha)$ we obtain $\langle 1,1\rangle\langle x,g\rangle = \langle x,g\rangle\langle 1,1\rangle = \langle x,g\rangle$, and $S \times (F,\alpha)$ is a monoid. That $G(S \times (F,\alpha)) = G(S) \times F_L$ follows from the next theorem and Lemma 3.12.

3.43. Theorem. *Let* $S \times (F_L,F_R,\alpha)$ *be a monoid. Then defining* F *by* $F(x) = F_L(x)$, *and* $F(u,x,v) = F_L(u,xv,1) \circ (\ )^{\alpha(x,v)} \circ F_R(1,x,v)$, *we obtain a left semi-functor from* $\mathbb{D}(S)$ *to* $\mathbb{G}r$, *such that* $S \times (F_L,F_R,\alpha) = S \times (F,\alpha)$. *Conversely, if* $S \times (F,\alpha)$ *is a monoid, then defining* (F_L,F_R) *by* $F_L(x) = F(x) = F_R(x)$, $F_L(u,x,1) = F(u,x,1)$, *and* $F_R(1,x,v) = (\ )^{-\alpha(x,v)} \circ F(1,x,v)$, *we obtain a compatible pair of functors such that* $S \times (F,\alpha) = S \times (F_L,F_R,\alpha)$. *Finally, the function* $(F,\alpha) \to (F_L,F_R,\alpha)$ *is the inverse of the function* $(F_L,F_R,\alpha) \to (F,\alpha)$.

Proof. The verification of the last statement is a sequence of straightforward calculations. We also leave the verification of the first statement to the reader, and prove the converse.

If $S \times (F,\alpha)$ is a monoid, then $F_L = F|\mathbb{L}(S)$ is clearly a functor. Moreover, using (ii) and (iii) of the above lemma,

$$\begin{aligned}
F_R(1,xv,\bar{v}) \circ F_R(1,x,v) &= (\ )^{-\alpha(xv,\bar{v})} \circ F(1,xv,\bar{v}) \circ (\ )^{-\alpha(x,v)} \circ F(1,x,v) \\
&= (\ )^{-\{\alpha(xv,\bar{v})F(1,xv,\bar{v})[\alpha(x,v)]\}} \circ F(1,xv,\bar{v}) \circ F(1,x,v) \\
&= (\ )^{-\{\alpha(x,v\bar{v})F(x,v\bar{v},1)[\alpha(v,\bar{v})]\}} \circ F(1,xv,\bar{v}) \circ F(1,x,v) \\
&= (\ )^{-\alpha(x,v\bar{v})} \circ F(1,x,v\bar{v}) \\
&= F_R(1,x,v\bar{v}) .
\end{aligned}$$

Hence F_R is a functor. Clearly (F_L,F_R) is a compatible pair of functors. If $\langle x,g\rangle,\langle y,h\rangle \in S \times F_L$, then

$$\begin{aligned}\langle x,g\rangle\langle y,h\rangle &= \langle xy,\alpha(x,y)F(1,x,y)[g]F(x,y,1)[h]\rangle \\ &= \langle xy,\alpha(x,y)F(1,x,y)[g]\alpha(x,y)^{-1}\alpha(x,y)F(x,y,1)[h]\rangle \\ &= \langle xy,F_R(1,x,y)[g]\alpha(x,y)F_L(x,y,1)[h]\rangle .\end{aligned}$$

Hence $S \times (F,\alpha) = S \times (F_L,F_R,\alpha)$.

3.44. Let $(\overline{S},p) \in \mathcal{H}(\underline{\underline{\text{Mon}}}_S)$, and let $x \to \overline{x}$ be a unitary lifting from S to $\overline{S}$. Using this lifting, we can construct a function, $F : \mathbb{D}(S) \to \underline{\underline{\text{Gr}}}$, defined by $F(x) = \Sigma(\overline{x})$ and $F(u,x,v) = \Sigma(\overline{u},\overline{x},\overline{v})$. Clearly, by condition $\mathcal{H}$VI of the FMT , $F|\mathbb{L}(S)$ is the functor, F_L . Using Corollary 2.14 we obtain:

$$\begin{aligned}F(u,xv,1) \circ F(1,x,v) &= \Sigma(\overline{u},\overline{xv},1) \circ \Sigma(1,\overline{x},\overline{v}) \\ &= \Sigma(\overline{u},\overline{x}\,\overline{v},1) \circ \Sigma(1,\overline{x},\overline{v}) \\ &= \Sigma(\overline{u},\overline{x},\overline{v}) \\ &= F(u,x,v) ,\end{aligned}$$

and similarly $F(1,ux,v) \circ F(u,x,1) = F(u,x,v)$. Moreover, if α is defined as before, we have:

$$\begin{aligned}(\overline{x}\cdot g)(\overline{y}\cdot h) &= \overline{x}\,\overline{y}\cdot(\Sigma(1,\overline{x},\overline{y})[g]\Sigma(\overline{x},\overline{y},1)[h]) \\ &= \overline{xy}\cdot(\alpha(xy)F(1,x,y)[g]F(x,y,1)[h]) .\end{aligned}$$

Hence $(\overline{S},p) \cong (S \times (F,\alpha),\pi)$ under the map $\overline{x}\cdot g \to \langle x,g\rangle$. If we use $x \to \overline{x}$ to compute the triple, (F_L,F_R,α) , in accord with the latter part of the FMT , we see that (F_L,F_R,α) corresponds with (F,α) under the above, (3.43), correspondence.

3.45. Let S be a group, and let $x \to \overline{x}$ be a unitary lifting from S to $\overline{S}$, whose induced left semifunctor is F . Then $C(1) = F(1)$, and for all $x \in S$,

$F(x) = F(1)$. Moreover, if $F(1) \times S \to F(1)$ is the map defined by $\langle g,x\rangle \to g^x = g^{\bar{x}} = \bar{x}^{-1} g \bar{x}$, then $F(u,x,v)[g] = g^v$ and equations (ii) and (iii) of (3.42) become:

$$\alpha(ux,v)\alpha(u,x)^v = \alpha(u,xv)\alpha(x,v) \text{ , and } (g^x)^v = (g^{xv})^{\alpha(x,v)} .$$

But these are just the equations arising in the unitary, right coset version of Schreier theory. Motivated by an important concept arising in group extension theory we introduce the following concept.

3.46. Let $x \to \bar{x}$ be a lifting from S to $\bar{S}$. $x \to \bar{x}$ is said to be a <u>semi-central</u> lifting if for all $x, y \in S$, $\bar{x}\bar{y} \in \overline{xy} \cdot Z(\Sigma(\overline{xy}))$, where $Z(\Sigma(\overline{xy}))$ is the center of $\Sigma(\overline{xy})$. Likewise we say that a pair, (F,α) , or a triple, (F_L,F_R,α) , is semicentral if for all $x, y \in S$, $\alpha(x,y) \in Z(F(xy))$ or $Z(F_L(xy))$. We say that a coextension, $S \times (F_L,F_R,\alpha) = S \times (F,\alpha)$, is semi-central if α is semicentral. We now present the appropriate analogue of a basic fact of group extension theory.

3.47. <u>Theorem</u>. <u>Let</u> $(S \times (F_L,F_R,\alpha),\pi) = (S \times (F,\alpha),\pi)$ <u>be an H-coextension of</u> S <u>such that for all</u> $x, y \in S$, $\alpha(x,y) \in Z(F_L(xy))$. <u>Then</u> $F : \mathbb{D}(S) \to \underline{\underline{Gr}}$ <u>is a functor which is the unique extension of</u> (F_L,F_R) <u>to</u> $\mathbb{D}(S)$.

<u>Proof</u>. Do to the hypothesis, $F|\mathbb{R}(S) = F_R$ by Theorem 3.43. Hence F is a functor and the result follows.

3.48. The question naturally arises as to whether or not converses to the above theorem exist. The example of 2.30 can easily be used to show that coextensions of the form, $(S \times (F,\alpha),\pi)$, exist such that F is a functor, but (F,α) is not semicentral. We now look at a well-known example which shows that is is possible to have a coextension, $(S \times (F_L,F_R,\alpha),\pi)$, where (F_L,F_R) can be extended to $\mathcal{D}(S)$, but (F_L,F_R,α) need not be semicentral.

3.49. Let G be a group and S a semigroup with 0. A semigroup, $\overline{S}$, is called an ideal extension of G by S if G is an ideal of $\overline{S}$, and S is the Rees factor semigroup, $\overline{S}/G$. In this case G is the minimum ideal of $\overline{S}$ and constitutes an $\mathcal{H}$-class of $\overline{S}$. Moreover, the canonical map $p : \overline{S} \to S$ induces a sub-$\mathcal{H}$ congruence on $\overline{S}$. We define a lifting $x \to \overline{x}$ by $\overline{x} = x$ if $x \neq 0$ and $\overline{0} = 1_G$. Under this lifting we have $F_R(x) = F_L(x) = 1$, the trivial group, if $x \neq 0$, and $F_R(0) = F_L(0) = G$. Moreover, $F_{LR}(u,x,v) = F_{RL}(u,x,v) =$ trivial homomorphism if $x \neq 0$. Since $u \cdot 0 = 1 \cdot 0$ and $0 \cdot v = 0 \cdot 1$ we have $F_L(u,0,1) = i_G$ and $F_R(1,0,v) = i_G$ so that $F_{RL}(u,0,v) = F_{LR}(u,0,v) = i_G$. Hence our lifting induces a functor, $F : \mathcal{D}(S) \to \underline{\underline{Gr}}$, for which $F(x) = 1$ if $x \neq 0$, $F(0) = G$; $F(u,x,v)$ is trivial if $x \neq 0$, and $F(u,0,v) = i_G$. Moreover, if α is the factor system induced by $x \to \overline{x}$, we have $\alpha(0,0) = 1_G$. If $x \neq 0$, condition $\mathcal{H}$Ia of (3.10) for $(0,x,0)$ yields $\alpha(x,0) = \alpha(0,x)$. If $x \neq 0, y \neq 0, xy = 0$, condition $\mathcal{H}$Ia for $(0,x,y)$ yields $\alpha(x,y) = \alpha(0,x)\alpha(0,y)$. Finally if $xy \neq 0$, then $\alpha(x,y) = 0$ and condition $\mathcal{H}$Ia for $(0,x,y)$ yields $\alpha(0,xy) = \alpha(0,x)\alpha(0,y)$. Hence every associative factor system, α, for F satisfying $\alpha(0,0) = 1_G$ is obtained by finding an element $f \in \mathrm{Hom}(S-0,G)$ the set of partial homomorphisms, and setting $\alpha(0,x) = \alpha(x,0) = f(x)$, $\alpha(x,y) = 1$ if $xy \neq 0$, and $\alpha(x,y) = f(x)f(y)$ if $x \neq 0$, $y \neq 0$, but $xy = 0$. Conversely every partial homomorphism gives rise to an associative factor system, α, with $\alpha(0,0) = 1$.

If $S = \mathcal{F}_X^0$, the free monoid on X with 0 adjoined, where $||X|| \geq 1$, and G is any nontrivial group with center equal to $\{1_G\}$, then $\mathrm{Hom}(\mathcal{F}_X, G)$ has nontrivial morphisms. Hence there must exist ideal extensions of G by S which do not split, and therefore are not semicentral coextensions of S.

We now look at several partial converses of (3.47).

3.50. <u>Theorem</u>. <u>Let</u> S <u>be regular, and let</u> $x \to \bar{x}$ <u>be a unitary lifting from</u> S <u>to</u> $(S,p) \in H(\underline{\underline{Mon}}_S)$. <u>If the induced pair,</u> (F,α) , <u>is such that</u> F <u>is a functor, then</u> (F,α) <u>is a semicentral pair</u>.

<u>Proof</u>. Since F is a functor, (3.42 iii) can be written as $\alpha(x,v)^{(u,xv,1)}g^{(1,u,xv)} = g^{(1,u,xv)}\alpha(x,v)^{(u,xv,1)}$. Now if $e \in E(S) \cap R(xv)$, we obtain $\alpha(x,v)g^{(1,e,xv)} = g^{(1,e,xv)}\alpha(x,v)$ since $exv = xv$. Now $\bar{e}\, R\, \overline{exv}\, R\, \bar{e}\,\overline{xv}$. Hence $F(1,e,xv) = \Sigma(1,\bar{e},\overline{xv})$ is an isomorphism, since $\Sigma \cong \Sigma'$ and $\Sigma'(1,\bar{e},\overline{xv}) = 1_{\Sigma'(\bar{e})}$. Hence every $h \in F(xv)$ is of the form $g^{(1,e,xv)}$ and $\alpha(x,v) \in Z(F(xv))$ follows.

3.51. $\underline{\underline{Gr}}^o$ denotes the subcategory of $\underline{\underline{Gr}}$ with the same set of objects, but whose morphisms are the epimorphisms of $\underline{\underline{Gr}}$. If $F : K \to \underline{\underline{Gr}}^o$ is a functor, we call F a <u>surjective functor</u>.

3.52. <u>Lemma</u>. <u>If</u> $F : K \to \underline{\underline{Gr}}^o$ <u>where</u> $K = \mathbb{D}(S)$, $\mathbb{R}(S)$, <u>or</u> $\mathbb{L}(S)$, <u>then</u> F <u>factors through</u> $\mathcal{D}(S)$, $R(S)$, <u>or</u> $L(S)$ <u>respectively</u>.

<u>Proof</u>. Let $K = \mathbb{R}(S)$. Then $Hom_{\mathbb{R}(S)}(1,y) = \{<1,1,y>\}$ for all $y \in S$. If $<1,x,v> \sim <1,x,v'>$, we obtain $F(1,x,v) \circ F(1,1,x) = F(1,x,v') \circ F(1,1,x)$, so that $F(1,x,v) = F(1,x,v')$ follows from the surjectivity of $F(1,1,x)$. Hence $F|\mathbb{R}(S)$ factors through $R(S)$ and dually, $F|\mathbb{L}(S)$ factors through $L(S)$.

3.53. <u>Theorem</u>. <u>If</u> $(S \times (F_L,F_R,\alpha),\pi) = (S \times (F,\alpha),\pi) \in \underline{\underline{Mon}}_S$, <u>and if</u> $(F_L,F_R) : (\mathbb{L}, \mathbb{R})(S) \to \underline{\underline{Gr}}^o$ (<u>i.e.</u>, <u>by</u> (3.43), $F : \mathbb{D}(S) \to \underline{\underline{Gr}}^o$) <u>then</u> $(S \times (F_L,F_R,\alpha),\pi) = (S \times (F,\alpha),\pi) \in H(\underline{\underline{Mon}}_S)$. <u>If furthermore</u> (F_L,F_R) <u>extends to all</u> $\mathbb{D}(S)$, <u>or if</u> F <u>is a functor, then</u> α <u>is a semicentral factor system</u>.

Proof. If (F_L, F_R) is a compatible pair of surjective functors, then for all $x \in S$, $F_L(x)^{(1,1,x)} = F_L(x) = F_L(x)^{(x,1,1)}$, so that condition $\mathcal{H}$II is satisfied, and the first statement of the theorem is proven. If furthermore (F_L, F_R) extends to a functor, $G : \mathbb{D}(S) \to \underline{\underline{Gr}}^{\circ}$, then condition $\mathcal{H}$I for $x = 1$ becomes:

$$\alpha(u,v)G(u,1,v)[g] = G(u,1,v)[g]\alpha(u,v) .$$

But since $G(u,1,v)$ takes $G(1)$ onto $G(uv)$ we obtain $\alpha(u,v) \in Z(G(uv))$. Suppose now that F is a functor. Then (3.42 iii) becomes $g^{(1,u,xv)} = (g^{(1,u,xv)})^{F(u,xv,1)[\alpha(x,v)]}$. Setting $u = 1$, this equation is equivalent to $\alpha(x,v)g^{(1,1,xv)} = g^{(1,1,xv)}\alpha(x,v)$. But since $F(1,1,xv)$ takes $F(1)$ onto $F(xv)$ we obtain $\alpha(x,v) \in Z(F(xv))$.

3.54. The case where (F_L, F_R) or F is $\underline{\underline{Gr}}^{\circ}$-valued is of sufficient interest to deserve a small discussion. So suppose that $(F_L, F_R) : (L,R)(S) \to \underline{\underline{Gr}}^{\circ}$ and that $(S \times (F_L, F_R, \alpha), \pi) \in \mathcal{H}(\underline{\underline{Mon}}_S)$. Then if we denote $F(1)$ by K, there exists an embedding $i : K \to S \times (F_L, F_R, \alpha)$ defined by $i(k) = \langle 1,k \rangle$. Moreover, for all $\langle x,g \rangle \in S \times F_L$ we have $K^i\langle x,g \rangle = \langle x,g \rangle K^i = \{x\} \times F_L(x)$ since $F_L(x,1,1)$ and $F_R(1,1,x)$ are epimorphisms. Hence K^i is a normal subgroup of $S \times (F_L, F_R, \alpha)$, and C_π is the coset congruence of K^i.

Chapter Four

Split $\mathcal{H}$-Coextensions of a Monoid

An $\mathcal{H}$-coextension of S, $(\overline{S},p)$, is said to split if there exists a homomorphism $f : S \to \overline{S}$ such that $p \circ f = i_S$. f is called a lifting homomorphism. $\mathrm{Hom}(S,(\overline{S},p))$ denotes the set of all lifting homomorphisms of $(\overline{S},p)$. Each such homomorphism maps 1 to 1. We begin with the splitting analogues of previous theorems and lemmas.

4.1. Let (F_L,F_R,α) be the triple determined by the lifting homomorphism, f. Since $f(xy) = f(x)f(y)$, $\alpha(x,y) = 1$ for all $x, y \in S$. By 3.47 we have $F_{RL} = F_{LR}$, i.e., (F_L,F_R) has a unique extension to all of $\mathcal{D}(S)$. In fact $(F_L,F_R) = (F|\mathcal{L}(S), F|\mathcal{R}(S))$, where $F = \Sigma_p \circ \mathcal{D}(f)$ and Σ_p is the right structure functor of $\mathcal{C}_p$. This leads to the next theorem.

4.2. Theorem. Every split $\mathcal{H}$-coextension of S, $(\overline{S},p)$, and every lifting homomorphism of $(\overline{S},p)$, f, induces a functor $F = \Sigma_p \circ \mathcal{D}(f) : \mathcal{D}(S) \to \underline{Gr}$ such that $(S \times F,\pi) \cong (\overline{S},p)$ under the map $\langle x,g\rangle \to f(x)\cdot g$. For all $x \in S$,

$$\bigcup_{u\in\ell(x)} F(u)^{(1,u,x)} = F(x) = \bigcup_{v\in r(x)} F(v)^{(x,v,1)} .$$

Conversely, if $F : \mathcal{D}(S) \to \underline{Gr}$ satisfies this equality, $(S \times F,\pi)$ is a split $\mathcal{H}$-coextension of S, and the map $x \to \langle x,1\rangle$ is in $\mathrm{Hom}(S,(S \times F,\pi))$.

4.3. Let $f, f_1 \in \mathrm{Hom}(S,(\overline{S},p))$. There exists $\psi \in C^1(F)$ such that for all $x \in S$, $f_1(x) = f(x)\cdot\psi(x)$. From $f_1(xy) = f(xy)\cdot\psi(xy)$ and $f_1(xy) = f_1(x)f_1(y) = (f(x)\cdot\psi(x))(f(y)\cdot\psi(y)) = f(x)f(y)\cdot F(1,x,y)[\psi(x)]F(x,y,1)[\psi(y)]$ we get $\psi \in Z^1(F) = \{\psi \in C^1(F) : \psi(xy) = \psi(x)^{(1,x,y)}\psi(y)^{(x,y,1)}\}$. Moreover, setting $F_1 = \Sigma_p \circ \mathcal{D}(f_1)$, 3.26 gives us $F_1(u,x,1) = F(u,x,1)$ for all $u, x \in S$, and $F_1(1,x,v) = (\ )^{\psi(xv)} \circ F(1,x,v) \circ (\ )^{-\psi(x)} = (\ )^{F(x,v,1)[\psi(v)]} \circ F(1,x,v)$ for all $x, v \in S$.

4.4. Lemma. Let $f, f_1 \in \mathrm{Hom}(S,(\overline{S},p))$, and let $F, F_1 : \mathcal{D}(S) \to \underline{\underline{Gr}}$ denote their induced functors. Then there exists $\psi \in Z^1(F)$ such that for all $x \in S$, $f_1(x) = f(x)\cdot\psi(x)$ and $F_1 = F^{\psi}$, i.e., $F_1(x) = F(x)$ for all $x \in S$, and $F_1(u,x,v) = (\ )^{F(ux,v,1)[\psi(v)]} \circ F(u,x,v)$. Also there exists a bijective correspondence between $Z^1(F)$ and $\mathrm{Hom}(S,(\overline{S},p))$ determined by $\psi \to (x \to f(x)\cdot\psi(x))$. Finally, if $F : \mathcal{D}(S) \to \underline{\underline{Gr}}$ is any functor and $\psi \in Z^1(F)$, then $F^{\psi}: \mathcal{D}(S) \to \underline{\underline{Gr}}$ is also a functor, and $\psi^* : (S \times F,\pi) \cong (S \times F^{\psi},\pi)$ where $\psi^*(x,g) = \langle x,\psi(x)^{-1}g\rangle$.

4.5. Lemma. Let $(S \times F,\pi), (S \times G,\pi) \in \mathcal{H}(\underline{\underline{Mon}}_S)$. For every $\sigma \in \mathrm{Hom}(F,G)$ define $\sigma^* : S \times F \to S \times G$ by $\sigma^*(x,g) = \langle x,\sigma(x)[g]\rangle$. Then σ^* is an $\mathcal{H}$-morphism and $\sigma^*(x,1) = \langle x,1\rangle$. Conversely if $f : \langle S \times F,\pi\rangle \to \langle S \times G,\pi\rangle$ is an $\mathcal{H}$-morphism such that $f(x,1) = \langle x,1\rangle$, then there exists $\sigma \in \mathrm{Hom}(F,G)$ such that $f = \sigma^*$.

Putting 4.4 and 4.5 together we obtain, by the analogous arguments, the following analogue of 3.33.

4.6. Theorem. Let $(S \times F,\pi)$, $(S \times G,\pi) \in \mathcal{H}(\underline{\underline{Mon}}_S)$. Then every morphism, $f : (S \times F,\pi) \to (S \times G,\pi)$, has the unique form, $f = (\phi^*)^{-1} \circ \sigma^*$, where $\phi \in Z^1(G)$ is defined implicitly by $f(x,1) = \langle x,\phi(x)\rangle$ and $\sigma : F \to G^{\phi}$. Moreover, for all $\langle x,g\rangle \in S \times F$, $f(x,g) = \langle x,\phi(x)\sigma(x)[g]\rangle$.

4.7. Lemma. For all $\langle x,g\rangle \in S \times F$ let us identify $\Sigma_\pi(x,g)$ with $F(x)$ under the map $[\langle u,a\rangle] \to a^{(x,u,1)}$ where $\langle u,a\rangle \in \mathcal{R}(x) \times F$ and $[\langle u,a\rangle]$ is the element of $\Sigma_\pi(x,g)$ to which $\langle u,a\rangle$ belongs. Then

$$\Sigma_\pi(\langle x,g\rangle,\langle y,h\rangle,\langle z,k\rangle) = (\ )^{F(xy,z,1)[k]} \circ F(x,y,z) \ .$$

Before proceeding further we mention the following generalization of a well-known fact in the theory of group extensions.

4.8. Theorem. Let $(S \times (F_L,F_R,\alpha),\pi) \in \mathcal{H}(\underline{\underline{Mon}}_S)$. Then $(S \times (F_L,F_R,\alpha),\pi)$ splits if and only if there exists $\psi \in C^1(F_R)$ such that

$$\alpha(x,y) = [\psi(x)^{-1}]^{(1,x,y)}\psi(xy)[\psi(y)^{-1}]^{(x,y,1)} \ .$$

4.9. Definition. $F : \mathcal{D}(S) \to \underline{\underline{Gr}}$ is called left $\mathcal{H}$-stable if for all $x, y \in S$ $x\mathcal{H}y$ implies $F(x) = F(y)$ and $F(u,x,1) = i_{F(x)}$ when $ux = y$.

4.10. Lemma. Let $F : \mathcal{D}(S) \to \underline{\underline{Gr}}$. In each $\mathcal{H}$-class of S choose a fixed element. For each $x \in S$, let x_0 denote this element in $\mathcal{H}(x)$. Set $\tau(x) = F(u,x_0,1)$ where $ux_0 = x$. Define $F_0 : \mathcal{D}(S) \to \underline{\underline{Gr}}$ by $F_0(x) = F(x_0)$ and $F_0(u,x,v) = \tau(uxv)^{-1} \circ F(u,x,v) \circ \tau(x)$. Then F_0 is left $\mathcal{H}$-stable and $\tau : F_0 \cong F$.

Hence, without loss of generality, we shall assume that all functors, $F : \mathcal{D}(S) \to \underline{\underline{Gr}}$, are left $\mathcal{H}$-stable throughout the remainder of this chapter. Clearly all functors obtained from lifting homomorphisms into $\mathcal{H}$-coextensions of S are left $\mathcal{H}$-stable.

4.11. Lemma. Let $F : \mathcal{D}(S) \to \underline{\underline{Gr}}$ be left $\mathcal{H}$-stable. Then if $x\mathcal{G}y$ in S , $F(u,x,v) = F(u,y,v)$ for all $u, v \in S$.

Proof. Let $y_{-1}x = y$. Then $[u,y,v] \circ [y_{-1},x,1] = [uy_{-1},x,v] = [uy_{-1},xv,1] \circ [1,x,v]$. Now $uy_{-1}xv \,\mathcal{H}\, uxv$. So let $\bar{u} \in S$ be such that $\bar{u}uy_{-1}xv = uxv$. Then we have $F(\bar{u},uy_{-1}xv,1) = i_{F(uxv)}$. Hence $F(uy_{-1},xv,1) = F(\bar{u},uy_{-1}xv,1) \circ F(uy_{-1},xv,1) = F(\bar{u}uy_{-1},xv,1) = F(u,xv,1)$. Therefore $F(u,y,v) = F(u,y,v) \circ F(y_{-1},x,1) =$

$F(u,xv,1) \circ F(1,x,v) = F(u,x,v)$.

4.12. Let $F, G : \mathcal{D}(S) \to \underline{\underline{Gr}}$. For each $x \in S$ let $\tau_x : F(x) \times G(x) \to F(x)$ be a right action of $G(x)$ on $F(x)$ such that for all $u,x,v \in S$ the following diagram commutes:

$$\begin{array}{ccc} F(x) \times G(x) & \xrightarrow{\quad \tau_x \quad} & F(x) \\ {\scriptstyle F(u,x,v)}\downarrow \quad \downarrow{\scriptstyle G(u,x,v)} & & \downarrow{\scriptstyle F(u,x,v)} \\ F(uxv) \times G(uxv) & \xrightarrow[\quad \tau_{uxv} \quad]{} & F(uxv) \end{array}$$

$\tau = \{\tau_x : x \in S\}$ is called a right action of G on F . We define the semi-direct product of F by G , $G \times {}_\tau F$, by $(G \times {}_\tau F)(x) = G(x) \times_{\tau_x} F(x)$ and $(G \times {}_\tau F)(u,x,v) = G(u,x,v) \times F(u,x,v)$. To show that $G \times {}_\tau F : \mathcal{D}(S) \to \underline{\underline{Gr}}$ is a functor we only need to show that $(G \times {}_\tau F)(u,x,v)$ is a homomorphism. So let $\langle g,f \rangle$, $\langle g',f' \rangle \in (G \times {}_\tau F)(x)$. Then

$$\begin{aligned} (\langle g,f\rangle \cdot \langle g',f'\rangle)^{(u,x,v)} &= \langle gg', f^{g'} f' \rangle^{(u,x,v)} \\ &= \langle (gg')^{G(u,x,v)}, (f^{g'}f')^{F(u,x,v)} \rangle \\ &= \langle g^{G(u,x,v)}(g')^{G(u,x,v)}, (f^{F(u,x,v)})^{G(u,x,v)[g']}(f')^{F(u,x,v)} \rangle \\ &= \langle g^{G(u,x,v)}, f^{F(u,x,v)} \rangle \cdot \langle (g')^{G(u,x,v)}, (f')^{F(u,x,v)} \rangle \\ &= \langle g,f \rangle^{(u,x,v)} \cdot \langle g',f' \rangle^{(u,x,v)} . \end{aligned}$$

Hence $G \times {}_\tau F : \mathcal{D}(S) \to \underline{\underline{Gr}}$ is a functor.

4.13. Theorem. Let $(S \times F, \pi) \in H(\underline{\underline{Mon}}_S)$ with F being left H-stable. For each $x \in S$ define $\tau_x : F(x) \times \Gamma_S(x) \to F(x)$ by $\tau_x(f,g) = F(1,x,\overline{g})[f]$ where $x\overline{g} = x \cdot g$. Then τ is a right action of Γ_S on F and $\Gamma|\mathcal{D}(S) \cong \Gamma_S \times {}_\tau F$.

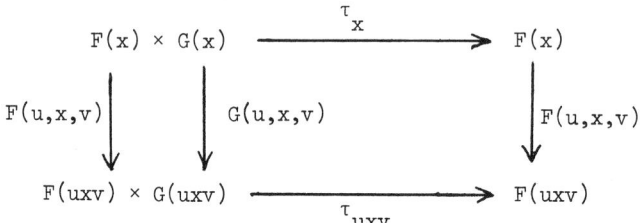

Remark. If we consider S a submonoid of $S \times F$ under the embedding $x \to \langle x,1\rangle$, $\mathcal{D}(S)$ is a subcategory of $\mathcal{D}(S \times F)$ because "S is a retract of $S \times F$" implies "$\mathcal{D}(S)$ is a retract of $\mathcal{D}(S \times F)$", and our formula makes sense.

Proof. Clearly $\tau_x(\ ,g)$ is an automorphism of $F(x)$. Moreover $\tau_x(\_,gh) = F(1,x,\overline{g}\,\overline{h}) = F(1,x,\overline{h}) \circ F(1,x,\overline{g}) = \tau_x(\_,h) \circ \tau_x(\_,g)$ by 4.11. Hence τ_x is a right action. We need to show that the following diagram commutes:

$$\begin{array}{ccc} F(x) \times \Gamma_S(x) & \xrightarrow{\quad\tau_x\quad} & F(x) \\ {\scriptstyle F(u,x,v)}\downarrow \quad \downarrow{\scriptstyle \Gamma_S(u,x,v)} & & \downarrow{\scriptstyle F(u,x,v)} \\ F(uxv) \times \Gamma_S(uxv) & \xrightarrow{\quad\tau_{uxv}\quad} & F(uxv) \end{array}$$

Let $(f,g) \in F(x) \times \Gamma_S(x)$. We must show that $F(u,x,v)[\tau_x(f,g)] = \tau_{uxv}(F(u,x,v)[f],\Gamma_S(u,x,v)[g])$. Now $F(u,x,v)[\tau_x(f,g)] = F(u,x,v) \circ F(1,x,\overline{g})[f] = F(u,x,\overline{g}v)[f]$, while

$$\begin{aligned} &\tau_{uxv}(F(u,x,v)[f],\Gamma_S(u,x,v)[g]) \\ &\quad = F(1,uxv,\overline{\Gamma_S(u,x,v)[g]}) \circ F(u,x,v)[f] \\ &\quad = F(u,x,v\overline{\Gamma_S(u,x,v)[g]})[f] . \end{aligned}$$

But $F(1,ux,\overline{g}v) = F(1,ux,v\overline{\Gamma_S(u,x,v)[g]})$. Hence the diagram commutes. If $\langle y,h\rangle \in G(x,1) = G_S(x) \times F$, let $\langle y,h\rangle^{\#}$ denote the element of $\Gamma(x,1)$ such that $\langle x,1\rangle \cdot \langle y,h\rangle^{\#} = \langle y,h\rangle$. Moreover $\tilde{y} \in \Gamma(x)$ is such that $\tilde{x} \cdot y = y$. Let us map $\Gamma(x,1)$ to $\Gamma_S(x) \times F(x)$ by $\langle y,h\rangle^{\#} \to \langle \tilde{y},h\rangle$. This map is clearly bijective. If $\langle z,k\rangle \in G(x,1)$ let $\langle v,b\rangle \in S \times F$ be such that $\langle x,1\rangle\langle v,b\rangle = \langle z,k\rangle$. Then $\langle x,1\rangle \cdot \langle y,h\rangle^{\#}\langle z,k\rangle^{\#} = \langle y,h\rangle \cdot \langle v,b\rangle = \langle yv,h^{(1,y,v)}b^{(y,v,1)}\rangle = \langle yv,h^{(1,x,v)}b^{(x,v,1)}\rangle = \langle yv,h^{(1,x,v)}k\rangle = \langle x \cdot \widetilde{\widetilde{yz}},\tau_x(h,\tilde{z})k\rangle$ by 4.11. Hence $\langle y,h\rangle^{\#}\langle z,k\rangle^{\#} \to \langle \widetilde{\widetilde{yz}},\tau_x(h,\tilde{z})k\rangle = \langle \tilde{y},h\rangle \cdot \langle \tilde{z},k\rangle$, and the map is an isomorphism.

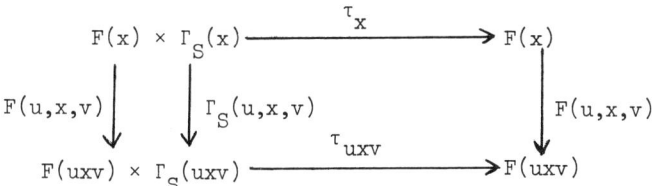

Under this identification, $\langle u,1\rangle(\langle x,1\rangle\cdot\langle\tilde{y},g\rangle)\langle v,1\rangle = \langle u,1\rangle\langle y,g\rangle\langle v,1\rangle = \langle uyv,F(u,y,v)[g]\rangle = \langle uxv\cdot\Gamma_S(u,x,v)[\tilde{y}],\ F(u,x,v)[g]\rangle$ by 4.11. Hence $\Gamma(\langle u,1\rangle,\langle x,1\rangle,\langle v,1\rangle)[\langle\tilde{y},g\rangle] = \langle\Gamma_S(u,x,v)[\tilde{y}],\ F(u,x,v)[g]\rangle$, and the theorem follows.

We now present a consequence of Theorem 4.13. If $(S \times F,\pi)$ is a split H-coextension then the lattice of congruences under $H_{S\times F}$ is isomorphic with the lattice of subfunctors of $\Gamma_{S\times F}$. Now the result of 4.13 could be used to derive a cumbersome computation of $\Gamma_{S\times F}$. However, the next theorem tells us that $\Gamma_S \times {}_\tau F$ is almost as good as $\Gamma_{S\times F}$.

4.14. Theorem. *Let* S *be a submonoid of the monoid,* $\overline{S}$, *and suppose that for each* G-*class,* K, *of* $\overline{S}$, $S \cap K$ *is a nonempty* G-*class of* S. *Then the lattice of congruences under* $H_{\overline{S}}$ *is isomorphic with the lattice of normal subfunctors of* $\Gamma_{\overline{S}}|\mathbb{D}(S)$. *In particular, if* $\overline{S} = S \times F$ *is an* H-*coextension of* S, $[C \leq H_{\overline{S}}]$ *is isomorphic with the lattice of normal subfunctors of* $\Gamma_S \times {}_\tau F$.

Proof. The map $\Sigma \to \Sigma|\mathbb{D}(S)$ is clearly a well defined lattice homomorphism from $[\Sigma \leq \Gamma_{\overline{S}}]$ to $[F \trianglelefteq \Gamma_{\overline{S}}|\mathbb{D}(S)]$. We need to show that it is bijective. If Σ, Σ' are such that $\Sigma(x) = \Sigma'(x)$ for all $x \in S$, then if $y \in \overline{S}$ with yGx, we must have $\Sigma(y) = \Sigma(x) = \Sigma'(x) = \Sigma'(y)$ by H-stability. Hence $\Sigma = \Sigma'$, and our restriction map is a monomorphism. Conversely, let $F \leq \Gamma_{\overline{S}}|\mathbb{D}(S)$. Let $y \in \overline{S}$ and $y = ux$ where $x \in S$ and xGy. $\Gamma(x) = \Gamma(y)$, and we set $F(y) = F(x)$. $F(y)$ is well defined since if $x, x' \in S$ with xGx', then xG_Sx' and there exists an $\mathbb{D}(S)$-morphism $\langle u,x,1\rangle : x \to x'$ such that $F(x) = \Gamma(u,x,1)[F(x)] \subseteq F(x')$. Similarly $F(x') \subseteq F(x)$ so that $F(x) = F(x')$. We need to show that if $\langle u,y,v\rangle$ is a $\mathbb{D}(S)$-morphism, we have $\Gamma(u,y,v)F(y) \subseteq F(uyv)$. Now $F(y) = F(x)$, $uxv\,G\,xyv$ so that $F(uxv) = F(uyv)$, and $\Gamma(u,y,v) = \Gamma(u,x,v)$ since Γ is

$\mathcal{H}$-stable. Hence we need to show that $\Gamma(u,x,v)F(x) \subseteq F(uxv)$. Let $u = au'$ and $v = v'b$ where $u', v' \in S$, $u\mathcal{G}u'$, and $v\mathcal{G}v'$. Then $\Gamma(u,x,v) = \Gamma(a,u'xv',b) \circ \Gamma(u',x,v')$. But $\Gamma(a,u'xv',1) = \Gamma(1,u'xv',1)$ and $\Gamma(1,au'xv',b) = \Gamma(1,u'xv',b)$ is merely conjugation of $\Gamma(u'xv')$ by the element of $\Gamma(u'xv')$ induced by b. Hence $\Gamma(a,u'xv',b)F(u'xv') = F(u'xv')$ and $\Gamma(u,x,v)F(x) = \Gamma(a,u'xv',b) \circ \Gamma(u',x,v')F(x) \subseteq \Gamma(a,u'xv,b)F(u'xv') = F(u'xv')$. But $u'xv' \mathcal{G} uxv$ implies $F(u'xv') = F(uxv)$, and $F(u,x,v)F(x) \subseteq F(uxv)$ follows. Hence F induces a functor, $F \leq \Gamma$, which induces in turn the original $F \trianglelefteq \Gamma | \mathbb{D}(S)$.

A second application of Theorem 4.13 is concerned with repeated split coextensions.

4.15. <u>Theorem</u>. <u>Let</u> $G, F : \mathcal{D}(S) \to \underline{\underline{Gr}}$ <u>be left</u> $\mathcal{H}$-<u>stable functors, and let</u> $\tau : F \times G \to F$ <u>be a right action. Define</u> $F_\tau : \mathbb{D}(S \times G) \to \underline{\underline{Gr}}$ <u>by</u>:

i. $F_\tau(x,g) = F(x)$.

ii. $F_\tau(\langle u,a\rangle,\langle x,g\rangle,\langle v,b\rangle)[h] = F(u,x,v)[h]^{G(ux,v,1)[b]}$.

<u>Then</u> F_τ <u>is a well defined left</u> $\mathcal{H}$-<u>stable functor, and</u> $(S \times G) \times F_\tau \cong S \times (G \times_\tau F)$ <u>under the map</u> $\langle\langle x,g\rangle,h\rangle \to \langle x,\langle g,h\rangle\rangle$. <u>Moreover, if</u> $S \times G \in \mathcal{H}(\underline{\underline{Mon}}_S)$, <u>then</u> $(S \times G) \times F_\tau \in \mathcal{H}(\underline{\underline{Mon}}_{S\times G})$ <u>if and only if</u> $S \times F \in \mathcal{H}(\underline{\underline{Mon}}_S)$. <u>Finally, if</u> $S \times G \in \mathcal{H}(\underline{\underline{Mon}}_S)$, <u>then for every left</u> $\mathcal{H}$-<u>stable functor</u>, $H : \mathcal{D}(S \times G) \to \underline{\underline{Gr}}$ <u>such that</u> $(S \times G) \times H \in \mathcal{H}(\underline{\underline{Mon}}_{S\times G})$, <u>there exists a left</u> $\mathcal{H}$-<u>stable functor</u> $F : \mathcal{D}(S) \to \underline{\underline{Gr}}$ <u>and a right action</u> $\tau : F \times G \to F$ <u>such that</u> $H = F_\tau$.

<u>Proof</u>. That F_τ is a well defined left $\mathcal{H}$-stable functor is seen by a straightforward calculation. Since $\langle u,a\rangle \in \mathcal{R}(x,g)$ implies $\langle u,1\rangle \in \mathcal{R}(x,g)$, and $\langle u,1\rangle \in \mathcal{R}(x,g)$ if and only if $u \in \mathcal{R}(x)$ we have:

$$\bigcup_{\langle u,a\rangle \in \mathcal{R}(x,g)} F(u,a)^{(\langle x,g\rangle,\langle u,a\rangle,1)} = \bigcup_{u \in \mathcal{R}_S(x)} F(u)^{(x,u,1)} .$$

Hence the right hand equality of 4.2 holds for F_τ if and only if it does for F. A similar argument holds for the left hand equality with the only formal difference being that the unions are acted upon by the automorphism, $\tau(\ ,g)$. Finally, let $H : \mathcal{D}(S \times G) \to \underline{\underline{Gr}}$ satisfy the above conditions. By 4.13 there exists an action $\tau_1 : H \times \Gamma_{S\times G} \to H$ such that $\Gamma_{(S\times G)\times H}|\mathcal{D}(S\times G) \cong \Gamma_{(S\times G)} \times_{\tau_1} H$.

Let $\tau : (H|\mathcal{D}(S)) \times G \to (H|\mathcal{D}(S))$ be the action defined by restriction of τ_1, first to $(H|\mathcal{D}(S)) \times (\Gamma_{S\times G}|\mathcal{D}(S))$, and then by chopping $(\Gamma_{S\times G}|\mathcal{D}(S))$ down to G. Let $F = H|\mathcal{D}(S)$. Since H is left H-stable, $H(x,g) = F(x) = F_\tau(x,g)$ for all $\langle x,g\rangle \in S \times G$. By 4.11

$$\begin{aligned} H(\langle u,a\rangle,\langle x,g\rangle,\langle v,b\rangle) &= H(\langle u,1\rangle,\langle x,1\rangle,\langle v,b\rangle) \\ &= (\ )^{G(ux,v,1)[b]} \circ H(\langle u,1\rangle,\langle x,1\rangle,\langle v,1\rangle) \\ &= (\ )^{G(ux,v,1)[b]} \circ F(u,x,v) \\ &= F_\tau(\langle u,a\rangle,\langle x,g\rangle,\langle v,b\rangle) . \end{aligned}$$

Hence in order to find all nice functors in $\mathrm{Hom}(\mathcal{D}(S \times G), \underline{\underline{Gr}})$ we need only find all nice functors in $\mathrm{Hom}(\mathcal{D}(S),\underline{\underline{Gr}})$ and all right actions of G upon them.

Chapter Five

Abelian Coextensions and Cohomology

We call an object $(\overline{S},p) \in \mathcal{H}(\underline{\underline{Mon}}_S)$ an abelian coextension of S if $\Sigma_p : \mathcal{D}(S) \to \underline{\underline{AG}}$, the category of abelian groups. We let $\mathcal{A}(\underline{\underline{Mon}}_S)$ denote the full subcategory of $\mathcal{H}(\underline{\underline{Mon}}_S)$ whose objects are abelian coextensions. Each such coextension has a representation of the form $(S \times (F_L,F_R,\alpha),\pi)$ where $(F_L,F_R) : (\mathcal{L},\mathcal{R})(S) \to \underline{\underline{AG}}$. But since F_{LR} and F_{RL} differ only by inner automorphisms of abelian groups, we have $F_{LR} = F_{RL}$ and there must exist a unique extension of (F_L,F_R) to a functor $F : \mathcal{D}(S) \to \underline{\underline{AG}}$. Hence we shall denote $(S \times (F_L,F_R,\alpha),\pi)$ by $(S \times (F,\alpha),\pi)$. If (F_L,F_R,α) is obtained from a lifting $x \to \overline{x}$ from S to $(\overline{S},p)$, then by 2.28 $\Sigma_p = F \circ \mathcal{D}(p)$ and $F(u,x,v) = \Sigma_p(\overline{u},\overline{x},\overline{v})$ for all $u, x, v \in S$. $\mathcal{A}(\underline{\underline{Mon}}_S)$ bears a relation to $\mathcal{H}(\underline{\underline{Mon}}_S)$ which is analogous to an important relation between $\underline{\underline{AG}}$ and $\underline{\underline{Gr}}$.

5.1. Theorem. There is a universal functor, $A : \mathcal{H}(\underline{\underline{Mon}}_S) \to \mathcal{A}(\underline{\underline{Mon}}_S)$, i.e., for each $(\overline{S},p) \in \mathcal{H}(\underline{\underline{Mon}}_S)$ there exists a morphism $\alpha_{(\overline{S},p)} : (\overline{S},p) \to A(\overline{S},p)$ such that for all $(\overline{S}_1,p_1) \in \mathcal{A}(\underline{\underline{Mon}}_S)$, $\alpha^*_{(\overline{S},p)} : \mathrm{Hom}(A(\overline{S},p),(\overline{S}_1,p_1)) \to \mathrm{Hom}((\overline{S},p),(S_1,p_1))$ is an isomorphism which is natural in both variables.

Proof. If $(\overline{S},p) \in \mathcal{H}(\underline{\underline{Mon}}_S)$ and $\Sigma = \Sigma_p$, define $K : \mathcal{D}(\overline{S}) \to \underline{\underline{Gr}}$ by $K(x) = (\Sigma(x))'$, the commutator subgroup of $\Sigma(x)$, and $K(u,x,v) = \Sigma(u,x,v)|K(x)$. Let $\mathcal{K}$ be the congruence corresponding to K. Clearly $\mathcal{K} \leq \mathcal{C}_p$. If $\pi : \overline{S} \to \overline{S}/\mathcal{K}$ is the canonical map, then a map, $\tilde{p} : \overline{S}/\mathcal{K} \to S$, is induced such that $\tilde{p} \circ \pi = p$. Moreover $\mathcal{C}_{\tilde{p}} = (\pi \times \pi)[\mathcal{C}] \leq \mathcal{H}_{\overline{S}/\mathcal{K}}$. If $\tilde{\Sigma}$ is the right structure functor of $\mathcal{C}_{\tilde{p}}$, then by 2.27, $\tilde{\Sigma} \circ \mathcal{D}(\pi) \cong \Sigma/K$. Hence $(\overline{S}/\mathcal{K},p) \in \mathcal{A}(\underline{\underline{Mon}}_S)$. If $(\overline{S},p_1) \in \mathcal{A}(\underline{\underline{Mon}}_S)$ and $f : (S,p) \to (\overline{S}_1,p_1)$ is a morphism, then the maps $f^x : H_R(x) \to H_R(f(x))$, $x \in S$, induce homomorphisms $f^x : \Sigma(x) \to \Sigma_1(x)$ where

Σ_1 is the right structure functor of $\mathcal{C}_{p_1}$. But $\Sigma_1(x)$ is abelian. Hence $K(x) \subseteq \mathrm{Ker}(f^x)$, and therefore $K \leq \mathcal{C}_f$. Hence a unique homomorphism $\tilde{f} : S/K \to \overline{S}_1$ is induced such that $\tilde{f} \circ \pi = f$. Hence $\pi : \overline{S} \to S/K$ is a universal morphism from $(\overline{S},p)$ to $A(\underline{\underline{\mathrm{Mon}}}_S)$. We define $A(\overline{S},p) = (\overline{S}/K,\tilde{p})$. If $f : (\overline{S},p) \to (\overline{S}_1,p_1)$ where $(\overline{S}_1,p_1)$ is now any object of $H(\underline{\underline{\mathrm{Mon}}}_S)$, then $A(f)$ is the unique morphism from $A(\overline{S},p)$ to $A(\overline{S}_1,p_1)$ induced by $\pi_1 \circ f$ where $\pi_1 : \overline{S}_1 \to \overline{S}_1/K_1$ is the universal morphism for $(\overline{S}_1,p_1)$. Finally we let $\alpha_{(\overline{S},p)} = \pi : (\overline{S},p) \to (\overline{S}/K,\tilde{p})$.

5.2. When S is a group, it is well known that the abelian coextension problem has intimate connections with group cohomology. The same is true in the general situation.

5.3. If K is any small category, then it is known that the category, $[K,\underline{\underline{AG}}]$, of functors from K to $\underline{\underline{AG}}$ together with their natural transformations, is a very well behaved abelian category with enough projectives and injectives. Hence there exists a sequence of bifunctors, $\mathrm{Ext}^n(\ ,\ ) : [K,\underline{\underline{AG}}] \times [K,\underline{\underline{AG}}] \to \underline{\underline{AG}}$, which is contravariant, connected in the first variable and covariant, connected in the second variable. If $\mathbb{Z}$ is the constant functor defined by $\mathbb{Z}(x) = \mathbb{Z}$ and $\mathbb{Z}(\mu) = id_{\mathbb{Z}}$ where $\mu : x \to y$ (of course $\mathbb{Z}$ = integers), then we define $H^n(F) = \mathrm{Ext}^n(\mathbb{Z},F)$. The sequence of the H^n together with their connecting homomorphisms can be described axiomatically as the minimal cohomological extension of the left exact functor, $\varprojlim : [K,\underline{\underline{AG}}] \to \underline{\underline{AG}}$.

5.4. Of particular interest is the case where $K = \mathbb{D}(S)$. By use of a certain standard resolution, H^n is computable from the following cochain complexes and cochain maps.

Given a functor $F : \mathbb{D}(S) \to \underline{\underline{AG}}$ we can construct a cochain complex of abelian groups, $\overline{C}(F) = 0 \to C^0(F) \xrightarrow{\delta^0} C^1(F) \xrightarrow{\delta^1} C^2(F) \xrightarrow{\delta^2} C^3(F) \xrightarrow{\delta^3} \cdots$

which is defined by:

(i) $C^0(F) = F(1)$ and for $n \geq 1$, $C^n(F) = \coprod_{\langle x_1,\ldots,x_n\rangle \in S^n} F(x_1 \ldots x_n)$.

(ii) $\delta^0 a(x) = a^{(x,1,1)} - a^{(1,1,x)}$,

$$\delta^1\psi(x,y) = \psi(y)^{(x,y,1)} - \psi(xy) + \psi(x)^{(1,x,y)},$$

$$\delta^2\alpha(x,y,z) = \alpha(y,z)^{(x,yz,1)} - \alpha(xy,z) + \alpha(x,yz) - \alpha(x,y)^{(1,xy,z)},$$

$$\delta^3\alpha(x,y,z,w) = \alpha(y,z,w)^{(x,yzw,1)} - \alpha(xy,z,w) + \alpha(x,yz,w) - \alpha(x,y,zw) + \alpha(x,y,z)^{(1,xyz,w)}, \text{ etc.}$$

Since $\delta^{n+1} \circ \delta^n = 0$ for all $n \geq 0$, we obtain $B^n(F) \subseteq Z^n(F)$ for all $n \geq 1$, where $Z^n(F) = \mathrm{Ker}(\delta^n)$ and $B^n(F) = \mathrm{Im}(\delta^{n-1})$. We define $H^0(F) = Z^0(F)$ and $H^n(F) = Z^n(F)/B^n(F)$ for $n \geq 1$.

If $G : \mathbb{D}(S) \to \underline{AG}$ and $\sigma : F \to G$ is a natural transformation, then σ induces a cochain map, $\overline{C}(\sigma) : \overline{C}(F) \to \overline{C}(G)$, defined by:

(iii) $C^0(\sigma) = \sigma(1)$, and

(iv) for $n \geq 1$, $C^n(\sigma)(\alpha)(x_1,\ldots,x_n) = \sigma(x_1 \ldots x_n)[\alpha(x_1,\ldots,x_n)]$.

Since $\overline{C}(\sigma)$ is a cochain map, for all $n \geq 0$ we have $C^n(\sigma)[Z^n(F)] \subseteq Z^n(G)$, and for all $n \geq 1$ we have $C^n(\sigma)[B^n(F)] \subseteq B^n(G)$. Hence for all $n \geq 0$, σ induces a homomorphism, $H^n(\sigma) : H^n(F) \to H^n(G)$, defined by: $H^n(\sigma)([\alpha]) = [C^n(\sigma)(\alpha)]$ for all $\alpha \in Z^n(F)$.

Clearly $C^1(F)$ and $Z^1(F)$ retain their original meanings, with the only difference being that an additive abelian group structure has been given to them. Moreover, $C^2(F) = \{\text{factor systems of } F\}$, and $Z^2(F) = \{\text{associative factor systems of } F\}$.

5.5. Theorem. *For each* $\alpha \in Z^2(F)$ *embed* $F(1)$ *into* $G(S \times (F,\alpha))$ *under the map,* $g \to \langle 1, g - \alpha(1,1)\rangle$. *Then* $Z^0(F) = Z(S \times (F,\alpha)) \cap F(1)$.

Proof. For each $\alpha \in F(1)$, $\langle x,g\rangle \in S \times (F,\alpha)$ we have $a^{-1}\langle x,g\rangle a = \langle x,(-a)^{(1,1,x)} + g + a^{(x,1,1)}\rangle = \langle x,g + \delta^0 a(x)\rangle$. Hence $a \in Z(S \times (F,\alpha))$ if and only if $\delta^0 a = 0$.

5.6. Definition. $\underline{\underline{AG}}^o = \underline{\underline{AG}} \cap \underline{\underline{Gr}}^o$.

5.7. Theorem. *If F factors through $J(S)$, then $H^0(F) = F(1)$. Conversely, if $F : \mathbb{D}(S) \to AG^o$, then the converse also holds.*

Proof. If F factors through $J(S)$, then $F(x,1,1) = F(1,1,x)$ for all $x \in S$. Hence $\delta^0 a = 0$ for all $a \in F(1)$. If $F : \mathbb{D}(S) \to \underline{\underline{AG}}^o$, then $H^0(F) = F(1)$ means $F(x,1,1) = F(1,1,x)$ for all $x \in S$. Hence for all $x, y \in S$, $F(x,1,y) = F(x,y,1) \circ F(1,1,y) = F(x,y,1) \circ F(y,1,1) = F(xy,1,1)$ so that for all $z \in S$, $F(\mathrm{Hom}(1,z))$ has only one morphism. Finally let $uxv = \bar{u}x\bar{v}$. Then $F(u,x,v) \circ F(x,1,1) = F(\bar{u},x,\bar{v}) \circ F(x,1,1)$, and since $F(x,1,1)$ is surjective, $F(u,x,v) = F(\bar{u},x,\bar{v})$. Hence for all $x, y \in S$, $F(\mathrm{Hom}(x,y))$ has at most one morphism, i.e., F can be factored through $J(S)$.

5.8. Lemma. *Let $\alpha \in Z^2(F)$. If $\mathcal{H}: Z^1(F) \to \mathrm{Aut}(S \times (F,\alpha))$ is defined by $\mathcal{H}(\psi)[\langle x,g\rangle] = \langle x,g - \psi(x)\rangle$, then $\mathcal{H}$ is a well defined monomorphism.*

Proof. $F^\psi = F$ and $\alpha^\psi = \alpha + \delta^1\psi = \alpha$. Hence, by Lemma 3.22, $\mathcal{H}(\psi)$ is an S-automorphism of $S \times (F,\alpha)$. Clearly $\mathcal{H}(\psi_1 + \psi_2) = \mathcal{H}(\psi_1) \circ \mathcal{H}(\psi_2)$, and $\mathcal{H}(\psi_1) = \mathcal{H}(\psi_2)$ implies $\psi_1 = \psi_2$.

5.9. If $F : \mathcal{D}(S) \to \underline{\underline{AG}}$, then we denote $H^n(F \circ \delta)$ by $H^n(F)$. This will cause no confusion since it is the $\mathbb{D}(S)$-cohomology of F which is of interest, and not the $\mathcal{D}(S)$-cohomology of F. In fact it is possible that $H^2_{\mathbb{D}}(F) \neq 0$ while $H^2_{\mathcal{D}}(F) = 0$. Throughout the remainder of this chapter, we shall assume that all functors under consideration are defined on $\mathcal{D}(S)$.

5.10. If Aut(F) denotes the group of natural automorphisms of F , then there exists a left action $\Delta : \mathrm{Aut}(F) \times Z^1(F) \to Z^1(F)$ defined by $\Delta(\sigma,\psi)(x) = \sigma(x)[\psi(x)]$ for all $x \in S$. We let $Z^1(F) \times {}_\Delta\mathrm{Aut}(F)$ be the resulting semidirect product. We define $\mathcal{H} : Z^1(F) \times {}_\Delta\mathrm{Aut}(F) \to \mathrm{Aut}_S(S \times F)$ by $\mathcal{H}(\psi,\sigma) = \psi^* \circ \sigma^*$.

5.11. <u>Theorem</u>. $\mathcal{H}$ <u>is a monomorphism</u>. <u>Moreover, if</u> $(S \times F,\pi) \in \mathcal{A}(\underline{\underline{\mathrm{Mon}}}_S)$, <u>then</u> $\mathcal{H}$ <u>is an isomorphism</u>.

<u>Proof</u>. That $\mathcal{H}$ is a monomorphism is a straightforward computation. If $(S \times F,\pi) \in \mathcal{A}(\underline{\underline{\mathrm{Mon}}}_S)$, then the fact that $\mathcal{H}$ is an isomorphism is a consequence of the split abelian analogues of 3.33 and 3.35. (Of course, in these analogues we only use those ϕ's which are in $Z^1(F)$.)

We now seek to give a description of those automorphisms of $(S \times (F,\alpha),\pi)$ which are of the form, $\mathcal{H}(\psi)$, when $(S \times (F,\alpha),\pi) \in \mathcal{A}(\underline{\underline{\mathrm{Mon}}}_S)$.

5.12. Let $(S \times (F,\alpha),\pi) \in \mathcal{A}(\underline{\underline{\mathrm{Mon}}}_S)$. We are interested in those automorphisms of $(S \times (F,\alpha),\pi)$ which in some way fix all of the $F(x)$. More precisely, if $f : (S \times (F,\alpha),\pi) \cong (S \times (F,\alpha),\pi)$, then for all $x \in S$, f induces an isomorphism $f^x : \Sigma_\pi(x,1) \cong \Sigma_\pi(f\langle x,1\rangle)$ which is defined implicitly by the equation, $f(\langle x,1\rangle \cdot a) = f(x,1) \cdot f^x(a)$. But since we have identified $\Sigma_\pi(x,1)$ and $\Sigma_\pi(f\langle x,1\rangle)$ with $F(x)$, it follows that f^x is an automorphism of $F(x)$. We are now interested in those automorphisms for which $f^x = 1_{F(x)}$ for all $x \in S$. Using the results of 3.33 and 3.35 we can describe each automorphism, f^x of $F(x)$. If $f \in \mathrm{Aut}(S \times (F,\alpha),\pi)$, then f has the unique decomposition of the form $(-\psi)^* \circ \sigma^*$ where $\sigma : (F,\alpha) \to (F,\alpha)^\psi = (F,\alpha^\psi)$, and f is defined by $f(x,a) = \langle x,\sigma(x)[a] + \psi(x)\rangle$. Since $f(x,1) = \langle x,\psi(x)\rangle$, we obtain $f^x = \sigma(x)$ for all $x \in S$. Hence $f^x = 1_{F(x)}$ for all $x \in S$, if and only if

$\sigma = 1_F$, and hence $\alpha^\psi = \alpha$. But $\alpha^\psi = \alpha + \delta^0\psi$. Hence $\delta^0\psi = 0$ and $f = \divideontimes(-\psi)$. Conversely, if $\psi \in Z^1(F)$, then $\divideontimes(-\psi)$ induces the identity automorphism on all the $F(x)$. Hence if we denote these special automorphisms by $\mathrm{Aut}_0(S \times (F,\alpha),\pi)$, we have $\mathrm{Aut}_0 = \mathrm{Im}(\divideontimes)$, so that $\mathrm{Aut}_0 \leq \mathrm{Aut}$ and $\divideontimes : Z^1(F) \cong \mathrm{Aut}_0(S \times (F,\alpha),\pi)$. A special subclass of Aut_0 are those automorphisms of $S \times (F,\alpha)$ which are obtained by conjugation by elements of $F(1)$. Indeed let $a \in F(1)$. Then $a \leftrightarrow \langle 1, a - \alpha(1,1)\rangle \in S \times (F,\alpha)$. Moreover, for all $\langle x,g\rangle \in S \times (F,\alpha)$ we have:

$$\begin{aligned} a^{-1}\cdot\langle x,g\rangle\cdot a &= \langle 1,-a-\alpha(1,1)\rangle\langle x,g\rangle\langle 1,a-\alpha(1,1)\rangle \\ &= \langle x,-a^{(1,1,x)}-\alpha(1,1)^{(1,1,x)} + \alpha(1,x) + g\rangle\langle 1,a-\alpha(1,1)\rangle \\ &= \langle x,-a^{(1,1,x)} + g\rangle\langle 1,a-\alpha(1,1)\rangle \\ &= \langle x,-a^{(1,1,x)} + g + \alpha(x,1) + a^{(x,1,1)} - \alpha(1,1)^{(x,1,1)}\rangle \\ &= \langle x,-a^{(1,1,x)} + a^{(x,1,1)} + g\rangle \\ &= \langle x,g + \delta^0 a(x)\rangle \end{aligned}$$

since $\alpha(x,1) = \alpha(1,1)^{(x,1,1)}$ and $\alpha(1,x) = \alpha(1,1)^{(1,1,x)}$. We have proven the following theorem.

5.13. <u>Theorem</u>. <u>Let</u> $(\overline{S},p) \in A(\underline{\mathrm{Mon}_S})$, <u>and let</u> $F : \mathcal{D}(S) \to \underline{AG}$ <u>be the unique functor such that</u> $F \circ \mathcal{D}(p) = \Sigma_p$. <u>If</u> $\mathrm{Aut}_0(\overline{S},p) = \{f \in \mathrm{Aut}(\overline{S},p) \mid f^x = 1_{\Sigma_p(x)}$ <u>for all</u> $x \in S\}$, <u>and</u> $I_0(\overline{S},p) = \{(\ )^a \mid a \in p^{-1}(1)\}$, <u>then</u> $I_0 \leq \mathrm{Aut}_0 \leq \mathrm{Aut}$ <u>is a sequence of subgroups</u>. <u>Moreover</u> Aut_0 <u>is an abelian group and</u> $H^1(F) \cong \mathrm{Aut}_0/I_0$.

5.14. If we call each object $(S \times (F,\alpha),\pi)$ a coextension of S by F , then not every such coextension is an H-coextension. If we let $Z^2_H(F)$ denote the subset of $Z^2(F)$ consisting of all factor systems giving rise to

H-coextensions, then $Z_H^2(F)$ need not be closed under addition. The next theorem tells us that $H_H^2(F) = Z_H^2(F)/B^2(F)$ makes sense.

5.15. Theorem. Every element of $H^2(F)$ is a class of equivalent coextensions of S by F. Moreover, if $\alpha \in Z_H^2(F)$, then $\alpha + B^2(F) \subseteq Z_H^2(F)$ and $-\alpha \in Z_H^2(F)$.

Proof. If $\psi \in C^1(F)$, then $F^\psi = F$ and $\alpha^\psi = \alpha + \delta^1\psi$. Hence $(S \times (F,\alpha),\pi)$ and $(S \times (G,\beta),\pi)$ are equivalent if and only if $F = G$ and $\beta - \alpha \in B^2(F)$. Moreover, membership in $H(\underline{\underline{Mon}}_S)$ is an equivalence property, so the second assertion is seen. Moreover, since $(F(u)^{(1,u,x)})^{-1} = F(u)^{(1,u,x)}$ and $(F(v)^{(x,v,1)})^{-1} = F(v)^{(x,v,1)}$ for all $u \in \ell_S(x)$, $v \in r_S(x)$, the equalities of 3.9 HII, upon application of the map $g \to g^{-1}$ for $g \in F(x)$, give $-\alpha \in Z^2(F)$.

5.16. Counterexample. Let F_X be the free monoid on X and let $\mathbb{Z}^+$ denote the positive integers under addition. Let $S = F_X \cup \mathbb{Z}^+$ with multiplication on S be defined by $\omega n = n\omega = n$ for $\omega \in F_X$ and $n \in \mathbb{Z}^+$ and the usual multiplication, or addition otherwise. Define $F : \mathbb{D}(S) \to \underline{\underline{AG}}$ by: $F(x) = \{0_A\}$ if $x \in F_X$, and $F(x) = A$ if $x \in \mathbb{Z}^+$; $F(u,x,v) = i_{\{0_A\}}$ if $x \in F_X$, and $F(u,x,v) = i_A$ if $x \in \mathbb{Z}^+$. We first show that $H^2(F) \cong \mathrm{Hom}(F_X,A)$. Let $\alpha \in Z^2(F)$. For all ω_1, $\omega_2 \in F_X$ and $n \in \mathbb{Z}^+$ we have $\alpha(\omega_1\omega_2,n) = \alpha(\omega_1,n) + \alpha(\omega_2,n)$ as $\alpha(\omega_1,\omega_2) = 0_A$ and $\omega_2 n = n$. Similarly for $m, n \in \mathbb{Z}^+$, $\omega \in F_X$ we have $\alpha(m,\omega) + \alpha(m,n) = \alpha(m,n) + \alpha(\omega,n)$, and hence $\alpha(m,\omega) = \alpha(\omega,n)$. Hence α induces a homomorphism $f : F_X \to A$ such that for all $\omega \in F_X$, $n \in \mathbb{Z}^+$ we have $\alpha(\omega,n) = \alpha(n,\omega) = f(\omega)$. Moreover, if we define $\psi : \mathbb{Z}^+ \to A$ by $\langle 1_{\mathbb{Z}}, 1_A\rangle^m = \langle m, -\psi(m)\rangle$, we get $-\psi(m+n) = \alpha(m,n) - \psi(m) - \psi(n)$, and hence $\alpha(m,n) = \delta^1\psi(m,n)$. Conversely it is easy to see that given $f \in \mathrm{Hom}(F_X,A)$

and $\psi : \mathbb{Z}^+ \to A$ we can uniquely determine a factor system, $\alpha \in Z^2(F)$, such that α induces f and ψ. Hence $Z^2(F) \cong \mathrm{Hom}(F_X, A) \oplus A^{\mathbb{Z}^+}$. Under this isomorphism all elements of $A^{\mathbb{Z}^+}$ are members of $B^2(F)$. Hence if $\alpha \in B^2(F)$, we may assume $\alpha | \mathbb{Z}^+ \times \mathbb{Z}^+ = 0_A$. Now if α is not the zero cochain, then $S \times (F,\alpha)$ does not split, since $\alpha(\omega,1) \neq 0$ for some $\omega \in F_X$ implies that no lifting $x \to \bar{x}$ from S to $S \times (F,\alpha)$ gives $\overline{\omega}\,\overline{1} = \overline{1}$. Hence $\alpha : S \times S \to 0_A$. Hence $B^2(F)$ is a direct summand of $Z^2(B)$, and $H^2(F) \cong \mathrm{Hom}(F_X, A)$ under the map $\alpha \to \alpha | F_X \times \{1_{\mathbb{Z}}\}]$. Now $\alpha \in Z^2_H(F)$ if and only if $\alpha[F_X \times \{1_{\mathbb{Z}}\}] = A$. Hence $S \times F$ is not an abelian coextension. Moreover, if $\alpha \in Z^2_H(F)$, then so does $-\alpha$. But $0 = \alpha + (-\alpha)$ does not. Hence $Z^2_H(F)$ is not closed under addition. Finally, let $||X|| = 1$, i.e., F_X is free cyclic. Then $Z^2_H(F)$ is not empty if and only if A is finite cyclic.

5.17. Example. If S is a semigroup with a zero and A is an abelian group, then every ideal extension of A by S is an abelian coextension of S by the functor $F : \mathbb{D}(S) \to \underline{\underline{AG}}$ defined as in the example of 3.49. Moreover, if $\psi \in C^1(F)$, then $\delta^1\psi(0,0) = \psi(0) - \psi(0) + \psi(0) = \psi(0)$. Hence $\delta^1\psi(0,0) = 0_A$ iff $\psi(0) = 0_A$, i.e., $\psi = 0_{C^1(F)}$. Hence by the argument in 3.49, $Z^2(F) = B^2(F) \oplus \mathrm{Hom}^*(S-0,A)$, and $H^2(F) \cong \mathrm{Hom}(S-0,A)$, where $\mathrm{Hom}^*(S-0,A)$ is the image of the embedding of $\mathrm{Hom}(S-0,A)$ into $Z^2(F)$.

5.18. Theorem. Suppose that either S is regular or $F : \mathcal{D}(S) \to \underline{\underline{AG}}^{\circ}$. Then $Z^2(F) = Z^2_H(F)$, so that $H^2(F) = H^2_H(F)$.

Proof. In either case, condition HII is automatically satisfied.

The following is the homological consequence of the Second Main Theorem.

5.19. Theorem. Let $F, G : \mathcal{D}(S) \to \underline{\underline{AG}}$, and let $\alpha \in Z^2_H(F)$ and $\beta \in Z^2_H(G)$.

Then $\mathrm{Hom}_S(S \times (F,\alpha),\ S \times (G,\beta))$ *is not empty if and only if there exists a natural transformation* $\sigma : F \to G$ *such that* $H^2(\sigma)[\alpha + B^2(F)] = \beta + B^2(G)$.

REFERENCES

1. Grillet, A. P., "Left Coset Extensions," Semigroup Forum, to appear.

2. Leech, J. E., "The $\mathcal{D}$-Category of a Monoid," submitted.

THE STRUCTURE OF A BAND OF GROUPS

by

Jonathan Leech

<u>Introduction</u>. In a semigroup, every subgroup is contained in a maximal subgroup and all maximal subgroups are disjoint. A semigroup which is a union of its subgroups is called a union of groups. A well-known theorem of Clifford states that every union of groups is the disjoint union of completely simple semigroups and the induced partition is a congruence partition with the quotient semigroup being a semilattice. Hence every union of groups is a semilattice of completely simple semigroups. In this paper, we study unions of groups which satisfy the restriction that the elementwise product of any pair of maximal subgroups is contained in a third subgroup. In this case the maximal subgroup partition is a congruence partition and the quotient semigroup is a band--a semigroup whose elements satisfy the identity: $x^2 = x$. Such a union of groups is called a band of groups.

Throughout this paper it will be convenient to think of a band of groups as a triple, $\langle S,\pi,\mathbb{B}$, where S is a semigroup, $\mathbb{B}$ is a band, and π is a homomorphism from S nto $\mathbb{B}$ such that $\pi^{-1}(a)$ is a subgroup of S for each $a \in \mathbb{B}$. We call $\langle S,\pi,\mathbb{B}\rangle$ a band of groups over $\mathbb{B}$. Clearly we can define a morphism between two bands of groups over $\mathbb{B}$, $f : \langle S,\pi,\mathbb{B}\rangle \to \langle S',\pi',\mathbb{B}\rangle$ to be a homomorphism $f : S \to S'$ such that $\pi' \circ f = \pi$. In the first section of this paper we give a representation of the category of bands of groups over $\mathbb{B}$ by means of a modified Schreier extension theory.

In the second section we list some of the important properties of bands; in particular, the fact that $\mathcal{D}$ is a congruence on $\mathbb{B}$ and the induced homomorphism $p : \mathbb{B} \to \mathbb{B}/\mathcal{D} = \tilde{\mathbb{B}}$ is a universal morphism from $\mathbb{B}$ to the category of semilattices.

In the third section we discuss central bands of groups and represent them by functors from $\mathbb{B}$ to the category of groups and by normal 2-cocycles which arise in a certain cochain complex described in the fourth section. This leads to the calculation of a certain second cohomology group, and in the fifth and sixth sections this group is calculated for special classes of bands.

In the final section, the splitting case is discussed.

Before beginning we make a few remarks concerning the relationship between this paper and previously obtained results. The first results of interest were obtained by Rees [4] and Clifford [1]. In 1940 Rees described the structure of a rectangular band of groups. In 1941 Clifford determined the structure of a semilattice of groups. In 1963 Yamada [5] gave a description of those bands of groups $(S,\pi,\mathbb{B})$ for which $E(S)$, the set of idempotents of S, is a subsemigroup of S, i.e., S splits over $\mathbb{B}$. He proved that every splitting band of groups is the fibred product of a band and a semilattice of groups over their underlying common (up to isomorphism) semilattice. In 1967 Petrich [3] characterized those bands of groups whose underlying bands are functorial (defined in Section Five) as those constructible by the Clifford construction with the only change being that one uses completely simple semigroup valued functors instead of group valued functors. Theorem 16 of this paper is the cohomological version of this result. It should be mentioned that functorial bands have elsewhere been called normal bands.

1. The General Case. Let us start by considering a fixed band of groups over $\mathbb{B}$, $\langle S,\pi,\mathbb{B}\rangle$.

Notation. For all $\alpha \in \mathbb{B}$, $G(\alpha) = \pi^{-1}(\alpha)$, and $\overline{\alpha} = 1_{G(\alpha)}$.

Definition. For all $\alpha,\beta \in \mathbb{B}$, let $I_{\alpha}^{\alpha\beta}\colon G(\alpha) \to G(\alpha\beta)$ and $J_{\beta}^{\alpha\beta}\colon G(\beta) \to G(\alpha\beta)$ be defined by $I_{\beta}^{\alpha\beta}(x) = x\cdot\overline{\alpha\beta}$, and $J_{\beta}^{\alpha\beta}(y) = \overline{\alpha\beta}\cdot y$, where $x \in G(\alpha)$ and $y \in G(\beta)$. Since $(G(\alpha)\cdot G(\alpha\beta)) \cup (G(\alpha\beta)\cdot G(\beta)) \subseteq G(\alpha\beta)$, $I_{\alpha}^{\alpha\beta}$ and $J_{\beta}^{\alpha\beta}$ are well defined.

Lemma 1. For all $\alpha,\beta \in \mathbb{B}$, $I_{\alpha}^{\alpha\beta}$ and $J_{\beta}^{\alpha\beta}$ are homomorphisms. Moreover, if $\gamma \in \mathbb{B}$, then $I_{\alpha\beta}^{(\alpha\beta)\gamma} \circ I_{\alpha}^{\alpha\beta} = I_{\alpha}^{\alpha(\beta\gamma)}$ and $J_{\beta\gamma}^{\alpha(\beta\gamma)} \circ J_{\gamma}^{\beta\gamma} = J_{\gamma}^{(\alpha\beta)\gamma}$. Finally $I_{\alpha}^{\alpha} = J_{\alpha}^{\alpha} = i_{G(\alpha)}$ for all $\alpha \in \mathbb{B}$.

Proof. Let $x_1,x_2 \in G(\alpha)$. Then we have

$$\begin{aligned} I_{\alpha}^{\alpha\beta}(x_1\cdot x_2) &= (x_1\cdot x_2)\cdot\overline{\alpha\beta} = x_1\cdot(x_2\cdot\overline{\alpha\beta}) \\ &= x_1\cdot I_{\alpha}^{\alpha\beta}(x_2) \\ &= x_1\cdot(\overline{\alpha\beta}\cdot I_{\alpha}^{\alpha\beta}(x_2)) = (x_1\cdot\overline{\alpha\beta})\cdot I_{\alpha}^{\alpha\beta}(x_2) \\ &= I_{\alpha}^{\alpha\beta}(x_1)\, I_{\alpha}^{\alpha\beta}(x_2)\,. \end{aligned}$$

If $\gamma \in \mathbb{B}$, then we have

$$\begin{aligned} x\cdot\overline{\alpha\beta\gamma} &= x\cdot 1_{\alpha\beta}^{\alpha\beta\gamma}(\overline{\alpha\beta}) = x\cdot(\overline{\alpha\beta}\cdot\overline{\alpha\beta\gamma}) \\ &= (x\cdot\overline{\alpha\beta})\cdot\overline{\alpha\beta\gamma}\ , \text{ or} \\ I_{\alpha}^{\alpha(\beta\gamma)}(x) &= I_{\alpha\beta}^{(\alpha\beta)\gamma} \circ I_{\alpha}^{\alpha\beta}(x)\,. \end{aligned}$$

That $I_{\alpha}^{\alpha} = i_{G(\alpha)}$ is obvious. The proofs of the corresponding facts about J are similar.

Corollary. For all $\alpha,\beta \in \mathbb{B}$, $\overline{\alpha}\cdot\overline{\alpha\beta} = \overline{\alpha\beta}\cdot\overline{\beta} = \overline{\alpha\beta}$.

Definition. Let $f : \mathbb{B} \times \mathbb{B} \to U\{G(\alpha) | \alpha \in \mathbb{B}\}$ be the function defined by $f(\alpha,\beta) = \overline{\alpha}\cdot\overline{\beta}$ for all $\alpha,\beta \in \mathbb{B}$. We call f the factor set of $<S,\pi,\mathbb{B}>$.

Lemma 2. For all $x \in G(\alpha)$ and $y \in G(\beta)$ we have

$$x\cdot y = I_\alpha^{\alpha\beta}(x)\cdot f(\alpha,\beta)\ J_\beta^{\alpha\beta}(y) .$$

Proof.
$$\begin{aligned} x\cdot y &= (x\cdot\overline{\alpha})\cdot(\overline{\beta}\cdot y) = x\cdot(\overline{\alpha}\cdot\overline{\beta})\cdot y \\ &= x\cdot f(\alpha,\beta)\cdot y \\ &= x\cdot(\overline{\alpha\beta}\cdot f(\alpha,\beta)\cdot\overline{\alpha\beta}\,)\cdot y \\ &= I_\alpha^{\alpha\beta}(x)\cdot f(\alpha,\beta)\cdot J_\beta^{\alpha\beta}(y) . \end{aligned}$$

Lemma 3. For all $\alpha,\beta\ \gamma \in \mathbb{B}$ and $y \in G(\beta)$ we have

$$\begin{aligned} &f(\alpha,\beta\gamma)\cdot[J_{\beta\gamma}^{\alpha\beta\gamma} \circ I_\beta^{\beta\gamma}(y)]\cdot J_{\beta\gamma}^{\alpha\beta\gamma}(f(\beta,\gamma)) \\ &\qquad = I_{\alpha\beta}^{\alpha\beta\gamma}(f(\alpha,\beta))\cdot[I_{\alpha\beta}^{\alpha\beta\gamma} \circ J_\beta^{\alpha\beta}(y)]\cdot f(\alpha\beta,\gamma) . \end{aligned}$$

Proof. This follows from Lemma 2, by multiplying out $\overline{\alpha}\cdot y\cdot\overline{\gamma}$, first as $\overline{\alpha}\cdot(y\cdot\overline{\gamma})$ and then as $(\overline{\alpha}\cdot y)\cdot\overline{\gamma}$.

Corollary. For all $\alpha,\beta,\gamma \in \mathbb{B}$ we have

$$f(\alpha,\beta\gamma)\cdot J_{\beta\gamma}^{\alpha\beta\gamma}(f(\beta,\gamma)) = I_{\alpha\beta}^{\alpha\beta\gamma}(f(\alpha,\beta))\cdot f(\alpha\beta,\gamma) .$$

Lemma 4. For all $\alpha,\beta \in \mathbb{B}$ we have $f(\alpha,\alpha\beta) = f(\alpha\beta,\beta) = 1_{G(\alpha\beta)}$. In particular we have $f(\alpha,\alpha) = 1_{G(\alpha)}$.

Theorem 1. Let $<S,\pi,\mathbb{B}>$ be a band of groups. Then $<S,\pi,\mathbb{B}>$ naturally induces a family of groups $G = <G(\alpha) | \alpha \in \mathbb{B}>$, two families of connecting homomorphisms $<I_\alpha^{\alpha\beta} : G(\alpha) \to G(\alpha\beta) | \alpha,\beta \in \mathbb{B}>$ and $<J_\beta^{\alpha\beta} : G(\beta) \to G(\alpha\beta) | \alpha,\beta \in \mathbb{B}>$, and a map $f : \mathbb{B} \times \mathbb{B} \to UG$ such that:

(i) For all $\alpha,\beta \in \mathbb{B}$, $f(\alpha,\beta) \in G(\alpha\beta)$.

(ii) *For all* $\alpha \in \mathbb{B}$, $f(\alpha,\alpha) = 1_{G(\alpha)}$.

(iii) *For all* $\alpha,\beta,\gamma \in \mathbb{B}$ *we have*

$$I^{\alpha\beta\gamma}_{\alpha\beta} \circ I^{\alpha\beta}_{\alpha} = I^{\alpha\beta\gamma}_{\alpha} \quad \text{and} \quad J^{\alpha\beta\gamma}_{\beta\gamma} \circ J^{\beta\gamma}_{\gamma} = J^{\alpha\beta\gamma}_{\gamma} .$$

(iv) *For all* $\alpha,\beta,\gamma \in \mathbb{B}$ *and* $y \in G(\beta)$ *we have*

$$f(\alpha,\beta\gamma)\cdot(J^{\alpha\beta\gamma}_{\beta\gamma} \circ I^{\beta\gamma}_{\beta})(y)\cdot J^{\alpha\beta\gamma}_{\beta\gamma}(f(\beta,\gamma)) = I^{\alpha\beta\gamma}_{\alpha\beta}(f(\alpha,\beta))\cdot(I^{\alpha\beta\gamma}_{\alpha\beta} \circ J^{\alpha\beta}_{\beta})(y)\cdot f(\alpha\beta,\gamma) .$$

Conversely, let $\mathbb{B}$ *be a band with each* $\alpha \in \mathbb{B}$; *let us associate a group* $G(\alpha)$ *so that the family* $\langle G(\alpha) | \alpha \in \mathbb{B}\rangle$ *is pairwise disjoint. Also suppose that there exist sets of homomorphisms* $\langle I^{\alpha\beta}_{\alpha} : G(\alpha) \to G(\alpha\beta) | \alpha,\beta \in \mathbb{B}\rangle$ *and* $\langle J^{\alpha\beta}_{\beta} : G(\beta) \to G(\alpha\beta) | \alpha,\beta \in \mathbb{B}\rangle$ *and a function* $f : \mathbb{B} \times \mathbb{B} \to \cup_{\alpha\in\mathbb{B}} G(\alpha)$ *which satisfy* (i) *through* (iv) *above. Then defining a composition,* $\circ_f$, *on the set* $\cup_{\alpha\in\mathbb{B}} G(\alpha)$ by $x \circ_f y = I^{\alpha\beta}_{\alpha}(x)\cdot f(\alpha,\beta)\cdot J^{\alpha\beta}_{\beta}(y)$ *where* $x \in G(\alpha)$ *and* $y \in G(\beta)$, *we thus obtain a band of groups* $\langle(\cup G(\alpha) , \circ_f),\pi,\mathbb{B}\rangle$ *where* π *is defined by* $\pi | G(\alpha) : G(\alpha) \to \{\alpha\}$. *Every band of groups is formed in just this manner.*

Let us now suppose that $\langle S_1,\pi_1,\mathbb{B}\rangle$ and $\langle S_2,\pi_2,\mathbb{B}\rangle$ are bands of groups over $\mathbb{B}$. Let $h : S_1 \to S_2$ be a $\mathbb{B}$-homomorphism. For all $\alpha \in \mathbb{B}$ let us denote $\pi_1^{-1}(\alpha)$ by $G_1(\alpha)$ and $\pi_2^{-1}(\alpha)$ by $G_2(\alpha)$. For all $\alpha \in \mathbb{B}$, $h_\alpha = h | G_1(\alpha)$ is a group homomorphism from $G_1(\alpha) \to G_2(\alpha)$. Let us denote the collection $\langle h_\alpha | \alpha \in \mathbb{B}\rangle$ by $h_{\mathbb{B}}$.

Theorem 2. Suppose $h : \langle S_1,\pi_1,\mathbb{B}\rangle \to \langle S_2,\pi_2,\mathbb{B}\rangle$ *is a* $\mathbb{B}$-*homomorphism. Then for all* $\alpha,\beta \in \mathbb{B}$ *the following diagram commutes*

$$\begin{array}{ccccc}
G_1(\alpha) & \xrightarrow{I^{\alpha\beta}_{\alpha}} & G_1(\alpha\beta) & \xleftarrow{J^{\alpha\beta}_{\beta}} & G_1(\beta) \\
\downarrow h_\alpha & & \downarrow h_{\alpha\beta} & & \downarrow h_\beta \\
G_2(\alpha) & \xrightarrow{I^{\alpha\beta}_{\alpha}} & G_2(\alpha\beta) & \xleftarrow{J^{\alpha\beta}_{\beta}} & G_2(\beta)
\end{array}$$

and $h_{\alpha\beta}(f_1(\alpha,\beta)) = f_2(\alpha,\beta)$ where $f_i(\alpha,\beta)$ is the factor set of $<S_i,\pi_i,\mathbb{B}>$ for $i = 1,2$. Conversely suppose that $h_{\mathbb{B}} = <h_\alpha : G_1(\alpha) \to G_2(\alpha) | \alpha \in \mathbb{B}>$ is a collection of homomorphisms satisfying the above conditions. Then the map $h : S_1 \to S_2$ defined by $h|G_1(\alpha) = h_\alpha$ for all $\alpha \in \mathbb{B}$ is a $\mathbb{B}$-homomorphism.

2. Some Remarks on the Category of Bands. In order to obtain sharper results concerning the structure of bands of groups we present some basic facts about the structure of bands.

Definition. By a left zero semigroup, Λ , we mean a nonempty set, Λ , together with the composition $\lambda_1\lambda_2 = \lambda_1$ for all $\lambda_1,\lambda_2 \in \Lambda$. By a right zero semigroup, P , we mean a nonempty set, P , together with the composition $\rho_1\rho_2 = \rho_2$ for all $\rho_1,\rho_2 \in P$. By a rectangular band we mean any semigroup isomorphic with the direct product of a left zero semigroup and a right zero semigroup, $\Lambda \times P$.

Definition. Let $\mathbb{B}$ be a band, and let $\alpha,\beta \in \mathbb{B}$. If $\alpha\mathbb{B} = \beta\mathbb{B}$ we say that α and β are right equivalent (they generate the same right ideal) and denote this by $\alpha\mathcal{R}\beta$. Similarly, if $\mathbb{B}\alpha = \mathbb{B}\beta$ we denote this by $\alpha\mathcal{L}\beta$. If $\mathbb{B}\alpha\mathbb{B} = \mathbb{B}\beta\mathbb{B}$ we denote this by $\alpha\mathcal{D}\beta$.

It should be easy to see that $\mathcal{R}$, $\mathcal{L}$, and $\mathcal{D}$ are equivalence relations, and for each $\alpha \in \mathbb{B}$ we denote by $\mathcal{R}(\alpha)$, $\mathcal{L}(\alpha)$, and $\mathcal{D}(\alpha)$ the respective equivalence classes of α .

McLean's Theorem. [2] For all $\alpha \in \mathbb{B}$, $\mathcal{R}(\alpha)$ is a right zero semigroup, $\mathcal{L}(\alpha)$ is a left zero semigroup, and $\mathcal{D}(\alpha)$ is a rectangular band which is the internal direct product of $\mathcal{R}(\alpha)$ and $\mathcal{L}(\alpha)$. $\mathcal{D}(\alpha) = \mathcal{L}(\alpha)\cdot\mathcal{R}(\alpha)$ and $\alpha = \mathcal{R}(\alpha)\cdot\mathcal{L}(\alpha)$. Moreover for all $\alpha \in \mathbb{B}$, $\mathcal{L}(\alpha)$ $(\mathcal{R}(\alpha),\mathcal{D}(\alpha))$ is a maximal left zero subsemigroup (*right zero subsemigroup*, *rectangular subband*) *of* $\mathbb{B}$. *Finally, the decomposition* $\tilde{\mathbb{B}} = \langle \mathcal{D}(\alpha) \mid \alpha \in \mathbb{B} \rangle$ *is a congruence decomposition such that* $\tilde{\mathbb{B}} = \mathbb{B}/\mathcal{D}$ *is a semilattice, and the canonical map* $p : \mathbb{B} \to \tilde{\mathbb{B}}$ *is a universal morphism from* $\mathbb{B}$ *to the category of semilattices.*

Definition. Let $\alpha, \beta \in \mathbb{B}$. We say that α *divides* β if and only if $\beta \in \mathbb{B}\alpha\mathbb{B}$. We denote this by $\alpha \mid \beta$.

Notation. $p(\alpha)$ will be denoted by $\tilde{\alpha}$ for all $\alpha \in \mathbb{B}$.

Notation. Let $\alpha, \beta \in \mathbb{B}$. We denote $\alpha\beta = \beta\alpha = \beta$ by $\alpha \geq \beta$.

Lemma 5. If $\alpha, \beta \in \mathbb{B}$, then the following are equivalent:

(a) $\alpha \mid \beta$ (b) $\beta\alpha\beta = \beta$ (c) $\tilde{\alpha} \geq \tilde{\beta}$.

3. *Central Bands of Groups.*

Definition. Let $\langle S, \pi, \mathbb{B} \rangle$ be a band of groups over $\mathbb{B}$, and suppose that $\alpha, \beta \in \mathbb{B}$. Then the maps $I^{\alpha\beta}_{\alpha}$ and $J^{\alpha\beta}_{\beta}$ are called *inner* homomorphisms. Suppose that $\alpha \mid \beta$. Then by an *algebraic homomorphism* $h : G(\alpha) \to G(\beta)$ we mean any homomorphism which is the composition of a finite number of inner homomorphisms.

Consider the following statements which may or may not hold about $\langle S, \pi, \mathbb{B} \rangle$.

I. There exists only one algebraic homomorphism between any two subgroups, $G(\alpha)$ and $G(\beta)$, where $\alpha \mid \beta$.

II. For all $\alpha, \beta, \gamma \in \mathbb{B}$ the following diagram commutes:

$$\begin{array}{ccc} G(\alpha\beta) & \xrightarrow{I^{\alpha\beta\gamma}_{\alpha\beta}} & G(\alpha\beta\gamma) \\ {\scriptstyle J^{\alpha\beta}_{\beta}}\uparrow & & \uparrow{\scriptstyle J^{\alpha\beta\gamma}_{\beta\gamma}} \\ G(\beta) & \xrightarrow{I^{\beta\gamma}_{\beta}} & G(\beta\gamma) \end{array}$$

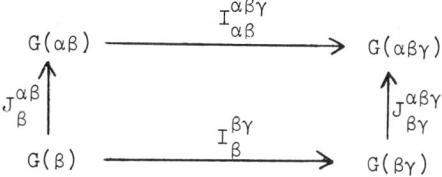

III. For all $\alpha,\beta \in \mathbb{B}$, $f(\alpha,\beta) \in Z(G(\alpha\beta))$, the center of $G(\alpha\beta)$.

Certainly I implies II. Also III implies II by Lemma 3. In the first part of this section it is shown that all three conditions are equivalent. If $\langle S,\pi,\mathbb{B}\rangle$ satisfies any one, and hence all, of these conditions, we call $\langle S,\pi,\mathbb{B}\rangle$ a <u>central band of groups</u>.

<u>Lemma 6</u>. <u>For all</u> $\alpha,\beta \in \mathbb{B}$, $f(\alpha\beta,\alpha) = f(\alpha,\beta\alpha) = 1_{G(\alpha\beta\alpha)}$.

<u>Proof</u>. That $f(\alpha\beta,\alpha) = 1_{G(\alpha\beta\alpha)}$ follows directly from $\overline{\alpha\beta}\cdot\overline{\alpha}\cdot\overline{\alpha\beta\alpha} = 1_{G(\alpha\beta\alpha)}$ using the corollary to Lemma 1. $f(\alpha,\beta\alpha) = 1_{G(\alpha\beta\alpha)}$ is seen in a similar manner.

<u>Lemma 7</u>. <u>For all</u> $\alpha,\beta \in \mathbb{B}$:

(a) $\alpha \geq \alpha\beta\alpha$, $\beta \geq \beta\alpha\beta$, <u>and</u> $\alpha\beta\alpha\mathcal{D}\beta\alpha\beta$.

(b) $f(\alpha,\beta) = f(\alpha\beta\alpha,\beta\alpha\beta)$.

<u>Proof</u>. $\overline{\alpha\beta\alpha}\cdot f(\alpha,\beta)\cdot\overline{\beta\alpha\beta} = \overline{\alpha\beta\alpha}\cdot\overline{\alpha\beta}\cdot f(\alpha,\beta)\cdot\overline{\alpha\beta}\cdot\overline{\beta\alpha\beta} = f(\alpha\beta\alpha,\alpha\beta)\cdot f(\alpha,\beta)\cdot f(\alpha\beta,\beta\alpha\beta) = f(\alpha,\beta)$ by Lemma 6. But we also have $\overline{\alpha\beta\alpha}\cdot f(\alpha,\beta)\cdot\overline{\beta\alpha\beta} = \overline{\alpha\beta\alpha}\cdot\overline{\alpha}\cdot\overline{\beta}\cdot\overline{\beta\alpha\beta} = \overline{\alpha\beta\alpha}\cdot\overline{\beta\alpha\beta} = f(\alpha\beta\alpha,\beta\alpha\beta)$, and (b) follows. The proof of (a) is trivial.

Hence showing that II implies III reduces to showing this whenever the α,β of III satisfy $\alpha\mathcal{D}\beta$.

<u>Lemma 8</u>. <u>Let</u> $\alpha,\beta \in \mathbb{B}$ <u>and suppose</u> $\alpha\mathcal{D}\beta$. <u>Then</u> $I_{\alpha}^{\alpha\beta}$ <u>and</u> $J_{\beta}^{\alpha\beta}$ <u>are isomorphisms</u>.

<u>Proof</u>. We have $I_{\alpha\beta}^{\alpha\beta\alpha} \circ I_{\alpha}^{\alpha\beta} = I_{\alpha}^{\alpha\beta\alpha} = I_{\alpha}^{\alpha} = i_{G(\alpha)}$ as $\alpha\beta\alpha = \alpha$. Also $I_{\alpha}^{\alpha\beta} \circ I_{\alpha\beta}^{\alpha\beta\alpha} = I_{\alpha\beta}^{\alpha\beta\alpha\beta} = I_{\alpha\beta}^{\alpha\beta} = i_{G(\alpha\beta)}$. Hence $I_{\alpha}^{\alpha\beta}$ is an isomorphism. Similarly $J_{\beta}^{\alpha\beta}$ is an isomorphism.

<u>Lemma 9</u>. <u>Let</u> $\alpha,\beta \in \mathbb{B}$ <u>and suppose</u> $\alpha\mathcal{D}\beta$. <u>Then</u> $f(\alpha,\beta) \in Z(G(\alpha\beta))$ <u>if</u> II <u>holds</u>.

Proof. Setting $\gamma = \alpha$ in Lemma 3 we see that

$$f(\alpha,\beta\alpha)[J^{\alpha\beta\alpha}_{\beta\alpha} \circ I^{\beta\alpha}_{\beta}(y)]J^{\alpha\beta\alpha}_{\beta\alpha}(f(\beta,\alpha)) = I^{\alpha\beta\alpha}_{\alpha\beta}(f(\alpha,\beta))\cdot[I^{\alpha\beta\alpha}_{\alpha\beta} \circ J^{\alpha\beta}_{\beta}(y)]\cdot f(\alpha\beta,\alpha) .$$

Since $f(\alpha,\beta\alpha) = f(\alpha\beta,\alpha) = 1_{G(\alpha)}$ and II holds, this becomes

$$J^{\alpha}_{\beta\alpha} \circ I^{\beta\alpha}_{\beta}(y)\cdot J^{\alpha}_{\beta\alpha}(f(\beta,\alpha)) = I^{\alpha}_{\alpha\beta}(f(\alpha,\beta))\cdot J^{\alpha}_{\beta\alpha} \circ I^{\beta\alpha}_{\beta}(y) .$$

Setting $y = 1$ gives us $J^{\alpha}_{\beta\alpha}(f(\beta,\alpha)) = I^{\alpha}_{\alpha\beta}(f(\alpha,\beta))$. Hence

$$J^{\alpha}_{\beta\alpha} \circ I^{\beta\alpha}_{\beta}(y)\cdot I^{\alpha}_{\alpha\beta}(f(\alpha,\beta)) = I^{\alpha}_{\alpha\beta}(f(\alpha,\beta))\cdot J^{\alpha}_{\beta\alpha} \circ I^{\beta\alpha}_{\alpha}(y) .$$

But $J^{\alpha}_{\beta\alpha} \circ I^{\beta\alpha}_{\beta}$ is an isomorphism and hence surjective so that $I^{\alpha}_{\alpha\beta}(f(\alpha,\beta) \in Z(G(\alpha))$, and since $I^{\alpha}_{\alpha\beta}$ is an isomorphism, $f(\alpha,\beta) \in Z(G(\alpha\beta))$ follows.

Theorem 3. Conditions II *and* III *are equivalent.*

Definition. A band of groups, $\langle S,\pi,\mathbb{B}\rangle$, splits if and only if the map $\alpha \to \bar{\alpha}$ is a homomorphism.

Lemma 10. Let $\langle S,\pi,\mathbb{B}\rangle$ *split over* $\mathbb{B}$. *If* $\alpha,\beta,\gamma \in \mathbb{B}$ *set* $\delta = \alpha\beta\gamma$. *Then for all* $x \in G(\beta)$ *we have* $\bar{\alpha}x\bar{\gamma} = \bar{\delta}x\bar{\delta}$.

Proof.

$$\begin{aligned}
\bar{\alpha}x\bar{\gamma} &= \bar{\delta}(\overline{\bar{\alpha}x\bar{\gamma}})\bar{\delta} = \overline{\delta\alpha}(\overline{x\beta\gamma\delta}) \\
&= \overline{\delta\alpha}(\overline{\beta\gamma\delta}\ I^{\beta\gamma\delta}_{\beta}(x)) = \bar{\delta}I^{\beta\gamma\delta}_{\beta}(x) \\
&= \overline{\delta\beta\gamma\delta}\ I^{\beta\gamma\delta}_{\beta}(x) \\
&= \bar{\delta}x\overline{\beta\gamma\delta} = \overline{\delta\beta}x\overline{\gamma\delta} = J^{\delta\beta}_{\beta}(x)\overline{\delta\beta\gamma\delta} \\
&= J^{\delta\beta}_{\beta}(x)\overline{\delta\beta\delta} \\
&= \overline{\delta\beta}x\bar{\delta} = \bar{\delta}x\bar{\delta} .
\end{aligned}$$

(In the above proof we constantly used Lemma 5).

Theorem 4. Conditions I *and* II *are equivalent.*

Proof. Let us first assume that $\langle S,\pi,\mathbb{B}\rangle$ splits. Assuming II holds we can use it to see that every algebraic homomorphism from $G(\beta)$ to $G(\delta)$ is equal to $J^{\alpha\beta\gamma}_{\beta\gamma} \circ I^{\beta\gamma}_{\beta}$ for some α,γ such that $\delta = \alpha\beta\gamma$. But by Lemma 10 $J^{\alpha\beta\gamma}_{\beta\gamma} \circ I^{\beta\gamma}_{\beta} = J^{\delta}_{\beta\delta} \circ I^{\beta\delta}_{\beta}$. Hence I follows. Assume now that $\langle S,\pi,\mathbb{B}\rangle$ does not necessarily split. If II holds then we can use the I and J homomorphisms to construct a band of groups which splits, but still has the same I and J homomorphisms. Hence since every algebraic homomorphism in the original band of groups is also an algebraic homomorphism in the new band of groups we see that condition I holds.

We shall now proceed to construct a certain class of central bands of groups and show that every central band of groups is equivalent to some member of this class.

Let $\mathbb{B}$ be a band, and let $\mathrm{Hom}(\tilde{\mathbb{B}},\underline{\underline{Gr}})$ denote the category whose objects are functors from $\tilde{\mathbb{B}}$ to $\underline{\underline{Gr}}$, the category of groups, and whose morphisms are natural transformations between functors. If $L \in \mathrm{Hom}(\tilde{\mathbb{B}},\underline{\underline{Gr}})$, then we let $H^2(\mathbb{B},L)$ denote the set of all maps $f : \mathbb{B} \times \mathbb{B} \to \cup_{a \in \tilde{\mathbb{B}}} L(a)$ such that:

(a) $\forall\ \alpha,\beta \in \mathbb{B}$, $f(\alpha,\beta) \in Z(L(\tilde{\alpha}\tilde{\beta}))$.

(b) $\forall\ \alpha \in \mathbb{B}$, $f(\alpha,\alpha) = 1_{L(\tilde{\alpha})}$.

(c) $\forall\ \alpha,\beta,\gamma \in \mathbb{B}$, $f(\alpha,\beta\gamma)L^{\alpha\beta\gamma}_{\beta\gamma}(f(\beta,\gamma)) = L^{\alpha\beta\gamma}_{\alpha\beta}(f(\alpha,\beta))f(\alpha\beta,\gamma)$.

(Here L^{v}_{u} denotes $L^{\tilde{v}}_{\tilde{u}}$ for $u,v \in \mathbb{B}$) .

$H^2(\mathbb{B},L)$ is an abelian group under pointwise multiplication, and we shall later see that $H^2(\mathbb{B},L)$ is isomorphic with the second cohomology group of a cochain complex associated with L .

Construction. Let $\mathbb{B}$ be a band, let $L \in \mathrm{Hom}(\tilde{\mathbb{B}},\underline{\underline{Gr}})$, and let $f \in H^2(\mathbb{B},L)$. Set $\mathbb{B} \times L = \{\langle\alpha,x\rangle \mid \alpha \in \mathbb{B}\ ,\ x \in L(\tilde{\alpha})\}$ and define a composition, $\circ_f$, on

$\mathbb{B} \times L$ by

$$\langle\alpha,x\rangle o_f \langle\beta,y\rangle = \langle\alpha\beta, L_\alpha^{\alpha\beta}(x)\cdot f(\alpha,\beta)\cdot L_\beta^{\alpha\beta}(y)\rangle \ .$$

We denote $\langle\mathbb{B} \times L, o_f\rangle$ by $\mathbb{B} \times_f L$ and the map $\langle\alpha,x\rangle \to \alpha$ from $\mathbb{B} \times_f L \to \mathbb{B}$ by π .

<u>Theorem 5</u>. $\langle\mathbb{B} \times_f L, \pi, \mathbb{B}\rangle$ <u>is a central band of groups</u>.

Let $\langle S,\pi,\mathbb{B}\rangle$ be a central band of groups over $\mathbb{B}$. Let $\ell : \tilde{\mathbb{B}} \to \mathbb{B}$ be a lifting map, i.e., if $p : \mathbb{B} \to \tilde{\mathbb{B}}$ is the canonical epimorphism, $p \circ \ell = i_{\tilde{\mathbb{B}}}$ even though ℓ may possibly not be a homomorphism. Corresponding to ℓ we define $L : \tilde{\mathbb{B}} \to \underline{\underline{Gr}}$ by

(a) $\forall\ \alpha \in \tilde{\mathbb{B}}$, $L(a) = G(\ell(a))$, and

(b) If $a|b$ in $\tilde{\mathbb{B}}$, then $L_a^b = J_{\ell(a)\ell(b)}^{\ell(b)} \circ I_{\ell(a)}^{\ell(a)\ell(b)}$.

Since condition I holds, $L \in \mathrm{Hom}(\tilde{\mathbb{B}},\underline{\underline{Gr}})$. We define $f_0 : \mathbb{B} \times \mathbb{B} \to \cup_{a\in\tilde{\mathbb{B}}} L(a)$ by

$$f_0(\alpha,\beta) = J_{\alpha\beta\ell(\widetilde{\alpha\beta})}^{\ell(\widetilde{\alpha\beta})} \circ I_{\alpha\beta}^{\alpha\beta\ell(\widetilde{\alpha\beta})}(f(\alpha,\beta)) \ .$$

Again since condition I holds it is easily seen that $f_0 \in H^2(\mathbb{B},L)$.

<u>Theorem 6</u>. $\langle S,\pi_S,\mathbb{B}\rangle \cong \langle\mathbb{B} \times_{f_0} L,\pi,\mathbb{B}\rangle$ <u>under the isomorphism</u>, ϕ , <u>defined by</u>

$$\phi(x) = \langle\pi_S(x), J_{\pi_S(x)\cdot\ell(\pi_S(s))}^{\ell(\pi_S(x))} \circ I_{\pi_S(x)}^{\pi_S(x)\cdot\ell(\pi_S(x))}(x)\rangle \ .$$

Let us now suppose that $L_1,L_2 \in \mathrm{Hom}(\tilde{\mathbb{B}},\underline{\underline{Gr}})$, and let $\sigma : L_1 \to L_2$ be a natural transformation. σ induces a map $\sigma^2 : \Pi_{\langle\alpha,\beta\rangle\in\mathbb{B}\times\mathbb{B}} L_1(\widetilde{\alpha\beta}) \to \Pi_{\langle\alpha,\beta\rangle\in\mathbb{B}\times\mathbb{B}} L_2(\widetilde{\alpha\beta})$ defined by $\sigma^2(f)(\alpha,\beta) = \sigma(\alpha\beta)[f(\alpha,\beta)]$ which is a homomorphism. σ also induces a map $\sigma^\# : \mathbb{B} \times L_1 \to \mathbb{B} \to L_2$ defined by $\sigma^\#(\langle\alpha,x\rangle) = \langle\alpha,\sigma(\tilde{\alpha})[x]\rangle$.

<u>Theorem 7</u>. <u>Let</u> $L_1,L_2 \in \mathrm{Hom}(\mathbb{B},\underline{\underline{Gr}})$, <u>let</u> $\sigma : L_1 \to L_2$ <u>be a natural transformation, and let</u> $f \in H^2(\mathbb{B},L_1)$ <u>and</u> $g \in H^2(\mathbb{B},L_2)$. <u>Then</u> $\sigma^\# : \mathbb{B} \times_f L_I \to \mathbb{B} \times_g L_2$

is a homomorphism over $\mathbb{B}$ *if and only if* $\sigma^2(f) = g$. *Moreover, if* $\phi : \mathbb{B} \times_f L_1 \to \mathbb{B} \times_g L_2$ *is a homomorphism over* $\mathbb{B}$ *there must exist a* $\sigma \in \mathrm{Nat}(L_1, L_2)$ *such that* $\phi = \sigma^{\#}$.

Proof. Only the proof of the last assertion is presented here. For each $\alpha \in \mathbb{B}$ we define a map $\sigma_0(\alpha) : L_1(\tilde{\alpha}) \to L_2(\tilde{\alpha})$ by the equation $\phi(<\alpha,x>) = <\alpha,\sigma_0(\alpha)[x]>$. If $\beta \mathcal{D} \alpha$ in $\mathbb{B}$, then since $<\beta,x> = <\beta\alpha,1> o_f <\alpha,x> o_f <\beta,1>$ and ϕ is a homomorphism we can show easily that $\sigma_0(\alpha) = \sigma_0(\beta)$. Hence for each $\alpha \in \tilde{\mathbb{B}}, \phi$ induces a homomorphism $\sigma(a) : L_1(a) \to L_2(a)$ which satisfies $\phi(<\alpha,x>) = <\alpha,\sigma(\tilde{\alpha})[x]>$ for all $\alpha \in \mathbb{B}$. Let $\alpha,\beta \in \mathbb{B}$. Then for all $x \in L_1(\alpha)$ we have $\phi(<\alpha,x> o_f <\beta,1>) = \phi(<\alpha\beta, f(\alpha,\beta)L_{1\alpha}^{\alpha\beta}(x)>) = <\alpha\beta, \sigma(\widetilde{\alpha\beta})[f(\alpha,\beta)]\cdot\sigma(\widetilde{\alpha\beta})[L_{1\alpha}^{\alpha\beta}(x)]>$ and $\phi(<\alpha,x> o_f <\beta,1>) = \phi(<\alpha,x>) o_g \phi(<\beta,1>) = <\alpha,\sigma(\tilde{\alpha})[x]> o_g <\beta,1> = <\alpha\beta, L_{2\tilde{\alpha}}^{\alpha\beta} o\sigma(\tilde{\alpha})[x]\cdot g(\alpha,\beta)>$. Hence for all $\alpha,\beta \in \mathbb{B}$, $x \in L_1(\tilde{\alpha})$ we have

$$\sigma(\widetilde{\alpha\beta})[f(\alpha,\beta)]\cdot\sigma(\widetilde{\alpha\beta})[L_{1\alpha}^{\alpha\beta}(x)] = g(\alpha,\beta)L_{2\alpha}^{\alpha\beta}(\sigma(\tilde{\alpha})[x]) .$$

Setting $x = 1$ we get $\sigma(\widetilde{\alpha\beta})[f(\alpha,\beta)] = g(\alpha,\beta)$. Then cancelling this from the above equation we get $\sigma(\widetilde{\alpha\beta}) o L_{1\alpha}^{\alpha\beta} = L_{2\alpha}^{\alpha\beta} o \sigma(\tilde{\alpha})$. Hence $\sigma : L_1 \to L_2$ is a natural transformation and $\phi = \sigma^{\#}$.

Of course $\mathbb{B} \times_f L$ splits over $\mathbb{B}$ if and only if $f = 1_{H^2(\mathbb{B},L)}$. In this case we denote $\mathbb{B} \times_f L$ by $\mathbb{B} \times L$.

Corollary. If $L_1, L_2 \in \mathrm{Hom}(\tilde{\mathbb{B}}, \underline{\underline{Gr}})$, then $\mathrm{Hom}_{\mathbb{B}}(\mathbb{B} \times L_1, \mathbb{B} \times L_2) \cong \mathrm{Nat}(L_1, L_2)$.

(The $\cong$ denotes a natural correspondence and not necessarily an algebraic isomorphism. If, however, $L_2(a)$ is abelian for all $a \in \tilde{\mathbb{B}}$, then both sets are naturally abelian groups and $\cong$ becomes an isomorphism).

Definition. Let $f \in H^2(\mathbb{B},L)$. Then for all $a \in \mathbb{B}$, $K_f(a)$ is the subgroup of $L(a)$ generated by the set $\{f(\alpha,\beta) \mid \tilde{\alpha} = \tilde{\beta} = a\}$. Of course $K_f(a) \subseteq Z(L(a))$.

Lemma 11. $L_a^b(K_f(a)) \subseteq K_f(b)$.

Proof. Let $\alpha,\beta \in \mathbb{B}_a$ and $\gamma \in \mathbb{B}_b$ where $a \geq b$. Then $L_a^b(f(\alpha,\beta)) = f(\alpha\beta,\gamma)^{-1}f(\alpha,\beta\gamma)f(\beta,\gamma)$, and using Lemma 7 we see that $L_a^b(f(\alpha,\beta)) \in K_f(b)$. Hence $L_a^b(K_f(a)) \subseteq K_f(b)$.

Hence we can define a quotient function L/K_f by setting $(L/K_f)(a) = L(a)/K_f(a)$ for all $a \in \tilde{\mathbb{B}}$ and letting $(L/K_f)_a^b$ be the homomorphism induced by $L_a^b : (L(a)\ ,\ K_f(a)) \to (L(b)\ ,\ K_f(b))$. Let $\sigma_f : L \to L/K_f$ be the induced natural transformation, i.e., $\sigma_f(a) : L(a) \to L(a)/K_f(a)$ is the canonical map. Certainly $\sigma_f^2(f) = 1$.

Theorem 8. $\sigma_f^{\#} : \mathbb{B} \times_f L \to \mathbb{B} \times (L/K_f)$ is a universal homomorphism from $\mathbb{B} \times_f L$ to the category of all bands of groups over $\mathbb{B}$ which split over $\mathbb{B}$.

Let $L \in \mathrm{Hom}(\mathbb{B},\underline{\underline{Gr}})$. We define the functor $Z(L)$ by:

(1) $Z(L)(a) = \bigcap_{b \leq a} (L_a^b)^{-1}[Z(L(b))]$.

(2) $Z(L)_a^b = L_a^b | Z(L)(a)$.

By Lemma 11 $Z(L) \in \mathrm{Hom}(\tilde{\mathbb{B}},\underline{\underline{AG}})$ where $\underline{\underline{AG}}$ is the category of abelian groups and $H^2(\mathbb{B},L) = H^2(\mathbb{B},Z(L))$. Hence we shall see in the next section that $H^2(\mathbb{B},L)$ is isomorphic to the second cohomology group of a certain cochain complex associated with $Z(L)$.

Definition. Let $L \in \mathrm{Hom}(\tilde{\mathbb{B}},\underline{\underline{Gr}})$. Then a differential transformation of L is any natural transformation $\theta : L \to L$ such that for all $a \in \tilde{\mathbb{B}}$ we have $\mathrm{Im}(\theta(a)) \subseteq \mathrm{Ker}(\theta(a))$.

Let $f \in H^2(\mathbb{B},L)$. By Diff(L,f) we mean the set of all differential transformations of L, θ, such that for all $a \in \tilde{\mathbb{B}}$, $\mathrm{Im}(\theta(a)) \subseteq K_f(a) \subseteq \mathrm{Ker}(\theta(a))$. Diff(L,f) is an abelian group under pointwise multiplication, i.e., $\theta * \Phi(a)[x] = \theta(a)[x]\Phi(a)[x]$.

We set $\mathrm{Aut}_f(L)$ to be the set of all functorial automorphisms of L, σ, such that for all $a \in \tilde{\mathbb{B}}$, $\sigma(a)|K_f(a) = i_{K_f(a)}$. Similarly we define $\mathrm{End}_f(L)$.

We now define an embedding $\chi : \mathrm{Diff}(L,f) \to \mathrm{Aut}_f(L)$ by $\theta^{\chi}(a)[x] = x \cdot \theta(a)[x]$. Since $K_f(a) \subseteq Z(L)(a)$ for all $a \in \tilde{\mathbb{B}}$, it is easy to see that $\theta^{\chi}(a)$ is an endomorphism. Moreover, if $\theta^{\chi}(a)[x] = \theta^{\chi}(a)[y]$, then $x\theta(a)[x] = y\theta(a)[y]$, and applying $\theta(a)$ to this equality we get $\theta(a)[x] = \theta(a)[y]$, and by cancelling we get $x = y$ so that θ^{χ} is a monomorphism. Since $\theta^{\chi}(a)[x\theta(a)[x^{-1}]]$ is x we see that $\theta^{\chi}(a) \in \mathrm{Aut}(L(a))$. Clearly $\theta^{\chi} \in \mathrm{Aut}_f(L)$ and χ is one-to-one. Finally we have

$$\begin{aligned}\theta^{\chi}(a) \circ \Phi^{\chi}(a)[x] &= x\Phi(a)[x]\theta(a)[x\Phi(a)[x]] \\ &= x\Phi(a)[x]\theta(a)[x] \\ &= x\theta(a)[x]\Phi(a)[x] \\ &= x(\theta * \Phi)(a)[x] \\ &= (\theta * \Phi)^{\chi}(a)[x] .\end{aligned}$$

Hence χ is an embedding. We denote the image of χ by $\mathcal{D}_f(L)$. It is easy to see that for all $\theta^{\chi} \in \mathcal{D}_f(L)$ and $\sigma \in \mathrm{End}_f(L)$ we have $\sigma \circ \theta^{\chi}(a)[x] = \sigma(a)[x]\theta(a)[x]$.

Theorem 9. Let $f \in H^2(\mathbb{B},L)$. Then $\mathrm{End}_f(L) \cong \mathrm{End}_{\mathbb{B}}(\mathbb{B} \times_f L)$, and $\mathrm{Aut}_f(L) \cong \mathrm{Aut}_{\mathbb{B}}(\mathbb{B} \times_f L)$ under the map $\sigma \to \sigma^{\#}$. Moreover there exists a monoid homomorphism $\psi : \mathrm{End}_f(L) \to \mathrm{End}(L/K_f)$ such that for all $\sigma \in \mathrm{End}_f(L)$ the following diagram commutes:

$$\begin{array}{ccc} L/K_f & \xrightarrow{\psi(\sigma)} & L/K_f \\ \uparrow{\scriptstyle \sigma_f} & & \uparrow{\scriptstyle \sigma_f} \\ L & \xrightarrow{\sigma} & L \end{array}$$

Finally, if $\sigma,\tau \in End_f(L)$, *then* $\psi(\sigma) = \psi(\tau)$ *if and only if there exists* $\theta^X \in \mathcal{D}_f(L)$ *such that* $\sigma = \tau\circ\theta^X$, *so that* $\mathcal{D}_f(L) \trianglelefteq Aut_f(L)$ *and* $Aut_f(L)$ *is a central group extension of* $\mathcal{D}_f(L)$ *by a subgroup of* $Aut(L/K_f)$.

Proof. If $\sigma \in End_f(L)$, define $\psi(\sigma)$ by

$$\psi(\sigma)(a)[x + K_f(a)] = \sigma(a)[x] + K_f(a) .$$

It is easy to see that the diagram commutes. If $\sigma = \tau\circ\theta^X$, then clearly $\psi(\sigma) = \psi(\tau)$. Conversely, if $\psi(\sigma) = \psi(\tau)$, then for all $a \in \tilde{\mathbb{B}}$, $x \in L(a)$ we must have $\sigma(a)[x] = \tau(a)[x]\cdot\theta(a)[x]$ where $\theta(a)[x] \in K_f(a)$. Since $\sigma(a)$ and $\tau(a)$ are endomorphisms, $\theta(a) : L(a) \to K_f(a)$, and $K_f(a) \subseteq Z(L(a))$, it is easy to see that $\theta(a)$ is an endomorphism of L . Moreover, since $\sigma(a)|K_f(a) = \tau(a)|K_f(a) = i_{K_f(a)}$, we have $Im(\theta(a)) \subseteq K_f(a) \subseteq Ker(\theta(a))$. That $\theta \in End(L)$ follows from $\sigma,\tau \in End(L)$. Hence $\theta = \{\theta(a) | a \in \tilde{\mathbb{B}}\} \in Diff(L,f)$. But $\tau\circ\theta^X(a)[x] = \tau(a)[x]\theta(a)[x]$ for all $a \in \tilde{\mathbb{B}}$ and $x \in L(a)$. Hence $\sigma = \tau\circ\theta^X$.

Before proceeding further it should be mentioned that the homomorphism $\omega : \mathbb{B} \times_f L \to \tilde{\mathbb{B}} \times L/K_f$ defined by $\omega(\langle\alpha,x\rangle) = \langle\tilde{\alpha}, x + K_f(\tilde{\alpha})\rangle$ is a universal homomorphism from $\mathbb{B} \times_f L$ to the category of semilattices of groups. The proof of this fact is a routine verification.

4. Cohomological Considerations. Suppose that $\langle S,\pi,\mathbb{B}\rangle$ is a band of groups all of whose subgroups are abelian. In this case we call $\langle S,\pi,\mathbb{B}\rangle$ a band of abelian groups. Of course every band of abelian groups is a central band of groups, and the category of bands of abelian groups over the band $\mathbb{B}$ can be

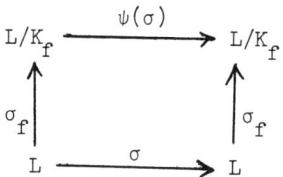

represented by functors $L \in \mathrm{Hom}(\tilde{\mathbb{B}},\underline{\underline{AG}})$, where $\underline{\underline{AG}}$ is the category of abelian groups, and by elements of $\mathcal{H}^2(\mathbb{B},L)$. In this case $\mathcal{H}^2(\mathbb{B},L)$ will be isomorphic with the second cohomology group of the following cochain complex.

For each $L \in \mathrm{Hom}(\tilde{\mathbb{B}},\underline{\underline{AG}})$ we define

$$C^n(\mathbb{B},L) = \begin{cases} \text{the trivial group if } n \leq 0 \\ \Pi_{<\alpha_1,\ldots,a_n> \in \mathbb{B}^n} L(\Pi\alpha_i) \text{ if } n \geq 1 \end{cases}$$

where the composition in $C^n(\mathbb{B},L)$ is pointwise multiplication. For all $n \geq 1$ we define $\partial^n : C^n(\mathbb{B},L) \to C^{n+1}(\mathbb{B},L)$ by $\partial^1 k(\alpha,\beta) = L^{\alpha\beta}_{\beta}(k(\beta))k^{-1}(\alpha\beta)L^{\alpha\beta}_{\alpha}(k(\alpha))$, $\partial^2 f(\alpha,\beta,\gamma) = L^{\alpha\beta\gamma}_{\beta\gamma}(f(\beta,\gamma))f(\alpha\beta,\gamma)^{-1}f(\alpha,\beta\gamma)\cdot L^{\alpha\beta\gamma}_{\alpha\beta}(f(\alpha,\beta)^{-1})$, etc. For $n \leq 0$ we let $\partial^n : C^n(\mathbb{B},L) \to C^{n+1}(\mathbb{B},L)$ be the trivial homomorphism. Of course we call $C^n(\mathbb{B},L)$ the nth-dimensional cochain group of the pair $(\mathbb{B},L)$, and we call ∂^n the nth-dimensional coboundary operator. For each integer n we set $S^n(\mathbb{B},L) = \mathrm{Ker}(\partial^n)$ and call it the group of n-dimensional cocycles, and we set $B^n(\mathbb{B},L) = \mathrm{Im}(\partial^{n-1})$ and call it the group of n-dimensional coboundaries. As usual, all coboundaries are cocyles, i.e., $B^n(\mathbb{B},L) \subseteq S^n(\mathbb{B},L)$ or $\partial^n \circ \partial^{n-1} = 0$, the trivial homomorphism, and we can form the group $H^n(\mathbb{B},L) = S^n(\mathbb{B},L)/B^n(\mathbb{B},L)$ which we call the nth-cohomology group of L. Of course, $\delta^2 \circ \delta^1 = 0$ is the only case of interest here.

Suppose that $L_1, L_2 \in \mathrm{Hom}(\tilde{\mathbb{B}},\underline{\underline{AG}})$ and that $\sigma : L_1 \to L_2$ is a natural transformation. For all $n \geq 1$, σ induces a map $\sigma^n : C^n(\mathbb{B},L_1) \to C^n(\mathbb{B},L_2)$ defined by

$$\sigma^n(f)(\alpha_1,\ldots\alpha_n) = \sigma(\Pi^n_{i=1}\tilde{\alpha}_i)[f(\alpha_1,\ldots\alpha_n)] .$$

For $n \leq 0$ we let σ^n be the trivial map. The collection of maps $\{\sigma^n\}$ is a cochain homomorphism so that for all n, σ naturally induces a homomorphism $H^n(\sigma) : H^n(\mathbb{B},L_1) \to H^n(\mathbb{B},L_2)$. Hence H^n is a covariant functor from

$\mathrm{Hom}(\tilde{\mathbb{B}},\underline{\underline{AG}})$ to $\underline{\underline{AG}}$.

Definition. If $f \in S^2(\mathbb{B},L)$ is such that $f(\alpha,\alpha) = 1_{L(\tilde{\alpha})}$ for all $\alpha \in \mathbb{B}$, then we say that f is a normal cocycle. Of course f is normal if and only if $f \in \mathcal{H}^2(\mathbb{B},L)$ and $\mathcal{H}^2(\mathbb{B},L)$ is a subgroup of $S^2(\mathbb{B},L)$.

Theorem 10. Let $L \in \mathrm{Hom}(\mathbb{B},\underline{\underline{AG}})$. Then

(a) $\partial^1 : C^1(\mathbb{B},L) \to C^2(\mathbb{B},L)$ is a monomorphism.

(b) $S^2(\mathbb{B},L) = B^2(\mathbb{B},L) \times \mathcal{H}^2(\mathbb{B},L)$.

(c) $H^1(\mathbb{B},L) \cong 0$ and $H^2(\mathbb{B},L) \cong \mathcal{H}^2(\mathbb{B},L)$.

Proof. Let $k \in C^1(B,L)$. For all $\alpha \in \mathbb{B}$ we have

$$\partial^1 k(\alpha,\alpha) = k(\alpha)\cdot(k(\alpha\cdot\alpha))^{-1}\cdot k(\alpha) = k(\alpha) .$$

Hence $\partial^1 k(\alpha,\alpha) = 1_{L(\tilde{\alpha})}$ iff $k(\alpha) = 1_{L(\tilde{\alpha})}$. Hence $\partial^1 k \in \mathcal{H}^2(\mathbb{B},L)$ if and only if $k = 1_{C^1(\mathbb{B},L)}$. In particular $\mathrm{Ker}(\partial^1) = 1_{C^1}$ so that ∂^1 is a monomorphism and $B^2(\mathbb{B},L) \cap \mathcal{H}^2(\mathbb{B},L) = 1_{C^2}$. To complete the proof we need only show that $S^2(\mathbb{B},L) = B^2(\mathbb{B},L) * \mathcal{H}^2(\mathbb{B},L)$. Let $f \in S^2$. Define $k \in C^1$ by $k(\alpha) = f(\alpha,\alpha)$. Then for all $\alpha \in \mathbb{B}$ we have $(\partial^1 k)(\alpha,\alpha) = k(\alpha) = f(\alpha,\alpha)$ and $f^N = f * (\partial^1 k)^{-1} = f * (\partial^1 k^{-1}) \in \mathcal{H}^2$ where $*$ is the composition of $C^2(\mathbb{B},L)$. But $f = (\partial^1 k) * f^N \in B^2(\mathbb{B},L) * \mathcal{H}^2(\mathbb{B},L)$, and hence $S^2(\mathbb{B},L) = B^2(\mathbb{B},L) * \mathcal{H}^2(\mathbb{B},L)$.

We now list some of the more important properties of normal 2-cocycles.

For all $\alpha,\beta \in \mathbb{B}$,

(1) $f(\alpha\beta,\alpha) = f(\alpha,\beta\alpha) = 1_{L(\alpha\beta)}$.

(2) $f(\alpha,\beta) = f(\beta,\alpha)$.

(3) $f(\alpha,\beta) = f(\alpha\beta,\beta\alpha)^{-1}$.

If $\alpha,\beta \in \mathcal{D}(\omega)$, then

(4) $$f(\alpha,\beta) = f(\omega,\alpha)f(\omega,\beta)f(\omega,\alpha\beta)^{-1}f(\omega,\beta\alpha)^{-1} .$$

The first equality is already known, the next two are easy consequences of the first and the fact that f is a normal cocycle, and the fourth is a direct consequence of the first three and the equality $\overline{\alpha}\cdot\overline{\beta} = \overline{\alpha}\cdot\overline{\omega\alpha}\cdot\overline{\beta\omega}\cdot\overline{\beta}$ in $\mathbb{B} \times_f L$.

5. The Case Where $\mathbb{B}$ is Functorial Over $\tilde{\mathbb{B}}$.

5i. Basic definitions and lemmas.

Lemma 11. *Let $\Lambda \times P$ and $\Lambda_1 \times P_1$ be rectangular bands. Then $\mathrm{Hom}(\Lambda,\Lambda_1) = \Lambda_1^{\Lambda}$, $\mathrm{Hom}(P,P_1) = P_1^{P}$, and $\mathrm{Hom}(\Lambda \times P,\Lambda_1 \times P_1) = \{f \times g | f \in \Lambda_1^{\Lambda}, g \in P_1^{P}\}$.*

Construction. Let $\underline{\underline{RB}}$ be the category of rectangular bands and their homomorphisms, and let $\mathbb{B}_0$ be a semilattice. If $\Gamma : \mathbb{B}_0 \to \underline{\underline{RB}}$ is a functor we define $\mathbb{B}_0 \times \Gamma$ to be the band whose underlying set is $\bigcup_{a\in \mathbb{B}_0} \{\alpha\} \times \Gamma(a)$ and whose composition is defined by $\langle a,\alpha\rangle \circ \langle b,\beta\rangle = \langle ab,\Gamma_a^{ab}(\alpha)\Gamma_b^{ab}(\beta)\rangle$. It is an easy check to see that $\mathbb{B}_0 \times \Gamma$ is a band whose $\mathcal{D}$-classes are the sets of the form $\{a\} \times \Gamma(a)$ and $\widetilde{\mathbb{B}_0 \times \Gamma} \cong \mathbb{B}_0$ in the obvious manner.

Definition. A band $\mathbb{B}$ is called *functorial* if there exists a functor $\Gamma : \tilde{\mathbb{B}} \to \underline{\underline{RB}}$ such that $\mathbb{B} \cong \tilde{\mathbb{B}} \times \Gamma$ as bands over $\tilde{\mathbb{B}}$.

Notation. Let $\alpha \in \mathbb{B}$ and $b \in \tilde{\mathbb{B}}$. $\mathrm{Ann}(\alpha,\mathbb{B}_b) = \{\beta \in \mathbb{B}_b | \alpha \geq \beta\}$.

Lemma 12. *$||\mathrm{Ann}(\alpha,\mathbb{B}_b)|| \geq 1$ if and only if $\tilde{\alpha} \geq b$. If $a \geq b$ in $\tilde{\mathbb{B}}$ and $\alpha_1,\alpha_2 \in \mathbb{B}_\alpha$, then $\mathrm{Ann}(\alpha_1,\mathbb{B}_b)\cdot\mathrm{Ann}(\alpha_2,\mathbb{B}_b) \subseteq \mathrm{Ann}(\alpha_1\alpha_2,\mathbb{B}_b)$.*

Theorem 11. *$\mathbb{B}$ is functorial over $\tilde{\mathbb{B}}$ if and only if for all $\alpha \in \mathbb{B}$, $b \in \tilde{\mathbb{B}}$, $||\mathrm{Ann}(\alpha,\mathbb{B}_b)|| \leq 1$. If the latter condition holds, define $\Gamma : \tilde{\mathbb{B}} \to \underline{\underline{RB}}$ by $\Gamma(a) = \mathbb{B}_a$ and $\Gamma_a^b(\alpha) = \alpha^b$ where $a \geq b$ and $\{\alpha^b\} = \mathrm{Ann}(\alpha,\mathbb{B}_b)$.*

Then $\mathbb{B} \cong \tilde{\mathbb{B}} \times \Gamma$ under the map $\alpha \to \langle\tilde{\alpha},\alpha\rangle$.

We are especially interested in functorial bands, $\mathbb{B}$, for which $\tilde{\mathbb{B}}$ is a retract of $\mathbb{B}$, i.e., there exists a homomorphism, $\ell : \tilde{\mathbb{B}} \to \mathbb{B}$, such that $p \circ \ell = i_{\tilde{\mathbb{B}}}$ where $p : \mathbb{B} \to \tilde{\mathbb{B}}$ is the canonical map. We call such a functorial band a pointed functorial band.

Definition. A pointed rectangular band is a pair $\langle\mathbb{B},\omega\rangle$ where $\mathbb{B}$ is a rectangular band and $\omega \in \mathbb{B}$. A morphism $h : \langle\mathbb{B},\omega\rangle \to \langle\mathbb{B}_1,\omega_1\rangle$ is a homomorphism $h : \mathbb{B} \to \mathbb{B}_1$ such that $h(\omega) = \omega_1$. It should be clear that $\underline{\underline{RB}}_0$, the category of pointed rectangular bands, is equivalent with $\underline{\underline{Ens}}_0 \times \underline{\underline{Ens}}_0$, where $\underline{\underline{Ens}}_0$ is the category of pointed sets, under the functor $T : \underline{\underline{RB}}_0 \to \underline{\underline{Ens}}_0 \times \underline{\underline{Ens}}_0$ defined by $T(\langle\mathbb{B},\omega\rangle) = (\langle L(\omega),\omega\rangle, \langle R(\omega),\omega\rangle)$ and $T(h) = (h|L(\omega) , h|R(\omega))$.

It is obvious that any pointed functorial band, $\mathbb{B}$, is isomorphic over $\tilde{\mathbb{B}}$ with $\tilde{\mathbb{B}} \times \Gamma$ for some $\Gamma : \tilde{\mathbb{B}} \to \underline{\underline{RB}}_0$, and conversely, for any $\Gamma : \tilde{\mathbb{B}} \to \underline{\underline{RB}}_0$, $\tilde{\mathbb{B}} \times \Gamma$ is a pointed rectangular band.

Let $\Gamma \in \mathrm{Hom}(\mathbb{B}_0,\underline{\underline{RB}}_0)$. We define $\Gamma_L,\Gamma_R \in \mathrm{Hom}(\mathbb{B}_0,\underline{\underline{RB}}_0)$ by $\Gamma_L(a) = \langle L(a_0),a_0\rangle$, ${\Gamma_L}^b_a = \Gamma^b_a|L(a_0)$, $\Gamma_R(a) = \langle R(a_0),a_0\rangle$, and ${\Gamma_R}^b_a = \Gamma^b_a|R(a_0)$, where a_0 is the distinguished element of $\Gamma(a)$. Clearly $\mathbb{B}_0 \times \Gamma$ is a semilattice of left zero semigroups while $\mathbb{B}_0 \times \Gamma_R$ is a semilattice of right zero semigroups and $\mathbb{B}_0 \times \Gamma \cong (\mathbb{B}_0 \times \Gamma_L) \times_{\mathbb{B}_0} (\mathbb{B}_0 \times \Gamma_R)$.

Construction. Let $\mathbb{B}$ be a band, and let $\alpha,\beta \in \mathbb{B}$. We write $\alpha A_0 \beta$ if and only if there exists $\delta \in \mathbb{B}$ such that $\delta \geq \alpha$ and $\delta \geq \beta$. Let A denote the congruence generated by A_0 . Let us denote $\mathbb{B}/A$ by $\Omega(\mathbb{B})$ and the canonical epimorphism from $\mathbb{B}$ to $\Omega(\mathbb{B})$ by Ω . Let S_0 be the relation on $\mathbb{B}$ defined by $\alpha S_0 \beta$ if and only if $\alpha A_0 \beta$ and $\alpha D \beta$. Let S denote the congruence generated by S_0, and let $\Sigma(\mathbb{B})$ denote the quotient band $\mathbb{B}/S$, and let $\Sigma : \mathbb{B} \to \Sigma(\mathbb{B})$ denote the canonical map.

Theorem 12. *Let* $\mathbb{B}$ *be a band. Then:*

(a) $\Omega(\mathbb{B})$ *is a rectangular band, and* $\Omega : \mathbb{B} \to \Omega(\mathbb{B})$ *is a universal homomorphism from* $\mathbb{B}$ *to the category* $\underline{\underline{RB}}$.

(b) $\tilde{\Sigma} : \tilde{\mathbb{B}} \to \widetilde{\Sigma(\mathbb{B})}$ *so that* $\Sigma(\mathbb{B})$ *is a band over* $\tilde{\mathbb{B}}$. *Moreover* $\Sigma(\mathbb{B})$ *is functorial over* $\tilde{\mathbb{B}}$, *and* $\Sigma : \mathbb{B} \to \Sigma(\mathbb{B})$ *is a universal homomorphism from* $\mathbb{B}$ *to the category of all bands functorial over their semilattice.*

5ii. *Coherent cohomology groups.* We say that a cochain $f \in C^n(\mathbb{B},L)$, where $n \geq 1$, is coherent if for all pairs of n-tuples $\langle\alpha_1,\ldots,\alpha_n\rangle$, $\langle\alpha'_1,\ldots,\alpha'_n\rangle$ such that $\alpha_i \geq \alpha'_i$ for each $i \leq n$, we have

$$f(\alpha'_1,\ldots,\alpha'_n) = L_{\Pi\alpha_i}^{\Pi\alpha'_i}(f(\alpha_1,\ldots,\alpha_n)) .$$

We set $\tilde{C}^n(\mathbb{B},L) = \{f \in C^n(\mathbb{B},L) \mid f \text{ is coherent}\}$ and define $\tilde{\partial}^n : C^n(\mathbb{B},L) \to \tilde{C}^{n+1}(\mathbb{B},L)$ by $\tilde{\partial}^n = \partial^n|\tilde{C}^n(\mathbb{B},L)$. For all $n \geq 1$, $\tilde{C}^n(\mathbb{B},L)$ is a subgroup of $C^n(\mathbb{B},L)$, and $\tilde{\partial}^n$ is a well defined homomorphism. Setting $\tilde{C}^n(\mathbb{B},L) = 0$ for $n \leq 0$ and defining $\tilde{\partial}^n$ accordingly, we can set up a coherent cochain complex and for all integers, n, define $\tilde{H}^n(\mathbb{B},L) = \tilde{S}^n(\mathbb{B},L)/\tilde{B}^n(\mathbb{B},L)$ where $\tilde{S}^n(\mathbb{B},L)$ and $\tilde{B}^n(\mathbb{B},L)$ have their obvious meanings. Moreover, if we let $\tilde{\mathcal{H}}^2(\mathbb{B},L)$ have its obvious meaning, the coherent form of Theorem 10 holds.

Theorem 13. *Let* $\mathbb{B}$ *be functorial over* $\tilde{\mathbb{B}}$, *and let* $L \in \mathrm{Hom}(\mathbb{B},\underline{\underline{AG}})$. *Then* $\mathcal{H}^2(\mathbb{B},L) = \tilde{\mathcal{H}}^2(\mathbb{B},L)$.

Proof. Let $f \in \mathcal{H}^2(\mathbb{B},L)$ and suppose that $\alpha,\alpha',\beta,\beta' \in \mathbb{B}$ are such that $\alpha \geq \alpha'$ and $\beta \geq \beta'$. Setting $\alpha'' = \alpha'\beta'\alpha'$ and $\beta'' = \beta'\alpha'\beta'$ we have $f(\alpha',\beta') =$

$f(\alpha'',\beta'')$ with $\alpha \geq \alpha''$, $\beta \geq \beta''$ $\alpha\beta \geq \alpha''\beta''$, and $\alpha''\mathcal{D}\beta''$. Since $\alpha'' = \alpha\beta''\alpha$, we have $f(\alpha'',\beta'') = f(\alpha,\beta'') = f(\alpha,\beta\beta'')f(\beta,\beta'') = L_{\alpha\beta}^{\alpha''\beta''}(f(\alpha,\beta))f(\alpha\beta,\beta'')$. But $f(\alpha\beta,\beta'') = f(\alpha\beta\beta''\alpha\beta,\beta''\alpha\beta\beta'') = f(\alpha''\beta'',\beta'') = 1$. Hence $f(\alpha'',\beta'') = L_{\alpha\beta}^{\alpha''\beta''}(f(\alpha,\beta))$.

Let $L \in \mathrm{Hom}(\tilde{\mathbb{B}},\underline{\underline{AG}})$. The canonical map $\Sigma : \mathbb{B} \to \Sigma(\mathbb{B})$ induces a monomorphism, $\Sigma^2 : H^2(\Sigma(\mathbb{B}),L) \to \tilde{H}^2(\mathbb{B},L)$ defined by $\Sigma^2(f)(\alpha,\beta) = f([\alpha],[\beta])$.

Theorem 14. *Σ^2 is an isomorphism and $\tilde{H}^2(\mathbb{B},L) \cong H^2(\Sigma(\mathbb{B}),L)$.*

Proof. Let $f \in \tilde{H}^2(\mathbb{B},L)$. If $\beta_1,\beta_2 \in \mathbb{B}$, then we let $\beta_1 \underset{f}{\sim} \beta_2$ denote:

$$\beta_1\mathcal{D}\beta_2 \quad \text{and} \quad f(\alpha,\beta_1) = f(\alpha,\beta_2) \quad \text{for all} \quad \alpha \in \mathbb{B} .$$

Certainly $\underset{f}{\sim}$ is an equivalence relation. Moreover, if $\gamma \in \mathbb{B}$ we have $\beta_1\gamma\mathcal{D}\beta_2\gamma$, and for all $\alpha \in \mathbb{B}$ we have

$$\begin{aligned} f(\alpha,\beta_1\gamma) &= f(\beta_1\gamma,\alpha) \\ &= f(\beta_1,\gamma\alpha)L_{\gamma\alpha}^{\beta_1\gamma\alpha}(f(\gamma,\alpha))L_{\beta_1\gamma}^{\beta_1\gamma\alpha}(f(\beta_1,\gamma)^{-1}) \\ &= f(\beta_2,\gamma\alpha)L_{\gamma\alpha}^{\beta_2\gamma\alpha}(f(\gamma,\alpha))L_{\beta_2\gamma}^{\beta_2\gamma\alpha}(f(\beta_2,\gamma)^{-1}) \\ &= f(\beta_2\gamma,\alpha) = f(\alpha,\beta_2\gamma) \end{aligned}$$

so that $\beta_1\gamma\underset{f}{\sim}\beta_2\gamma$. Similarly $\gamma\beta_1\underset{f}{\sim}\gamma\beta_2$ and hence $\underset{f}{\sim}$ is a congruence relation. Suppose that $[\alpha] = [\beta]$. Then there exists a sequence $\alpha_0,\alpha_1,\ldots,\alpha_n$ such that $\alpha_0 = \alpha,\alpha_n = \beta$ and for each pair $\langle\alpha_{i-1},\alpha_i\rangle$ there exists $\delta_i,\varepsilon_i,\gamma_i,\gamma_i'$ such that $\gamma_i S_0 \gamma_i'$, $\alpha_{i-1} = \delta_i\gamma_i\varepsilon_i$ and $\alpha_i = \delta_i\gamma_i'\varepsilon_i$. Since f is coherent this means that $\gamma_i\underset{f}{\sim}\gamma_i'$, and since $\underset{f}{\sim}$ is a congruence relation, $\alpha_{i-1} \underset{f}{\sim}\alpha_i$. Hence $\alpha\underset{f}{\sim}\beta$. Hence we can construct $f_0 \in C^2(\Sigma(\mathbb{B}),L)$ by $f_0([\alpha],[\beta]) = f(\alpha,\beta)$. By what we have shown, f_0 is well defined. That $f_0 \in H^2(\Sigma(\mathbb{B}),L)$ should be clear. Moreover, $\Sigma^2(f_0) = f$. Hence Σ^2 is surjective.

Theorem 15. *Let K be an abelian group, $\mathbb{B}$ be a band, and $K^{\#} : \tilde{\mathbb{B}} \to \underline{\underline{AG}}$ be the* constant functor. If $\Omega^2 : H^2(\Omega(\mathbb{B}),K^{\#}) \to \tilde{H}^2(\mathbb{B},K^{\#})$ is defined by $\Omega^2 f(\alpha,\beta) = f(\Omega(\alpha),\Omega(\beta))$, then Ω^2 is an isomorphism.

Proof. That Ω^2 is a monomorphism is immediate. To see that it is onto we need only use an argument similar to that of Theorem 14.

5iii. Some calculations of $H^2(\mathbb{B},L)$. We present a series of theorems concerning the computation $H^2(\mathbb{B},L)$ where $\mathbb{B}$ is functorial. The first theorem holds for arbitrary functorial bands. In the second and third theorems of this subsection we obtain sharper results by making the additional assumptions that $\mathbb{B}$ be pointed and that $\mathbb{B}$ be nonsingular separable, respectively. First we present a lemma which is the H^2 version of Lemma 7.

Lemma 13. Let $\mathbb{B}$ be a band, and let $L \in \mathrm{Hom}(\tilde{\mathbb{B}},\underline{AG})$. Then $H^2(\mathbb{B},L)$ can be isomorphically embedded in $\Pi_{a\in\tilde{\mathbb{B}}} H^2(\mathbb{B}_a,L(a)^{\#})$.

Proof. For each $a \in \tilde{\mathbb{B}}$ we define $\phi_a : H^2(\mathbb{B},L) \to H^2(\mathbb{B}_a,L(a)^{\#})$ to be the restriction map, i.e., $\phi_a(f) = f|\mathbb{B}_a \times \mathbb{B}_a$. Clearly ϕ_a is a well defined homomorphism. Define $\phi : H^2(\mathbb{B},L) \to \Pi_a H^2(\mathbb{B}_a,L(a)^{\#})$ by $\phi = \Pi_{a\in\tilde{\mathbb{B}}}\phi_a$. Suppose now that $\phi(f) = \phi(g)$. Then for all $\alpha,\beta \in \mathbb{B}$ we have $f(\alpha,\beta) = f(\alpha\beta\alpha,\beta\alpha\beta) = g(\alpha\beta\alpha,\beta\alpha\beta) = g(\alpha,\beta)$. Hence $f = g$.

Definition. If $\langle f_a\rangle \in \Pi_{a\in\tilde{\mathbb{B}}} H^2(\mathbb{B}_a,L(a)^{\#})$, we say it is coherent when for all $\alpha,\beta \in \mathbb{B}_a$ and $\alpha',\beta' \in \mathbb{B}_b$, if $\alpha \geq \alpha'$ and $\beta \geq \beta'$, then $L_a^b(f_a(\alpha,\beta)) = f_b(\alpha',\beta')$.

Definition. If $\langle f_a\rangle \in \Pi_{a\in\tilde{\mathbb{B}}} H^2(\mathbb{B}_a,L(a)^{\#})$, we define $\langle f_a\rangle^0 : \mathbb{B} \times \mathbb{B} \to \bigcup_{a\in\tilde{\mathbb{B}}} L(a)$ by $\langle f_a\rangle^0(\alpha,\beta) = f_{\widetilde{\alpha\beta}}(\alpha\beta\alpha,\beta\alpha\beta)$.

Theorem 16. $\langle f_a\rangle^0 \in \tilde{H}^2(\mathbb{B},L)$ if and only $\langle f_a\rangle$ is coherent. In particular, *if* $\mathbb{B}$ *is a functorial band,* $\mathcal{H}^2(\mathbb{B},L) = \{\langle f_a\rangle^0 | \langle f_a\rangle$ *is coherent*$\}$.

Proof. ($\longrightarrow$) is obvious. ($\longleftarrow$) is a straightforward computation.

Let $\mathbb{B}_0$ be a semilattice and suppose that $\Gamma \in \mathrm{Hom}(\mathbb{B}_0,\underline{\underline{RB}})$. If $L \in \mathrm{Hom}(\mathbb{B}_0,\underline{\underline{AG}})$, we will let $C^n(\Gamma,L)$ denote $C^n(\mathbb{B}_0 \times \Gamma,L)$ with $S^n(\Gamma,L)$, $B^n(\Gamma,L)$, and $H^n(\Gamma,L)$ denoting their corresponding groups. Similar notations hold for their coherent counterparts as well as $\mathcal{H}^2(\Gamma,L)$ and $\tilde{\mathcal{H}}^2(\Gamma,L)$ standing for $\mathcal{H}^2(\mathbb{B}_0 \times \Gamma,L)$ and $\tilde{\mathcal{H}}^2(\mathbb{B}_0 \times \Gamma,L)$.

Let $\Gamma \in \mathrm{Hom}(\mathbb{B}_0,\underline{\underline{RB}}_0)$ and $L \in \mathrm{Hom}(\mathbb{B}_0,\underline{\underline{AG}})$. $\# : \tilde{C}^1(\Gamma_L,L) \times \tilde{C}^1(\Gamma_R,L) \to \tilde{C}^1(\Gamma,L)$ is defined by $(\nu_L\#\nu_R)(\langle a,\langle\lambda_a,\rho_a\rangle\rangle) = \nu_L(\lambda_a)\nu_R(\rho_a)$ where $\langle\lambda_a,\rho_a\rangle \in \Gamma(a)$. We denote the image of $\#$ by $\tilde{C}^1(\Gamma_L,L)\#\tilde{C}^1(\Gamma_R,L)$ and if $\nu \in \tilde{C}^1(\Gamma_L,L)\#\tilde{C}^1(\Gamma_R,L)$ we say that ν splits.

Theorem 17. Let $\Gamma \in \mathrm{Hom}(\mathbb{B}_0,\underline{\underline{RB}}_0)$, *and let* $\varepsilon : \tilde{C}^1(\Gamma,L) \to \mathcal{H}^2(\Gamma,L)$ *be defined by* $\varepsilon(\nu)(\alpha,\beta) = L_\alpha^{\alpha\beta}(\nu(\alpha))L_\beta^{\alpha\beta}(\nu(\beta))\nu(\alpha\beta)^{-1}\nu(\beta\alpha)^{-1}$ *for all* $\nu \in \tilde{C}^1(\Gamma, L)$ *and* $\alpha,\beta \in \mathbb{B}_0 \times \Gamma$. *Then* ε *is an epimorphism whose kernel is* $\tilde{C}^1(\Gamma_L,L)\#\tilde{C}^1(\Gamma_R,L)$ (*i.e.*, $\mathbb{B} \times_{\varepsilon(\nu)} L$ *splits if and only if* ν *splits*), *and*

$$\mathcal{H}^2(\Gamma,L) \cong \frac{\tilde{C}^1(\Gamma,L)}{\tilde{C}^1(\Gamma_L,L)\#\tilde{C}^1(\Gamma_R,L)} .$$

Proof. That $\varepsilon(\nu) \in \mathcal{H}^2(\Gamma, L)$ is a routine calculation, and so is the fact that ε is a homomorphism. Suppose now that $f \in \mathcal{H}^2(\Gamma,L)$. Define $\nu \in \tilde{C}^1(\Gamma,L)$ by $\nu(\alpha) = f((\tilde{\alpha})_0,\alpha)$ where for all $a \in \mathbb{B}_0$, $a_0 = p^{-1}(a)$ with $p : \mathbb{B}_0 \times \Gamma \to \mathbb{B}_0$ being the projection and p^{-1} its natural right inverse. For all $\alpha,\beta \in \mathbb{B}_0 \times \Gamma$ setting $\alpha' = \alpha\beta\alpha$ and $\beta' = \beta\alpha\beta$ we get

$$f(\alpha,\beta) = f(\alpha',\beta')$$

$$= \nu(\alpha')\nu(\beta')\nu(\alpha'\beta')^{-1}\nu(\beta'\alpha')^{-1}$$

$$= L_{\alpha}^{\alpha\beta}(\nu(\alpha))L_{\beta}^{\alpha\beta}(\nu(\beta))\nu(\alpha\beta)^{-1}\nu(\beta\alpha)^{-1}$$

$$= \varepsilon(\nu)(\alpha,\beta) .$$

Hence the map ε is surjective. Surely $\nu \in \mathrm{Ker}(\varepsilon)$ if and only if $L_{\alpha}^{\alpha\beta}(\nu(\alpha))L_{\beta}^{\alpha\beta}(\nu(\beta))\nu(\alpha\beta)^{-1}\nu(\beta\alpha)^{-1} = 1_{L(\widetilde{\alpha\beta})}$ for all $\alpha,\beta \in \mathbb{B}_0 \times \Gamma$. It is a straightforward computation to show that $\tilde{C}^1(\Gamma_L,L)\#\tilde{C}^1(\Gamma_R,L) \subseteq \mathrm{Ker}(\varepsilon)$. Suppose that $\nu \in \mathrm{Ker}(\varepsilon)$. Then for all $\alpha \in \mathbb{B}_a$ we have

$$\nu(\alpha)\nu(a_0)\nu(\alpha a_0)^{-1}\nu(a_0\alpha)^{-1} = 1 \quad \text{or}$$

$$\nu(\alpha) = \nu(\alpha a_0)\nu(a_0\alpha)\nu(a_0)^{-1}$$

Setting $\nu_L(\alpha a_0) = \nu(\alpha a_0)$ and $\nu_R(a_0\alpha) = \nu(a_0\alpha)\nu(a_0)^{-1}$ we have $\nu_L \in \tilde{C}^1(\Gamma_L,L)$, $\nu_R \in \tilde{C}^1(\Gamma_R,L)$, and $\nu = \nu_L\#\nu_R$. Hence $\mathrm{Ker}(\varepsilon) = \tilde{C}^1(\Gamma_L,L)\#\tilde{C}^1(\Gamma_R,L)$.

Corollary. *If K is an abelian group and $K^{\#} : \mathbb{B}_0 \to \underline{\underline{AG}}$ is the constant functor induced by K, then let $\varepsilon : K^{\Lambda\times P} \to H^2(\Lambda \times P, K^{\#})$ be defined by $\varepsilon(\nu)(\alpha,\beta) = \nu(\alpha)\nu(\beta)\nu(\alpha\beta)^{-1}\nu(\beta\alpha)^{-1}$ for all $\nu \in K^{\Lambda\times P}$ and $\alpha,\beta \in \Lambda \times P$. Then ε is an epimorphism whose kernel is $K^{\Lambda}\#K^{P}$ and $H^2(\Lambda \times P, K^{\#}) \cong \frac{K^{\Lambda\times P}}{K^{\Lambda}\#K^{P}}$.*

Definition. A rectangular band is called *singular* if it is either a right zero or a left zero semigroup. A functor $\Gamma : \mathbb{B}_0 \to \underline{\underline{RB}}$ is said to be *nonsingular* if for all $a \in \mathbb{B}_0$, $\Gamma(a)$ is nonsingular. A nonsingular functor $\Gamma : \mathbb{B}_0 \to \underline{\underline{RB}}_0$ is called *separable* if for all pairs $a \geq b$ in $\mathbb{B}_0$, if $\Gamma_a^b(\alpha) = b_0$, then $\alpha = a_0$. In this case we can form a functor $\Gamma_S : \mathbb{B}_0 \to \underline{\underline{RB}}$ defined by: $\Gamma_S(a) = \Gamma(a) - (L(a_0) \cup R(a_0))$ and $\Gamma_{Sa}^b = \Gamma_a^b|\Gamma_S(a)$. It follows from the peculiar properties of rectangular bands and their homomorphisms that Γ_S is a well defined functor.

Theorem 18. *Let* $\Gamma : \mathbb{B}_0 \to \underline{\underline{RB}}_0$ *be nonsingular and separable. Then for all* $L \in \mathrm{Hom}(\mathbb{B}_0, \underline{\underline{AG}})$, $\tilde{H}^2(\Gamma, L) \cong \tilde{C}^1(\Gamma_S, L)$ *under the map,* $E : H^2 \to \tilde{C}^1$, *defined by* $E(f)(\alpha) = f((\tilde{\alpha})_0, \alpha)$.

Proof. Since f is coherent, E(f) must be coherent, and E is a well defined monomorphism. Suppose that $\nu \in \tilde{C}^1(\Gamma_S, L)$. We can extend it to a cochain $\nu^0 \in \tilde{C}^1(\Gamma, L)$ by setting

$$\nu^0(\alpha) = \begin{cases} \nu(\alpha) & \text{if } \alpha \in \{\tilde{\alpha}\} \times \Gamma_S(\tilde{\alpha}), \text{ or} \\ 1 & \text{otherwise}. \end{cases}$$

Then $E(\varepsilon(\nu^0)) = \nu$. Hence E is onto.

Definition. If $\mathbb{B} = \Lambda \times P$ is a nonsingular rectangular band (i.e., $||\Lambda|| \geq 2$ and $||P|| \geq 2$), then if $\omega_0 = \langle\lambda_0, \rho_0\rangle \in \mathbb{B}$, $\mathbb{B}/\omega_0$ denotes the band $(\Lambda - \{\lambda_0\}) \times (P - \{\rho_0\}) = B - (L(\omega_0) \cup R(\omega_0))$.

Corollary 1. *Let* $\mathbb{B}$ *be a nonsingular rectangular band, and let* $\omega \in \mathbb{B}$. *Let* $E : H^2(\mathbb{B}, K^{\#}) \to K^{\mathbb{B}/\omega}$ *be the map defined by* $E(f)(\alpha) = f(\omega, \alpha)$ *for all* $\alpha \in \mathbb{B}/\omega$. *Then* E *is an isomorphism.*

Corollary 2. *Let* $\mathbb{R}$ *be a rectangular band, let* $\mathbb{B} = \mathbb{B}_0 \times \mathbb{R}$, *and let* $L \in \mathrm{Hom}(\mathbb{B}_0, \underline{\underline{AG}})$. *Then* $H^2(\mathbb{B}, L) \cong H^2(\mathbb{R}, \tilde{C}^1(\mathbb{B}_0, L)^{\#})$.

Proof. If $\mathbb{R}$ is singular, the isomorphism is obvious. Otherwise let $\omega_0 \in \mathbb{R}$ and set $\mathbb{B}_S = \mathbb{B}_0 \times \mathbb{R}/\omega_0$. Let Γ, Γ_S be the functors such that $\mathbb{B} = \mathbb{B}_0 \times \Gamma$ and $\mathbb{B}_S = \mathbb{B}_0 \times \Gamma_S$. Since Γ, Γ_S satisfy the conditions of Theorem 18 we have $H^2(\mathbb{B}, L) \cong \tilde{C}^1(\Gamma_S, L)$. Now, $\tilde{C}^1(\Gamma_S, L) \cong (\tilde{C}^1(\mathbb{B}_0, L))^{\mathbb{R}/\omega_0}$ under the map F defined by $F(\nu)(\alpha)(a) = \nu(\langle a, \alpha\rangle)$. But $(\tilde{C}^1(\mathbb{B}_0, L))^{\mathbb{R}/\omega_0} \cong H^2(\mathbb{R}, \tilde{C}^1(\mathbb{B}_0, L)^{\#})$.

6. The Case When $\mathbb{B}$ is an A-band.

Suppose that a band $\mathbb{B}$ satisfies the following condition:

(A) If $a > b$ in $\tilde{\mathbb{B}}$, then for all $\alpha \in \mathbb{B}_a$, $\mathrm{Ann}(\alpha, \mathbb{B}_b) = \mathbb{B}_b$. For lack of a better word we will refer to $\mathbb{B}$ as an A-band.

Definition. $a \in \tilde{\mathbb{B}}$ is called irreducible if and only if $a = bc$ implies $a = b$ or $a = c$. Otherwise a is reducible. $\mathrm{Irr}(\mathbb{B})$ denotes the set of all irreducible elements of $\tilde{\mathbb{B}}$.

Lemma 14. If $a \in \mathbb{B}$ is reducible, and $\mathbb{B}$ is an A-band, then $||\mathbb{B}_a|| = 1$.

Structure Theorem for A-bands. Let $\mathbb{B}_0$ be a semilattice, and let $\chi : \mathrm{Irr}(\mathbb{B}_0) \to \mathrm{Objects}\ (\underline{\underline{RS}})$ be a pairwise disjoint function. Set $\mathbb{B} = (\mathbb{B}_0 - \mathrm{Irr}(\mathbb{B}_0)) \cup \mathrm{U}\{\chi(a) | a \in \mathrm{Irr}(\mathbb{B}_0)\}$. $\mathbb{B}$ becomes an A-band if we define a product on it by

$$\alpha\beta = \begin{cases} \tilde{\alpha}\tilde{\beta} & \text{if } \tilde{\alpha} \text{ and } \tilde{\beta} \text{ are incomparable.} \\ \beta & \text{if } \tilde{\alpha} > \tilde{\beta}\,. \\ \alpha & \text{if } \tilde{\beta} > \tilde{\alpha}\,. \\ \alpha\beta & \text{if } \tilde{\alpha} = \tilde{\beta}\,. \end{cases}$$

Moreover all A-bands are formed in this manner.

The following theorem is a corollary of Lemma 13, and its easy verification is left to the reader.

Theorem 19. Let $\mathbb{B}$ be an A-band and $L \in \mathrm{Hom}(\tilde{\mathbb{B}}, \underline{\underline{AG}})$. Then

$$H^2(\mathbb{B}, L) \underset{\phi}{\cong} \Pi_{a \in \tilde{\mathbb{B}}} H^2(\mathbb{B}_a, [\textstyle\bigcap_{b<a} \mathrm{Ker}(L_a^b)]^{\#})\,.$$

7. Remarks on the Splitting Case.

Lemma 15. Let $\mathbb{B}$ be a left zero or a right zero semigroup. Then every band of groups over $\mathbb{B}$ splits and hence is isomorphic to $\mathbb{B} \times G$ where G is a group.

Proof. This is an immediate consequence of Lemmas 4 and 6, and Theorem 6.

Notation. If $\langle S,\pi,\mathbb{B}\rangle$ is a band of groups over $\mathbb{B}$, then for all $a \in \tilde{\mathbb{B}}$, $S_a = \pi^{-1}(\mathbb{B}_a)$.

Lemma 16. $\langle S,\pi,\mathbb{B}\rangle$ splits if and only if for all $a \in \tilde{\mathbb{B}}$, $\langle S_a,\pi|S_a,\mathbb{B}_a\rangle$ splits.

Proof. This follows from Lemma 7.

Definition. A band $\mathbb{B}$ is *purely singular* if for all $a \in \tilde{\mathbb{B}}$, $\mathbb{B}_a$ is singular.

Theorem 20. If $\mathbb{B}$ is a purely singular band, then every band of groups over $\mathbb{B}$ splits.

Theorem 21. Let $\mathbb{B}$ be functorial over $\mathbb{B}$. Then every band of groups over $\mathbb{B}$ splits if and only if $\mathbb{B}$ is purely singular.

Proof. We need to show $\longrightarrow$. Hence suppose that $\mathbb{B}$ is not purely singular and that $\mathbb{B}_a$ is nonsingular where $a \in \tilde{\mathbb{B}}$. Let $\mathbb{B}^a = \cup_{b\geq a}\mathbb{B}_b$. Clearly $\mathbb{B}^a$ is a subband of $\mathbb{B}$, and $\mathbb{B}_a = \Omega(\mathbb{B}^a)$. Let K be any nontrivial abelian group. We define $L : \tilde{\mathbb{B}} \to \underline{\underline{AG}}$ by $L|\mathbb{B}^a = K^{\#}$, $L(b)$ is the trivial group when $b \not\geq a$, and L_b^c is the trivial map when $c \not\geq a$. It should be clear that $H^2(\mathbb{B},L) \cong H^2(\mathbb{B}^a,L|\mathbb{B}^a) \cong H^2(\mathbb{B}_a,K^{\#})$ where the last isomorphism comes from Theorem 15. But since $\mathbb{B}_a$ is assumed to be nonsingular, Theorem 18, Corollary 1 tells us that $H^2(\mathbb{B}_a,K^{\#})$ is nontrivial. Contradiction. Hence $\mathbb{B}$ is purely singular.

Using Theorem 19, the following theorem is clear.

<u>Theorem 22</u>. <u>Let</u> $\mathbb{B}$ <u>be</u> <u>an</u> A-<u>band</u>. <u>Then</u> <u>every</u> <u>band</u> <u>of</u> <u>groups</u> <u>over</u> $\mathbb{B}$ <u>splits</u> <u>if</u> <u>and</u> <u>only</u> <u>if</u> $\mathbb{B}$ <u>is</u> <u>purely</u> <u>singular</u>.

We close the section and the paper by constructing a band which, although not purely singular, still has every band of groups over it splitting. In fact, to within isomorphism or antiisomorphism, this is the smallest band having this property.

Let $\Lambda = \{a,b\}$ and $P = \{1,2\}$. Let $\omega_r : P \to P$ map P onto $\{1\}$, and let $\omega = i_\Lambda \times \omega_r$. Let $\mathbb{B} = (\Lambda \times P) \cup \{\omega\}$ and extend the multiplication from $\Lambda \times P$ to all of $\mathbb{B}$ by

$$\omega\cdot\omega = \omega \ ; \ \omega\cdot\alpha = \omega(\alpha)\alpha \ ; \ \alpha\cdot\omega = \alpha\omega(\alpha)$$

where $\alpha \in \Lambda \times P$. Under this multiplication $\mathbb{B}$ is a band whose $\mathcal{D}$-classes are $\Lambda \times P$ and $\{\omega\}$. Since the general case is only notationally harder, we only show that every central band of groups over $\mathbb{B}$ splits. Hence let $L \in \mathrm{Hom}(\tilde{\mathbb{B}},\underline{\underline{Gr}})$, and let $f \in H^2(\mathbb{B},L)$. We need to show that $f(\langle a,1\rangle, \langle b,2\rangle) = 1_{L(\widetilde{\langle a,1\rangle})}$. But

$$\begin{aligned} f(\langle a,1\rangle , \langle b,2\rangle) &= f(\langle a,2\rangle,\omega)^{-1} f(\langle a,1\rangle , \langle b,2\rangle\omega) f(\langle b,2\rangle,\omega) \\ &= f(\langle a,2\rangle , \langle a,1\rangle)^{-1} f(\langle a,1\rangle , \langle b,1\rangle) f(\langle b,2\rangle , \langle b,1\rangle) \\ &= 1 . \end{aligned}$$

Hence $f(\langle a,1\rangle,\alpha) = 1$ for all $\alpha \in \Lambda \times P$ so that $f|(\Lambda \times P) \times (\Lambda \times P) \equiv 1$ and $f = 1_{H^2(\mathbb{B},L)}$.

REFERENCES

1. A. H. Clifford, "Semigroups Admitting Relative Inverses," Annals of Math., 42(1941), 1037-1049.

2. D. McLean, "Idempotent Semigroups," Amer. Math. Monthly, 61(1954), 110-113.

3. M. Petrich, "Topics in Semigroups," Lecture Notes, Pennsylvania State University (1967).

4. D. Rees, "On Semigroups," Proc. Cambridge Phil. Soc., 36(1940), 387-400.

5. M. Yamada, "Inversive Semigroups I," Proc. Japan Acad., 39(1963), 100-103.

Department of Mathematics

The University of Tennessee

Knoxville, Tennessee 37916

Memoirs of the American Mathematical Society

Number 158

Leslie Cohn

The dimension of spaces of automorphic forms on a certain two-dimensional complex domain

Published by the

AMERICAN MATHEMATICAL SOCIETY

Providence, Rhode Island

VOLUME 1 · ISSUE 2 · NUMBER 158 (end of volume) · MARCH 1975

ABSTRACT

In this paper, we apply the Selberg trace formula to derive a formula for the dimensions of certain spaces of automorphic forms. We also obtain the explicit value for the volume of a certain fundamental domain.

AMS (MOS) subject classifications (1970). Primary 32N10; Secondary 22E40.

ISBN 0-8218-1858-9.

TABLE OF CONTENTS

INTRODUCTION

The object of this paper is to compute the dimension of certain spaces of automorphic forms on the complex domain $D = \{(z,u) \in \mathbb{C}^2 \mid 2\mathrm{Im}z - |u|^2 > 0\}$, which is of interest in the theory of automorphic forms as the simplest example of a bounded symmetric domain which is not a "tube domain". Hopefully, the information obtained on the dimension of these spaces will facilitate the further study of automorphic forms on such domains, which differ from forms on the domains so far studied in that they have Fourier coefficients which are not constants, but are theta functions.

The main result which we derive is the following dimension formula, obtained by means of Selberg's trace formula:

$$\begin{aligned}\dim A_m(\Gamma) &= \frac{(3m-1)(3m-2)}{3\times 64} - \frac{1}{12}[3\mathrm{tr}_{Q[i]/Q}(\frac{i^{3m}}{1+i}) + 3(-1)^{3m} + 1] \\ &+ \frac{1}{3\times 128}[10\mathrm{tr}_{Q[i]/Q}\{(3m-1)i^{3m-1} - (3m-2)i^{3m-2}\} + 17(-1)^{3m}(6m-3)] \\ &+ \frac{17}{128} + \frac{7}{32}\mathrm{tr}_{Q[i]/Q}(\frac{i^{3m}}{1-i}) + \frac{1}{16}\mathrm{tr}_{Q[\zeta]/Q}(\frac{\zeta^{3m}}{1-\zeta^{-1}}) + \frac{1}{36}\mathrm{tr}_{Q[\rho]/Q}(\rho^{3m}) \\ &+ \frac{1}{12}\mathrm{tr}_{Q[i]/Q}(i^{3m})\mathrm{tr}_{Q[\rho]/Q}(\frac{\rho^{3m}}{1-\rho}) \\ &- \frac{1}{12}\mathrm{tr}_{Q[i]/Q}(i^{3m+1})\mathrm{tr}_{Q[\rho]/Q}(\frac{\rho^{3m+1}}{1-\rho}) \\ &+ \frac{(-1)^{3m}}{12}\mathrm{tr}_{Q[\rho]/Q}(\frac{\rho^{3m-1}}{1-\rho}) + \frac{1}{9}\,.\end{aligned}$$

Here Γ is a certain discrete group of analytic automorphisms of the domain D; $A_m(\Gamma)$ is the space of "cusp forms for Γ of weight m" (see below for the definition); ρ and ζ denote primitive cube and 8-th roots of unity;

Received by the editor November 16, 1973.

and $tr_{k|Q}$ denotes the trace map of the field extension k of Q. We also find an explicit value, $\frac{\pi^2}{3\times 64}$, for the volume of a fundamental domain for Γ acting in D.

This research was done as a NASA fellow at the University of Chicago, 1967-69. The author wishes to thank Professor Walter Baily, Jr., for suggesting the topic of this work, and for his constant encouragement and advice. Conversations with Professor Paul Sally were also particularly helpful. The author would also like to express his thanks to Professor Robert Langlands for pointing out an error in, and suggesting a proof of, Lemma 6 of Chapter VIII.

CHAPTER I

SELBERG'S DIMENSION FORMULA

Selberg's method of finding the dimension of spaces of automorphic forms on a bounded complex domain D, outlined in [8], consists essentially in the following. Suppose that D admits a transitive group of analytic automorphisms G, and let $j(g,P)$ be the jacobian of the mapping $g \in G$ at $P \in D$. It turns out that the space of holomorphic functions on D which are square integrable on D with respect to ordinary Lebesque measure is a Hilbert space $\mathcal{H}$ and that the evaluation maps $f \to f(P)$ $(P \in D)$ are continuous linear functionals on this Hilbert space (see Helgason, [3]); hence, there exists a unique function $k(P,Q)$ on $D \times D$ (the Bergman kernel function) such that, as a function of Q, $\overline{k(P,Q)}$ is in $\mathcal{H}$ and

$$f(P) = \int_D k(P,Q)f(Q)dQ \tag{1}$$

for all $f \in \mathcal{H}$. Expanding $\overline{k(P,Q)}$ in terms of an orthonormal basis of $\mathcal{H}$, one finds that $k(P,Q) = \overline{k(Q,P)}$; and checking that $j(g,P)d(gP,gQ)\overline{j(g,Q)}$ also satisfies (1) for all $g \in G$, one see that

$$j(g,P)k(gP,gQ)\overline{j(g,Q)} = k(P,Q) \tag{2}$$

for all $g \in G$. In fact, the transitivity of G implies that (2) determines $k(P,Q)$ uniquely up to a constant. Moreover, $d\omega(P) = k(P,P)dP$ is clearly the G-invariant volume element on D.

If one now considers the space $\mathcal{H}_m$ of holomorphic functions f on D such that

$$\int_D |f(P)|^2 \frac{d\omega(P)}{k(P,P)^m} < \infty ,$$

one again gets a Hilbert space, which is non-empty if $m \geq 1$ (it contains the constant functions), such that the evaluation maps $f \to f(P)$ are continuous. Hence, as before, there exists a kernel function $k_m(P,Q)$ such that

$$f(P) = \int_D k_m(P,Q) f(Q) \frac{d\omega(Q)}{k(Q,Q)^m} \tag{3}$$

for all $f \in \mathcal{H}_m$ and

$$j(g,P)^m k_m(gP,gQ)\overline{j(g,Q)}^m = k_m(P,Q) \tag{4}$$

for $g \in G$. As before, (4) determines $k_m(P,Q)$ up to a constant, so $k_m(P,Q) = a(m)k(P,Q)^m$, where

$$\frac{1}{a(m)} = \int_D \frac{|k(P,Q)|^m}{k(P,P)^m} \frac{d\omega(Q)}{k(Q,Q)^m} . \tag{5}$$

(Apply (3) with $f(Q) = k_m(Q,P) = \overline{k_m(P,Q)}$.) According to Selberg, (3) also holds for the space of holomorphic functions f on D such that $k(P,P)^{\frac{-m}{2}} f(P)$ is bounded on D if $m \geq 2$.

Now, if Γ is a discrete subgroup of G such that the orbit space $\Gamma\backslash D$ has finite volume (with respect to the invariant measure $d\omega(P)$), we say that a holomorphic function f on D is a automorphic form of weight m for Γ if $j(\gamma,P)^m f(\gamma P) = f(P)$ for all $\gamma \in \Gamma$. Here, m is an integer or a real number such that $j(g,P)^m$ is a single-valued function on G, and m is greater than or equal to 2. We let $A_m(\Gamma)$ denote the space of automorphic forms of weight m for Γ which have the property that $k(P,P)^{\frac{-m}{2}} f(P)$ is bounded on a fundamental domain F (i.e. a measurable set of orbit representations) of Γ acting in D. We note that if the automorphic form f belongs to $A_m(\Gamma)$, f satisfies (3), so

$$f(P) = a(m)\int_F K_m^\Gamma(P,Q) f(Q) \frac{d\omega(Q)}{k(Q,Q)^m} , \tag{6}$$

where

$$K_m^\Gamma(P,Q) = \Sigma_\Gamma k(\gamma P,Q)^m j(\gamma,P)^m.$$

The operator I_m^Γ: $f \to a(m)\int_F K_m^\Gamma(P,Q)f(Q)\frac{d\omega(Q)}{k(Q,Q)^m}$ is clearly Hermitian and has only automorphic forms of weight m as eigenfunctions, the functions in $A_m(\Gamma)$ having eigenvalue 1. Furthermore, since $\mathrm{vol}(F) < \infty$, $A_m(\Gamma)$ is contained in the Hilbert space of automorphic forms f such that

$$\int_F |f(Q)|^2 \frac{d\omega(Q)}{k(Q,Q)^m} < \infty.$$

Suppose that this Hilbert space is finite dimensional and that all the eigenfunctions of I_m^Γ are in $A_m(\Gamma)$. Then, if $f_1,\ldots,f_d$ is an orthonormal basis of $A_m(\Gamma), a(m)K_m^\Gamma(P,Q) = \Sigma f_i(P)\overline{f_i(Q)}$; and so $\dim A_m(\Gamma)$ equals the trace of I_m^Γ -that is

$$\dim A_m(\Gamma) = a(m)\int_F K_m^\Gamma(P,P)\frac{d\omega(P)}{k(P,P)^m}$$

$$= a(m)\int_F \Sigma_\Gamma k(P,\gamma P)^m \overline{j(\gamma,P)}^m \frac{d\omega(P)}{k(P,P)^m}. \qquad (7)$$

In our case, D is the domain $\{(z,u) \in \mathbb{C} | 2\mathrm{Im}z - |u|^2 > 0\}$. The group G is the special unitary group of the Hermitian form

$$H = \begin{pmatrix} & & i \\ & -1 & \\ -i & & \end{pmatrix}-$$

that is, $G = \{g \in SL(3,\mathbb{C}) | {}^t\bar{g}Hg = H\}$ - which acts on D via

$$\begin{pmatrix} a_1 & a_2 & a_3 \\ b_1 & b_2 & b_3 \\ c_1 & c_2 & c_3 \end{pmatrix} (z,u) = \left(\frac{a_1z+a_2u+a_3}{c_1z+c_2u+c_3}, \frac{b_1z+b_2u+b_3}{c_1z+c_2u+c_3}\right).$$

As a short computation convinces us, $j(g,P) = (c_1z+c_2u+c_3)^{-3}$ if

$$g = \begin{pmatrix} a_1 & a_2 & a_3 \\ b_1 & b_2 & b_3 \\ c_1 & c_2 & c_3 \end{pmatrix} \in G$$

and $P = (z,u) \in D$. As our discrete group Γ, we take $G \cap SL(3,Z[i])$, the set of matrices in G with Gaussian integer coordinates. Then, in order to evaluate the expression (7), we need to find the Bergman kernel function $k(P,Q)$ and the constant $a(m)$. But in fact, as an examination of (5) readily convinces us, it suffices merely to find any function $k(P,Q)$ on $D \times D$ holomorphic in P which satisfies (2). Another short computation convinces us that if $P = (z,u)$, $Q = (z',u') \in D$, the function

$$k(P,Q) = [i(\bar{z}'-z) - \bar{u}'u]^{-3}$$

works. Hence, if $dP = dxdydu_1du_2$ for $P = (z,u) = (x+iy, u_1+iu_2) \in D$ (x,y,u_1,u_2 real), then

$$d\omega(P) = k(P,P)dP = (2y-|u|^2)^{-3}dxdydu_1du_2$$

is an invariant volume element (henceforth to be denoted simply by dP).

Clearly, G can be regarded as the group of real points of a reductive algebraic Q-group $\underline{\underline{G}}$; and Γ is an arithmetic subgroup. Furthermore, if $K = G \cap SU(3)$, K is a maximal compact subgroup of G and $G/K \cong D$ (see Chapter VIII, where this is shown explicitly). Hence, the space of automorphic forms of weight m for Γ is finite dimensional (see the Harish-Chandra notes [4]; the automorphic forms of weight m can be identified with a space $(G/\Gamma,\sigma,\chi)$, where $\sigma(k) = j(k,(i,0))^m$ and χ is a certain character of $\mathfrak{Z}$, the algebra of bi-invariant differential operators on G).

LEMMA 1. $a(m)^{-1} = \dfrac{\pi^2}{(3m-1)(3m-2)}$

PROOF. From formula (5) of Chapter I, we have

$$a(m)^{-1} = \int_D \frac{|k(P,P')|^{2m} dP}{k(P,P)^m k(P',P')^m},$$

where the integral is independent of $P' \in D$. Therefore, taking $P' = (\frac{i}{2},0)$, we get

$$a(m)^{-1} = \int_D \frac{(2y-|u|^2)^{3m-3}}{|\frac{1}{2} - ix + y|^{6m}} dxdy|du| = \frac{1}{2}\int_{\substack{y>0\\ x\in R\\ u\in\mathbb{C}}} \frac{y^{3m-3}dxdy|du|}{|\frac{1}{2} + \frac{y+|u|^2}{2} - ix|^{6m}}$$

$$= 4^{3m-1}\int_{\substack{y>0\\ x\in R\\ u\in\mathbb{C}}} \frac{y^{3m-3}dxdy|du|}{[(1 + y + |u|^2)^2 + x^2]^{3m}}.$$

If $u = u_1 + iu_2 = re^{i\theta}$, $|du| = du_1du_2 = rdrd\theta = \frac{dr^2d\theta}{2}$; so

$$a(m)^{-1} = 2\times 4^{3m-1}\pi \int_{x,y,r>0} \frac{y^{3m-3}dxdydr}{[x^2+(1+y+|u|^2)^2]^{3m}}$$

$$= 2\times 4^{3m-1}\pi \int_{y,r=0}^{\infty} \frac{y^{3m-3}dydr}{(1+y+r)^{6m-1}} \int_{x=0}^{\infty} \frac{dx}{(1+x^2)^{3m}}$$

$$= 4^{3m-1}\pi \int_{y=0} y^{3m-3}\left(\int_{r=0}^{\infty} \frac{dr}{(1+y+r)^{6m-1}}\right) dy \int_{x=0}^{\infty} \frac{dx}{(1+x^2)^{3m}}$$

$$= 4^{3m-1}\frac{\pi}{6m-2}\int_{y=0}^{\infty} \frac{y^{3m-3}dy}{(y+1)^{6m-2}} \int_{t=0}^{\infty} \frac{t^{-1/2}dt}{(1+t)^{3m}}$$

$$= 4^{3m-1}\frac{\pi}{6m-2} B(3m-2,3m)\ B(1/2,3m-1/2)$$

$$= 4^{3m-1}\frac{\pi}{6m-2}\frac{\Gamma(3m-2)\Gamma(3m)\Gamma(1/2)\Gamma(3m-1/2)}{\Gamma(6m-2)\Gamma(3m)}$$

$$= 4^{3m-1}\frac{\pi}{6m-2}\frac{\Gamma(3m-2)\Gamma(1/2)\Gamma(3m-1/2)}{\Gamma(2(3m-1))}$$

$$= 4^{3m-1}\,\frac{\pi^2}{2(3m-1)}\,\frac{\Gamma(3m-2)\Gamma(3m-1/2)}{2^{6m-3}\Gamma(3m-1)\Gamma(3m-1/2)} \qquad \text{(by the Duplication Formula)}$$

$$= \frac{\pi^2}{(3m-1)(3m-2)}\ .$$

Hence, if our assumption on the eigenfunctions of the operator I_m^Γ holds,

$$\dim A_m(\Gamma) = \frac{(3m-1)(3m-2)}{\pi^2}\int_F \Sigma_\Gamma\, k(P,\gamma P)^m \overline{j(\gamma,P)}^m \frac{dP}{k(P,P)^m}.$$

To proceed further, we must investigate the validity of our assumption and the manner of convergence of this integral.

CHAPTER II

REDUCTION THEORY AND CONVERGENCE LEMMAS

The following lemma appears without proof in Shimizu's paper [9].

LEMMA 1. For all $r > 1$, there exits $K > 0$ such that if $g(t) = (t^2 + \alpha^2)^{-r}$ with $|\alpha| > 2$, then $\sum_{n\varepsilon Z} g(t+n) \leq \frac{K}{|\alpha|^{2r-1}}$ for all real t.

PROOF. Let $t' = t + s$ with $s \geq 0$. Then $\frac{g(t)}{g(t')} = \left(\frac{t'^2 + \alpha^2}{t^2 + \alpha^2}\right)^r = \left(\frac{t^2 + \alpha^2 + 2st + s^2}{t^2 + \alpha^2}\right)^r = \left(1 + \frac{2st}{t^2 + \alpha^2} + \frac{s^2}{t^2 + \alpha^2}\right)^r$. But $t^2 + \alpha^2 > t^2 + 1 \geq 2t$; so $\frac{2t}{t^2 + \alpha^2} \leq 1$ and $\frac{1}{\alpha^2 + t^2} \leq \frac{1}{2}$. Hence, $\frac{g(t)}{g(t')} \leq (1+s+\frac{s^2}{2})^r < e^{rs}$; so $g(t) < e^{rs} g(t+s)$ for $s \geq 0$. Therefore, $g(t)\int_0^1 e^{-rs} ds = g(t)(\frac{1-e^{-r}}{r}) \leq \int_0^1 g(t+s)ds$; so $\{\sum_{n\varepsilon Z} g(t+n)\}\{\frac{1-e^{-r}}{r}\} \leq \sum_{n\varepsilon Z} \int_0^1 g(t+n+s)ds = \sum_{n\varepsilon Z} \int_{s=n}^{n+1} g(t+s)ds = \int_{-\infty}^{\infty} g(t+s)ds = \int_{-\infty}^{\infty} g(s)ds = \int_{-\infty}^{\infty} \frac{ds}{(s^2 + \alpha^2)^r} = \frac{1}{|\alpha|^{2r-1}} \int_{-\infty}^{\infty} \frac{ds}{(s^2 + 1)^r}$. Thus, if $K = \frac{r}{1 - e^{-r}} \int_{-\infty}^{\infty} \frac{ds}{(s^2 + 1)^r}$, $\sum_{n\varepsilon Z} g(t+n) \leq \frac{K}{|\alpha|^{2r-1}}$.

COROLLARY. For all $r > 1$, there exists $K' > 0$ such that if $h(u) = (|u|^2 + |\alpha|^2)^{-r}$ with $|\alpha|^2 > 2$, then $\sum_{a\varepsilon Z[i]} h(u+a) \leq \frac{K'}{(|\alpha|^2)^{r-1}}$ for all $u \varepsilon \mathbb{C}$.

PROOF. $\sum_{a\varepsilon Z[i]} h(u+a) = \sum_{k,j\varepsilon Z} \frac{1}{[(\mathrm{Re}u+j)^2 + (\mathrm{Im}u+k)^2 + |\alpha|^2]^r}$

$$\leq K \sum_{k\varepsilon Z} \frac{1}{[(\mathrm{Im}u+k)^2 + |\alpha|^2]^{r-\frac{1}{2}}} \leq \frac{K^2}{(|\alpha|^2)^{r-1}} .$$

We now define the (parabolic) subgroup P of G to be the group of upper triangular matrices belonging to G - that is

$$P = \left\{ \begin{pmatrix} a_1 & a_2 & a_3 \\ 0 & b_2 & b_3 \\ 0 & 0 & c_3 \end{pmatrix} \middle| a_j, b_j, c_j \in \mathbb{C} \right\} \cap G .$$

The unipotent radical P_u of P is the set of matrices of the form

$$\begin{pmatrix} 1 & a & i\frac{|a|^2}{2}+j \\ & 1 & i\bar{a} \\ & & 1 \end{pmatrix} = [a,j] \quad (a\in\mathbb{C}, j\in R).$$

We have the multiplication rule $[a,j][a',j'] = [a+a',j+j'-\mathrm{Im}\,a\bar{a}']$; P_u acts on D by the rule $[a,j](z,u) = (z+au+\frac{|a|^2}{2}i+j, u+i\bar{a})$ $((z,u) \in D)$. We also define the subgroups $\Gamma_\infty^{(1)} = \Gamma \cap P$ and $\Gamma_\infty = \Gamma \cap P_u$ of Γ; $\Gamma_\infty = \{[a,j]\in P \mid a\in(1+i)Z[i], j\in Z\}$ and $\Gamma_\infty^{(1)} = \{g_o^k[a,j] \mid [a,j]\in\Gamma_\infty, k\in Z\} \cong \Gamma_\infty \times Z/(4)$, where $g_o = \mathrm{diag}\{i,-1,-i\}\in\Gamma$. Finally, we define, for $L > 0$, the subset $V_\infty(L)$ of D to be

$$\{(z,u)\in D \mid |\mathrm{Re}\,z| \le \tfrac{1}{2},\ 2\,\mathrm{Im}\,z-|u|^2 \ge L,\ |\mathrm{Re}\,u| \le \tfrac{\sqrt{2}}{2}, |\mathrm{Im}\,u| \le \tfrac{\sqrt{2}}{2}\}.$$

$V_\infty(L)$ is the image in D (under the projection $g \to g(i,0)$ of G onto D) of a Siegel domain in G (see Borel [1] and Chapter VIII, where this is shown explicitly); so $\mathrm{vol}(V_\infty(L))$ is finite. More exactly, $\mathrm{vol}(V_\infty(L)) =$

$$\iint_{u_1,u_2=-\frac{\sqrt{2}}{2}}^{\frac{\sqrt{2}}{2}} du_1du_2 \int_{y=L+\frac{|u|^2}{2}}^{\infty} \frac{dy}{(2y-|u|^2)^3} = \frac{1}{2}\iint_{u_1,u_2=-\frac{\sqrt{2}}{2}}^{\frac{\sqrt{2}}{2}} du_1du_2 \int_{y=L}^{\infty} \frac{dy}{y^3} = \int_{y=L}^{\infty} \frac{dy}{y^3} = \frac{1}{2L^2}.$$

We now investigate the convergence of $\dfrac{K_m^{\Gamma}(P,P)}{k(P,P)^m}$ on the sets $V_\infty(L)$.

LEMMA 2. There exists $K' > 0$ such that $\sum_{[a,j]\in\Gamma_\infty} |k(P,[a,j]P')^m| \le \dfrac{K'}{(\operatorname{Im} z-|u|^2)^{3m-2}}$ for $P = (z,u) \in V_\infty(2\sqrt{2})$ and any $P' = (z',u') \in D$.

PROOF.

$$k(P,[a,j]P')^m = [i(\bar{z}'+\bar{a}\bar{u}' - i|\tfrac{a}{2}|^2+j-z)-(\bar{u}'-ia)u]^{-3m}$$

$$= [i(\bar{z}'-z)-\bar{u}'u+i(\bar{a}\bar{u}'+au)+|\tfrac{a}{2}|^2+ji]^{-3m}.$$

Therefore,

$$\sum_{[a,j]\in\Gamma_\infty} |k(P,[a,j]P')|^m \le K\sum_{a\in Z[i]} \frac{1}{[\operatorname{Im} z-|\frac{u}{2}|^2+\operatorname{Im} z'-|\frac{u'}{2}|^2+|ia+\frac{(u'-u)}{1+i}|^2]^{\frac{3m}{2}-1}}$$

(by Lemma 1)

$$\le \frac{K'}{[\operatorname{Im} z-|\frac{u}{2}|^2+\operatorname{Im} z'-|\frac{u'}{2}|^2]^{3m-2}}$$

(by the Corollary to Lemma 1)

$$\le \frac{K'}{(\operatorname{Im} z-|\frac{u}{2}|^2)^{3m-2}} \text{ for } P = (z,u)\in V_\infty(2\sqrt{2}).$$

LEMMA 3. If $m \ge 1$ and $L > 2\sqrt{2}$,

$$\int_{V_\infty(L)} \sum_{\gamma\in\Gamma_\infty^{(1)}} \frac{k(P,\gamma P)^m}{k(P,P)^m}\overline{j(\gamma,P)}^m dP = \lim_{s\to+0} \sum_{\gamma\in\Gamma_\infty^{(1)}} \int_{V_\infty(L)} \frac{k(P,\gamma P)^m}{k(P,P)^{m(1-s)}}\overline{j(\gamma,P)}^m dP .$$

PROOF. $k(P,P)^{ms} = (2\operatorname{Im} z-|u|^2)^{-3ms} \le L^{-3ms}$ on $V_\infty(L)$, so the left hand side equals

$$\lim_{s\to+0}\int_{V_\infty(L)} \sum_{\gamma\in\Gamma_\infty^{(1)}} \frac{k(P,\gamma P)^m}{k(P,P)^{m(1-s)}} \overline{j(\gamma,P)}^m dP.$$

Let $S(P) = \sum_{\gamma\in\Gamma_\infty^{(1)}} |k(P,\gamma P)|^m$. It suffices to show that $\frac{S(P)}{k(P,P)^{m(1-s)}}$ is integrable on $V_\infty(L)$ for $s > 0$. But

$$\sum_{\Gamma_\infty\backslash\Gamma_\infty^{(1)}} \sum_{[a,j]\varepsilon\Gamma_\infty} |k(P,[a,j]\gamma(P)|^m \leq \frac{4K'}{(2\operatorname{Im}z-|u|^2)^{3m-2}} \quad \text{if } P\,\varepsilon\, V_\infty(2\sqrt{2}).$$

Therefore, $\frac{S(P)}{k(P,P)^{m(1-s)}} \leq 4K'(2\operatorname{Im}z-|u|^2)^{2-3ms}$. But $\int_{V_\infty(L)} (2\operatorname{Im}z-|u|^2)^{2-3ms}dP =$

$\int_{V_\infty(L)} \frac{dxdydu_1du_2}{(2y-|u|^2)^{1+3ms}} = \text{cons.} \int_{y=L}^{\infty} \frac{dy}{y^{1+3ms}} < \infty$. Hence, $\frac{S(P)}{k(P,P)^{m(1-s)}}$ is integrable on $V_\infty(L)$.

LEMMA 4. If $m \geq 2$ and L is sufficiently large,

$\int_{V_\infty(L)} \sum_{\Gamma-\Gamma_\infty^{(1)}} \frac{k(P,\gamma P)^m}{k(P,P)^m} \overline{j(\gamma,P)}^m dP$ is termwise integrable.

PROOF. It is enough to show that $S(P) = \sum_{\Gamma-\Gamma_\infty^{(1)}} \left|\frac{k(P,\gamma P)^m}{k(P,P)^m} \overline{j(\gamma,P)}^m\right|$ is integrable on $V_\infty(L)$ for large L. But $P = (z,u)\varepsilon V_\infty(L)$ implies that

$$\sum_{\Gamma-\Gamma_\infty^{(1)}} \left|\frac{k(P,\gamma P)}{k(P,P)} \overline{j(\gamma,P)}\right|^m = \sum_{\Gamma_\infty\backslash\Gamma-\Gamma_\infty^{(1)}} \sum_{[a,j]\varepsilon\Gamma_\infty} |k(P,[a,j]\gamma P)|^m \frac{|j(\gamma,P)|^m}{k(P,P)^m}$$

$$\leq \frac{K'}{(2\operatorname{Im}z-|u|^2)^{3m-2}} \sum_{\Gamma_\infty\backslash\Gamma-\Gamma_\infty^{(1)}} \frac{|j(\gamma,P)|^m}{k(P,P)^{m/2}} \times \frac{1}{k(P,P)^{m/2}}$$

$$\leq \frac{K'}{L^{\frac{3m-2}{2}}} \sum_{\Gamma_\infty\backslash\Gamma-\Gamma_\infty^{(1)}} k(\gamma P,\gamma P)^{-m/2} \quad \text{for } L > 2\sqrt{2}.$$

But since $V_\infty(L)$ is a Siegel domain in D, it satisfies the Siegel finiteness property - that is, $\{\gamma\varepsilon\Gamma|\gamma(V_\infty(L)) \cap V_\infty(L) \neq \phi\}$ is finite, say of order $k \geq 1$. Then

$$\int_{V_\infty(L)} \sum_{\Gamma\backslash\Gamma-\Gamma_\infty^{(1)}} k(\gamma P,\gamma P)^{-m/2} dP \leq k \int_{\bigcup_{\gamma\varepsilon T}\gamma(V_\infty(L))} k(P,P)^{-m/2} dP, \tag{1}$$

where T is a set of representatives for $\Gamma_\infty\backslash\Gamma-\Gamma_\infty^{(1)}$. But we can choose T so that

$$\bigcup_{\gamma\varepsilon T}\gamma(V_\infty(L))$$

$$\subset \{(z,u)\varepsilon D|\,|\mathrm{Re} z| \leq \tfrac{1}{2}, |\mathrm{Re} u| \leq \tfrac{\sqrt{2}}{2}, |\mathrm{Im} u| \leq \tfrac{\sqrt{2}}{2},\ 0 \leq 2\mathrm{Im} z-|u|^2 \leq L\}.$$

Therefore, the right side of (1) is bounded by

$$k\int_{x=-\frac{1}{2}}^{\frac{1}{2}} \int_{u_1,u_2=-\frac{\sqrt{2}}{2}}^{\frac{\sqrt{2}}{2}} \int_{2y=|u|^2}^{|u|^2+L} (2y-|u|^2)^{\frac{3m-3}{2}}\, dxdydu_1du_2 = k\int_0^L y^{\frac{3m-3}{2}}\, dy = \frac{2kL^{\frac{3m-2}{2}}}{3m-4} < \infty.$$

Let G and P be as before, and let $G_{Q[i]} = G \cap SL(3,Q[i]), P_{Q[i]} = P \cap G_{Q[i]}$. Then Borel's main theorem of reduction theory [1] implies that $\Gamma\backslash G_{Q[i]}/P_{Q[i]}$ is finite and if Σ is a subset of $G_{Q[i]}$, $\Gamma\Sigma V_\infty(L) = D$ for some L if and only if $G_{Q[i]} = \Gamma\Sigma P_{Q[i]}$.

Let $V_\infty^o(L) = \{(z,u)\varepsilon V_\infty(L)|\mathrm{Re}\, u \geq 0,\ \mathrm{Im}\, u \geq 0\}$.

LEMMA 5. If $G_{Q[i]} = \Gamma P_{Q[i]}$, then there exists $L > 0$ such that $V_\infty^o(L)$ contains a fundamental domain F for Γ.

PROOF. By Borel's Theorem, there exists $L > 0$ such that $D = \Gamma V_\infty(L)$. But $V_\infty(L) = \bigcup_{\varepsilon=0}^{3} g_o^\varepsilon(V_\infty^o(L))$; so $D = \Gamma V_\infty^o(L)$ as well. Let $\{\gamma_1,\ldots,\gamma_k\} =$

$\{\gamma\varepsilon\Gamma | \gamma(V_\infty(L)) \cap V_\infty(L) \neq \phi\}$, and let $F = V_\infty^o(L) \cap \{P\varepsilon D | |j(\gamma_i,P)|^2 \leq 1$ for $i=1,2,\ldots,k\}$. We claim that F is a fundamental domain for Γ. For if $P,\gamma_i(P)\varepsilon F$, $1 = |j(\gamma_i^{-1},\gamma_i P)|^2 |j(\gamma_i,P)|^2 \leq |j(\gamma_i^{-1},\gamma_i P)|^2$. But since $\gamma_i^{-1} = \gamma_j$ for some $j=1,\ldots,k$, $|j(\gamma_i^{-1},\gamma_i P)|^2 = |j(\gamma_i,P)|^2 = 1$; and so, P and $\gamma_i(P)$ lie on the boundary of F. Therefore, since every point of D can be transformed under the action of Γ into $V_\infty^o(L)$, it suffices to show that the points of $V_\infty^o(L)$ can be transformed into F.

But if $P \varepsilon V_\infty^o(L)$, there exists $P'\varepsilon\Gamma\cdot P$ with $k(P',P') \leq k(\gamma P',\gamma P')$ for all $\gamma\varepsilon\Gamma$. For if not, let $K = \operatorname{glb}_{\gamma\varepsilon\Gamma} k(\gamma P,\gamma P)$. Clearly $0 \leq K \leq L^{-3}$; so there exists δ_ν such that $L^{-3} \geq k(\delta_\nu P,\delta_\nu P)$ and $\lim_{\nu\to\infty} k(\delta_\nu P,\delta_\nu P) = K$. But then there exists $\delta_\nu' \varepsilon \Gamma_\infty^{(1)}$ such that $\delta_\nu'\delta_\nu P\varepsilon V_\infty^o(L)$. Hence $\delta_\nu'\delta_\nu = \gamma_{i(\nu)}$ for some $i(\nu)$; so $k(\delta_\nu'\delta_\nu P,\delta_\nu'\delta_\nu P) = k(\delta_\nu P,\delta_\nu P)$ takes on only finitely many values, hence equals K for infinitely many ν.

Clearly, we may assume as well that $P'\varepsilon V_\infty^o(L)$, hence equals $\gamma_i(P)$ for some i. But then $k(P',P') \leq k(\gamma_j P',\gamma_j P') = |j(\gamma_j,P')|^2 k(P',P')$ and $|j(\gamma_j,P')|^2 \leq 1$ for $j=1,\ldots,k$; so $P' = \gamma_i P\varepsilon F$, as required.

REMARK. For our Γ, we shall soon see (Corollary 2 of Lemma 2, Chapter IV) that $G_{Q[i]} = \Gamma P_{Q[i]}$; but if Γ' is any arithmetic subgroup of G and if $G_{Q[i]} = \Gamma\Sigma P_{Q[i]}$ with Σ finite, the above proof shows that $\Sigma V_\infty(L)$ will contain a fundamental domain of Γ' for appropriate $L > 0$.

COROLLARY 1. Lemmas 3 and 4 hold for F in place of $V_\infty(L)$.

PROOF. If $L' < L$, $V_\infty(L') - V_\infty(L)$ is compact. Hence, Lemmas 3 and 4 hold for any $L > 0$. But by Lemma 5, we may assume $F \subset V_\infty(L)$ for some L.

LEMMA 6. If $\gamma,\delta\varepsilon G$ and $P\varepsilon D$, $\dfrac{k(P,\delta^{-1}\gamma\delta P)}{k(P,P)} \overline{j(\delta^{-1}\gamma\delta,P)} = \dfrac{k(\delta P,\gamma\delta P)}{k(\delta P,\delta P)} \overline{j(\gamma,\delta P)}$.

PROOF. Trivial.

COROLLARY If our assumption on the operator I_m^Γ holds, if $G_{Q[i]} = \Gamma P_{Q[i]}$, and if $3m \in Z$ and $m \geq 2$, then $a(m)^{-1}\dim A_m(\Gamma) = \mathrm{vol}(F)$

$$+ \sum_{\gamma\in\Gamma-\Gamma_\infty^{(1)}} \int_F \frac{k(P,\gamma P)^m}{k(P,P)^m} \overline{j(\gamma,P)}^m dP + \lim_{s\to+0} \sum_{\gamma\in\Gamma_\infty^{(1)}} \int_F \frac{k(P,\gamma P)^m}{k(P,P)^{m(1-s)}} \overline{j(\gamma,P)}^m dP$$

$$= \mathrm{vol}(F) + \sum_{[\gamma]_\Gamma : [\gamma]_\Gamma \cap \Gamma_\infty^{(1)} = \phi} \int_{F_\gamma} \frac{k(P,\gamma P)^m}{k(P,P)^m} \overline{j(\gamma,P)}^m dP$$

$$+ \lim_{s\to+0} \sum_{[\gamma]_\Gamma : [\gamma]_\Gamma \cap \Gamma_\infty^{(1)} \neq \phi} \Big[\int_{F_\gamma - \bigcup_{\delta\gamma\delta^{-1}\in\Gamma_\infty^{(1)}} \delta^{-1}F} \frac{k(P,\gamma P)^m}{k(P,P)^m} \overline{j(\gamma,P)}^m dP$$

$$+ \sum_{\delta\gamma\delta^{-1}\in\Gamma_\infty^{(1)}} \int_{\delta^{-1}F} \frac{k(P,\gamma P)^m}{k(P,P)^{m(1-s)}} \frac{\overline{j(\gamma,P)}^m dP}{|j(\delta,P)|^{2ms}} \Big] .$$

Here, $[\gamma]_\Gamma = \{\delta^{-1}\gamma\delta \mid \delta\in\Gamma\}$ is the Γ-conjugacy class of γ; G_γ is the centralizer of γ in G; $\Gamma_\gamma = G_\gamma \cap \Gamma$ is the centralizer of γ in Γ; and $F_\gamma = \bigcup_{\delta\in\Gamma/\Gamma_\gamma} \delta^{-1}F$ is a fundamental domain in D for Γ_γ.

PROOF. Apply Lemma 6 and Corollary 1 of Lemma 5.

NOTATIONS. We shall denote by $I_m(\gamma;s)$ the integral

$$\int_{F_\gamma} \frac{k(P,\gamma P)^m}{k(P,P)^{m(1-s)}} \overline{j(\gamma,P)}^m dP$$

($\gamma\in\Gamma$, $s \geq 0$). We write $I_m(\gamma)$ for $I_m(\gamma;0)$.

We now prove that, under the (inessential) assumption that $G_{Q[i]} = \Gamma P_{Q[i]}$, the eigenfunctions of the operator I_m^Γ belong to $A_m(\Gamma)$.

First, note that if $f(z,u)$ is an automorphic form of weight m for Γ,

$$f([a,j](z,u)) = f(z+au+i\frac{|a|^2}{2}+j, u+i\bar{a}) = f(z,u)$$

for $[a,j]\varepsilon\Gamma_\infty$; and in particular, $f(z+n,u) = f(z,u)$ for $n\varepsilon Z$. Hence, in the usual way, we can write

$$f(z,u) = \sum_{n=n_0}^{\infty} \theta_n(u)e^{2\pi inz}, \tag{2}$$

where $\theta_n(u)$ is an entire function such that

$$\theta_n(u+ia)e^{2\pi in[au+i\frac{|a|^2}{2}]} = \theta_n(u)$$

for $a\varepsilon(1+i)Z[i]$, i.e., a theta function. But by the theory of theta functions of one variable (Weber, [10]), $\theta_n(u) = 0$ for $n < 0$ and $\theta_o(u)$ is a constant. If $\theta_o = 0$, we call f a cusp form. Since $f(g_o(z,u)) = f(z,iu) = i^{3m}f(z,u)$, $\theta_n(iu) = i^{3m}\theta_n(u)$; so in particular, $\theta_o = i^{3m}\theta_o$. Thus, if $3m$ is not divisible by 4, every automorphic form of weight m is a cusp form. However, if 4 divides $3m$, there exist automorphic forms of weight m which are not cusp forms, for example the "Eisenstein series"

$$g_m(z,u) = \sum_{\Gamma_\infty^{(1)}\backslash\Gamma} j(\gamma,P)^m .$$

In fact, in this case the cusp forms of weight m form a subspace of co-dimension one of the space of all automorphic forms of weight m for Γ.

LEMMA 7. $A_m(\Gamma)$ is the space of cusp forms of weight m with respect to Γ.

PROOF. If $f(z,u) = \Sigma_o^\infty\theta_n(u)e^{2\pi inz}$ is an automorphic form of weight m for Γ,

$$(y-\frac{|u|^2}{2})^{\frac{3m}{2}}f(z,u) = (y-\frac{|u|^2}{2})^{\frac{3m}{2}}\theta_o+\Sigma_1^\infty\theta_n(u)(y-\frac{|u|^2}{2})^{\frac{3m}{2}}e^{2\pi inz} .$$

But

$$|\Sigma_1^\infty \theta_n(u)(y-\frac{|u|^2}{2})^{\frac{3m}{2}} e^{2\pi inz}|$$

$$\leq \Sigma_1^\infty |\theta_n(u)| e^{-\pi n|u|^2} (y-\frac{|u|^2}{2})^{\frac{3m}{2}} e^{-2\pi n(y-\frac{|u|^2}{2})}, \tag{3}$$

the right hand side converging uniformly on compact subsets of D to a continuous function, since the convergence of (2) is absolute and uniform on compact subsets of D.

But $\frac{d}{dy} y^k e^{-2\pi ny} = y^{k-1}(k-2\pi ny)e^{-2\pi ny} < 0$ for $y > \frac{k}{2\pi}$, so $y^k e^{-2\pi ny}$ is decreasing for y sufficiently large. Therefore, the right side of (2) is bounded by

$$L^{3m}\Sigma_1^\infty |\theta_n(u)| e^{-\pi n|u|^2} e^{-2\pi nL}$$

on $V_\infty(L)$ for large L, the latter being continuous as a function of u; and so the right side of (3) is bounded on $V_\infty(L)$ for large L, hence for all positive L.

Thus, $(y-\frac{|u|^2}{2})^{\frac{3m}{2}} f(z,u)$ is bounded on $V_\infty(L)$ if and only if f is a cusp form.

We shall next show that $K_m^\Gamma(P,Q)$ as a function of P is a cusp form, hence in $A_m(\Gamma)$. This will imply the validity of our dimension formula; for if $K_m^\Gamma(P,Q)$ is in $A_m(\Gamma)$ for all $Q, a(m)K_m^\Gamma(P,Q)$ must be the reproducing kernel for the Hilbert space $A_m(\Gamma)$ - i.e., the unique function $K(P,Q)$ in $A_m(\Gamma)$ such that for all $f \epsilon A_m(\Gamma)$, $\int_F K(P,Q)f(Q)\frac{dQ}{k(Q,Q)^m} = f(P)$; and so

$$\dim A_m(\Gamma) = \int_F K_m^\Gamma(P,P)\frac{dP}{k(P,P)^m} .$$

LEMMA 8. As a function of P, $K_m^\Gamma(P,Q) = \Sigma_\Gamma k(\gamma P,Q)^m j(\gamma,P)^m$ is a cusp form.

PROOF. We recall that if $k \geq 2$,

$$\Sigma \frac{1}{(z+n)^k} = \frac{(-2\pi i)^k}{(k-1)!} \Sigma_1^\infty n^{k-1} e^{2\pi i n z} ,$$

the convergence of both sides being absolute and uniform on compact subset of $\{z \in \mathbb{C} \mid z \neq n \in Z\}$. Hence if $\Gamma_o = \{[o,n] \mid n \in Z\}$, $k([o,n]P,Q) = [i(\bar{z}'-z-n)-\bar{u}'u]^{-3}$ $(P=(z,u), Q=(z',u'))$; and so

$$K_m^\Gamma(P,Q) = \sum_{\gamma \in \Gamma_o \backslash \Gamma} \sum_{[o,n] \in \Gamma_o} k([o,n]\gamma P,Q)^m j(\gamma,P)^m$$

$$= \text{cons. } \Sigma_1^\infty n^{3m-1} \sum_{\gamma \in \Gamma_o \backslash \Gamma} j(\gamma,P)^m e^{2\pi i n[\gamma_1(P)-\gamma_2(P)\bar{u}'i-\bar{z}']}$$

$$= \text{cons. } \Sigma_1^\infty n^{3m-1} \sum_{\gamma \in \Gamma_o \backslash \Gamma} \overline{j(\gamma,Q)}^m e^{-2\pi i n[\overline{\gamma_1(Q)}+\gamma_2(Q)ui-z]}$$

$$= \text{cons. } \Sigma_1^\infty n^{3m-1} \{ \sum_{\gamma \in \Gamma_o \backslash \Gamma} \overline{j(\gamma,Q)}^m e^{-2\pi i n(\gamma_1(Q)+\gamma_2(Q)ui)} \} e^{2\pi i n z}$$

(where $\gamma(z,u) = (\gamma_1(z,u), \gamma_2(z,u))$. Therefore, $K_m^\Gamma(P,Q)$ is a cusp form.

CHAPTER III

CLASSIFICATION OF CONJUGACY CLASSES IN G

In order to evaluate the integrals appearing in the Corollary to Lemma 6 of the last Chapter, we need to find representatives for the conjugacy classes in Γ and the centralizers in Γ of such representatives. We begin the task of evaluating the integrals by classifying the conjugacy classes in G.

First, we need to introduce some further notations and definitions. We let V be the vector space of column matrices

$$\begin{pmatrix} a_1 \\ a_2 \\ a_3 \end{pmatrix}$$

with $a_j \varepsilon \mathbb{C}$ and let

$$e_1 = \begin{pmatrix} 1 \\ 0 \\ 0 \end{pmatrix}, \quad e_2 = \begin{pmatrix} 0 \\ 1 \\ 0 \end{pmatrix}, \quad e_3 = \begin{pmatrix} 0 \\ 0 \\ 1 \end{pmatrix}$$

be the standard basis of V. For any ring extension A of the rational integers Z, we let $V_A = \sum_1^3 Ae_i$, the space of column vectors

$$\begin{pmatrix} a_1 \\ a_2 \\ a_3 \end{pmatrix}$$

with $a_j \varepsilon A$; we denote the lattice $V_{Z[i]}$ by L. The matrix

$$H = \begin{pmatrix} & & i \\ & -1 & \\ -i & & \end{pmatrix}$$

determines a Hermitian form (,) on $V \times V$ by the rule $(x,y) = {}^t\bar{y}Hx$, where

$${}^t\overline{\begin{pmatrix} a_1 \\ a_2 \\ a_3 \end{pmatrix}} = (\bar{a}_1, \bar{a}_2, \bar{a}_3);$$

the basis elements e_1, e_2, e_3 thus satisfy

$$(e_1,e_1) = (e_3,e_3) = (e_1,e_2) = (e_3,e_2)$$
$$= 0, \ (e_2,e_2) = -1, \ (e_3,e_1) = i.$$

Our group G can then be characterized as the group of matrices $g \in SL(3,\mathbb{C})$ such that $(gx,gy) = (x,y)$ for all $x,y \in V$; Γ is the subgroup of elements γ such that $\gamma(L) \subset L$. Finally, we say that the non zero vector $x \in V$ is positive, isotropic, or negative according as (x,x) is positive, zero, or negative.

We can now give the following classification of conjugacy classes in G.

DEFINITION. If $g \in G$ and $g \neq 1$, we say that g is <u>elliptic</u> if g has a positive eigenvector and has no isotropic eigenvector; <u>hyperelliptic</u> if there exists a hyperbolic plane $W \subset V$ (i.e. a two-dimensional non-degenerate subspace containing an isotropic vector) such that $g|W$ is multiplication by a scalar (of absolute value 1); <u>hyperbolic</u> if there exist linearly independent isotropic vectors x_1 and x_2 in V such that $g(x_i) = \gamma_i x_i (1=1,2)$ with

$\lambda_i \epsilon \mathbb{C}, \lambda_1 \neq \lambda_2$; or parabolic if g has an isotropic eigenvector and is neither hyperelliptic nor hyperbolic.

Clearly, if $g = hg'h^{-1}$ with $g', h \epsilon G$, then if g' is one of the above, so is g. Therefore, we may speak of elliptic, hyperelliptic, hyperbolic, or parabolic conjugacy classes (either in G or in Γ). Clearly, also, a conjugacy class cannot belong to more than one of the above types.

LEMMA 1. If $g \epsilon G$ and $g \neq 1$, then g belongs to one of the above types of conjugacy classes.

PROOF. i) Assume that g has a positive eigenvector $x \epsilon V$ and that $g(x) = \lambda x$ $(\lambda \epsilon \mathbb{C})$. Then g leaves invariant $\{x\}^{\perp} = \{y \epsilon V | (x,y) = 0\}$, so g has an eigenvector $y \epsilon \{x\}^{\perp}$. But since V is clearly the orthogonal direct sum of a hyperbolic plane and a negative definite line, invariance of the signature implies that $\{x\}^{\perp} \cap \{y\}^{\perp}$ is then one-dimensional negative - definite, hence can be written $\mathbb{C}z$ with z negative, and is again invariant under g - that is, is an eigenvector of g. Therefore, $V = \mathbb{C}x \oplus \mathbb{C}y \oplus \mathbb{C}z$ with x,y,z eigenvectors of g, x positive, y and z negative, x,y,z pairwise orthogonal, $g(x) = \lambda x, g(y) = \lambda' y, g(z) = \lambda'' z (\lambda, \lambda', \lambda'' \epsilon \mathbb{C})$. The spaces $\mathbb{C}x \oplus \mathbb{C}y$ and $\mathbb{C}x \oplus \mathbb{C}z$ are hyperbolic planes; so if $\lambda = \lambda'$ or $\lambda = \lambda''$, g is hyperelliptic. Assume then that $\lambda \neq \lambda'$ and $\lambda \neq \lambda''$. Then the eigenvectors of g are the elements of $\mathbb{C}x, \mathbb{C}y$, and $\mathbb{C}z$ if $\lambda' \neq \lambda''$ or the elements of $\mathbb{C}x$ and $\mathbb{C}y \oplus \mathbb{C}z$ if $\lambda' = \lambda''$; hence g has no isotropic eigenvector and so is elliptic. Therefore, if g has a positive eigenvector, g is elliptic or hyperelliptic.

ii) Assume that g has no positive eigenvector but has a negative eigenvector x with $g(x) = \lambda x$ $(\lambda \epsilon \mathbb{C})$. Then g leaves invariant $\{x\}^{\perp}$ and so has an eigenvector y with $y \perp x$. If y is negative, the space

$\{x\}^{\perp} \cap \{y\}^{\perp}$ equals $\mathbb{C}z$ with z positive (invariance of the signature) and an eigenvector of g. Therefore, y is isotropic. If $\{x\}^{\perp} = \mathbb{C}y \oplus \mathbb{C}y'$ with y' another eigenvector of g, y' is also isotropic by the same argument. Therefore, if $g(y) = \mu y$ and $g(y') = \mu' y'$ with $\mu, \mu' \in \mathbb{C}$, we must have $\mu \neq \mu'$ and g hyperbolic. Otherwise, y is the only eigenvector of g in $\{x\}^{\perp}$ and g is parabolic.

iii) If g has no positive or negative eigenvectors, g is parabolic as it must have at least one eigenvector, which must be isotropic.

The vector space V is related to the domain D in the following way. Let $P(V)$ be the projective space based on V - that is, $V-\{0\}/\sim$, where $v \sim v'$ if and only if there exists $t \in \mathbb{C}^*$ such that $v = tv'$ - and let $\pi: V-\{0\} \to P(V)$ be the natural projection. Also, let

$$U = \left\{ \begin{pmatrix} a \\ b \\ c \end{pmatrix} \in P(V) \,\middle|\, c \neq 0 \right\} \subset P(V).$$

Then if ρ is the homeomorphism

$$\rho : \mathbb{C}^2 \to U$$
$$(z,u) \to \begin{pmatrix} z \\ u \\ 1 \end{pmatrix},$$

$\rho(D) = \pi(\{v \in V | v$ is positive $\})$ and $\rho(\bar{D}-D) = \pi(\{v \in V | v$ is isotropic, $v \notin \mathbb{C}e_1\})$ (note that if $v \in V$ is positive or isotropic, $\pi(v) \in V$ unless $v \in \mathbb{C}e_1$). We denote $\pi(e_1)$ as ∞, the "infinitely distant point of the boundary of D". Then if $g \in G$, and $g \neq 1$, g is <u>elliptic</u> if g has precisely one fixed point in D; hyperelliptic if g fixes a one-dimensional (complex) submanifold of D; <u>hyperbolic</u> if g fixes two distinct boundary points of D, but no interior points of D; and parabolic if g fixes precisely one boundary point of D

and no interior points of D.

The next four chapters will be devoted to the classification of the Γ-conjugacy classes of each of these four types and to the evaluation of the integrals appearing in the Corollary to Lemma 6 of the preceding chapter.

CHAPTER IV

PARABOLIC CONJUGACY CLASSES

LEMMA 1. If $\gamma\varepsilon\Gamma$ is parabolic, the eigenvalues of γ are Gaussian integers.

PROOF. Let $f_\gamma(x) = \det(\gamma - x\cdot id) = \prod_{j=1}^{3}(x-\gamma_j)\varepsilon\ Z[i][x]$ be the characteristic polynomial of γ. If the roots of $f_\gamma(x)$ are distinct, γ has three linearly independent eigenvectors, which is impossible if γ is parabolic. Therefore, $f_\gamma(x)$ is not irreducible over Q[i]; and so $f_\gamma(x) = f_1(x)f_2(x)$ with f_1, f_2 monic polynomials over Z[i] of degree 1 and 2 respectively. In particular, $f_1(x) = x-\alpha$ with $\alpha\varepsilon Z[i]$. But then, since f_γ has a double root, α is a root of f_2 if f_2 has distinct roots. Hence, since $\alpha\varepsilon Z[i], f_2$ is not irreducible over Z[i].

REMARKS. 1) If M is any module over a principal ideal domain R and if $v\varepsilon M$ is primitive, v can be embedded in a basis of M. Also, every submodule or quotient module of M is a direct summand. This is well-known. (Recall that $v\varepsilon M$ is primitive if $a\varepsilon R$ and $v\varepsilon aM$ implies that a is a unit of R.)

2) If M is a module over Z[i] with a Hermitian form $(\ ,\ ): M\times M \to Z[i]$, then if $\{e_j\}$ and $\{f_j\}$ are two bases for M over Z[i], $\det((e_j,e_k)) = \det((f_j,f_k))$. The same holds if we replace Z[i] with the ring of integers of an imaginary quadratic extension of a totally real algebraic number field. We say that M is unimodular if $\det((e_j,e_k)) = 1$.

3) If M is a unimodular module over Z[i] or over the ring of integers of an imaginary quadratic extension R of a totally real algebraic number field such that R has class number 1, then if $v\varepsilon M$ is primitive,

there exists $w \in M$ with $(w,v) = 1$. For by Remark 1, we can embed v in a basis $v = v_1, v_2, \ldots, v_n$ of M; and since by Remark 2 $\det((v_j,v_k)) = 1$, we can find $a_j \in R$ with $\Sigma a_j(v_j,v_1) = 1$. Then if $w = \Sigma a_j v_j$, $(w,v) = 1$.

LEMMA 2. Every primitive isotropic vector $v \in L$ can be embedded in a basis $\{v,y,v'\}$ of L with v' isotropic, $(v',v) = i$, $(y,y) = -1$, and $y \perp v,v'$.

PROOF. By Remark 3, we can find $v \in L$ such that $(v',v) = i$. The map $L \to Z[i]$ of left $Z[i]$-modules sending $x \in L$ to $(x,v) \in Z[i]$ is therefore onto, so it splits - that is, $L = Z[i]v' \oplus L \cap \{v\}^\perp$. By Remark 1, $L \cap \{v\}^\perp$ has a basis $\{v,x\}$; and by Remark 2,

$$\det\begin{pmatrix} (v',v') & (v',v) & (v',x) \\ (v,v') & (v,v) & (v,x) \\ (x,v') & (x,v) & (x,x) \end{pmatrix} = \det\begin{pmatrix} (v',v') & i & (v',x) \\ -i & 0 & 0 \\ (x,v') & 0 & (x,x) \end{pmatrix} = -(x,x) = 1;$$

so $(x,x) = -1$. Also, $(v'+ax+bv,v) = (v',v) = i$, $(v'+ax+bv,x) = (v',x)-a$, and $(v'+ax+bv,v'+ax+bv) = (v'+ax,v'+ax) + 2\mathrm{Im}b$; so replacing v' by $v'+ax+bv$ for appropriate $a,b \in Z[i]$, we may assume that $(v',x) = 0$, $(v',v) = i$, and $(v',v') = 0$ or 1. If $(v',v') = 0$, we are done (take $y=x$). If $(v',v') = 1$, let $y = x+v$ and $\tilde{v}' = v'+ix$. Then $(\tilde{v}',\tilde{v}') = 0, (\tilde{v}',y) = 0$, and $(y,y) = -1$, so $\{v,y,\tilde{v}'\}$ is a basis of the required type.

COROLLARY 1. $\gamma \in \Gamma$ parabolic implies that $[\gamma]_\Gamma \cap P \neq \phi$. In particular, the eigenvalues of γ are $\{i^\varepsilon,(-1)^\varepsilon,i^\varepsilon\}(\varepsilon \in Z)$.

PROOF. Since the eigenvalues λ_j of γ belong to $Z[i]$, the eigen-spaces $V_{\lambda_j} = \{x \in V | \lambda(x) = \lambda_j x\}$ are defined over $Q[i]$ - that is, they have a basis contained in $V_{\lambda_j} \cap V_{Q[i]}$, and hence have a basis consisting of primitive vectors of L. Therefore, there exists v primitive isotropic in L such that $\gamma(v) = \lambda v$ with $\lambda \varepsilon Z[i]$. Let $\{v,y,v'\}$ be a basis of L as in Lemma 2, and let $\delta(e_1) = v, \delta(e_2) = i^{\nu}y$, and $\delta(e_3) = v'$. Then if ν is chosen so that $\det \delta = 1, \delta\varepsilon\Gamma$ and $\delta^{-1}\gamma\delta(e_1) = \delta^{-1}\gamma(v) = \lambda\delta^{-1}(v) = \lambda e_1$. Hence, $\delta^{-1}\gamma\delta\varepsilon\Gamma \cap P$.

COROLLARY 2. We have $G_{Q[i]} = \Gamma P_{Q[i]}$.

PROOF. Suppose that $g\varepsilon G_{Q[i]}$ and that $v = g(e_1)$. Then $v\varepsilon V_{Q[i]}$ is isotropic and there exists $a\varepsilon Z[i]$ such that av is primitive isotropic in L. By Lemma 2, there exists $\gamma\varepsilon\Gamma$ such that $\gamma(av) = a\gamma(v) = e_1$. Then $\gamma g(e_1) = \gamma(v) = a^{-1}e_1$; and so $\gamma g\varepsilon P_{Q[i]}$ and $g\varepsilon\Gamma P_{Q[i]}$.

REMARK. By Corollary 1, every parabolic Γ-conjugacy class $[\gamma]_\Gamma$ has a representative in $\Gamma_\infty^{(1)} = \Gamma \cap P$. But if $\gamma, \gamma' \varepsilon \Gamma_\infty^{(1)}$ are parabolic and conjugate in Γ, they are conjugate in $\Gamma \cap P$. For if $\gamma' = \delta^{-1}\gamma\delta$ with $\delta\varepsilon G$, $\delta\gamma'(\infty) = \delta(\infty) = \gamma(\delta(\infty))$; so $\delta(\infty)$ is a fixed point of γ lying on the boundary of D. Thus, $\delta(\infty) = \infty$ and $\delta\varepsilon P$. Therefore, it suffices to find representatives for the $\Gamma_\infty^{(1)}$-conjugacy classes of parabolic elements of $\Gamma_\infty^{(1)}$.

Similarly, if $\gamma\varepsilon\Gamma_\infty^{(1)}$ is parabolic, $G_\gamma \subset P$ and hence $\Gamma_\gamma \subset \Gamma \cap P = \Gamma_\infty^{(1)}$. For if $\delta\varepsilon G_\gamma$, $\gamma\delta(\infty) = \delta\gamma(\infty) = \delta(\infty)$ so $\delta(\infty) = \infty$ and $\delta\varepsilon P$.

LEMMA 3. A complete set of representatives for the $\Gamma_\infty^{(1)}$-conjugacy classes of elements of $\Gamma_\infty^{(1)}$ is given by the following: type i) $[(1+i)ja,k]$ $(a\neq 0\ Z[i], (\mathrm{Re}\,a, \mathrm{Im}\,a) = 1,\ j\neq 0\varepsilon Z, k=0,1,\ldots,2j-1)$; type ii) $[0,k]$ $(k\neq 0\varepsilon Z)$; type iii) $g_0^{\varepsilon}[0,k], g_0^{\varepsilon}[i^{\varepsilon}-1, k+\mathrm{Im}\,i^{\varepsilon}], g_0^{2}[i-1,k]$ $(\varepsilon=1,2,$ or 3, $k\varepsilon Z)$. We have $g_0^{\varepsilon}[i^{\varepsilon}-1,k+\mathrm{Im}\,i^{\varepsilon}] = \sigma g_0^{\varepsilon}[0,k]\sigma^{-1}$ and $g_0^{2}[i-1,k] = \tau g_0^{2}[0,k]\tau^{-1}$ with $\sigma = [1,0], \tau = [\frac{1-i}{2},0]$.

PROOF. This is an immediate consequence of the fact that

$g_o^{-\varepsilon}[a,k]g_o^{\varepsilon} = [i^{\varepsilon}a,k]$ and that $[b,k]g_o^{\varepsilon}[a,j][-b,-k] = g_o^{\varepsilon}[a+(i^{\varepsilon}-1)b, j+\mathrm{Im}(i^{\varepsilon}+1)b\bar{a}-|b|^2\mathrm{Im}i^{\varepsilon}] = [a,j+2\mathrm{Im}b\bar{a}]$ if $i^{\varepsilon} = 1, g_o^{\varepsilon}[a-(1\mp i)b, j\mp|b|^2 + \mathrm{Im}(1\pm i)b\bar{a}]$ if $i^{\varepsilon} = \pm\, i$, or $g_o^2[a-2b,j]$ if $i^{\varepsilon} = -1$.

LEMMA 4. If $w\varepsilon L$ is negative with $(w,w) = -1$, $L\cap\{w\}^{\perp}$ contains an isotropic vector. If $v\varepsilon L\cap\{w\}^{\perp}$ is primitive isotropic, there exists $x\varepsilon L\cap\{w\}^{\perp}$ such that $L\cap\{w\}^{\perp} = Z[i]v\oplus Z[i]x$ with $(x,v) = i$ and $(x,x) = 0$ or -1. If $(x,x) = -1$, $L\cap\{w\}^{\perp} = Z[i]x\oplus Z[i]y$ with $x\perp y$, $(x,x) = (y,y) = -1$.

PROOF. Choose $v\varepsilon L\cap\{w\}^{\perp}$ such that $|(v,v)| = j$ is minimal > 0. Then v is primitive in $L\cap\{w\}^{\perp}$, so let $L\cap\{w\}^{\perp} = Z[i]v\oplus Z[i]x$. Since $x+av,v$ is a basis of $L\cap\{w\}^{\perp}$ for any $a\varepsilon Z[i]$ and $(x+av,v) = (x,v) + a(v,v)$, we may assume $|(x,v)|^2 < (v,v)^2 = j^2$. But $\det\begin{pmatrix}(x,x) & (x,v)\\(v,x) & (v,v)\end{pmatrix} = (x,x)(v,v) - |(x,v)|^2 = -1$, so $(x,x)(v,v) = |(x,v)|^2 - 1 > 1$. If $(x,x)(v,v)= -1$, $j=1 = \pm\,(v,v) = \mp\,(x,x)$ and $x+v$ is isotropic. If $(x,x)(v,v) = 0$, $(x,x) = 0$. If $(x,x)(v,v) > 0$, $|(x,x)| > 0$ so $|(x,x)| \geq j$ and $j^2 - 1 > |(x,v)|^2 -1 \geq j^2$, which is impossible.

Now suppose that v is any primitive isotropic vector of $L\cap\{w\}^{\perp}$ and that $\{v,x\}$ is a basis of $L\cap\{w\}^{\perp}$. Then $\det\begin{pmatrix}(x,x) & (x,v)\\(v,x) & (v,v)\end{pmatrix} = \det\begin{pmatrix}(x,x) & (x,v)\\(v,x) & 0\end{pmatrix} = -\,|(v,x)|^2 = -1$, so $|(v,x)|^2 = 1$ and, as we may assume, $(v,x) = i$. But $(x+av,x+av) = (x,x) - 2\mathrm{Im}a$ and $\{x+av,v\}$ is a basis of $L\cap\{w\}^{\perp}$, so we may assume $(x,x) = 0$ or -1. If $(x,x) = -1$, let $y = v+ix$. If $(x,x) = 0$, (u,u) is even for all $u\varepsilon L\cap\{w\}^{\perp}$, so the two cases are mutually exclusive.

LEMMA 5. If $w\varepsilon L$ is primitive negative with $(w,w) = -2$, $L\cap\{w\}^{\perp}$ contains an isotropic vector. If $v = L\cap\{w\}^{\perp}$ is primitive isotropic, L has a basis $\{v,x,u\}$ with u isotropic, $(u,v) = i$, $(x,x) = -1$, $(v,x) = 1+i$, $(u,x) = 0$, and $w = 2iu+v+(1+i)x$.

PROOF. Choose $v \in L \cap \{w\}^{\perp}$ such that $|(v,v)| = j$ is minimal > 0, and choose a basis $\{x,v\}$ of $L \cap \{w\}^{\perp}$ with $|(x,v)|^2 < |(v,v)|^2 = j^2$ as in Lemma 4. Also, choose $u \in L$ with $(w,u) = -i$ and $L = Z[i]u \oplus L \cap \{w\}^{\perp}$, and suppose that $w = au+bv+cx$ with a,b,c relatively prime in $Z[i]$, $(w,w) = ai = -2$. Then if $(w_1,w_2,w_3) = (w,v,x)$ and $(\tilde{w}_1,\tilde{w}_2,\tilde{w}_3) = (u,v,x)$, $\det((w_j,w_k)) = \det A((\tilde{w}_j,\tilde{w}_k))^t\bar{A} = |\det A|^2 \det((\tilde{w}_j,\tilde{w}_k)) = |\det A|^2 = 4$,

where $A = \begin{pmatrix} 2i & b & c \\ 0 & 1 & 0 \\ 0 & 0 & 1 \end{pmatrix}$; but $\det((w_j,w_k)) = \det \begin{pmatrix} (w,w) & 0 & 0 \\ 0 & (v,v) & (v,x) \\ 0 & (x,v) & (x,x) \end{pmatrix} =$

$-2\,(v,v)(x,x)-|(v,x)|^2$. Hence $(v,v)(x,x)-|(v,x)|^2 = -2$ and $(x,x)(v,v) = |(x,v)|^2 -2 \geq -2$. If $(x,x)(v,v) = -2$, $j=1 = \pm (v,v)$, $(x,x) = \mp 2$, $(x,v) = 0$, and $x + (1+i)v$ is isotropic. If $(x,x)(v,v) = -1$, $j=1$, $(x,x) = \pm 1$, $(v,v) = \mp 1$, $(x,v) = i^{\varepsilon}$, and $i^{-\varepsilon}x+iv$ is isotropic. If $(x,x)(v,v) = 0$, x is isotropic. If $(x,x)(v,v) > 0$, $(x,x)(v,v) \geq j^2$ and $j^2-2 > |(x,v)|^2 -2 = (x,x)(v,v) \geq j^2$, which is impossible.

Now suppose that v is any isotropic primitive vector of $L \cap \{w\}^{\perp}$ and that $\{v,x\}$ is a basis of $L \cap \{w\}^{\perp}$. Again, suppose that $w = au+bv+cx$ with a,b,c relatively prime in $Z[i]$, that $\{w_1,w_2,w_3\} = \{w,v,x\}$, and that $\{\tilde{w}_1,\tilde{w}_2,\tilde{w}_3\} = \{u,v,x\}$. Then, as before, $-2 = (v,v)(x,x)-|(v,x)|^2 = -|(v,x)|^2$; so, as we may assume, $(v,x) = 1+i$. But $(v,x+av) = 1+i$ and $(x+av,x+av) = (x,x)+2\mathrm{Re}(1+i)a$; so we may assume that $(x,x) = 0$ or -1. But if $(x,x) = 0$,

$$1 = \det \begin{pmatrix} (u,u) & (u,v) & (u,x) \\ (v,u) & 0 & 1+i \\ (x,u) & 1-i & 0 \end{pmatrix} \equiv 0 \bmod 1+i, \text{ which is impossible. Hence,}$$

$(x,x) = -1$.

Now let $y = v+(1+i)x$. Then $L\cap\{w\}^{\perp} = Z[i]y \oplus Z[i]x, y\perp x, (y,y) = -2$, and $(x,x) = -1$. Also, $(u+px+qy,w) = (u,u) = i$, $(u+px+qy,x) = (u,x)-p$, and $(u+px+qy,y) = (u,y)+2q$. Hence, choosing appropriate $p,q\varepsilon Z[i]$, we may assume $(u,x) = 0$ and $(u,y) = 0,i,1+i$, or $1-i$. But since

$$\det\begin{pmatrix}(u,u) & 0 & (u,y)\\ 0 & -1 & 0\\ (y,u) & 0 & -2\end{pmatrix} = -2(u,u)+|(u,y)|^2 = 1,$$ we have $(u,y) = (u,v) = i$

and $(u,u) = 0$. Then if $w = 2iu+bv+cx, -bi=(w,u) = -i$ and $0 = (w,x) = b(1+i) -c$; so $w = 2iu+v+(1+i)x$, as claimed.

COROLLARY. If $w\varepsilon L$ is primitive negative with $(w,w) = -2$, L has a basis $\{u,x,y\}$ with $L\cap\{w\}^{\perp} = Z[i]x \oplus Z[i]y, (x,x) = -1, (y,y) = -2, (u,u) = 0, (u,x) = 0, (u,y) = i, (x,y) = 0$, and $w = 2iu+y$.

REMARKS. 1) The conjugacy classes listed in Lemma 3 are all parabolic, except for $g_o^{\varepsilon}, \sigma g_o^{\varepsilon}\sigma^{-1}$, and $\tau g_o^{2}\tau^{-1}$ (ε=1,2,or 3). For if $\gamma = [(1+i)ja,k]$ is of type i), we see easily that e_1 is the only eigenvector of γ, so γ is parabolic. Similarly, if $\gamma = [0,k]$ is of type ii), $\gamma(z,u) = (z+k,u) = (z,u)$ only if $(z,u) = \infty\varepsilon\overline{D}-D$, so again γ is parabolic. Finally, if $\gamma = g_o^{\varepsilon}[0,k]$ with ε=1,2, or 3, $\gamma(e_1) = i^{\varepsilon}e_1, \gamma(e_2) = (-1)^{\varepsilon}e_2$, and $\gamma(e_3) = i^{\varepsilon}(e_3+ke_1)$. Hence, if $k\neq 0$, γ has only one isotropic eigenvector e_1, whereas if $k=0$, $\gamma = i^{\varepsilon}\times \mathrm{id}$ on the hyperbolic plane $V\cap\{e_2\}^{\perp}$. Thus, γ is parabolic if $k\neq 0$, hyperelliptic if $k=0$. Since $\sigma g_o^{\varepsilon}[0,k]\sigma^{-1}$ and $\tau g_o^{\varepsilon}[0,k]\tau^{-1}$ are G-conjugate to $g_o^{\varepsilon}[0,k]$, the same result holds for these elements.

2) The elements $g_o^{\varepsilon}[0,k], \sigma g_o^{\varepsilon}[0,k]\sigma^{-1}$ have primitive negative eigenvectors $w=e_2$ and $w=e_1+e_2$, respectively. In both cases, $(w,w) = -1$.

$L\cap\{w\}^{\perp} = Z[i]e_1 \oplus Z[i]x$ as in Lemma 4 with $x=e_3$ isotropic or $x = e_3+ie_2$ satisfying $(x,x) = -1$. The elements $\tau g_o^{\,2}[0,k]\tau^{-1}$ have the primitive negative eigenvector $w=e_1+(1+i)e_2$ with $(w,w) = -2$.

LEMMA 6. 1) If $\gamma = [(1+i)ja,k]$ is parabolic of type i), $\Gamma_\gamma = \{[(1+i)ra,s] \mid r,s\varepsilon Z\}$. 2) If $\gamma = [0,k]$ is parabolic of type ii, $\Gamma_\gamma = \Gamma\cap P$. 3) If $\varepsilon=1,2$, or 3 and $k\neq 0$, $\Gamma_{g_o^\varepsilon[0,k]} = \{g_o^{\,\nu}[0,j] \mid j,\nu\varepsilon Z\}$ and $\Gamma_{\sigma g_o^\varepsilon[0,k]\sigma^{-1}} = \sigma\Gamma_{g_o^\varepsilon[0,k]}\sigma^{-1}$. Also, $\Gamma_{\tau g_o^2[0,k]\tau^{-1}} = \{\tau g_o^{\,2}[0,j]\tau^{-1} \mid j,\nu\varepsilon Z\}$ and is of index 2 in $\tau\Gamma_{g_o^2[0,k]}\tau^{-1}$.

PROOF. 1) and 2) are obvious from the fact that $\Gamma_\gamma \subset \Gamma\cap P$ and the multiplication rule in $\Gamma\cap P$. If $\gamma = g_o^{\,\varepsilon}[0,k], \sigma g_o^{\,\varepsilon}[0,k]\sigma^{-1}$, or $\tau g_o^{\,2}[0,k]\tau^{-1}$ with $\varepsilon=1,2$, or 3 and $k\neq 0$, then γ has a primitive negative eigenvector $w=e_2, e_1+e_2$, or $e_1+(1+i)e_2$ respectively. Hence, if $\delta\varepsilon\Gamma_\gamma, \delta(e_1) = i^\nu e_1$; and $\gamma\delta(w) = \delta\gamma(w) = i^\varepsilon\delta(w)$ implies that $\delta(w) = i^u w$. Clearly, $i^u = (-1)^\nu$; and thus, since $g_o^{\,\nu}[a,j]$ has the eigenvector $w=e_1, e_1+e_2$, or $e_1+(1+i)e_2$ if and only if $g_o^{\,\nu}[a,j] = g_o^{\,\nu}[0,j], \sigma g_o^{\,\nu}[0,j]\sigma^{-1}$, or $\tau g_o^{\,2\nu}[0,j]\tau^{-1}, \delta = g_o^{\,\nu}[0,j], \sigma g_o^{\,\nu}[0,j]\sigma^{-1}$, or $\tau g_o^{\,2\nu}[0,j]\tau^{-1}$ with $\nu,j\varepsilon Z$.

Computation of the Integrals for Parabolic Conjugacy Classes

We now calculate the contribution to the dimension of $A_m(\Gamma)$ of the parabolic conjugacy classes. We recall that if γ and $\delta\gamma\delta^{-1}\varepsilon\Gamma_\infty^{(1)} = \Gamma\cap P$ are parabolic with $\delta\varepsilon\Gamma$, then $\delta\varepsilon\Gamma\cap P$. Hence, $|j(\delta,P)|^2 = 1$. Therefore, the contribution of the parabolic conjugacy classes is given by

$$I_{par} = \lim_{s\to+0} \sum_{[\gamma]_\Gamma \text{ parabolic}} \Big[\int_{\substack{F_\gamma - \cup\delta^{-1}F \\ \delta\gamma\delta^{-1}\varepsilon\Gamma_\infty^{(1)}}} \frac{k(P,\gamma P)^m}{k(P,P)^m} \overline{j(\gamma,P)}^{\,m} dP$$

$$+ \int_{\substack{\bigcup \delta^{-1}F \\ \delta\gamma\delta^{-1}\varepsilon\Gamma_\infty^{(1)}}} \frac{k(P,\gamma P)^m}{k(P,P)^{m(1-s)}} \overline{j(\gamma,P)}^m dP \Big]$$

$$= \lim_{s\to+0} \sum_{[\gamma]_\Gamma \text{ parabolic}} I_m(\gamma;s) \quad \text{(see the Corollary to Lemma 6, Chapter III).}$$

We will require the following

LEMMA 7. If γ_o and γ are elements of Γ with $\gamma = \rho\gamma_o\rho^{-1}$ for some $\rho\varepsilon G$, and if $r_1 = [\Gamma_{\gamma_o} : \Gamma_{\gamma_o} \cap \rho^{-1}\Gamma\rho]$, $r_2 = [G_{\gamma_o} \cap \rho^{-1}(\Gamma)\rho : \Gamma_{\gamma_o} \cap \rho^{-1}(\Gamma)\rho]$, then if $s=0$ or if $\rho\varepsilon N$ and γ and γ_o are parabolic elements of $\Gamma_\infty^{(1)}$, we have

$$I_m(\gamma;s) = \frac{r_1}{r_2} I_m(\gamma_o;s).$$

PROOF. By Lemma 6 of Chapter II,

$$I_m(\gamma;s) = \int_{F_\gamma} \frac{k(\rho^{-1}P,\gamma\rho^{-1}P)^m}{k(\rho^{-1}P,\rho^{-1}P)^m} \overline{j(\gamma,\rho^{-1}P)}^m \frac{k(\rho^{-1}P,\rho^{-1}P)^{ms}}{|j(\rho,\rho^{-1}P)|^{2ms}} dP,$$

and the right side is equal to

$$\int_{\rho^{-1}(F_\gamma)} \frac{k(P,\gamma P)^m}{k(P,P)^{m(1-s)}} \overline{j(\gamma,P)}^m dP$$

if either $s=0$ or γ and γ_o are parabolic elements of $\Gamma_\infty^{(1)}$ with $\rho\varepsilon N = P_u$. But $\rho^{-1}(F_\gamma)$ is a fundamental domain for $\rho^{-1}(\Gamma_\gamma)\rho = G_{\gamma_o} \cap \rho^{-1}(\Gamma)\rho$, which contains $\Gamma_{\gamma_o} \cap \rho^{-1}(\Gamma)\rho$ as a subgroup of index r_2.
Therefore, if $\tilde{F}$ is a fundamental domain for $\Gamma_\gamma \cap \rho^{-1}(\Gamma)\rho$, $\tilde{F}$ is an essentially disjoint union of r_1 fundamental domains for Γ_{γ_o} and r_2

fundamental domains for $G_{\gamma_o} \cap \rho^{-1}(\Gamma)\rho$; so

$$\int_{\tilde{F}} \frac{k(P,\gamma_o P)^m}{k(P,P)^{m(1-s)}} \overline{j(\gamma_o P)}^m dP = r_1 \int_{F_{\gamma_o}} \frac{k(P,\gamma_o P)^m}{k(P,P)^{m(1-s)}} \overline{j(\gamma_o, P)}^m dP$$

$$= r_2 \int_{F_\gamma} \frac{k(P,\gamma P)^m}{k(P,P)^{m(1-s)}} \overline{j(\gamma,P)}^m dP$$

and $I_m(\gamma;s) = \frac{r_1}{r_2} I_m(\gamma_o;s)$ as claimed, if s is as above.

LEMMA 8. If $[\gamma]_\Gamma$ is parabolic of type i), then $I_m(\gamma;s) = 0$.

PROOF. We can assume that $\gamma = [(1+i)a,r]$ with $a\varepsilon Z[i], a\neq 0, r\varepsilon Z$. Then $\Gamma_\gamma = \{[j(1+i)\underset{\alpha}{a},s] \mid j,s\varepsilon Z\}$ with $\alpha = (\mathrm{Re}a, \mathrm{Im}a)$; hence for F_γ we may take

$$\{(z,u)\varepsilon D \mid \ |\mathrm{Re}z| \leq \tfrac{1}{2}, |\mathrm{Re}(1+i)\overline{a}u| \leq \frac{|a|^2}{\alpha}\}.$$

Then $\gamma(P) = \gamma(z,u) = (z+(1+i)au+|a|^2 i+r, u+(1+i)\bar{a})$; so

$$k(P,\gamma P)^m = [2y-|u|^2+|a|^2+i\{2\mathrm{Im}(1+i)\overline{au}+r\}]^{-3m}.$$

Therefore,

$$I_m(\gamma;s) =$$

$$\frac{1}{2}\int_{y=0}^{\infty}\int_{\mathrm{Re}(1+i)\overline{au}=-\frac{|a|^2}{\alpha}}^{\frac{|a|^2}{\alpha}}\int_{\mathrm{Im}(1+i)\overline{au}=-\infty}^{\infty} \frac{y^{3m(1-s)-3}dy|du|}{[y+|a|^2+ir+2i\mathrm{Im}(1+i)\overline{au}]^{3m}} = 0.$$

LEMMA 9. If $[\gamma]_\Gamma$ is parabolic of type ii), then

$$I_m(\gamma;s) = \frac{1}{4|r|^{2+3ms}} \int_{x=0}^{\infty} \frac{x^{3m(1-s)-3}dx}{[x+(\mathrm{sgn}r)i]^{3m}}$$

if $[\gamma]_\Gamma = [[0,r]]_\Gamma$ with $r \neq 0$, $r \in Z$.

PROOF. We can assume $\gamma = [0,r]$. Then $\Gamma_\gamma = \Gamma \cap P$; so for F_γ, we can take

$$\{ (z,u) \in D \mid |\mathrm{Re}z| \le \frac{1}{2},\ 0 \le u_i \le \frac{\sqrt{2}}{2}\ (i=1,2,)\}.$$

Then

$$I_m(\gamma;s) = \frac{1}{2} \int_{u_1,u_2=0}^{\frac{\sqrt{2}}{2}} \int_{x=0}^{\infty} \frac{x^{3m(1-s)-3}}{[x+ir]^{3m}} dx du_1 du_2$$

$$= \frac{1}{4} \int_{x=0}^{\infty} \frac{x^{3m(1-s)-3} dx}{[x+ir]^{3m}}$$

$$= \frac{1}{4|r|^{2+3ms}} \int_{x=0}^{\infty} \frac{x^{3m(1-s)-3} dx}{[x+(\mathrm{sgn} r)i]^{3m}} .$$

LEMMA 10. If $[\gamma]_\Gamma$ is parabolic of type iii) with $w \in L$ a primitive negative eigenvector of γ, then

$$I_m(\gamma;s) = \frac{(w,w)}{3m-1} \frac{\pi}{8} \frac{i^{3m\varepsilon}}{|r|^{1+3ms}} \frac{1}{1-i^{-\varepsilon}} \int_{x=0}^{\infty} \frac{x^{3m(1-s)-3} dx}{[x+(\mathrm{sgn} r)i]^{3m-1}}$$

if $[\gamma]_\Gamma = [\rho g_o^{\varepsilon}[0,r]\rho^{-1}]_\Gamma$ with $r \neq 0$, $r \in Z, \rho = 1, \sigma$, or τ, $\varepsilon = 1, 2$, or 3, and $\varepsilon = 2$ if $\rho = \tau$.

PROOF. We can assume that $\gamma = \rho g_o^{\varepsilon}[0,r]\rho^{-1}$. By Lemmas 10 and 11, $I_m(\gamma;s) = (w,w) I_m(\gamma_o;s)$ if $\gamma_o = g_o^{\varepsilon}[0,r]$. Therefore, it suffices to assume that $\gamma = g_o^{2}[0,r]$. Then $\Gamma_\gamma = \{ g_o^{\nu}[0,k] \mid \nu, k \in Z\}$; and so for F_γ we may take

$$\{ (z,u) \in D \mid |\mathrm{Re}z| \le \frac{1}{2},\ \mathrm{Re}u \ge 0,\ \mathrm{Im}u \ge 0\}.$$

Thus,

$$I_m(\gamma_o;s) = \frac{i}{2}^{3m\varepsilon}\int_{x,u_i=0}^{\infty} \frac{x^{3m(1-s)-3}dx|du|}{[x+(1-i^{-\varepsilon})|u|^2+ir]^{3m}}$$

$$= \frac{\pi i}{8}^{3m\varepsilon}\int_{x,v=0}^{\infty} \frac{x^{3m(1-s)-3}dxdv}{[x+(1-1^{-\varepsilon})v+ir]^{3m}}$$

$$= \frac{\pi i^{3m\varepsilon}}{8|r|^{1+3ms}}\int_{x,v=0}^{\infty} \frac{x^{3m(1-s)-3}dxdv}{[x+(1-i^{-\varepsilon})v+(\mathrm{sgn} r)i]^{3m}} .$$

But if $a\neq 0$,

$$\int_{x,v=0}^{\infty} \frac{x^{3m(1-s)-3}dxdv}{(x+av\pm i)^{3m}} = \frac{1}{a^{3m}}\int_{x=0}^{\infty}\left[\int_{v=0}^{\infty} \frac{dv}{\left(v+\frac{x\pm i}{a}\right)^{3m}}\right]x^{3m(1-s)-3}dx$$

$$= \frac{1}{(3m-1)a^{3m}}\int_{x=0}^{\infty} \frac{x^{3m(1-s)-3}dx}{\left(\frac{x\pm i}{a}\right)^{3m-1}} = \frac{1}{(3m-1)a}\int_{x=0}^{\infty} \frac{x^{3m(1-s)-3}dx}{(x\pm i)^{3m}} .$$

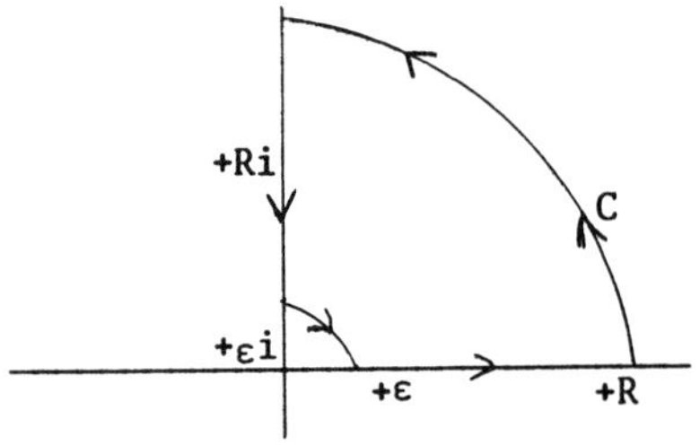

Fig. 1. The Contour C

LEMMA 11. If $\eta = 0$ or 1, $\displaystyle\int_{x=0}^{\infty} \frac{x^{3m(1-s)-3}dx}{(x+i)^{3m-\eta}}$

$$= e^{\frac{-\pi i(3ms+2-\eta)}{2}}\int_{x=0}^{\infty} \frac{x^{3m(1-s)-3}dx}{(x+1)^{3m-\eta}} .$$

PROOF. If C is the contour shown in Figure 1,

$$\oint_C \frac{z^{3m(1-s)-3}dx}{(z+i)^{3m-\eta}} = 0$$

by the residue theorem. Hence

$$\int_{x=\varepsilon}^{R} \frac{x^{3m(1-s)-3}dx}{(x+i)^{3m-\eta}} + \int_{\theta=0}^{\frac{\pi}{2}} \frac{(Re^{i\theta})^{3m(1-s)-3}Rie^{i\theta}d\theta}{(Re^{i\theta}+i)^{3m-\eta}}$$

$$+ \int_{x=R}^{\varepsilon} \frac{(e^{\frac{\pi i}{2}}x)^{3m(1-s)-3}e^{\frac{\pi i}{2}}\,dx}{(e^{\frac{\pi i}{2}}x+i)^{3m-\eta}}$$

$$+ \int_{\theta=\frac{\pi}{2}}^{0} \frac{(\varepsilon e^{i\theta})^{3m(1-s)-3}\varepsilon ie^{i\theta}d\theta}{(\varepsilon e^{i\theta}+i)^{3m-\eta}} = 0.$$

Letting $\varepsilon \to 0$ and $R \to \infty$ and noting that the second and fourth integrals converge to 0, we get

$$\int_{x=0}^{\infty} \frac{x^{3m(1-s)-3}dx}{(x+i)^{3m-\eta}} = e^{\frac{-\pi i}{2}(3ms+2-\eta)} \int_{x=0}^{\infty} \frac{x^{3m(1-s)-3}dx}{(x+1)^{3m-\eta}},$$

as claimed.

REMARK. If $\eta=0$ or 1, $\displaystyle\int_{x=0}^{\infty} \frac{x^{3m-3}dx}{(x+1)^{3m-\eta}} = B(3m-2,2-\eta)$

$$= \frac{\Gamma(3m-2)\Gamma(2-\eta)}{\Gamma(3m-\eta)} = \frac{(3m-3)!}{(3m-\eta-1)!} = \begin{cases} \dfrac{1}{(3m-1)(3m-2)} & \text{if } \eta=0; \\ \dfrac{1}{3m-2} & \text{if } \eta=1. \end{cases}$$

Let $tr_{k/Q}$ denote the trace map of the field k into Q.

PROPOSITION 1. $a(m)^{-1}\ I_{par} = \frac{-1}{12}[1+3tr_{Q[i]/Q}(\frac{i^{3m}}{1+i})+3(-1)^{3m}]$.

PROOF. By the preceeding, we have

$$I_{par} = \lim_{s\to+0}[\tfrac{1}{4}(\Sigma_1^{\infty}\frac{1}{r^{2+3ms}})(2\mathrm{Re}\int_{x=0}^{\infty}\frac{x^{3m(1-s)-3}dx}{(x+i)^{3m}})]$$

$$+\lim_{s\to+0}[\frac{\pi}{4(3m-1)}\{tr_{Q[i]/Q}(\frac{i^{3m}}{1+i})+(-1)^{3m}\}\{\Sigma_{r=1}^{\infty}\frac{1}{r^{1+3ms}}\}$$

$$\times\{2\mathrm{Re}\int_{x=0}^{\infty}\frac{x^{3m(1-s-3)}dx}{(x+i)^{3m-1}}\}]$$

$$=\frac{\zeta(2)}{2}\lim_{s\to+0}[\mathrm{Re}\ e^{\frac{-\pi i}{2}(3ms-2)}\int_{x=0}^{\infty}\frac{x^{3m(1-s)-3}dx}{(x+1)^{3m}}]$$

$$+\frac{\pi}{2(3m-1)}[tr_{Q[i]/Q}(\frac{i^{3m}}{1+i})+(-1)^{3m}]\lim_{s\to+0}[(\Sigma_{r=1}^{\infty}\frac{1}{r^{1+3ms}})$$

$$\times\ (\mathrm{Re}\ e^{\frac{-\pi i}{2}(3ms+1)}\int_{x=0}^{\infty}\frac{x^{3m(1-s)-2}dx}{(x+1)^{3m-1}})]$$

$$=-\frac{\zeta(2)}{2}\int_{x=0}^{\infty}\frac{x^{3m-3}dx}{(x+1)^{3m}}$$

$$-\frac{\pi}{2(3m-1)}\int_{x=0}^{\infty}\frac{x^{3m-3}dx}{(x+1)^{3m-1}}[tr_{Q[i]/Q}(\frac{i^{3m}}{1+i})+(-1)^{3m}]$$

$$\times\lim_{s\to+0}\zeta(1+3ms)\mathrm{Re}\ ie^{\frac{-\pi i3ms}{2}},$$

where $\zeta(s) = \Sigma_{r=1}^{\infty}\frac{1}{r^s}$ is the Riemann zeta function. But $\mathrm{Re}\ ie^{\frac{-\pi i3ms}{2}} =$

$$-\frac{i}{2}\{\exp(\frac{3ms\pi i}{2})-\exp(\frac{-3ms\pi i}{2})\} = -\frac{i}{2}\pi i3ms + \text{higher powers of } s.$$

Hence,

$$\lim_{s\to+0} \zeta(1+3ms)\mathrm{Re}\, ie^{\frac{-\pi i 3ms}{2}} = \frac{\pi}{2}\lim_{s\to+0} 3ms\,\zeta(1+3ms) = \frac{\pi}{2}.$$

Therefore, $I_{par} = \dfrac{-\zeta(2)}{2(3m-1)(3m-2)} + \dfrac{-\pi^2}{4(3m-1)(3m-2)}[\mathrm{tr}_{Q[i]/Q}(\dfrac{i^{3m}}{1+i})+(-1)^{3m}]$.

Since $\zeta(2) = \dfrac{\pi^2}{6}$, our assertion follows.

CHAPTER V

HYPERELLIPTIC CONJUGACY CLASSES

LEMMA 1. Suppose that $\gamma\varepsilon\Gamma$ is hyperelliptic and let $W \subset V$ be a hyperbolic plane such that $\gamma|W = \lambda\times id$ with $\lambda\varepsilon\mathbb{C}$. Then $\lambda=i^{\varepsilon}(\varepsilon=1,2,$ or $3)$ and $\gamma(x) = (-1)^{\varepsilon}x$ for $x\varepsilon W^{\perp}$.

PROOF. Since $\det\gamma=1, \gamma(x) = \lambda^{-2}x$ for $x\varepsilon W^{\perp}$. Therefore, $f_{\gamma}(x) = \det(\gamma-x\cdot id) = (x-\lambda)^2(x-\lambda^{-2})\varepsilon Z[i][x]$ and is not irreducible over $Q[i]$; so $g(x) = (x-\lambda)(x-\lambda^{-2}) = x^2 - (\lambda+\lambda^{-2})x + \lambda^{-1}\varepsilon Z[i][x]$. But then λ^{-1} and $\lambda\varepsilon Z[i]$, so $\lambda=i^{\varepsilon} \neq 1$.

LEMMA 2. Suppose that $w\varepsilon L$ is primitive non-isotropic and suppose that $\gamma(w) = (-1)^{\varepsilon}w$ and $\gamma(x) = i^{\varepsilon}x$ for $x\varepsilon L\cap\{w\}^{\perp}$ with $\varepsilon=1,2,$ or 3. Then if $\gamma\varepsilon\Gamma, (w,w) = \pm 1$ or ± 2, and if $(w,w) = \pm 2$, then $\varepsilon=2$.

PROOF. Choose $x\varepsilon L$ with $(w,w) = 1$ and $L = Z[i]x \oplus L\cap\{w\}^{\perp}$. Then $w=ax+y$ with $y\varepsilon L\cap\{w\}^{\perp}$ and $a=(w,w)\varepsilon Z$, $a\neq 0$. Therefore, $\gamma(w) = (-1)^{\varepsilon}w = a(-1)^{\varepsilon}x+(-1)^{\varepsilon}y = a\gamma(x)+i^{\varepsilon}y$; so $a(\gamma(x) - (-1)^{\varepsilon}x) = ((-1)^{\varepsilon}-i^{\varepsilon})y$. If $y=0$, $a=(w,w)$ is a unit of $Z[i]$ and $a\varepsilon Z$; so $a=(w,w) = \pm 1$. So assume $y\neq 0$ and let $y=by'$ with $b\varepsilon Z[i]$ and y' primitive in L. Then a and b are relatively prime since w is primitive; hence, the integer a divides $(-1)^{\varepsilon}-i^{\varepsilon} = -(1+i), 2,$ or $-(1-i)$ in $Z[i]$. Therefore, $a = \pm 1$ or ± 2, and if $a = \pm 2$, $\varepsilon=2$.

REMARK. If γ and w are as in Lemma 2 and if $\Gamma_w = \{\delta\varepsilon\Gamma|\delta(w)\varepsilon Z[i]w\}$, $\Gamma_w = \Gamma_{\gamma}$.

By Lemmas 4 and 5 of Chapter IV and Lemmas 1 and 2, we see that there are seven Γ- conjugacy classes of hyperelliptic elements of Γ.

Clearly, the elements $g_o^{\varepsilon}, \sigma g_o^{\varepsilon}\sigma^{-1}$, and $\tau g_o^{2}\tau^{-1} (\varepsilon=1,2,3)$ in $\Gamma_{\infty}^{(1)}$ are non-conjugate hyperelliptic elements of Γ; hence, they form a set of representatives for these conjugacy classes. Therefore, since the number of Γ-conjugacy classes of hyperelliptic elements of Γ is the same as the number of $\Gamma_{\infty}^{(1)}$-conjugacy classes of hyperelliptic elements of $\Gamma_{\infty}^{(1)}$, if two hyperelliptic elements γ_1 and γ_2 of $\Gamma_{\infty}^{(1)}$ are conjugate in Γ, they are already conjugate in $\Gamma_{\infty}^{(1)}$.

The purpose of the following is to find the indices $[\Gamma_{g_o} : \Gamma_{g_o} \cap \rho^{-1}\Gamma\rho]$ and $[G_{g_o} \cap \rho^{-1}\Gamma\rho : \Gamma_{g_o} \cap \rho^{-1}\Gamma\rho]$ for $\rho=\sigma$ and $\rho=\tau$.

We introduce the following notations and definitions. If W is a vector space over $\mathbb{C}$ with non-degenerate Hermitian form (,), we let $G(W) = \{ g \varepsilon GL(W) \mid (g(x), g(y)) = (x,y) \forall\, x,y \varepsilon W\}$. If $P \subset W$ is a $Z[i]$-submodule of W such that $(x,y) \varepsilon Z[i]$ for all x and y in P, we let $GL(P) = \{ g \varepsilon GL(W) \mid g(P) = P\}$ and $\Gamma(P) = GL(P) \cap G(W)$. Choose a negative vector w_1 in L with $(w_1,w_1) = -1$ and suppose that $L \cap \{w\}^{\perp} = Z[i]u_1 \oplus Z[i]v_1$ with $u_1 \perp v_1, (u_1,u_1) = -(v_1,v_1) = 1$. Similarly, choose a primitive negative vector w_2 in L with $(w_2,w_2) = -2$; and, as in the Corollary to Lemma 5 of Chapter IV, choose a basis $\{x_2,u_2,v_2\}$ of L with $L \cap \{w_2\}^{\perp} = Z[i]u_2 \oplus Z[i]v_2, u_2 \perp v_2, (u_2,u_2) = 2, (v_2,v_2) = -1, (x_2,v_2) = 0, (x_2,x_2) = 0, (x_2,u_2) = i$, and $w_2 = 2ix_2+u_2$. Define the linear transformations σ and τ of V onto itself by

$$\sigma(e_1) = \frac{u_1+v_1}{1+i} \qquad \tau(e_1) = \frac{u_2+(1+i)v_2}{2}$$

$$\sigma(e_2) = w_1 \qquad \tau(e_2) = \frac{w_2}{1+i}$$

$$\sigma(e_3) = \frac{i(u_1-v_1)}{1+i} \qquad \tau(e_3) = \frac{i[u_2-(1+i)v_2]}{2} .$$

Then σ and τ are clearly in $G(V)$ and have determinant i^{ν}; so we may assume, by multiplying u_j, v_j, and w_j by $i^{-\nu}$, that $\det\sigma = \det\tau = 1$ and hence that σ and τ are in G. Also, we have

$$\sigma^{-1}(u_1) = \frac{e_3+ie_1}{1+i} \qquad \tau^{-1}(u_2) = e_1-ie_3$$

$$\sigma^{-1}(w_1) = e_2 \qquad \tau^{-1}(w_2) = (1+i)e_2$$

$$\sigma^{-1}(v_1) = \frac{(e_3-ie_1)}{1+i} \qquad \tau^{-1}(v_2) = \frac{e_1+ie_3}{1+i} .$$

(These transformations are not, of course, the same as the transformations $\sigma = [1,0]$ and $\tau = [\frac{1-i}{2},0]$ considered previously, in general. But clearly the indices $[\Gamma g_0 : \Gamma g_0 \cap \sigma^{-1}\Gamma\sigma]$, etc, are.) Consider next the vector space $W = \mathbb{C}e_1 \oplus \mathbb{C}e_3$ and the $Z[i]$-submodules $M = Z[i]e_1 \oplus Z[i]e_3$, $N = \sigma^{-1}(L \cap \{w_1\}^{\perp})$, and $\tilde{N} = \tau^{-1}(L \cap \{w\}^{\perp})$. Finally, let $\Gamma^{o}(\tilde{N}) = \{g \in \Gamma(\tilde{N}) \mid g\tau^{-1}(u_2) - (\det g)^{-1}\tau^{-1}(u_2) \in 2N\}$.

LEMMA 3. The restriction map $R: G_{e_2} \to G(W)\,(g \mapsto g|W)$ is an isomorphism inducing isomorphisms of Γ_{e_2} onto $\Gamma(M)$, of $G_{e_2} \cap \sigma^{-1}(\Gamma)\sigma$ onto $\Gamma(N)$, and of $G_{e_2} \cap \tau^{-1}(\Gamma)\tau$ onto $\Gamma^{o}(\tilde{N})$.

PROOF. R is clearly a well-defined group homomorphism. If $g \in G_{e_2}$ and $g|W = \mathrm{id}$, then since $\det g = 1$, $g = \mathrm{id}$ on $W^{\perp} = \mathbb{C}e_2$. Therefore, $g = \mathrm{id}$ and R is 1-1. If $\tilde{g} \in G(W)$, $|\det\tilde{g}|^2 = 1$; so if $g = \tilde{g}$ on W and $g(e_2) = \det(\tilde{g})^{-1}e_2$, $g \in G_{e_2}$ and $R(g) = \tilde{g}$. Therefore, R is an isomorphism.

Clearly, $R(\Gamma_{e_2}) \subset \Gamma(M)$, $R(G_{e_2} \cap \sigma^{-1}\Gamma\sigma) \subset \Gamma(N)$, and $R(G_{e_2} \cap \tau^{-1}(\Gamma)\tau) \subset \Gamma(\tilde{N})$. But if $\tilde{\gamma} \in \Gamma(M)$, $\det\tilde{\gamma} = i^{\nu}$; hence, $R^{-1}(\tilde{\gamma}) \in \Gamma_{e_2}$. Similarly, if $\tilde{\gamma} \in \Gamma(N)$, $\det\tilde{\gamma} = i^{\nu}$,

$\sigma R^{-1}(\tilde{\gamma})\sigma^{-1}(L\cap\{w_1\}^{\perp}) = \sigma R^{-1}(\tilde{\gamma})N = \sigma\tilde{\gamma}(N) = \sigma N = L\cap\{w_1\}^{\perp}$, and $\sigma R^{-1}(\tilde{\gamma})\sigma^{-1}(w_1)$ $= \sigma R^{-1}(\tilde{\gamma})e_2 = \det\tilde{\gamma}^{-1}\sigma(e_2) = i^{\nu}w_1$; so $\sigma R^{-1}(\tilde{\gamma})\sigma^{-1}\varepsilon\Gamma$ and $R^{-1}(\tilde{\gamma})\varepsilon G_{e_2}\cap\sigma^{-1}(\Gamma)\sigma$. Finally, if $\tilde{\gamma}\varepsilon\Gamma(\tilde{N}), \det(\tilde{\gamma})=i^{\nu}, \tau R^{-1}(\tilde{\gamma})\tau^{-1}(L\cap\{w_2\}^{\perp}) = L\cap\{w_2\}^{\perp}$, and $\tau R^{-1}(\tilde{\gamma})\tau^{-1}(w_2) = (1+i)\tau R\gamma^{-1}e_2 = (1+i)i^{-\nu}\tau(e_2) = i^{-\nu}w_2$. Therefore, $-2i\tau^{-1}R(\tilde{\gamma})\tau(x_2) + \tau R^{-1}(\tilde{\gamma})\tau^{-1}(u_2) = -2ii^{-\nu}x_2+i^{-\nu}u_2$; so $2i(\tau^{-1}R^{-1}(\tilde{\gamma})\tau(x_2)-i^{\nu}x_2)$ $= \tau R^{-1}(\tilde{\gamma})\tau^{-1}(u_2)-i^{-\nu}u_2\varepsilon L\cap\{w_2\}^{\perp}$. Clearly, then, $\tau R^{-1}(\tilde{\gamma})\tau^{-1}\varepsilon\Gamma$ if and only if $\tau\tilde{\gamma}\tau^{-1}(u_2)-i^{-\nu}u_2\varepsilon 2L\cap\{w_2\}^{\perp}$ - that is, if and only if $\tilde{\gamma}\tau^{-1}(u_2)-i^{-\nu}\tau^{-1}(u_2)\varepsilon 2N$.

LEMMA 4. We have a short exact sequence $1 \to SL(2,R) \to G(W) \xrightarrow{\det} \pi' \to 1$.

PROOF. Suppose that $g\varepsilon G(W)$ and that $\det g=1$. Then $g(e_1) = ae_1+be_3$ and $g(e_3) = ce_1+de_3$ with $a\bar{b}$ and $c\bar{d}$ real, $a\bar{d}-b\bar{c} = 1$, and $ad-bc = 1$. Then $1 = a\mathrm{Re}d-b\mathrm{Re}c$ and $b = b\bar{a}\mathrm{Re}d-|b|^2\mathrm{Re}c\varepsilon R$; and similarly, $a,c,d\varepsilon R$. Hence, $\begin{pmatrix} a & b \\ c & d \end{pmatrix}\varepsilon SL(2,R)$, and conversely, if $\begin{pmatrix} a & b \\ c & d \end{pmatrix}\varepsilon SL(2,R)$, $g\varepsilon G(W)$ and $\det g=1$. Furthermore, if g = multiplication by $\exp\frac{i\theta}{2}$, $g\varepsilon G(W)$ and $\det g = \exp i\theta$. Thus, det is onto and has kernel equal to $SL(2,R)$.

COROLLARY 1. The above sequence induces $1 \to SL(2,Z) \to \Gamma(M) \xrightarrow{\det} \{\pm 1\} \to 1$.

PROOF. Clearly, if $\gamma\varepsilon\Gamma(M)$ and $\det =1, \gamma\varepsilon SL(2,Z)$. Suppose $\gamma\varepsilon\Gamma(M)$ and $\det\gamma=i$. Then $\gamma(e_1) = ae_1+be_3$ and $\gamma e_3 = ce_1+de_3$ with $a,b,c,d\varepsilon Z[i], a\bar{b}, c\bar{d}\varepsilon Z, a\bar{d}-b\bar{c}=1$, and $\det\gamma=ad-bc=i$. Then $1+i = 2(a\mathrm{Re}d-b\mathrm{Re}c) \equiv 0 \bmod 2$, which is absurd. Therefore, since $\gamma = i\times id$ has determinant -1, the image of det is $\{\pm 1\}$.

COROLLARY 2. We have an induced short exact sequence $1 \to \Delta \to \Gamma(W) \to \{\pm 1, \pm i\} \to 1$ with $\Delta = \{g\varepsilon SL(2,Z) | g \equiv \begin{pmatrix} 1 & 0 \\ 0 & 1 \end{pmatrix}$ or $\begin{pmatrix} 0 & 1 \\ 1 & 0 \end{pmatrix} \bmod 2\}$.

PROOF. Suppose that $g = \begin{pmatrix} a & b \\ c & d \end{pmatrix}\varepsilon SL(2,R)$. Then $g\sigma^{-1}(u_1) =$

$\frac{[(a+d)+(b-c)i]}{2}\sigma^{-1}(u_1) + \frac{[(a-d)-(b+c)i]}{2}\sigma^{-1}(v_1)$ and $g\sigma^{-1}(v_1) =$

$\frac{[(a-d)+(b+c)i]}{2}\sigma^{-1}(u_1) + \frac{[(a+d)+(c-b)i]}{2}\sigma^{-1}(v_1)$. Therefore, $g\varepsilon\Gamma(N)$ if and only if $a,b,c,d\varepsilon Z$, $a\equiv d \pmod 2$, and $b\equiv c \pmod 2$ - that is, if and only if $g \equiv \begin{pmatrix}1 & 0\\ 0 & 1\end{pmatrix}$ or $\begin{pmatrix}0 & 1\\ 1 & 0\end{pmatrix}$ mod 2.

Also, note that if $\gamma\sigma^{-1}(u_1) = i\sigma^{-1}(u_1)$ and $\gamma\sigma^{-1}(v_1) = \sigma^{-1}(v_1)$, then $\gamma\varepsilon\Gamma(N)$ and $\det \gamma = i$.

COROLLARY 3. The induced sequence $1 \to \Delta \to \Gamma(N)\cap\Gamma(M) \to \{\pm 1\} \to 1$ is exact; and so $[\Gamma(M):\Gamma(M)\cap\Gamma(N)] = 3$ and $[\Gamma(N):\Gamma(N)\cap\Gamma(M)] = 2$.

PROOF. The exactness follows from Corollaries 1 and 2; the nine-lemma implies that $[\Gamma(M):\Gamma(M)\cap\Gamma(N)] = [SL(2,Z):\Delta]$ and $[\Gamma(N):\Gamma(N)\cap\Gamma(M)] = [Z/(4):Z/(2)] = 2$. But $SL(2,Z) \to SL(2,Z/(n))$ is onto for $n \geq 2$; so we see easily that $[SL(2,Z):\Delta] = 3$.

COROLLARY 4. The induced sequence $1 \to \tilde{\Delta} \to \Gamma^o(\tilde{N}) \to \{\pm 1\} \to 1$ is exact, where $\tilde{\Delta} = \{ g = \begin{pmatrix}a & b\\ c & d\end{pmatrix} \varepsilon SL(2,Z) \mid a\equiv d\equiv \pm 1 \pmod 4$ and $b\equiv c\equiv 0$ or $2 \pmod 4\}$.

PROOF. Suppose that $g = \begin{pmatrix}a & b\\ c & d\end{pmatrix} \varepsilon SL(2,R)$. Then

$$g\tau^{-1}(u_2) = \frac{[(a+d)+(b-c)i]}{2}\tau^{-1}(u_2) + \frac{[(a+b+c+d)+(a-b-c-d)i]}{2}\tau^{-1}(v_2),$$

and

$$g\tau^{-1}(v_2) = \frac{[(a-d)+(b+c)i]}{2(1+i)}\tau^{-1}(u_2) + \frac{[(a-d)+(c-b)i]}{2}\tau^{-1}(v_2).$$

Also,

$$g \cdot \tau^{-1}(u_2) = \frac{[(a+d-2)+(b-c)i]}{2}\tau^{-1}(u_2) + \frac{[(a+b+c+d)+(a-b-c-d)i]}{2}\tau^{-1}(v_2) + \tau^{-1}(u_2).$$

Hence, $g\varepsilon\Gamma^o(\tilde{N})$ if and only if $g\varepsilon SL(2,Z)$, $b\equiv c$ mod 4, $a+d\equiv 2$ mod 4, and

$a\equiv d \bmod 2$. But if $g\epsilon SL(2,Z)$, $1=ad-bc$; so if $g\epsilon\Gamma^o(\tilde{N})$ and $a\equiv d\equiv 0 \bmod 2$, then $ad\equiv 0 \bmod 4$, $bc\equiv -1$, and $b\not\equiv c \bmod 4$. Hence if $g\epsilon\Gamma^o(\tilde{N})$, $a\equiv d\equiv 1 \bmod 2$ and so $a\equiv d\equiv\pm 1 \bmod 4$. But then $bc\equiv 0 \bmod 4$, so $b\equiv c\equiv 0$ or $2 \bmod 4$. Thus $\Gamma^o(\tilde{N})\cap SL(2,Z) = \tilde{\Delta}$.

Also, if $\gamma\cdot\tau^{-1}(u_2)=i\tau^{-1}(u_2)$ and $\gamma\cdot\tau^{-1}(v_2)=\tau^{-1}(v_2), \gamma\epsilon\Gamma(\tilde{N})$ and $\det\gamma=i$. Moreover, $\gamma\cdot\tau^{-1}(u_2)-i^{-1}\tau^{-1}(u_2) = 2i\tau^{-1}(u_2)\epsilon 2\tilde{N}$ so $\gamma\epsilon\Gamma^o(\tilde{N})$. Hence the image of det is $\{\pm 1,\pm i\}$.

COROLLARY 5. The sequence $1\rightarrow\tilde{\Delta}\rightarrow\Gamma^o(\tilde{N})\cap\Gamma(M)\rightarrow\pm 1\rightarrow 1$ is exact; hence $[\Gamma^o(\tilde{N}):\Gamma^o(\tilde{N})\cap\Gamma(M)] = 2$ and $[\Gamma(M):\Gamma(M)\cap\Gamma^o(\tilde{N})] = [SL(2,Z):\tilde{\Delta}] = \frac{1}{4}[SL(2,Z/(4)):1] = \frac{48}{4} = 12$.

PROOF. It suffices to show that if γ is the map constructed in Corollary 4 such that $\gamma\tau^{-1}(u_2) = i\tau^{-1}(u_2)$ and $\gamma\tau^{-1}(v_2) = \tau^{-1}(v_2)$, then $\gamma^2\epsilon\Gamma(M)$. But $\gamma^2(e_1)=ie_3$ and $\gamma^2(e_3) = -ie_1$, so $\gamma^2\epsilon\Gamma(M)$ and has determinant -1.

From Lemma 3 and Corollaries 3 and 5, we see that $[\Gamma g_o:\Gamma g_o\cap\sigma^{-1}\Gamma\sigma] = 3$, $[Gg_o\cap\sigma^{-1}\Gamma\sigma:\Gamma g_o\cap\sigma^{-1}\Gamma\sigma] = 2, [\Gamma g_o:\Gamma g_o\cap\tau^{-1}\Gamma\tau] = 12$, and $[Gg_o\cap\tau^{-1}\Gamma\tau:\Gamma g_o\cap\tau^{-1}\Gamma\tau] = 2$.

REMARK. Suppose that $w\epsilon L$ is primitive negative with $(w,w) = -1$ or -2 and that $\gamma\epsilon\Gamma_w$. Then if $(w,w) = -1$ and (x,x) is even for all $x\epsilon L\cap\{w\}^{\perp}$, $\gamma|L\cap\{w\}^{\perp}$ has determinant $=\pm 1$ (Corollary 1); if $(w,w) = -1$ and $L\cap\{w\}^{\perp}$ contains an element x with (x,x) odd, then $\gamma|L\cap\{w\}^{\perp}$ has even trace if it has determinant 1 (Corollary 2); and if $(w,w) = -2$, $\gamma|L\cap\{w\}^{\perp}$ has trace $\equiv 2 \bmod 4$ if it has determinant 1 (Corollary 4).

Evaluation of the Integrals for Hyperelliptic Conjugacy Classes

We have seen that if γ and γ' are hyperelliptic elements of $\Gamma_{\infty}^{(1)}$

with $\delta^{-1}\gamma\delta = \gamma'$ for some $\delta\varepsilon\Gamma, \delta^{-1}\Gamma_{\gamma'}\delta \cap \Gamma_\infty \neq \phi$. Therefore, in the integrals for the hyperelliptic conjugacy classes, we may assume that $|j(\delta,P)|^2 = 1$ and since there are only finitely many such classes, that $s=0$. Therefore, using Lemma 7 of Chapter IV, we see that the contribution $I_{h.e.}$ of the hyperelliptic conjugacy classes is given by

$$I_{h.e.} = \Sigma_{[\gamma]_\Gamma h.e.} I_m(\gamma) = \frac{5}{2} I_m(g_o) + \frac{17}{2} I_m(g_o^{\,2}) + \frac{5}{2} I_m(g_o^{\,3}).$$

LEMMA 5. $I_m(g_o^{\,\varepsilon}) = \dfrac{a(m)}{(1-i^\varepsilon)^2} \dfrac{i^{(3m+1)\varepsilon}}{48} [(3m-1)i^\varepsilon - (3m-2)]$

(ε=1,2, or 3).

PROOF. We have $k(P, g_o^{\,\varepsilon}P)^m = [i(\bar{z}-z)-(-i)^\varepsilon |u|^2]^{-3m}$, $\overline{j(g_o^{\,\varepsilon}P)^{3m}} = i^{3m\varepsilon}$, and $F_{g_o} = \{P=(z,u)\varepsilon D \,|\, |\mathrm{Re} z| \leq \frac{1}{2},\ |z|^2 \geq 1,\ \mathrm{Re} u, \mathrm{Im} u \geq 0\}$. Therefore,

$$I_m(g_o^{\,\varepsilon}) = \int_{F_{g_o}} \frac{(2y-|u|^2)^{3m-3}|dz|du|}{[2i^{-\varepsilon}y-(-1)^\varepsilon|u|^2]^{3m}} = \frac{\pi}{4} \int_{\substack{y\geq 0 \\ |x|\leq\frac{1}{2} \\ x^2+y^2\geq 1}} dxdy \int_{r=0}^{\sqrt{2y}} \frac{(2y-r^2)^{3m-3}2rdr}{[2i^{-\varepsilon}y-(-1)^\varepsilon r^2]^{3m}}$$

$$= \frac{\pi}{4} \int_{\substack{y\geq 0 \\ |x|\leq\frac{1}{2} \\ x^2+y^2\geq 1}} dxdy \int_{r=0}^{2y} \frac{(2y-r)^{3m-3}dr}{[2i^{-\varepsilon}y-(-1)^\varepsilon r]^{3m}}$$

$$= \frac{\pi}{16} \int_{x=-\frac{1}{2}}^{\frac{1}{2}} dx \int_{y=\sqrt{1-x^2}}^{\infty} \frac{dy}{y^2} \int_0^1 \frac{(1-s)^{3m-3}ds}{[i^{-\varepsilon}-(-1)^\varepsilon s]^{3m}} .$$

But $\int_{x=-\frac{1}{2}}^{\frac{1}{2}} dx \int_{y=\sqrt{1-x^2}}^{\infty} \frac{dy}{y^2} = \int_{x=-\frac{1}{2}}^{\frac{1}{2}} \frac{dx}{\sqrt{1-x^2}} = \int_{-\frac{\pi}{6}}^{\frac{\pi}{6}} d\theta = \frac{\pi}{3}$; and

$$\frac{1}{(1-\zeta)^2(3m-1)(3m-2)} \frac{d}{ds}\{ \frac{(1-s)^{3m-2}[s+(3m-2)-(3m-1)\zeta]}{(\zeta-s)^{3m-1}}\}$$

$$= \frac{(1-s)^{3m-3}}{(\zeta-s)^{3m}} .$$

Therefore,

$$\int_0^1 \frac{(1-s)^{3m-3}ds}{(i^{\varepsilon}-s)^{3m}} = \frac{(3m-1)i^{\varepsilon}-(3m-2)}{(3m-1)(3m-2)(1-i^{\varepsilon})^{3m+1}} .$$

Thus,

$$I_m(g_o^{\varepsilon}) = \frac{\pi^2}{3\times 16} \frac{(i^{3m+1})\varepsilon}{(1-i^{\varepsilon})^2} [\frac{(3m-1)i^{\varepsilon}-(3m-2)}{(3m-1)(3m-2)}] \text{ as claimed.}$$

CHAPTER VI

HYPERBOLIC CONJUGACY CLASSES

LEMMA 1. Let γ be a hyperbolic element of Γ and let v_1, v_2, v_3 be linearly independent eigenvectors of γ with v_2 negative, v_1 and v_3 isotropic and $\gamma(v_j) = \lambda_j v_j (j=1,2,3, \lambda_j \in \mathbb{C})$. Then $(\lambda_1, \lambda_2, \lambda_3) = (ui^\varepsilon, (-1)^\varepsilon, u^{-1}i^\varepsilon)$ or $(r(1+i)i^\varepsilon, (-1)^\varepsilon i, \frac{r^{-1}}{2}(1+i)i^\varepsilon)$ with $\varepsilon \in Z, u \neq 1$ a positive unit of real quadratic number field $Q(\sqrt{d})$ of norm 1, and r^{-1} a positive integer of $Q(\sqrt{d})$ such that $2r^2$ is a positive unit $u \neq 1$ of $Q(\sqrt{d})$ of norm 1.

PROOF. We see easily that $|\lambda_2|^2 = 1, \lambda_1\bar{\lambda}_3 = 1$, and $\lambda_1 \neq \lambda_3$. Clearly, then, λ_1 and λ_3 are not roots of unity; for otherwise $\lambda_3^{-1} = \bar{\lambda}_3$ and $\lambda_1 = \lambda_3$. But if $f_\gamma(x) = \det(\gamma - x \cdot id) = \prod_{j=1}^{3} (x-\lambda_j)$ is irreducible over $Q[i]$, then λ_1 and λ_3 are roots of unity since λ_2 is; and likewise, if $f_\gamma(x)$ splits completely over $Q[i]$, then $\det \gamma = \lambda_1\lambda_2\lambda_3 = 1$ implies the same result.

Therefore, $f_\gamma(x)$ has an irreducible factor $g(x)$ of degree 2. If λ_2 were a root of $g(x)$, its other root, λ_1 or λ_3, would be a root of unity; hence $g(x) = (x-\lambda_1)(x-\lambda_3) \in Z[i][x]$ and $\lambda_1\lambda_3 \in Z[i]$. But $\lambda_1\bar{\lambda}_3 = 1$; so $\lambda_3 = \bar{\lambda}_1^{-1}$ and $\lambda_1\bar{\lambda}_1^{-1}$ is a root of unity in $Z[i]$, hence equals i^η for some $\eta \in Z$. Therefore, $\lambda_1 = i^\eta \bar{\lambda}_1$ and so $\lambda_1 = ri^\varepsilon$ or $r(1+i)i^\varepsilon$ with $\varepsilon \in Z$ and $r \in R$, $r > 0$.

If $\lambda_1 = ri^\varepsilon$, then $\lambda_3 = r^{-1}i^\varepsilon$ and $g(x) = x^2 + (r+r^{-1})i^\varepsilon x + (-1)^\varepsilon \in Z[i][x]$. Therefore, $r+r^{-1} \in Z$ and $\tilde{g}(x) = (x-r)(x-r^{-1}) = x^2-(r+r^{-1})x + 1 \in Z[x]$. Hence, $r=u$ is a positive unit of a real quadratic number field $Q[\sqrt{d}]$ with $u \neq 1$ and $N_{Q[\sqrt{d}]/Q}(u) = 1$.

Similarly, if $\lambda_1 = r(1+i)i^{\varepsilon}, \lambda_3 = r^{-1}(1-i)^{-1}i^{\varepsilon} = \frac{r^{-1}}{2}(1+i)i^{\varepsilon}$ and $g(x) = x^2-(\frac{r+r^{-1}}{2})(1+i)i^{\varepsilon}x + (-1)^{\varepsilon}i\varepsilon Z[i][x]$. Hence, $r+\frac{r^{-1}}{2} \varepsilon Z$ and $\tilde{g}(x) = (x-2r)(x-r^{-1}) = x^2 - (2r+r^{-1})x + 2\varepsilon Z[x]$. Therefore, since $\tilde{g}(x)$ is clearly irreducible over Q, r^{-1} is an integer of a real quadratic number field $Q(\sqrt{d})$ of norm 2 and even trace. But if $r^{-1} = \frac{\alpha+\beta\sqrt{d}}{2}$ with $\alpha,\beta\varepsilon Z$, we have α even and $\alpha^2-d\beta^2 = 8$. Hence, if d is odd, β is even and so $r^{-1} = \alpha+\beta\sqrt{d}$ with $\alpha,\beta\varepsilon Z, \alpha^2-d\beta^2 = 2$ (recall that the ring of integers of $Q(\sqrt{d})$ is $Z \oplus Z\sqrt{d}$ if d is square-free and $\equiv 2$ or $3 \bmod 4$). Clearly, if $d\equiv 1 \bmod 4$, the equation $x^2-dy^2 = 2$ has no solution in integers x,y as $x^2-dy^2\equiv x^2-y^2 \equiv 0,1,$ or $-1 \bmod 4$. Therefore, if d is square-free, $d\equiv 2$ or $3 \bmod 4$ and the discriminant of $Q(\sqrt{d})$ is $4d$. But then ([11], Theorem 6-2-1, page 235) the ideal (2) is the square of a prime ideal of $Q(\sqrt{d})$. Hence, $(2) = (r^{-1})^2 = (r^{-2})$ and $2r^2$ is a unit of $Q(\sqrt{d})$. Also, if σ is the non-trivial element of $gal(Q(\sqrt{d})/Q)$, $(r^{-1})^{\sigma} = 2r$. Hence,

$$u = \frac{(r^{-1})^{\sigma}}{r^{-1}}$$ has norm 1.

REMARKS. 1) If r,r' are two elements of $Q(\sqrt{d})$ with r^{-1},r'^{-1} integers of $Q(\sqrt{d})$ and $2r^2,2r'^2$ units, $(2) = (r^{-1})^2 = (r'^{-1})^2$; so $(r^{-1}) = (r'^{-1})$ and $r^{-1} = ur'^{-1}$ with u a unit of $Q(\sqrt{d})$.

2) Γ contains many hyperbolic elements. For example, if $\gamma = \begin{pmatrix} a & b \\ c & d \end{pmatrix}$ $\varepsilon SL(2,Z)$ is hyperbolic with $\gamma(x_j) = x_j$ $(j=1,2)$, where x_j lies on the boundary of $H^1 = \{z\varepsilon\mathbb{C} \mid Imz>0\}$, then $\tilde{\gamma} = \begin{pmatrix} a & 0 & b \\ 0 & 1 & 0 \\ c & 0 & d \end{pmatrix} \varepsilon\Gamma$ and $\tilde{\gamma}$ fixes

$(x_j,0)\varepsilon\bar{D}-D$ for $j=1$ or 2. Moreover, these are the only fixed points of $\tilde{\gamma}$ in D so $\tilde{\gamma}$ is hyperbolic.

COROLLARY 1. $[\gamma]_\Gamma$ hyperbolic implies that $[\gamma]_\Gamma \cap P = \phi$.

PROOF. The elements of $[\gamma]_\Gamma \cap P \subset \Gamma \cap P$ have eigenvalues $(i^\varepsilon,(-1)^\varepsilon,i^\varepsilon)$.

COROLLARY 2. If $\gamma\varepsilon\Gamma$ is hyperbolic, γ is conjugate in G to $\mathrm{diag}\{i^\varepsilon u,(-1)^\varepsilon,i^\varepsilon u^{-1}\}$ or $\mathrm{diag}\{(1+i)i^\varepsilon r,(-1)^\varepsilon i,(1+i)i^\varepsilon \frac{r^{-1}}{2}\}$ with u,r,ε as in Lemma 1.

PROOF. Changing v_1,v_2, and v_3 by appropriate scalars if necessary, we may assume that $(v_3,v_1) = i$ and $(v_2,v_2) = 1$. Then if $h(v_1) = e_1, h(v_2) = \lambda e_2$, and $h(v_3) = e_3$, we have $h\varepsilon G$ for suitable λ such that $|\lambda|^2 = 1$; and $h\gamma h^{-1}$ is diagonal of the required form.

NOTATIONS. We let $g_u = \mathrm{diag}\{u,1,u^{-1}\}$ and $h_r = \mathrm{diag}\{(1+i)r,i,(1+i)\frac{r^{-1}}{2}\}(u,r,>0)$. Note that if $2r^2 = u$, $(h_r)^2 = g_o g_u$ and that $h_{ru} = h_r g_u$.

LEMMA 2. Let $\gamma = h^{-1}g_o{}^\varepsilon g_u h$ or $h^{-1}g_o{}^\varepsilon h_r h$ be a hyperbolic element of Γ $(u,r\varepsilon Q(\sqrt{d})$ as in Lemma 1, $h\varepsilon G)$. Then if $\delta\varepsilon\Gamma_\gamma$, we have $h\delta h^{-1} = g_o{}^\nu$, $g_o{}^\nu g_{u'}$, or $g_o{}^\nu h_{r'}$ with $\nu\varepsilon Z, u'$ or r' elements of $Q(\sqrt{d})$ as in Lemma 1.

PROOF. Since the eigenvalues of γ belong to $Q[i,\sqrt{d}]$, γ has eigenvectors v_1,v_2,v_3 with v_1,v_3 isotropic, v_2 negative, $v_j\varepsilon V_{Q(i,\sqrt{d})}$. If $\delta\neq 1\varepsilon\Gamma_\gamma, \delta(v_j) = \lambda_j v_j$ with $\lambda_j\varepsilon\mathbb{C}$ $(j=1,2$, or $3)$. Thus, δ is hyperbolic if $\lambda_1\neq\lambda_3$, hyperelliptic if $\lambda_1=\lambda_3$. Also, since $\delta\varepsilon\Gamma, \delta(V_{Q[i,\sqrt{d}]})\subset V_{Q[i,\sqrt{d}]}$ and so $\lambda_j\varepsilon Q[i,\sqrt{d}]$. If $\delta=1$ or is hyperelliptic, $\lambda_1 = \lambda_3 = i^\nu$ and $\lambda_2 = (-1)^\nu$ with $\nu\varepsilon Z$ (by Lemma 1 of Chapter V) and, since $h\delta h^{-1}$ is diagonal, $h\delta h^{-1} = g_o{}^\nu$. If δ is hyperbolic, Lemma 1 implies in the same say that $h\delta h^{-1} =$

$g_o{}^{\nu}g_{u'}$ or $g_o{}^{\nu}h_{r'}$ with $u',r'\varepsilon Q[\sqrt{d}]$.

COROLLARY. If $\gamma = h^{-1}g_o{}^{\varepsilon}g_u h$ or $h^{-1}g_o{}^{\varepsilon}h_r h$ is a hyperbolic element of Γ with $u,r\varepsilon Q[\sqrt{d}]$ as in Lemma 1 and $h\varepsilon G, h(\Gamma_\gamma)h^{-1}$ is commensurable with the group $G_o = \{g_o{}^{\nu}g_u | \nu\varepsilon Z,\ u$ a positive unit of $Q(\sqrt{d})\}$.

PROOF. First note that the set of elements $g_o{}^{\nu}, g_o{}^{\nu}g_{u'}, g_o{}^{\nu}h_{r'}$ with $\nu\varepsilon Z, u'$ a positive unit of $Q[\sqrt{d}]$, and r^{-1} a positive integer of $Q[\sqrt{d}]$ with $2r^2$ a positive unit of norm 1 of $Q[\sqrt{d}]$ forms a group G_o' containing G_o. This is clear if $Q[\sqrt{d}]$ contains no integers r^{-1} with $2r^2$ a positive unit of norm 1. On the other hand, if r_1^{-1} and r_2^{-1} are two such integers, $r_1 = vr_2$ with v a positive unit of $Q(\sqrt{d})$, so $h_{r_1}h_{r_2} = g_v(h_{r_2})^2 = g_o g_{v\times 2r_2{}^2}\ \varepsilon G_o$. Thus G_o' is a group; and clearly, $[G_o':G_o] = 1$ or 2. Also, $G_o \cong Z \times Z/(4)$. But $h\gamma^n h^{-1}\varepsilon h(\Gamma_\gamma)h^{-1}$ for all $n\varepsilon Z$, so $h\Gamma_\gamma h^{-1}$ is infinite. Thus, $h(\Gamma_\gamma)h^{-1}\cap G_o$ is an infinite subgroup of G_o, hence of finite index in G_o and in $h\Gamma_\gamma h^{-1}$.

PROPOSITION 1. The contribution I_{hyp} of the hyperbolic conjugacy classes of Γ is 0.

PROOF. By Corollary 1 of Lemma 1, $I_{hyp} = \Sigma_{[\gamma]_\Gamma \text{ hyperbolic}} I_m(\gamma)$. Therefore, it suffices to show that if $\gamma\varepsilon\Gamma$ is hyperbolic, $I_m(\gamma) = 0$.

Let $\gamma = h^{-1}g_o{}^{\varepsilon}g_v h$ or $h^{-1}g_o{}^{\varepsilon}h_r h$ be a hyperbolic element of Γ with $v,r\varepsilon Q[\sqrt{d}]$ as in Lemma 1, and let $\gamma_o = g_o{}^{\varepsilon}g_v$ or $g_o{}^{\varepsilon}h_r$. By Lemma 6 of Chapter II,

$$I_m(\gamma) = \int_{h(F_\gamma)} \frac{k(P,\gamma_o P)^m}{k(P,P)^m}\ \overline{j(\gamma_o,P)^m}\,dP,$$

where F_γ is a fundamental domain of Γ_γ. But then $h(F_\gamma)$ is a fundamental domain for $h(\Gamma_\gamma)h^{-1}$, hence by the preceeding Corollary and Lemma 2 of Chapter IV, it suffices to prove that $J_m(\gamma_o) = 0$, where $J_m(\gamma_o) = \int_{F_o} \frac{k(P,\gamma_o P)^m}{k(P,P)^m} \overline{j(\gamma_o,P)}^m dP$, if F_o is a fundamental domain for G_o acting in D.

Let v_o be a fundamental unit of $Q[\sqrt{d}]$ $(v_o>1)$. Then, if $\delta = g_o{}^\nu g_{\tilde{v}} \varepsilon G_o$ and $P = (z,u)\varepsilon D$, $\tilde{v} = v_o{}^n$ for some $n\varepsilon Z$ and $\delta(z,u) = (\tilde{v}^2 z, i^\nu \tilde{v}u) = (v_o{}^{2n}z, i^\nu v_o{}^n u)$. Then $\log(2\mathrm{Im}v^2z-|vu|^2) = 2n\log v_o + \log(2\mathrm{Im}z-|u|^2)$. Therefore, for F_o we may take $\{(z,u)\varepsilon D|\mathrm{Re}u,\mathrm{Im}u \geq 0,\ 0 \leq \log(2\mathrm{Im}z-|u|^2) \leq 2\log v_o\} = \{(z,u)\varepsilon D|\mathrm{Re}u,\mathrm{Im}u \geq 0, 1 \leq 2\mathrm{Im}z-|u|^2 \leq v_o{}^2\}$.

Moreover, if $\gamma_o = g_o{}^\varepsilon g_v$ or $g_o{}^\varepsilon h_r$ and $P = (z,u) = (x+iy,u)\varepsilon D, \gamma_o(z,u) = (tz,au)$ with $t > 1$, $a\neq 0\varepsilon\mathbb{C}$. Thus, $\overline{j(\gamma_o,P)}^m = (t\bar{a})^m$ and $k(P,\gamma_o P)^m = [i(t\bar{z}-z) - \bar{a}|u|^2]^{-3m} = [i(t-1)x + (\frac{t+1}{2})(2y-|u|^2) + (\frac{t+1}{2} - \bar{a})|u|^2]^{-3m}$. Hence

$$J_m(\gamma_o) = (t\bar{a})^m \int_{F_o} \frac{(2y-|u|^2)^{3m-3}dxdy|du|}{[i(t-1)x + \frac{(t+1)}{2}(2y-|u|^2) + \frac{t+1}{2} - \bar{a}|u|^2]^{3m}}$$

$$= \frac{\pi}{4}(t\bar{a})^m \int_{y=1}^{v_o^2} \int_{u=0}^{\infty} \int_{x=-\infty}^{\infty} \frac{y^{3m-3}dydudx}{[i(t-1)x + \frac{(t+1)}{2}y + (\frac{t+1}{2}-\bar{a})u]^{3m}} = 0,$$

since $\int_{-\infty}^{\infty} \frac{dx}{(x-A)^{3m}} = 0$ if $\mathrm{Im}\ A\neq 0$.

CHAPTER VII

ELLIPTIC CONJUGACY CLASSES

Suppose that $\gamma\varepsilon\Gamma$ is elliptic and that $\gamma(v_j) = \lambda_j v_j (j=1,2,3)$ with v_1 positive, v_2 and v_3 negative, $v_j \perp v_k (j\neq k), |\lambda_j|^2 = 1, \lambda_1\lambda_2\lambda_3 = 1, \lambda_1\neq\lambda_2$, $\lambda_1\neq\lambda_3$.

LEMMA 1. γ belongs to one of the following four types:

i) $\lambda_2=\lambda_3, \lambda_j \varepsilon Z[i], (\lambda_1,\lambda_2,\lambda_3) = ((-1)^\varepsilon, i^\varepsilon, i^\varepsilon)$ (ε=1,2, or 3);

ii) $\lambda_j \varepsilon Z[i], \lambda_j\neq\lambda_k (j\neq k), \{\lambda_1,\lambda_2,\lambda_3\} = \{i,-i,1\}$;

iii) $\lambda_j \varepsilon Q(\exp 2\pi i/8), \{\lambda_1,\lambda_2,\lambda_3\} = \{\zeta,-\zeta,\zeta^2\}$ (ζ a primitive 8-th root of unity);

iv) $\lambda_j \varepsilon Q(\exp 2\pi i/12), \{\lambda_1,\lambda_2,\lambda_3\} = \{i^\varepsilon\rho, i^\varepsilon\rho^2, (-1)^\varepsilon\}$ ($\rho=\exp 2\pi i/3, \varepsilon=0,1,2$ or 3).

PROOF. Let $f_\gamma(x) = \prod_{j=1}^{3} (x-\lambda_j) \varepsilon Z[i][x]$, and assume $f_\gamma(x)$ is irreducible over $Q[i]$. Then $Q[i,\lambda_j]$ is a cyclotomic field of degree 6 over Q. Let ϕ be the Euler function $\phi(\prod_{j=1}^{k} p_j^{n_j}) = \prod_{j=1}^{k} (p_j-1)p_j^{n_j-1}$ (p_j distinct primes) and suppose that $Q[i,\lambda_j] = Q[\zeta]$ with ζ a primitive n-th root of unity. Then $\phi(n) = 6$, so clearly, $n = 7$ or 9. But $Q[i]$ is not a subfield of $Q(\exp 2\pi i/7)$ or $Q(\exp 2\pi i/9)$. Therefore, $f_\gamma(x)$ is not irreducible over $Q[i]$.

Assume, then, that $f_\gamma(x)$ has a factor $g(x)$ of degree 2 which is irreducible over $Q[i]$. Then $g(x) = (x-\lambda_{\sigma(1)})(x-\lambda_{\sigma(2)})$, where $\sigma\varepsilon S_3$ (permutations of three letters); and $\lambda_{\sigma(3)} = \lambda_{\sigma(1)}^{-1}\lambda_{\sigma(2)}^{-1} \varepsilon Z[i]$. Therefore,

$[Q[\lambda_{\sigma(1)},i]:Q[i]] = 2$ and $[Q[\lambda_{\sigma(1)},i]:Q] = 4$; so $Q[\lambda_{\sigma(1)},i]$ is a cyclotomic field of degree 4 over Q. But $\phi(n) = 4$ holds only for $n=5,8$, or 12, and $Q[i]\not\subset Q(\exp 2\pi i/5)$. Hence, $Q[\lambda_{\sigma(1)},i] = Q(\exp 2\pi i/8)$ or $Q(\exp 2\pi i/12)$; $\lambda_{\sigma(1)}$ is a primitive 8-th, cube, 6-th, or 12-th root of unity; and $\lambda_{\sigma(2)} = \lambda_{\sigma(1)}^5$ is the conjugate of $\lambda_{\sigma(1)}$ over $Q[i]$. Therefore, $\{\lambda_1,\lambda_2,\lambda_3\} = \{\zeta,-\zeta,\zeta^2\}$ or $\{i^\varepsilon\rho, i^\varepsilon\rho^2, (-1)^\varepsilon\}$; the eigenvalues of γ are distinct; and γ is of type iii) or iv).

Suppose, finally, that $f_\lambda(x)$ splits completely over $Q[i]$. If the λ_j are distinct, $\lambda_j \neq 1 (j=1,2,\text{ or } 3)$. For if $\lambda_{\sigma(1)} = -1$ ($\sigma\varepsilon S_3$), then $\lambda_{\sigma(2)} = -\lambda_{\sigma(3)}^{-1}$. Therefore, $\lambda_{\sigma(3)} = 1$ implies $\lambda_{\sigma(2)} = -1 = \lambda_{\sigma(1)}$; $\lambda_{\sigma(3)} = i$ implies $\lambda_{\sigma(2)} = i = \lambda_{\sigma(3)}$; and $\lambda_{\sigma(2)} = -i$ implies $\lambda_{\sigma(2)} = -i = \lambda_{\sigma(3)}$. Hence, $\{\lambda_1,\lambda_2,\lambda_3\} = \{i,-i,1\}$; and γ is of type ii). If the λ_j are not distinct, we must have $\lambda_2 = \lambda_3 = i^\varepsilon$, $\lambda_1 = (-1)^\varepsilon$; and γ is of type i).

LEMMA 2. Let $M = Z[i]v_1 \oplus Z[i]v_2$ be a rank 2 module over $Z[i]$ and let $(\ ,\ )$ be a negative-definite $Z[i]$-valued Hermitian form on M which is unimodular. Then M has a basis $\{w_1,w_2\}$ with $(w_1,w_2) = 0$, $(w_j,w_j) = -1$ $(j=1,2)$.

PROOF. Pick $w\neq 0$ in M such that $k = |(w,w)|$ is minimal, and pick $x\varepsilon M$ such that $M = Z[i]w \oplus Z[i]x$. Since $(x+aw,w) = (x,w)+a(w,w)$ and $\{x+aw,w\}$ is a basis of M, we may assume that $|(x,w)|^2 < k^2$. M unimodular implies that $1 = (x,x)(w,w)-|(w,w)|^2 = |(x,x)|k-|(w,x)|^2 > |(x,x)|k-k^2$, so $|(x,x)| \leq k$. But $k = |(w,w)|$ minimal implies that $k = |(x,x)|$, $|(w,x)|^2 = k^2 -1$. Also, $(w+i^\varepsilon x,w+i^\varepsilon x) = (w,w) + 2\text{Re}\, i^\varepsilon(x,w) + (x,x) = -2k + \text{Re}\, i^\varepsilon(x,w) \leq -k$. Hence, $2|\text{Re}(x,w)| \leq k$ and $2|\text{Im}(x,w)| \leq k$; so $4(\text{Re}(x,w))^2 +$

$4(\mathrm{Im}(x,w))^2 = 4|(x,w)|^2 \leq 2k^2$. Thus, $2|(x,w)|^2 = 2k^2-2 \leq k^2$, so $k^2 \leq 2$ and $k=1$. But then $(w,w) = (x,x) = -1$ and $(w,x) = 0$, as claimed.

LEMMA 3. Suppose that $w \in L$ is primitive positive with $(w,w) = 2$. Then there exist vectors $x,y,v \in L$ such that $L = Z[i]v \oplus Z[i]x \oplus Z[i]y$ and $L \cap \{w\}^{\perp} = Z[i]x \oplus Z[i]y$, with $(x,y) = (x,v) = 0$, $(y,v) = (w,v) = 1$, $(x,x) = -1$, $(y,y) = -2$, and $(v,v) = 0$. In particular, $w = y+2v$.

PROOF. Pick $x \in L \cap \{w\}^{\perp}$ such that $k = |(x,x)|$ is minimal, and let $\{x,y\}$ be a basis of $L \cap \{w\}^{\perp}$. As above, we can assume that $|(x,y)|^2 < k^2$. Choose $v \in L$ such that $(v,w) = 1$, so $L = Z[i]v \oplus L \cap \{w\}^{\perp}$; and let $w = av+bx+ay$ with $a,b,c \in Z[i]$ and $a = (w,w) = 2$. Letting $\{u_1,u_2,u_3\} = \{v,x,y\}$ and $\{\tilde{u}_1,\tilde{u}_2,\tilde{u}_3\} = \{w,x,y\}$, we get $2\{(x,x)(y,y)-|(x,y)|^2\} = \det((\tilde{u}_i,\tilde{u}_j)) = \det A((u_i,u_j))\,{}^t\bar{A} = |\det A|^2|\det(u_i,u_j) = |\det A|^2 = 4$, where $A =$

$$\begin{pmatrix} 2 & b & c \\ 0 & 1 & 0 \\ 0 & 0 & 1 \end{pmatrix}$$

; so $k|(y,y)| - |(x,y)|^2 = 2$. But then $k^2 > |(x,y)|^2 = k|(y,y)| -2$ and $k^2 \geq k|(y,y)| -1$; so $|(y,y)| \leq k$ or $k|(y,y)| = k^2+1$.

Assume that $|(y,y)| \leq k$. Then by minimality, $|(y,y)| = k$, $|(x,y)|^2 = k^2-2 \geq 0$, and $k \geq 2$. Therefore, $(x+i^{\varepsilon}y,x+i^{\varepsilon}y) = -2k+2\mathrm{Re}\,i^{\varepsilon}(y,x) \leq -k$ $(\varepsilon \in Z)$, $2|\mathrm{Re}(x,y)| \leq k$, $2|\mathrm{Im}(x,y)| \leq k$, $2|(x,y)|^2 \leq k^2 = 2 + |(x,y)|^2$, and $|(x,y)|^2 = k^2-2 \leq 2$. Hence, $k=2$, $(x,x) = (y,y) = -2$, and $|(x,y)|^2 = 2$.

Assuming, as we may, that $(x,y) = 1+i$, we get that

$$1 = \det \begin{pmatrix} (v,v) & (v,x) & (v,y) \\ (x,v) & -2 & 1+i \\ (y,v) & 1-i & -2 \end{pmatrix} \equiv 0 \bmod 1 + i,$$

which is impossible. Hence, $k\,|(y,y)| = k^2+1$. But then k divides k^2+1; so $k=1$, $(y,y) = -2$, and $(x,y) = 0$.

If $v' = v+dx+ey, (v',w) = 1, (v',x) = (v,x)-d$, and $(v',y) = (v,y)-2e$. Therefore, choosing appropriate $d,e\varepsilon Z[i]$, we may assume that $(v,x) = 0$ and $|(v,y)|^2 < 4$. But then $\det((u_i,u_j)) = 2(v,v) + |(v,y)|^2 = 1$; so $|(v,y)|^2 = (v,y) = i^u$, and $(v,v) = 0$. Changing y by a power of i, we can get $(v,y) = 1$.

LEMMA 4. If M is as in Lemma 2, $\Gamma(M)$ is finite of order 32.

PROOF. Choose $\gamma\varepsilon\Gamma(M)$ and w_1,w_2 as in Lemma 2. Then $\gamma(w_1) = aw_1+bw_2$ and $\gamma(w_2) = cw_1+dw_2$ with $a,b,c,d\varepsilon Z[i]$, $|a|^2+|b|^2 = |c|^2+|d|^2 = 1$, $a\bar{c}+b\bar{d} = 0$. But then $\begin{pmatrix} a & b \\ c & d \end{pmatrix} = \begin{pmatrix} i^\varepsilon & 0 \\ 0 & i^\nu \end{pmatrix}$ or $\begin{pmatrix} 0 & i^\varepsilon \\ i^\nu & 0 \end{pmatrix}$ with $\varepsilon,\nu=0,1,2,$ or 3; and these elements are all in $\Gamma(M)$.

LEMMA 5. If w is as in Lemma 3 and $\Gamma_w = \{\gamma\varepsilon\Gamma|\gamma(w)\varepsilon Z[i]w\}$, then Γ_w is finite of order 8.

PROOF. If $\gamma\varepsilon\Gamma w$, $\gamma(L\cap\{w\}^\perp) = L\cap\{w\}^\perp$; so if x,y,v are as in Lemma 3, $\gamma(x) = ax+by$ and $\gamma(y) = cx+dy$ with $(\gamma(x),\gamma(x)) = (x,x) = -1 = -|a|^2-2|b|^2, (\gamma(y),\gamma(y)) = (y,y) = -2 = -|c|^2-2|d|^2$, $(\gamma(x),\gamma(y)) = (x,y) = -a\bar{c}-2b\bar{d} = 0, a,b,c,d\varepsilon Z[i]$. Therefore, $b = c = 0$, $a=i^\varepsilon$, $d=i^\nu$, $\begin{pmatrix} a & b \\ c & d \end{pmatrix} = \begin{pmatrix} i^\varepsilon & 0 \\ 0 & i^\nu \end{pmatrix}$. But $w = 2v+y$. Therefore, $\gamma(w) = i^\lambda w = 2i^\lambda v+i^\lambda y = 2\gamma(v) + i^\nu y$; so $2(\gamma(v) - i^\lambda v) = (i^\lambda-i^\nu)y$. Hence, 2 divides $1-i^{\nu-\lambda}$ and $\nu\equiv\lambda \bmod 2$. But $\nu+\varepsilon+\lambda\equiv 0 \bmod 4$, so ε is even, $\varepsilon=2\mu$, $\begin{pmatrix} a & b \\ c & d \end{pmatrix} = \begin{pmatrix} -1^\mu & 0 \\ 0 & i^\nu \end{pmatrix}$. Moreover, if $\gamma(x) = (-1)^\mu x, \gamma(y) = i^\nu y, \gamma(w) = i^{2\mu-\nu}w$, then γ is in Γ_w so $^{\#}\Gamma_w = 8$.

LEMMA 6. There are four Γ-conjugacy classes of elliptic elements of type i), represented by $\gamma_o = \begin{pmatrix} 0 & 0 & i \\ 0 & -1 & 0 \\ -i & 0 & 0 \end{pmatrix}$ and $\gamma_\varepsilon =$

$$A\begin{pmatrix} i^\varepsilon & 0 & 0 \\ 0 & i^\varepsilon & 0 \\ 0 & 0 & (-1)^\varepsilon \end{pmatrix} A^{-1} \ (\varepsilon=1,2, \text{ or } 3), \text{ where } A = \begin{pmatrix} 1 & 0 & 1 \\ 1 & i & 1 \\ 0 & 1 & -i \end{pmatrix}.$$

$\#\Gamma_{\gamma_o} = 8$ and $\#\Gamma_{\gamma_\varepsilon} = 32$ for $\varepsilon=1,2$, or 3. γ_o is conjugate to γ_2 in G (but not in Γ).

PROOF. Suppose that $\gamma\varepsilon\Gamma$ is elliptic of type i), $\gamma(w)\varepsilon Z[i]w$, $w\varepsilon L$ is primitive positive such that $\gamma|L\cap\{w\}^\perp = i^\varepsilon\times\text{id}$, $\gamma(w) = (-1)^\varepsilon w$. By Lemma 2 of Chapter V, $(w,w) = 1$ or 2 and if $(w,w) = 2$, $\varepsilon=2$.

If $(w,w) = 1$, let $M = L\cap\{w\}^\perp$. Then the restriction map $\Gamma_\gamma \to \Gamma(M)$ is an isomorphism, so $\#\Gamma_\gamma = 32$ by Lemma 4. Also, if $\gamma'\varepsilon\Gamma$ is elliptic of type i) with $\gamma(w')\varepsilon Z[i]w'\subset L, (w',w') = 1$, $\gamma'|L\cap\{w\}^\perp = i^\varepsilon\times\text{id}$, then $\gamma'\varepsilon[\gamma]_\Gamma$. For by Lemma 2, $L = Z[i]w\oplus Z[i]x\oplus Z[i]y = Z[i]w'\oplus Z[i]x'\oplus Z[i]y'$ with $x\perp y\perp w(x'\perp y'\perp w')$ and $(x,x) = (y,y) = -1$ $((x',x') = (y',y') = -1)$; so if $\delta(w)=i^\nu w', \delta(x) = x'$, and $\delta(y) = y'$, then $\delta\varepsilon\Gamma$ if ν is chosen so that $\det\delta = 1$ and $\delta\gamma\delta^{-1} = \gamma'$. Moreover, $\gamma_\varepsilon = A\,\text{diag}\{i^\varepsilon,i^\varepsilon,(-1)^\varepsilon\}A^{-1}$ (A as above) is of this type with $x = e_1+e_2$, $y = ie_2+e_3, w = e_1+e_2-ie_3$ (e_1,e_2,e_3 the standard basis of L), so γ is conjugate to one of the γ_ε.

If $(w,w) = 2$, then $\Gamma_\gamma = \Gamma_w$ so $\#\Gamma_\gamma = 8$ by Lemma 5. Again, if $\gamma'\varepsilon\Gamma$ is elliptic of type i) with $\gamma(w')\varepsilon Z[i]w', w'$ primitive in L, $(w',w') = 2$, and $\gamma|L\cap\{w\}^\perp = -\text{id}$, then $\gamma'\varepsilon[\gamma]_\Gamma$. For by Lemma 3 (changing w,w' by units if necessary), $L = Z[i]v\oplus Z[i]x\oplus Z[i]y = Z[i]v'\oplus Z[i]x'\oplus Z[i]y'$ with $(v,v) = (v,x) = (x,y) = 0$, $(v,y) = -(x,x) = (v,w) = 1$, $(y,y) = -2$, $w\perp x,y$, and similarly for the primed letters. Hence, if $\delta(v) = v', \delta(x) = i^{\lambda}x', \delta(y) = y'$ for appropriate $\lambda, \delta\varepsilon\Gamma$ and $\delta\gamma\delta^{-1} = \gamma'$. Since in particular $\gamma_o = \begin{pmatrix} 0 & 0 & i \\ 0 & -1 & 0 \\ -1 & 0 & 0 \end{pmatrix}$ is elliptic of this type with $w = e_1 - ie_3$, γ is Γ-conjugate to γ_o.

LEMMA 7. Suppose that $\gamma\varepsilon\Gamma$ is elliptic of type ii) and that $w\varepsilon L$ is primitive positive with $\gamma(w)\varepsilon Z[i]w$. Then $(w,w) = 1$ or 2.

PROOF. Let $\gamma(w) = \lambda_1 w$, and choose u,v primitive in L such that $\gamma(u) = \lambda_2 u$, $\gamma(v) = \lambda_3 v$, $\{\lambda_1,\lambda_2,\lambda_3\} = \{i,-i,1\}$. If $\lambda_1 = 1$, $\gamma^2 = id$ on $L \cap \{w\}^{\perp}$; so by Lemma 2 of Chapter V, $(w,w) = 1$ or 2. Hence we may suppose that $\lambda_1 = \pm i, \lambda_3 = \mp i$, $\lambda_2 = 1$; and then $(u,u) = -1$ or -2. But $\gamma | L \cap \{u\}^{\perp}$ has determinant 1 and trace 0, so by the remark following Corollary 5 of Chapter V, $(u,u) = -1$. Hence if $\delta = \mp i \times id$ on $L \cap \{u\}^{\perp}$ and $\delta(u) = -u, \delta\varepsilon\Gamma$. But then $\delta\gamma\varepsilon\Gamma, \delta\gamma(w) = w, \delta\gamma(u) = -u$, and $\delta\gamma(v) = -v$; so again, $(w,w) = 1$ or 2.

LEMMA 8. There are six Γ-conjugacy classes of elliptic elements of Γ of type ii), represented by $\tilde{\gamma}_1 = A \begin{pmatrix} i & 0 & 0 \\ 0 & -i & 0 \\ 0 & 0 & 1 \end{pmatrix} A^{-1}$,

$$\tilde{\gamma}_2 = A \begin{pmatrix} 1 & 0 & 0 \\ 0 & -i & 0 \\ 0 & 0 & i \end{pmatrix} A^{-1}, \quad \tilde{\gamma}_3 = A \begin{pmatrix} i & 0 & 0 \\ 0 & 1 & 0 \\ 0 & 0 & -i \end{pmatrix} A^{-1}, \quad \tilde{\gamma}_4 = A \begin{pmatrix} 0 & -1 & 0 \\ 1 & 0 & 0 \\ 0 & 0 & 1 \end{pmatrix} A^{-1},$$

$$\tilde{\gamma}_5 = A \begin{pmatrix} 0 & 0 & -1 \\ 0 & 1 & 0 \\ 1 & 0 & 0 \end{pmatrix} A^{-1}, \quad \text{and} \quad \tilde{\gamma}_6 = \begin{pmatrix} 0 & 0 & 1 \\ 0 & 1 & 0 \\ -1 & 0 & 0 \end{pmatrix} \qquad (A \text{ as in Lemma 6}).$$

$\#\Gamma_{\tilde{\gamma}_j} = 16$ $(j=1,2,3)$; $\#\Gamma_{\tilde{\gamma}_j} = 8$ $(j=4,5,6)$. We have $\tilde{\gamma}_1 \sim \tilde{\gamma}_4, \tilde{\gamma}_2 \sim \tilde{\gamma}_5, \tilde{\gamma}_3 \sim \tilde{\gamma}_6$ in G, but not in Γ .

PROOF. Suppose that $\gamma\varepsilon\Gamma$ is elliptic of type ii) with w as in Lemma 7 and $(w,w) = 1$. Let $M = L\cap\{w\}^{\perp} = Z[i]x\oplus Z[i]y$ as in Lemma 2. By Lemma 4, $\gamma|L\cap\{w\}^{\perp} = \begin{pmatrix} i^{\varepsilon} & 0 \\ 0 & i^{\nu} \end{pmatrix}$ or $\begin{pmatrix} 0 & i^{\varepsilon} \\ i^{\nu} & 0 \end{pmatrix}$ with respect to the basis $\{x,y\}$. Assume that $\gamma(w) = w$. Then $\gamma|L\cap\{w\}^{\perp} = \begin{pmatrix} \pm i & 0 \\ 0 & \pm i \end{pmatrix}$ or $\begin{pmatrix} 0 & i^{\varepsilon} \\ -i^{-\varepsilon} & 0 \end{pmatrix} = \begin{pmatrix} 0 & \pm i \\ \pm i & 0 \end{pmatrix}$ or $\begin{pmatrix} 0 & \pm 1 \\ \mp 1 & 0 \end{pmatrix}$; and by interchanging x and y or using the basis $i^{\varepsilon}x,y$, we may assume that $\gamma|L\cap\{w\}^{\perp} = \begin{pmatrix} i & 0 \\ 0 & -i \end{pmatrix}$ or $\begin{pmatrix} 0 & -1 \\ 1 & 0 \end{pmatrix}$.

On the other hand, if $\gamma(w) = \pm iw$, $\gamma|L\cap\{w\}^{\perp}$ has trace $1 \mp i$; so we see that $\gamma|L\cap\{w\}^{\perp} = \begin{pmatrix} 1 & 0 \\ 0 & \mp i \end{pmatrix}$ or $\begin{pmatrix} \mp i & 0 \\ 0 & 1 \end{pmatrix}$, and again we may assume $\gamma|L\cap\{w\}^{\perp} = \begin{pmatrix} 1 & 0 \\ 0 & \mp i \end{pmatrix}$. Thus, if $(w,w) = 1$, $\gamma = \begin{pmatrix} i & 0 & 0 \\ 0 & -i & 0 \\ 0 & 0 & 1 \end{pmatrix}$, $\begin{pmatrix} 0 & -1 & 0 \\ 1 & 0 & 0 \\ 0 & 0 & 1 \end{pmatrix}$, or $\begin{pmatrix} 1 & 0 & 0 \\ 0 & i & 0 \\ 0 & 0 & -i \end{pmatrix}$ with respect to the basis $\{x,y,w\}$ of L. In particular, if $x = e_1+e_2$, $y = ie_2+e_3$, and $w = e_1+e_2-ie_3$, γ has matrix $\tilde{\gamma}_1,\tilde{\gamma}_2,\tilde{\gamma}_3$, or $\tilde{\gamma}_4$ with respect to the standard basis; and in general, γ is conjugate to one of these elements.

Clearly, $\Gamma_{\tilde{\gamma}_1} = \Gamma_{\tilde{\gamma}_2} = \Gamma_{\tilde{\gamma}_3} = \{A\,\mathrm{diag}\{i^{\eta_1}, i^{\eta_2}, i^{-\eta_1-\eta_2}\}A^{-1} \mid \eta_1,\eta_2 = 0,1,2 \text{ or } 3\}$.

Similarly, $\Gamma_{\tilde{\gamma}_4} = \{A\,\mathrm{diag}\{i^\eta, i^\eta, (-1)^\eta\}A^{-1} | \eta = 0, 1,2$ or $3\} \cup \{A \begin{pmatrix} 0 & i^\eta & 0 \\ -i^\eta & 0 & 0 \\ 0 & 0 & (-1)^\eta \end{pmatrix} A^{-1} | \eta = 0, 1,2$ or $3\}$. Hence, $\#\Gamma_{\tilde{\gamma}_j} = 16$ ($j=1,2$, or 3) and $\#\Gamma_{\tilde{\gamma}_4} = 8$.

If $(w,w) = 2$, assume $L = Z[i]v \oplus Z[i]x \oplus Z[i]y$ as in Lemma 3. By Lemma 5, $\gamma(x) = (-1)^\mu x, \gamma(y) = i^\nu y, \gamma(w) = i^{2\mu-\nu}w$. But $\gamma(x) = -x$ is impossible, so $\gamma(x) = x$, $\gamma(y) = \pm iy$, and $\gamma(w) = \mp iw$. In particular, if $w = e_1 - ie_3, x = e_2$, and $y = e_1 + ie_3, \gamma = \tilde{\gamma}_5$ or $\tilde{\gamma}_6$; and in general, γ is conjugate to one of these elements. Clearly, $\Gamma_\gamma = \Gamma_w$, so $\#\Gamma_{\tilde{\gamma}_5} = \#\Gamma_{\tilde{\gamma}_6} = 8$.

Suppose that $\gamma \varepsilon \Gamma$ is elliptic of type iii) with eigenvalues $\{\zeta, -\zeta, \zeta^2\}$ (ζ a primitive 8-th root of unity), and pick $w \varepsilon L$ primitive such that $\gamma(w) = \zeta^2 w$. γ^2 is elliptic or hyperelliptic with eigenvalues $\{\zeta^2, \zeta^2, \zeta^4\} = \{\pm i, \pm i, -1\}$, and $\gamma^2(w) = \zeta^4 w = -w$; so by Lemma 2, Chapter V, $(w,w) = \pm 1$.

LEMMA 9. There are 2 Γ-conjugacy classes of elliptic, type iii) elements $\gamma \varepsilon \Gamma$, with $(w,w) = 1$ (w primitive in L, $\gamma(w) = \zeta^2 w$), represented by

$$\hat{\gamma}_1 = A \begin{pmatrix} 0 & i & 0 \\ 1 & 0 & 0 \\ 0 & 0 & i \end{pmatrix} A^{-1} \text{ and } \hat{\gamma}_2 = A \begin{pmatrix} 0 & i & 0 \\ -1 & 0 & 0 \\ 0 & 0 & i \end{pmatrix} A^{-1}. \quad \#\Gamma_{\hat{\gamma}_j} = 8.$$

PROOF. As in Lemma 8, let $L \cap \{w\}^\perp = Z[i]x \oplus Z[i]y$. Then $\gamma | L \cap \{w\}^\perp = \begin{pmatrix} i^\varepsilon & 0 \\ 0 & i^\nu \end{pmatrix}$ or $\begin{pmatrix} 0 & i^\varepsilon \\ i^\nu & 0 \end{pmatrix}$ with respect to the basis $\{x,y\}$ and has trace 0 and determinant $-\zeta^2 = \pm i$; so $\gamma | L \cap \{w\}^\perp = \begin{pmatrix} 0 & \pm i \\ (-1)^\varepsilon & 0 \end{pmatrix}$ or $\begin{pmatrix} 0 & (-1)^\varepsilon \\ \pm i & 0 \end{pmatrix}$. Interchanging x and y if necessary, we may assume

$\gamma|L\cap\{w\}^{\perp} = \begin{pmatrix} 0 & \pm i \\ (-1)^{\varepsilon} & 0 \end{pmatrix}$; and then substituting $-x$ for x, we may assume

$\gamma|L\cap\{w\}^{\perp} = \begin{pmatrix} 0 & i \\ (-1)^{\varepsilon} & 0 \end{pmatrix}$. Then the matrix of γ with respect to the basis

$\{x,y,w\}$ of L is $\begin{pmatrix} 0 & i & 0 \\ 1 & 0 & 0 \\ 0 & 0 & i \end{pmatrix}$ or $\begin{pmatrix} 0 & i & 0 \\ -1 & 0 & 0 \\ 0 & 0 & -i \end{pmatrix}$; so as before

(Lemmas 6 and 8) γ is conjugate to $\hat{\gamma}_1$ or $\hat{\gamma}_2$.

Clearly, $\Gamma_{\hat{\gamma}_j} = \{\hat{\gamma}_j^{\,n}|n=0,1,\dots 7\}$ has order $8(j=1,2)$.

LEMMA 10. There are 4 Γ-conjugacy classes of elliptic, type iii) elements $\gamma\varepsilon\Gamma$ with $(w,w) = -1$ (w primitive in L, $\gamma(w) = \zeta^2 w$), represented by the elements $\gamma_{\zeta} = A\begin{pmatrix} \zeta^2 & 0 & 0 \\ 0 & -\sqrt{2}\zeta & \zeta^2 \\ 0 & -1 & \sqrt{2}\zeta \end{pmatrix}A^{-1}$, $\zeta = \frac{1+i}{\sqrt{2}}\, i^{\nu}$, $\nu=0,1,2$ or 3. $\#\Gamma_{\gamma_{\zeta}} = 8$.

PROOF. First, note that $\gamma|L\cap\{w\}^{\perp}$ has determinant $\zeta^{-2} = \pm i$. Hence, by the remark following Corollary 5 of Lemma 4, Chapter V, $L\cap\{w\}^{\perp}$ has a basis v_1,v_2 with $v_1\perp v_2, (v_1,v_1) = 1, (v_2,v_2) = -1$. Let $M = L\cap\{w\}^{\perp}$, $\tilde{M} = Z[\zeta]v_1 \oplus Z[\zeta]v_2 = M\oplus\zeta M$, and choose x primitive positive in $\tilde{M}$ such that $\gamma(x) = \zeta x$. If σ is the non-trivial automorphism of $Q[\zeta]/Q[i]$ ($\zeta^{\sigma} = -\zeta$), σ extends to an automorphism of $\tilde{M}$; and since γ is defined over $Q[i]$, $\gamma(x^{\sigma}) = \zeta^{\sigma}x^{\sigma} = -\zeta x^{\sigma}$. Therefore, $x^{\sigma}\perp x$. Write $x=x_1+\zeta x_2$ with $x_j\varepsilon M$. Then $x^{\sigma} = x_1-\zeta x_2$ and $(x_1+\zeta x_2,x_1-\zeta x_2) = \{(x_1,x_1)-(x_2,x_2)\} + \zeta\{(x_2,x_1) - \zeta^{-2}(x_1,x_2)\} = 0$. Hence, since (x_i,x_j) and $\zeta^{-2}\varepsilon Z[i], (x_1,x_1) = (x_2,x_2)$ and $(x_1,x_2) = \zeta^2(x_2,x_1)$. Also, $(x,x) = (x_1,x_1)+(x_2,x_2)+\zeta(x_2,x_1)+\zeta(x_1,x_2) = 2\{(x_1,x_1)+\zeta(x_2,x_1)\}\varepsilon 2Z[\sqrt{2}]$ (recall that $Z[\sqrt{2}]$ is the ring of integers of $Q(\sqrt{2}) = R\cap Q[\zeta]$). Hence, 2 divides (x,x) in $Z[\sqrt{2}]$.

We claim that $(x,x) = 2u$, with u a unit of $Z[\sqrt{2}]$. For since $Q[\zeta]$ has class number one ([2], p.570), x primitive in $\tilde{M}$ implies that there exists $y \varepsilon \tilde{M}$ such that $(y,x) = 1$, so $\tilde{M} = Z[\zeta]y \oplus \tilde{M} \cap \{x\}^{\perp}$. But $x^{\sigma} \perp x$ and x^{σ} primitive implies that $\tilde{M} \cap \{x\}^{\perp} = Z[\zeta]x^{\sigma}$, so $\tilde{M} = Z[\zeta]y \oplus Z[\zeta]x^{\sigma}$. Then $x = ay+bx^{\sigma}$ with a,b relatively prime in $Z[\zeta]$, $a = (x,x) \varepsilon Z[\sqrt{2}]$, $\gamma(x) = \zeta x = a\zeta y + b\zeta x^{\sigma} = a\gamma(y) - b\zeta x^{\sigma}, a(\gamma(y)-\zeta y) = 2b\zeta x^{\sigma}$. If $\gamma(y) = \zeta y, b=0$ and a is a unit of $Z[\zeta]$. But 2 divides $a = (x,x)$, so this is impossible. Hence, $\gamma(y)-\zeta y \neq 0$ and a divides 2; so $a = (x,x) = 2u$ with u a unit of $Z[\sqrt{2}]$. Since x is positive, $u > 0$.

As is well known, the units of $Z[\sqrt{2}]$ are the numbers $\pm(1+\sqrt{2})^k$ $(k \varepsilon Z)$; so $u = (1+\sqrt{2})^k (k \varepsilon Z)$. If $k=2\ell+\varepsilon$ with $\varepsilon=0$ or 1, $((\sqrt{2}-1)^{\ell}x, (\sqrt{2}-1)^{\ell}x) = 2(1+\sqrt{2})^{\varepsilon}$; so replacing x by $(\sqrt{2}-1)x$, we may assume $(x,x) = 2$ or $2(1+\sqrt{2})$. But if $(x,x) = 2, (x^{\sigma},x^{\sigma}) = (x,x)^{\sigma} = 2$; so since $x \perp x^{\sigma}$, M is positive definite, which is a contradiction. Therefore, $(x,x) = 2(1+\sqrt{2})$; so $(x_1,x_1)+\zeta(x_2,x_1) = 1+\sqrt{2}, (x_1,x_1)-\zeta(x_2,x_1) = 1-\sqrt{2}, (x_1,x_1) = (x_2,x_2) = 1$, and $\zeta(x_2,x_1) = \sqrt{2}$.

We claim that x_1,x_2 is actually a basis of M. First, if $x_1 = \lambda x_2$ with $\lambda \varepsilon Q[i]$, $x = (\lambda+\zeta)x_2$ and $\gamma(x_2) = \zeta x_2 \varepsilon M$, which is absurd. So x_1 and x_2 are independent over $Q[i]$; and since $(x_1,x_1) = (x_2,x_2) = 1$, they are primitive in M. Suppose then, that $x_k = \sum_{j=1}^{2} a_{kj}v_j (k=1,2)$. Then $a_{kj} \varepsilon Z[i]$ and

$$\det((x_k,x_j)) = \det \begin{pmatrix} 1 & \sqrt{2}\zeta \\ \sqrt{2}\bar{\zeta} & 1 \end{pmatrix} = 1-2 = -1 = |\det(a_{jk})|^2 \det((v_j,v_k)) = -|\det(a_{jk})|^2;$$

so $\det(a_{jk}) = i^{\lambda}$ and $\{x_1,x_2\}$ is a basis of M, as claimed. Since $\gamma(x) = \zeta x$ and $\gamma(x^{\sigma}) = -\zeta x^{\sigma}$, $\gamma(x_1) = \zeta^2 x_2$ and $\gamma(x_2) = x_1$.

Let $v_1' = x_1, v_2' = x_2 - \sqrt{2}\,\bar{\zeta}x_1$. Then $(v_1', v_1') = 1$, and $v_1' \perp v_2'$. So without loss of generality, $x_1 = v_1, x_2 = v_2 + \sqrt{2}\,\bar{\zeta}v_1, \gamma(v_1) = \sqrt{2}\,\zeta v_1 + \zeta^2 v_2, \gamma(v_2) = -v_1 - \sqrt{2}\,\zeta v_2$, and $\gamma(w) = \zeta^2 w$. Then if $w^o = e_1 + e_2, v_1{}^o = e_1 + e_2 - ie_3, v_2{}^o = ie_2 + e_3$ (e_j the standard basis of L) and if $\delta(v_j) = v_j{}^o (j=1,2)$ and $\delta(w) = i^\nu w^o$ for appropriate $\nu \varepsilon Z$, then $\delta\gamma\delta^{-1}(w^o) = \zeta^2 w^o, \delta\gamma\delta^{-1}(v_1{}^o) = \sqrt{2}\,\zeta v_1{}^o + \zeta^2 v_2{}^o$, and $\delta\gamma\delta^{-1}(v_2{}^o) = -v_1{}^o - \sqrt{2}\,\zeta v_2{}^o$. Hence, γ is conjugate in Γ to γ_ζ, as claimed.

If $\delta \varepsilon \Gamma_\gamma, \delta(x) = \lambda x$ and $\delta(x^\sigma) = \lambda^\sigma x^\sigma$ with $\lambda \varepsilon Z[\zeta], |\lambda|^2 = 1$; so $\lambda = \zeta^k$ with $k \varepsilon Z$. Hence, $\sigma = \gamma_\zeta^k$ and $\#\Gamma_{\gamma_\zeta} = 8$.

Suppose that $\gamma \varepsilon \Gamma$ is elliptic of type iv) with eigenvalues $\{i^\varepsilon\rho, i^\varepsilon\rho^2, (-1)^\varepsilon\}$ (where $\rho = \frac{-1}{2} + \frac{\sqrt{-3}}{2}$ is a primitive cube root of unity), and pick w primitive in L such that $\gamma(w) = (-1)^\varepsilon w$. Assume that $\varepsilon \neq 0 \bmod 4$. Then γ^3 is elliptic or hyperelliptic with eigenvalues $\{i^{-\varepsilon}, i^{-\varepsilon}, (-1)^\varepsilon\}$; so Lemma 2 of Chapter V implies again that $(w,w) = \pm 1$ or ± 2. If $(w,w) = +1$ or $+2$, γ^3 is elliptic of type i) and $\gamma \varepsilon \Gamma_{\gamma^3}$; but by Lemmas 4 and 5, the elements of Γ_{γ^3} have eigenvalues belonging to $Z[i]$ or $Z[\exp 2\pi i/8]$, which is a contradiction. Similarly, if $(w,w) = -2$, $\gamma^4 | L \cap \{w\}^\perp$ has determinant 1 and trace -1, which contradicts the remark following Corollary 5 of Chapter V. Hence, if $\varepsilon \neq 0 \bmod 4$, $(w,w) = -1$.

LEMMA 11. If $\gamma \varepsilon \Gamma$ is elliptic of type iv) with eigenvalues $\{i^\varepsilon\rho, i^\varepsilon\rho^2, (-1)^\varepsilon\} (\varepsilon \varepsilon Z)$ and if $\gamma(w) = (-1)^\varepsilon w$ with $w \varepsilon L$ and $(w,w) = -1$,

γ is conjugate in Γ to $\gamma_{\pm,\varepsilon} = \begin{pmatrix} 0 & 0 & \pm i^\varepsilon \\ 0 & -1^\varepsilon & 0 \\ \mp i^\varepsilon & 0 & -i^\varepsilon \end{pmatrix}$ (ε=0,1,2, or 3). Also,

$\#\Gamma_\gamma = 12$.

PROOF. Since $\gamma^4 | L\cap\{w\}^\perp$ has trace -1 and determinant 1, the above mentioned remark and Lemma 4 of Chapter IV imply that $L\cap\{w\}^\perp$ has a basis v_1, v_2 with v_j isotropic, $(v_2,v_1) = i$. Let $M = L\cap\{w\}^\perp$, $\tilde{M} = Z[i,\rho]v_1 \oplus Z[i,\rho]v_2 = M \oplus \rho M$; and choose $x \in \tilde{M}$ primitive positive such that $\gamma(x) = i^\varepsilon \rho x$. If σ is the non-trivial element of $Gal(Q[\rho,i]/Q[i])(\rho^\sigma = \rho^2)$, σ extends to an automorphism of $\tilde{M}, \gamma(x^\sigma) = i^\varepsilon \rho^2 x^\sigma$, and $x^\sigma \perp x$. Hence, writing $x = x_1 + \rho x_2 (x_j \in M)$, we have $0 = (x,x^\sigma) = (x_1+\rho x_2, x_1+\rho^2 x_2) = (x_1,x_1) + \rho\{(x_2,x_1)+(x_1,x_2)\}+\rho^2(x_2,x_2) = (x_1,x_1)-(x_2,x_2) + \rho\{(x_1,x_2)+(x_2,x_1)-(x_2,x_2)\}$ (as $\rho+\rho^2 = -1$); so $(x_1,x_1) = (x_2,x_2) = 2Re(x_1,x_2)$. Also, $(x,x) = (x_1+\rho x_2, x_1+\rho x_2) = (x_1,x_1)+(x_2,x_2)+\rho(x_2,x_1)+\rho^2(x_1,x_2) = 3Re(x_1,x_2) - (2\rho+1)iIm(x_1,x_2) = \sqrt{3}(\sqrt{3}\,Re(x_1,x_2) \mp Im(x_1,x_2))$. Therefore, since $(x_1,x_2) \in Z[i]$, (x,x) is divisible by $\sqrt{3}$ in $Z[\sqrt{3}]$ (the ring of integers of $Q(\exp 2\pi i/12) \cap R = Q(\sqrt{3})$).

Since $Z[\rho,i]$ is a principal ideal domain ([2] p. 570), x primitive in $\tilde{M}$ implies that there exists $y \in \tilde{M}$ such that $(y,x) = 1$ and $\tilde{M} = Z[i,\rho]y \oplus \tilde{M} \cap \{x\}^\perp$. But x^σ is primitive and $x^\sigma \perp x$, so $\tilde{M} \cap \{x\}^\perp = Z[\rho,i]x^\sigma, \tilde{M} = Z[\rho,i]y \oplus Z[\rho,i]x^\sigma$, and $x = ay+bx^\sigma$ (a,b relatively prime in $Z[\rho,i]$, $a = (x,x) \in Z[\sqrt{3}]$). Then $\gamma(x) = i^\varepsilon \rho x = ai^\varepsilon \rho y + bi^\varepsilon \rho x^\sigma = a\gamma(y) + bi^\varepsilon \rho^2 x^\sigma$, so $a(\gamma(y) - i^\varepsilon \rho y) = bi^\varepsilon(\rho-\rho^2)x^\sigma$. If $b=0$, $a=(x,x)$ is a unit of $Z[i,\rho]$, but a is divisible by $\sqrt{3}$, which is prime in $Z[\sqrt{3}]$ (and in $Z[i,\rho]$), so a is not a unit. Hence, $b \neq 0$ and a divides $\rho-\rho^2 = \pm\sqrt{-3}$ in $Z[i,\rho]$. Therefore, $a = \sqrt{3}u$ where u is a unit of $Z[i,\rho]$ and so a unit of $Z[\sqrt{3}]$. But the positive units of $Z[\sqrt{3}]$ are the numbers $(2+\sqrt{3})^k (k \in Z)$, so $(x,x) = \sqrt{3}\,(2+\sqrt{3})^{2j+\nu} (\nu=0$ or 1, $j \in Z)$; and replacing x by $(2-\sqrt{3})^j x$, we may assume that $(x,x) = \sqrt{3}$ or $\sqrt{3}(2+\sqrt{3})$. But also, $|i+\rho|^2 = 2\pm\sqrt{3}$, so if

$(x,x) = \sqrt{2}(2+\sqrt{3})$, we may replace x by $(1+\rho i)x$ or $(1+\rho i)^{-1}x$ to get $(x,x) = \sqrt{3}$. But then $(x_2,x_1) = \pm 1$ and x_1,x_2 are isotropic. Clearly, then, $\{x_1,x_2\}$ form a basis for M. Let $\delta(w) = i^{\nu}e_2, \delta(x_1) = \pm e_1$, and $\delta(x_2) = e_3$. Then $\delta\varepsilon\Gamma$ for appropriate $\nu\varepsilon Z, \delta\gamma\delta^{-1}(e_2) = (-1)^{\varepsilon}e_2$, and $\delta\gamma\delta^{-1}(x^o) = i^{\varepsilon}\rho x^o$, where $x^o = \pm e_1+\rho e_3$. Thus, the Γ- conjugacy class of γ contains an element $\gamma_{\rho,\varepsilon}$ with $\gamma_{\rho,\varepsilon}(e_2) = (-1)\, e_2$ and $\gamma_{\rho,\varepsilon}(x^o) = i^{\varepsilon}\rho x^o$.

Also, if $x = x_1+\rho x_2$ with $x_1,x_2\varepsilon L$ and $\gamma(x) = i^{\varepsilon}\rho x, \gamma(x_1)+\rho\gamma(x_2) = i^{\varepsilon}\rho x_1+i^{\varepsilon}\rho^2 x_2$ and $\gamma(x_1) + \rho^2\gamma(x_2) = i^{\varepsilon}\rho^2 x_1+i^{\varepsilon}\rho x_2$; so $\gamma(x_1) = -i^{\varepsilon}x_1$, and $\gamma(x_2) = i^{\varepsilon}(x_1-x_2)$. In particular, $\gamma_{\rho,\varepsilon}(e_1) = \mp i^{\varepsilon}e_3, \gamma_{\rho,\varepsilon}(e_2) = (-1)^{\varepsilon}e_2$, and $\gamma_{\rho,\varepsilon}(e_3) = i^{\varepsilon}(\pm e_1-e_3)$. Hence, each elliptic γ of type iv) with $(w,w) = -1$ is conjugate to some $\gamma_{\rho,\varepsilon}$; and clearly, no two $\gamma_{\rho,\varepsilon}$ are conjugate.

Clearly, if $\delta\varepsilon\Gamma_{\gamma}, \delta(x) = \lambda x, \delta(x^{\sigma}) = \lambda^{\sigma}x^{\sigma}$, and $\delta(w) = (\lambda\lambda^{\sigma})^{-1}w$ with $\lambda\varepsilon Z[i,\rho]$ and $|\lambda|^2 = 1$. Hence, $\lambda = i^{\nu}\rho^{\mu}$ for some μ,ν and $\#\Gamma_{\gamma} = 12$.

LEMMA 12. Suppose that $\gamma\varepsilon\Gamma$ is elliptic of type iv) with eigenvalues $\{1,\rho,\rho^2\}$(ρ a primitive cube root of unity), and suppose that $w\varepsilon L$ is a primitive vector such that $\gamma(w) = w$. Then $(w,w) = -1$ or ± 3.

PROOF. Choose $x\varepsilon L$ such that $(x,w) = 1$ and $L = Z[i]x \oplus L\cap\{w\}^{\perp}$, and let $M = L\cap\{w\}^{\perp}$, $\tilde{M} = M \oplus \rho M$. Choose $y\varepsilon\tilde{M}$ primitive such that $\gamma(y) = \rho y$, and let $\{y,v\}$ be a basis of $\tilde{M}$ over $Z[i,\rho]$. Then $w=ax+by+cv$ with $(w,w) = a\varepsilon Z$ and a,b,c relatively prime elements of $Z[i,\rho]$. Since $\gamma(w) = w$, $a\gamma(x)+b\rho y+c\gamma(v) = ax+by+cv$ and $a(\gamma(x)-x) = b(1-\rho)y-c(\gamma(v)-v)$. But $\gamma(v) = \rho^2 v+dy$ with $d\varepsilon Z[i,\rho]$; hence, $a(\gamma(x)-x) = [b(1-\rho)-cd]y+c(1-\rho^2)v$. Suppose that $a=3^{\nu}a'$ with $a'\varepsilon Z, (a',3) = 1$. Then a' divides b and c in $Z[i,\rho]$ so $a' = \pm 1$. Similarly, suppose that $\nu > 1$. Then since $3=(1-\rho)(1-\rho^2)$, 3 divides c and $1-\rho^2$ divides b in $Z[i,\rho]$, which is impossible. Thus, $\nu \leq 1$ and $a = (w,w) = \pm 1$ or ± 3. Of course, by the proof of Lemma 4, $(w,w) = +1$ is impossible.

LEMMA 13. If $w \in L$ is primitive with $(w,w) = -3$, then $L \cap \{w\}^{\perp} = Z[i]v \oplus Z[i]y$ with $(v,v) = 1, (y,y) = -3$. In particular, $L \cap \{w\}^{\perp}$ contains no isotropic vectors.

PROOF. Choose $x \in L$ such that $(w,x) = 1$, $L = Z[i]x \oplus L \cap \{w\}^{\perp}$. Pick $v \in L \cap \{w\}^{\perp}$ such that $k = |(v,v)|$ is minimal, and let $L \cap \{w\}^{\perp} = Z[i]v \oplus Z[i]y$. Then the usual argument (see Lemma 5 of Chapter IV) shows that $-3 = (v,v)(y,y) - |(v,y)|^2$. Also, we may assume that $|(v,y)|^2 < (v,v)^2 = k^2$. Then $k^2 - 3 > (v,v)(y,y)$. If $(v,v)(y,y) > 0$, $(v,v)(y,y) \geq k^2$, which is impossible. If $(v,v)(y,y) = 0$, $|(v,y)|^2 = 3$, which is also impossible. Hence, $(v,v)(y,y) < 0$, and so $(v,v)(y,y) \leq -k^2$, $-(v,v)(y,y) \geq k^2$, $3 = |(v,y)|^2 - (v,v)(y,y) \geq k^2$, $k=1, (v,y) = 0, (v,v) = \pm 1$, and $(y,y) = \mp 3$. If $(v,v) = -1$ and $(y,y) = 3$, then if $v' = y + (1+i)v$, $(v',v') = +1$. Hence, we may assume that $(v,v) = 1$, $(v,y) = 0$, and $(y,y) = -3$, as claimed.

If $x \in L \cap \{w\}^{\perp}$ is isotropic, $x = av + by$ with $a,b \in Z[i]$ and $|a|^2 - 3|b|^2 = 0$. But since 3 remains prime in $Z[i]$, such an equation is impossible.

LEMMA 14. If $w \in L$ is primitive with $(w,w) = +3$, then $L \cap \{w\}^{\perp} = Z[i]v \oplus Z[i]y$ with either $x \perp y, (v,v) = -1, (y,y) = -3$ (case 1) or $(v,y) = -1, (v,v) = (y,y) = -2$ (case 2). In the second case, changing v and y by powers of i if necessary, we can find $x \in L$ such that $L = Z[i]x \oplus L \cap \{w\}^{\perp}$, $(x,y) = 0$, $(x,v) = 1+i, (x,x) = -1$, and $w = 3x - (1+i)(y-2v)$.

PROOF. As in Lemma 13, choose $x \in L$ such that $(w,x) = 1$, $v \in L \cap \{w\}^{\perp}$ such that $k = |(v,v)|$ is minimal, and y such that $L \cap \{w\}^{\perp} = Z[i]v \oplus Z[i]y$ and $|(v,y)|^2 < k^2$. Then $3 = (v,v)(y,y) - |(v,y)|^2$; so $k^2 \leq (v,v)(y,y) = k|(y,y)| = 3 + |(v,y)|^2 < 3 + k^2$ and $k|(y,y)| = k^2, k^2+1,$

or k^2+2.

If $k|(y,y)| = k^2+2$, k divides 2 so $k=1$ or 2. If $k=1, (y,y) = -3$, $(v,v) = -1$, and $(v,y) = 0$. This is the first case of the Lemma. If $k=2, (y,y) = -3, (v,v) = -2$, and $|(v,y)|^2 = 3$, which is impossible.

If $k|(y,y)| = k^2+1$, $k=1, (y,y) = -2, (v,y) = 0$, and $2(v,v) = 3$, which is impossible.

If $k|(y,y)| = k^2, (y,y) = -k = (v,v)$ and $(v+i^{\varepsilon}y, v+i^{\varepsilon}y) = -2k+2\mathrm{Re}\, i^{\varepsilon}(y,v) \leq -k$, so $2\mathrm{Re}\, i^{\varepsilon}(y,v) \leq k, 2|\mathrm{Re}(y,v)| \leq k, 2|\mathrm{Im}(y,v)| \leq k$, and $4|(y,v)|^2 \leq 4k^2-12 \leq 12k^2$. Hence, $k^2 \leq 6$ and $k=1$ or 2. If $k=1, (y,y) = -1 = (v,v)$ and $|(v,y)|^2 = -2$, which is impossible. Hence $k=2, (y,y) = (v,v) = -2, |(v,y)|^2 = 1$, and, as we may assume, $(v,y) = -1$. This is the second case.

In the second case, let $x' = x+qv$. Then $(x',w) = 1$ and $(x',y) = (x,y)-q$ so we may assume that $(x,y) = 0$. Let $x'' = x+p(y-2v)$. Then $(x'',w) = 1, (x'',y) = 0$, and $(x'',v) = (x,v) +3p$. Thus, we may assume that $(x,v) = 0, i^{\varepsilon}$, or $(1+i)i^{\varepsilon}$ with $\varepsilon \in Z$. Also, since

$$1 = \det\begin{pmatrix} (x,x) & 0 & (x,v) \\ 0 & -2 & -1 \\ (v,x) & -1 & -2 \end{pmatrix} = 3(x,x)+2|(x,v)|^2,$$

$|(x,v)|^2 = 2 \bmod 3$. Thus, $(x,v) = (1+i)i^{\varepsilon}$. Multiplying v and y by the same power of i, we may then assume that $(x,v) = 1+i, (x,y) = 0$, and then $(x,x) = -1$, as claimed.

If $w = ax+by+cv, (w,w) = 3=a, (w,y) = 0 = -2b-c$, and $(w,v) = 0 = (1+i)a-b-2c = 3(1+i) + 3b$. So $b = -(1+i), c = 2(1+i)$, and $w = 3x - (1+i)(y-2v)$.

Note that in the second case (u,u) is even for all $u \in L \cap \{w\}^{\perp}$, so the two cases are mutually exclusive.

LEMMA 15. Suppose that $\gamma\varepsilon\Gamma$ is elliptic of type iv) with eigenvalues $\{1,\rho,\rho^2\}$ and that w is a primitive vector of L such that $\gamma(w) = w$. Then if $(w,w) = \pm 3$, there exists a primitive vector $y\varepsilon\tilde{L} = L \oplus \rho L$ such that $\gamma(y) = \rho y$ and $|(y,y)| = \sqrt{3}$ if $(w,w) = -3, (y,y) = -3$ if $(w,w) = 3$.

PROOF. Since $\gamma(y^\sigma) = \rho^2 y^\sigma, y^\sigma \perp y$. Therefore, as in Lemma 11, $\sqrt{3}$ divides (y,y) in $Z[\sqrt{3}]$. Choose $\tilde{y}\varepsilon L$ such that $(\tilde{y},y) = 1, \tilde{L} = Z[i,\rho]\tilde{y} \oplus L \cap \{y\}^\perp$; and let $\tilde{L} \cap \{y\}^\perp = Z[i,\rho]w \oplus Z[i,\rho]u$. Then $\gamma(u) = \rho^2 u + dw$ with $d\varepsilon Z[i,\rho]$ and $y = a\tilde{y}+bw+cu$ with a,b,c relatively prime in $Z[i,\rho]$ and $a=(u,w)\varepsilon Z[\sqrt{3}]$. Then since $\gamma(y) = \rho y, a\gamma(\tilde{y})+bw+c\rho^2 u+cdw = a\rho\tilde{y}+b\rho w+cu$ and $a(\gamma(\tilde{y})-\rho\tilde{y}) = [b(\rho-1)-cd]w+c(1-\rho^2)u$. Let $a = (\sqrt{3})^k a'$ with $a'\varepsilon Z[\sqrt{3}],(a',3) = 1$. Then a' divides c and b in $Z[i,\rho]$, hence is a unit of $Z[\sqrt{3}]$. Suppose $k \geq 3$. Then since $\rho-\rho^2 = \sqrt{3}\varepsilon$ and $\rho-1 = \sqrt{3}\varepsilon'$ with ε,ε' units, 3 divides c and $\sqrt{3}$ divides b in $Z[i,\rho]$. But then a,b, and c are not relatively prime, which is a contradiction. Hence, $k=1$ or 2.

If $(w,w) = -3$, $L\cap\{w\}^\perp$ is indefinite; hence (y,y) and $(y^\sigma,y^\sigma) = (y,y)^\sigma$ have opposite sign. But the units of $Z[\sqrt{3}]$ have norm 1; so if $(y,y) = \sqrt{3}^k a', (y,y)^\sigma = (-\sqrt{3})^k a'^{-1}$ and $(y,y),(y^\sigma,y^\sigma)$ have opposite sign if and only if $k=1$. Similarly, if $(w,w) = +3$, (y,y) and (y^σ,y^σ) are both negative, so $k=2$.

Since $\varepsilon_0 = 2+\sqrt{3}$ is a fundamental unit of $Q[\sqrt{3}], a' = \pm\varepsilon_0^{2\nu+\delta}$ with $\nu\varepsilon Z$ and $\delta=0$ or 1. Then $y' = \varepsilon_0^{-k}y$ is a primitive eigenvector of γ with $(y',y') = \pm(\sqrt{3})^k\varepsilon_0^{\delta}$. Also, $\varepsilon_0 = |1+\rho i|^2$ or $|1+\rho^2 i|^2$; so if $\delta=1, y'' = (1+\rho i)^{-1}y'$ or $(1+\rho^2 i)^{-1}y'$ is primitive and satisfies $(y'',y'') = \pm(\sqrt{3})^k$.

COROLLARY. If γ is as in Lemma 15, $(w,w) \neq -3$.

PROOF. Suppose $(w,w) = -3$ and take y as in the Lemma, $(y,y) = \pm\sqrt{3}$. Then $(y^\sigma,y^\sigma) = (y,y)^\sigma = \mp\sqrt{3}$ and $(y,y^\sigma) = 0$, so $y+y^\sigma \varepsilon L \cap \{w\}^\perp$ and is isotropic. But this contradicts Lemma 13.

LEMMA 16. If $\gamma \varepsilon \Gamma$ is elliptic of type iv) with eigenvalues $\{1,\rho,\rho^2\}$ and if $w \varepsilon L$ is primitive and satisfies $(w,w) = 3$ and $\gamma(w) = w$, then γ is conjugate in Γ to

$$\gamma_\rho = B \begin{pmatrix} 1 & 0 & 0 \\ -1-i & 0 & 1 \\ 1+i & -1 & -1 \end{pmatrix} B^{-1}, \text{ where } B = \begin{pmatrix} 1-i & i & 1+2i \\ 2-i & 1+i & 1+3i \\ -2i & -i & 2 \end{pmatrix}.$$

Also, $\#\Gamma_{\gamma_\rho} = 3$.

PROOF. By Lemma 15, we can find $y \varepsilon \tilde{L} = L \oplus \rho L$ such that $(y,y) = -3 = (y^\sigma,y^\sigma)$, $\gamma(y) = \rho y, \gamma(y^\sigma) = \rho^2 y^\sigma$, and $y \perp y^\sigma, w$. Let $y = y_1 + \rho y_2$ with

$y_1, y_2 \varepsilon L \cap \{w\}^\perp$. Then $y_1 = \dfrac{\rho^2 y - \rho y^\sigma}{\rho^2 - \rho}$ and $y_2 = \dfrac{y - y^\sigma}{\rho - \rho^2}$, so $(y_1,y_1) = -2 = (y_2,y_2)$ and $(y_1,y_2) = -1$. Hence, $L \cap \{w\}^\perp = Z[i]y_1 \oplus Z[i]y_2$ and $\{y_1,y_2\}$ is a basis of $L \cap \{w\}^\perp$ of the second type (Lemma 14). Then, by Lemma 14, multiplying y by a power of i if necessary, we can find $x \varepsilon L$ such that $L = Z[i]x \oplus Z[i]y_2, (x,x) = -1, (x,y_1) = 0, (x,y_2) = 1+i$, and $w = 3x-(1+i)(y_1-2y_2)$.

Also, since $\gamma(w) = w$, $\gamma(y) = \rho y$, and $\gamma(y^\sigma) = \rho^2 y^\sigma, \gamma(y_1) = -y_2, \gamma(y_2) = y_1 - y_2$, and $\gamma(x) = x+(1+i)(y_2-y_1)$.

In particular, if $\{e_1,e_2,e_3\}$ is the standard basis of L and if $w = 2(1+i)e_1+(2+3i)e_2+(3-i)e_3, (w,w) = 3$ and $x = (1-i)e_1+(2-i)e_2-2ie_3$,

$y_1=ie_1+(1+i)e_2-ie_3$, and $y_2=(1+2i)e_1+(1+3i)e_2+2e_3$ is a basis of the above form. The matrix of γ with respect to the standard basis is then given by

$$\gamma_\rho = B\begin{pmatrix} 1 & 0 & 0 \\ -1-i & 0 & 1 \\ 1+i & -1 & -1 \end{pmatrix} B^{-1}, \text{ with } B \text{ as above.}$$

Now suppose that γ' is another element of Γ of the same type and choose w', y_1', y_2', and x' in the same way as before. Then if $\delta(y_1) = i^\nu y_1', \delta(y_2) = i^\nu y_2'$, and $\delta(x) = i^\nu x', \delta\varepsilon\Gamma$ for suitable ν and $\delta^{-1}\gamma'\delta = \gamma$. Hence any two elements satisfying the conditions of the Lemma are conjugate in Γ.

If $\delta\varepsilon\Gamma_\gamma, \delta(w) = \lambda_1 w, \delta(y) = \lambda_2 y$, and $\delta(y^\sigma) = \lambda_2{}^\sigma y^\sigma$ with λ_1, λ_2 roots of unity, $\lambda_1\varepsilon Z[i], \lambda_2\varepsilon Z[i,\rho]$. Hence, $\lambda_2 = i^\varepsilon\rho^\delta, \lambda_2{}^\sigma = i^\varepsilon\rho^{2\delta}$, and $\lambda_1 = (-1)^\varepsilon$. But then $\delta^3(w) = (-1)^\varepsilon w$ and $\delta^3 = i^{-\varepsilon}\times id$ on $L\cap\{w\}^\perp$. Hence, by Lemma 2 of Chapter V, $\varepsilon=0 \bmod 4$ since $(w,w) \neq \pm 1$ or ± 2. Thus, $\delta=\gamma,\gamma^2$, or $\gamma^3=id$, and $\#\Gamma_\gamma = 3$.

We can summarize our results on elliptic conjugacy classes in Table 1.

TABLE 1

REPRESENTATIVES OF ELLIPTIC Γ-CONJUGACY CLASSES

	representative γ of $[\gamma]_\Gamma$	λ_1	$\{\lambda_2,\lambda_3\}$	$\dfrac{\lambda_1^{3m}}{(1-\lambda_1\lambda_2^{-1})(1-\lambda_1\lambda_3^{-1})}$	$\#\Gamma_\gamma$
type i)	γ_o	1	$\{-1,-1\}$	$\frac{1}{4}$	8
	$\gamma_\varepsilon\ (\varepsilon=1,2,3)$	$(-1)^\varepsilon$	$\{i^\varepsilon, i^\varepsilon\}$	$\dfrac{(-1)^{3m\varepsilon}}{(1-i^\varepsilon)^2}$	32
type ii)	$\tilde{\gamma}_1$	1	$\{i,-i\}$	$\frac{1}{2}$	16
	$\tilde{\gamma}_4$	1	$\{i,-i\}$	$\frac{1}{2}$	8
	$\tilde{\gamma}_2$	i	$\{1,-i\}$	$\frac{1}{2}(\frac{i^{3m}}{1-i})$	16
	$\tilde{\gamma}_5$	i	$\{1,-i\}$	$\frac{1}{2}(\frac{i^{3m}}{1-i})$	8
	$\tilde{\gamma}_3$	-i	$\{1,i\}$	$\frac{1}{2}(\frac{(-i)^{3m}}{1+i})$	16
	$\tilde{\gamma}_6$	-i	$\{1,i\}$	$\frac{1}{2}(\frac{(-i)^{3m}}{1+i})$	8

	$\hat{\gamma}_1$	i	$\{(\frac{1+i}{\sqrt{2}}), -(\frac{1+i}{\sqrt{2}})\}$	$\frac{i^{3m}}{1-i}$	8
type iii)	$\hat{\gamma}_2$	$-i$	$\{\frac{(1-i)}{\sqrt{2}} \frac{(1-i)}{\sqrt{2}}\}$	$\frac{(-i)^{3m}}{1+i}$	8
	$\gamma_\zeta(\zeta=\frac{(1+i)i^\nu}{\sqrt{2}})$	ζ	$\{\zeta^2,-\zeta\}$	$\frac{\zeta^{3m}}{2(1-\zeta^{-1})}$	8
	$\gamma_{\pm,\varepsilon}=\gamma_{\rho,\varepsilon}$ (ε=0,1,2, or 3,	$i^\varepsilon\rho=\lambda$	$\{i^\varepsilon\rho^2,(-1)^\varepsilon\}=$ $\{\lambda^5,\lambda^6\}$	$\frac{(i^\varepsilon\rho)^{3m}}{(1-\rho^2)(1-i^{-\varepsilon}\rho)} = \frac{\lambda^{3m}}{(1-\lambda^7)(1-\lambda^8)}$	12
type iv)	$\rho = -\frac{1}{2} \pm \frac{\sqrt{-3}}{2}$)				
	γ_ρ	1	$\{\rho,\rho^2\}$	$\frac{1}{3}$	3

λ_1 is the eigenvalue belonging to a positive eigenvector of γ;

λ_2 and λ_3 are the eigenvalues belonging to two linearly independent negative eigenvectors of Γ.

Computation of the Integrals for Elliptic Conjugacy Classes

As we have just seen, if $\gamma\varepsilon\Gamma$ is elliptic, Γ_γ is finite and $[\gamma]_\Gamma \cap P \neq \phi$. Therefore, the contribution of the elliptic conjugacy classes is

$$I_{el} = \Sigma_{[\gamma]_\Gamma \text{elliptic}} \frac{1}{\#\Gamma_\gamma} \int_D \frac{k(P,\gamma P)^m}{k(P,P)^m} \overline{j(\gamma,P)^m} dP.$$

Moreover, from Lemma 6 of Chapter II, we see that $I(\gamma) =$

$\int_D \frac{k(P,\gamma P)^m}{k(P,P)^m} \overline{j(\gamma,P)^m} dP$ depends only on the G-conjugacy class of γ, hence only on the eigenvalue λ_1 of a positive eigenvector of γ and the eigenvalues $\{\lambda_2,\lambda_3\}$ of two linearly independent negative eigenvectors. Therefore, we may assume that $\gamma = A\text{diag}\{\lambda_3,\lambda_2,\lambda_1\}A^{-1}$ with $A = \begin{pmatrix} 1 & 0 & 1 \\ 1 & i & 1 \\ 0 & 1 & -i \end{pmatrix}$ and $A^{-1} = \begin{pmatrix} 0 & 1 & -i \\ i & -i & 0 \\ 1 & -1 & i \end{pmatrix}$ as above.

The matrix A determines a bijective biholomorphic transformation A of the bounded domain $S = \{(u_1,u_2)\varepsilon\mathbb{C}^2 \mid |u_1|^2+|u_2|^2 < 1\}$ onto D by $A(u_1,u_2) = (\frac{u_1+1}{u_2-i}, \frac{u_1+iu_2+1}{u_2-i})$; we have $A^{-1}(z,u) = (\frac{u-i}{z-u+i}, \frac{(z-u)i}{z-u+i})$ and $j(A,(u_1,u_2)) = (u_2-i)^{-3}$. (Recall that if $g = (a_{ij})\varepsilon GL(3,\mathbb{C})$ acts on $\mathbb{C}^2$ as a projective linear transformation, $j(g,(u_1,u_2)) = \frac{\det g}{(a_{31}u_1+a_{32}u_2+a_{33})^3}$

Therefore, if $\tilde{\gamma} = \text{diag}\{\lambda_3,\lambda_2,\lambda_1\}$ and d_L denotes Lebesque measure,

$$I(\gamma) = \int_D \frac{k(P,A\tilde{\gamma}A^{-1}P)}{k(P,P)^m} \overline{j(A\tilde{\gamma}A^{-1},P)}^m dP$$

$$= \int_S \frac{k(AQ,A\tilde{\gamma}Q)^m}{k(AQ,AQ)^m} \overline{j(A\tilde{\gamma}A^{-1},AQ)}^m dAQ.$$

But, since $j(A\tilde{\gamma}A^{-1},AQ) = \frac{j(A,\tilde{\gamma}Q)}{j(A,Q)} j(\tilde{\gamma},Q)$, we have

$$I(\gamma) = \int_S \frac{j(A,Q)^m}{j(A,Q)^m} \frac{k(AQ,A\tilde{\gamma}Q)^m}{k(AQ,AQ)^m} \frac{\overline{j(A,\tilde{\gamma}Q)}^m}{\overline{j(A,Q)}^m} \overline{j(\tilde{\gamma},Q)}^m k(AQ,AQ)d_L AQ$$

$$= \int_S \frac{\tilde{k}(Q,\tilde{\gamma}Q)^m}{\tilde{k}(Q,Q)^m} \overline{j(\tilde{\gamma},Q)}^m dQ,$$

where $\tilde{k}(Q,Q') = j(A,Q)k(AQ,AQ')\overline{j(A,Q')}$ is the Bergman kernel function on S and $dQ = \tilde{k}(Q,Q)d_L Q$.

LEMMA 17. If $Q = (u_1,u_2)$ and $Q' = (u_1',u_2')$, then $\tilde{k}(Q,Q') = (1-u_1\bar{u}_1'-u_2\bar{u}_2')^{-3}$

PROOF. We have $k(P,P') = [i(\bar{z}'-z)-\bar{u}'u]^{-3}$ for $P = (z,u)$ and $P' = (z',u')$ in D; so $j(A,Q)k(AQ,AQ')\overline{j(A,Q')} = (u_2-i)^{-3}(\bar{u}_2'+i)^{-3}[i(\frac{\bar{u}_1'+1}{\bar{u}_2'+i} - \frac{u_1+1}{u_2-i}) - (\frac{\bar{u}_1'-i\bar{u}_2'+1}{\bar{u}_2'+i})(\frac{u_1+iu_2+1}{u_2-i})]^{-3} = [(\bar{u}_1'+1)(iu_2+1)-(u_1+1)(i\bar{u}_2'-1)-(\bar{u}_1'+1)(u_1+iu_2+1) +i\bar{u}_2'u_1+i\bar{u}_2'(iu_2+1)]^{-3} = [u_1+1-iu_1\bar{u}_2'-i\bar{u}_2'-u_1u_1'-u_1+i\bar{u}_2'u_1+i\bar{u}_2'-\bar{u}_2'u_2]^{-3} = [1-u_1\bar{u}_1'-u_2\bar{u}_2']^{-3} = \tilde{k}(Q,Q')$.

LEMMA 18. $a(m)I(\gamma) = \dfrac{\lambda_1^{3m}}{(1-\lambda_1\lambda_2^{-1})(1-\lambda_1\lambda_3^{-1})}$

PROOF. Let $u_j = r_j e^{i\theta_j} (j=1,2)$ be coordinates on S. Since

$d_L P = \frac{1}{4}|dz\wedge d\bar{z}\wedge du\wedge d\bar{u}|$, $d_L Q = \frac{1}{4}|du_1\wedge d\bar{u}_1\wedge du_2\wedge d\bar{u}_2|$. But $du_j\wedge d\bar{u}_j = 2ir_j d\theta_j\wedge dr_j$,

so $d_L Q = r_1 r_2 dr_1 dr_2 d\theta_1 d\theta_2$. Therefore,

$$I(\gamma) = \frac{\lambda_1^{3m}}{4}\int_S \frac{(1-|u_1|^2-|u_2|^2)^{3m-3}\ |du_1\wedge d\bar{u}_1\wedge du_2\wedge d\bar{u}_2|}{(1-\overline{\lambda_3\lambda_1}^{-1}|u_1|^2-\overline{\lambda_2\lambda_1}^{-1}|u_2|^2)^{3m}}$$

$$= \frac{\lambda_1^{3m}}{4}\int_S \frac{(1-|u_1|^2-|u_2|^2)^{3m-3}\ |du_1\wedge d\bar{u}_1\wedge du_2\wedge d\bar{u}_2|}{(1-\lambda_1\lambda_3^{-1}|u_1|^2-\lambda_1\lambda_2^{-1}|u_2|^2)^{3m}}$$

$$= \lambda_1^{3m}4\pi^2 \int_{r_1^2+r_2^2\leq 1} \frac{(1-r_1^2-r_2^2)^{3m-3}\quad r_1 r_2 dr_1 dr_2}{(1-\lambda_1\lambda_3^{-1}r_1^2-\lambda_1\lambda_2^{-1}r_2^2)^{3m}}$$

$$= \lambda_1^{3m}\pi^2 \int_{0\leq r_1+r_2\leq 1} \frac{(1-r_1-r_2)^{3m-3}\quad dr_1 dr_2}{(1-\lambda_1\lambda_3^{-1}r_1-\lambda_1\lambda_2^{-1}r_2)^{3m}}$$

$$= \lambda_1^{3m}\pi^2 \int_{s=0}^{1}\int_{r_2=0}^{s} \frac{dr_2}{[1-\lambda_1\lambda_3^{-1}s+(\lambda_1\lambda_3^{-1}-\lambda_1\lambda_2^{-1})r_2]^{3m}}(1-s)^{3m-3}ds.$$

Therefore, if $\lambda_2\neq\lambda_3$,

$$I(\gamma) = \frac{\lambda_1^{3m}\pi^2}{(3m-1)(\lambda_1\lambda_3^{-1}\lambda_1\lambda_2^{-1})}\left[\int_{s=0}^{1}\frac{(1-s)^{3m-3}ds}{(1-\lambda_1\lambda_3^{-1}s)^{3m-1}} - \int_{s=0}^{1}\frac{(1-s)^{3m-3}ds}{(1-\lambda_1\lambda_2^{-1}s)^{3m-1}}\right];$$

whereas if $\lambda_2 = \lambda_3$,

$$I(\gamma) = \lambda_1^{3m}\pi^2 \int_{s=0}^{1} \frac{(1-s)^{3m-3} s\,ds}{(1-\lambda_1\lambda_3^{-1}s)^{3m}} .$$

But $\dfrac{(1-s)^{3m-3}ds}{(1-\zeta s)^{3m-1}} = \dfrac{-1}{(3m-2)(1-\zeta)}\, d\left[\left(\dfrac{1-s}{1-\zeta s}\right)^{3m-2}\right]$;

so $\displaystyle\int_0^1 \frac{(1-s)^{3m-s}ds}{(1-\zeta s)^{3m-1}} = \frac{1}{(3m-2)(1-\zeta)}$ and $\displaystyle\int_0^1 \frac{(1-s)^{3m-3}s\,ds}{(1-\zeta s)^{3m}} = \frac{1}{(3m-1)(3m-2)(1-\zeta)^2}$.

Therefore, $a(m)I(\gamma) = \dfrac{\lambda_1^{3m}}{\lambda_1\lambda_3^{-1}-\lambda_1\lambda_2^{-1}}\left[\dfrac{1}{1-\lambda_1\lambda_3^{-1}} - \dfrac{1}{1-\lambda_1\lambda_2^{-1}}\right]$

$$= \frac{\lambda_1^{3m}}{(1-\lambda_1\lambda_3^{-1})(1-\lambda_1\lambda_2^{-1})} \quad \text{if } \lambda_2 \neq \lambda_3 ;$$

$$= \frac{\lambda_1^{3m}}{(1-\lambda_1\lambda_3^{-1})^2} \quad \text{if } \lambda_2 = \lambda_3.$$

Hence, in either case, $a(m)I(\gamma) = \dfrac{\lambda_1^{3m}}{(1-\lambda_1\lambda_2^{-1})(1-\lambda_1\lambda_3^{-1})}$,

as claimed.

PROPOSITION 1. $a(m)\, I_{el} = \frac{17}{128} + \frac{7}{32} \operatorname{tr}_{Q[i]/Q}(\frac{i^{3m}}{1-i}) +$

$$+ \frac{1}{16} \operatorname{tr}_{Q[\zeta]/Q}(\frac{\zeta^{3m}}{1-\zeta^{-1}}) + \frac{1}{36} \operatorname{tr}_{Q[\rho]/Q}(\rho^{3m})$$

$$+ \frac{1}{12} \operatorname{tr}_{Q[i]/Q}(i^{3m}) \operatorname{tr}_{Q[\rho]/Q}(\frac{\rho^{3m}}{1-\rho}) - \frac{1}{12} \operatorname{tr}_{Q[i]/Q}(i^{3m+1}) \operatorname{tr}_{Q[\rho]/Q}(\frac{\rho^{3m-1}}{1-\rho})$$

$$+ \frac{(-1)^{3m}}{12} \operatorname{tr}_{Q[\rho]/Q}(\frac{\rho^{3m-1}}{1-\rho}) + \frac{1}{9} ,$$

where ρ is a primitive cube root of unity and ζ is a primitive 8-th root of unity.

PROOF. Let $I_{el}^{i)}$, $I_{el}^{ii)}$, $I_{el}^{iii)}$, and $I_{el}^{iv)}$, be the contribution of the elliptic conjugacy classes of type i), ii), iii), and iv), respectively. Then from Table 1 we get that

$$a(m)I_{el}^{i)} = \frac{1}{32} + \frac{1}{32}\Big[\sum_{\varepsilon=1}^{3} \frac{(-1)^{3m\varepsilon}}{(1-i^{\varepsilon})^2} \Big] = \frac{1}{32} \Big[1 - \frac{(-1)^{3m}}{2i} + \frac{1}{4} + \frac{(-1)^{3m}}{2i} \Big] = \frac{5}{128} ;$$

$$a(m)I_{el}^{ii)} = \frac{3}{32} + \frac{3}{32}\, tr_{Q[i]/Q}\Big(\frac{i^{3m}}{1-i}\Big);$$

$$a(m)I_{el}^{iii)} = \frac{1}{8} tr_{Q[i]/Q}\Big(\frac{i^{3m}}{1-i}\Big) + \frac{1}{16}\, tr_{Q[\zeta]/Q}\Big(\frac{\zeta^{3m}}{1-\zeta^{-1}}\Big); \quad \text{and}$$

$$a(m)I_{el}^{iv)} = \frac{1}{12}\Big[\sum_{\lambda^{12}=1,\lambda^4\neq 1} \frac{\lambda^{3m}}{(1-\lambda^7)(1-\lambda^8)} \Big] + \frac{1}{9}$$

$$= \frac{1}{12}\, tr_{Q[\rho,i]/Q} \left\{ \frac{(\rho i)^{3m}}{(1-\rho^2)(1+\rho i)} \right\} + \frac{1}{12}\, tr_{Q[\rho]/Q} \left\{ \frac{(-\rho)^{3m}}{\rho-\rho^2} \right\}$$

$$+ \frac{1}{12}\, tr_{Q[\rho]/Q} \left\{ \frac{\rho^{3m}}{(1-\rho)(1-\rho^2)} \right\} + \frac{1}{9} .$$

But since $tr_{Q[\rho,i]/Q} = tr_{Q[\rho]/Q} \circ tr_{Q[\rho,i]/Q[\rho]}$ and $\rho+\rho^2 = -1$,

$$tr_{Q[\rho,i]/Q} \left\{ \frac{(\rho i)^{3m}}{(1+\rho i)(1-\rho^2)} \right\} = tr_{Q[\rho]/Q} \left\{ \frac{(\rho i)^{3m}}{(1+\rho i)(1-\rho^2)} + \frac{(-\rho i)^{3m}}{(1-\rho i)(1-\rho^2)} \right\}$$

$$= tr_{Q[\rho]/Q} \left\{ \frac{\rho^{3m} tr_{Q[i]/Q}(i^{3m}) - \rho^{3m+1} tr_{Q[i]/Q}(i^{3m+1})}{(1-\rho^2)(1+\rho^2)} \right\}$$

$$= tr_{Q[i]/Q}(i^{3m})\, tr_{Q[\rho]/Q}\Big(\frac{\rho^{3m}}{1-\rho}\Big) - tr_{Q[i]/Q}(i^{3m+1})\, tr_{Q[\rho]/Q}\Big(\frac{\rho^{3m+1}}{1-\rho}\Big).$$

Therefore,

$$a(m)I_{el}^{iv)} = \frac{1}{12}\,tr_{Q[i]/Q}(i^{3m})\,tr_{Q[\rho]/Q}\left(\frac{\rho^{3m}}{1-\rho}\right)$$

$$-\frac{1}{12}\,tr_{Q[i]/Q}(i^{3m+1})\,tr_{Q[\rho]/Q}\left(\frac{\rho^{3m+1}}{1-\rho}\right)$$

$$+\frac{(-1)^{3m}}{12}\,tr_{Q[\rho]/Q}\left(\frac{\rho^{3m}}{1-\rho}\right) + \frac{1}{36}\,tr_{Q[\rho]/Q}(\rho^{3m}) + \frac{1}{9}\,.$$

Adding, we get the required result.

CHAPTER VIII

THE VOLUME OF THE FUNDAMENTAL DOMAIN

We follow the results and methods of Langlands' Boulder Conference papers (1) "Eisenstein Series" and (2) "The Volume of the Fundamental Domain for some Arithmetical Subgroups of Chevalley Groups".

Notations and Conventions

As usual, $G = \{ g \in SL(3,\mathbb{C}) \mid {}^t\bar{g}Hg = H\}$, where

$$H = \begin{pmatrix} & & i \\ & -1 & \\ -i & & \end{pmatrix},$$

and P is the parabolic group of upper triangular matrices belonging to G. We have the Langlands decomposition $P = NAM$, where

$$A = \left\{ \begin{pmatrix} \delta & & \\ & 1 & \\ & & \delta^{-1} \end{pmatrix} \,\middle|\, \delta > 0 \right\},\quad M = \left\{ \begin{pmatrix} \beta & & \\ & \beta^{-2} & \\ & & \beta \end{pmatrix} \,\middle|\, |\beta|^2 = 1 \right\},\quad N = \left\{ \begin{pmatrix} 1 & \gamma & \frac{|\gamma|^2}{2}i+r \\ & 1 & i\bar{\gamma} \\ & & 1 \end{pmatrix} \right.$$

$$\left. = [\gamma, r] \,\middle|\, \gamma \in \mathbb{C}, r \in R \right\}.$$ We let $L(A)$, $L(M)$ and $L(N)$ be the Lie algebras of A, M, and N respectively. Clearly,

$$L(A) = \left\{ \begin{pmatrix} \delta & & \\ & 0 & \\ & & -\delta \end{pmatrix} \,\middle|\, \delta \in R \right\},\quad L(M) = \left\{ \begin{pmatrix} \beta i & & \\ & -2\beta i & \\ & & \beta i \end{pmatrix} \,\middle|\, \beta \in R \right\},\quad L(N) =$$

$$\left\{ \begin{pmatrix} 0 & \gamma & r \\ 0 & 0 & i\bar{\gamma} \\ 0 & 0 & 0 \end{pmatrix} \,\middle|\, \gamma \in \mathbb{C}, r \in R \right\}.$$ Also, if $a = \begin{pmatrix} \delta & & \\ & 0 & \\ & & -\delta \end{pmatrix} \in L(A)$ and

$$n = \begin{pmatrix} 0 & \gamma & r \\ 0 & 0 & i\bar{\gamma} \\ 0 & 0 & 0 \end{pmatrix} \varepsilon L(N), \ \operatorname{ada}(n) = \delta \begin{pmatrix} 0 & \gamma & 0 \\ 0 & 0 & i\bar{\gamma} \\ 0 & 0 & 0 \end{pmatrix} + 2\delta \begin{pmatrix} 0 & 0 & r \\ 0 & 0 & 0 \\ 0 & 0 & 0 \end{pmatrix}. \quad \text{Hence,}$$

$$L(N) = L(N)_\alpha \oplus L(N)_{2\alpha}, \quad \text{where} \quad \alpha(a) = \delta \quad \text{if} \quad a = \begin{pmatrix} \delta & & \\ & 0 & \\ & & -\delta \end{pmatrix} \varepsilon L(A),$$

$$L(N)_\alpha = \left\{ \begin{pmatrix} 0 & \gamma & 0 \\ 0 & 0 & i\bar{\gamma} \\ 0 & 0 & 0 \end{pmatrix} \middle| \gamma \varepsilon \mathbb{C} \right\} = Re_1 \oplus Re_2, L(N)_{2\alpha} = \left\{ \begin{pmatrix} 0 & 0 & r \\ 0 & 0 & 0 \\ 0 & 0 & 0 \end{pmatrix} \middle| r \varepsilon R \right\} = Re_3,$$

$$e_1 = \begin{pmatrix} 0 & 1 & 0 \\ 0 & 0 & i \\ 0 & 0 & 0 \end{pmatrix}, \quad e_2 = \begin{pmatrix} 0 & i & 0 \\ 0 & 0 & 1 \\ 0 & 0 & 0 \end{pmatrix}, \quad \text{and} \quad e_3 = \begin{pmatrix} 0 & 0 & 1 \\ 0 & 0 & 0 \\ 0 & 0 & 0 \end{pmatrix}.$$

If $K = G \cap SU(3) = \{ g \varepsilon G | {}^t\bar{g}g = 1\}$, K is a maximal compact subgroup of G containing M and $G = NAMK = NAK$ (Iwasawa Decomposition). Also, if D is, as usual, the domain $\{ (z,u) \varepsilon \mathbb{C}^2 | 2\operatorname{Im} z - |u|^2\} > 0$, $K = \pi^{-1}(i,0)$, where $\pi : G \to D$ is the projection sending $g \varepsilon G$ to $g(i,0) \varepsilon D$; thus, π induces a bijection $G/K \overset{\sim}{\to} D$.

We have a normalized Haar measure on G defined as follows. dk is the Haar measure on K such that $\int_K dk = 1$; $da = \delta^{-1} d\delta$ is a Haar measure on A $(a = \operatorname{diag}\{\delta, 1, \delta^{-1}\} \varepsilon A)$; and $dn = \frac{1}{2} dr\, d\gamma_1 d\gamma_2$ is a Haar measure on N $(n = [\gamma_1 + i\gamma_2, r] \varepsilon N, \gamma_1, \gamma_2, r \varepsilon R)$. Then if ρ = one-half the sum of the positive roots (counting multiplicities) of $L(A) = \frac{1}{2}(4\alpha) = 2\alpha$,

$$\int_G \phi(g) dg = \int_{N \times A \times K} \phi(nak) e^{-2\rho(\log a)} dn\, da\, dk$$

is the normalized Haar measure. Note that $e^{-2\rho(\log a)} da = \delta^{-5} d\delta$ if $a = \operatorname{diag}\{\delta, 1, \delta^{-1}\} \varepsilon A$.

LEMMA 1. If dP is the measure on D defined in Chapter I,

$$\int_D \phi(P)dP = \frac{1}{2}\int_G \phi\cdot\pi(g)dg$$

if ϕ is continuous of compact support on D.

PROOF. Since $\int_D \phi(P)dP$ and $\int_G \phi\cdot\pi(g)dg$ are both G invariant integrals on D, $\int_D \phi(P)dP = K\int_G \phi\cdot\pi(g)dg$ with $K > 0$.

Let $S_t = \{ g=nak\varepsilon G \mid \log\delta=\alpha(\log a)>t, n=[\gamma,r]\varepsilon N, |r|\leq\frac{1}{2}, |\mathrm{Re}\,t|\leq\frac{\sqrt{2}}{2}, |\mathrm{Im}\gamma|\leq\frac{\sqrt{2}}{2} \}$

where $a = \mathrm{diag}\{\delta,1,\delta^{-1}\}$. Then if $g = nak\varepsilon S_t, \pi(g) = nak(i,0) = na(i,0) = n(\delta^2 i,0) = (\delta^2 i+|\frac{\gamma}{2}|^2 i+r, i\bar{\gamma})$. Hence if $\pi(g) = (z,u)$, $2\mathrm{Im}z-|u|^2 = 2(\delta^2+|\frac{\gamma}{2}|^2)-|\gamma|^2 = 2\delta^2 > 2e^{2t}, |\mathrm{Re}z| = |r| \leq \frac{1}{2}, |\mathrm{Re}u| \leq \frac{\sqrt{2}}{2}, |\mathrm{Im}u| \leq \frac{\sqrt{2}}{2}$, and $\pi(g)\varepsilon V_\infty(2e^{2t})$. Conversely, it is clear that if $\pi(g)\varepsilon V_\infty(2e^{2t}), g\varepsilon S_t$. Therefore,

$$\int_{V_\infty(2e^{2t})} dP = \frac{1}{2(2e^{2t})^2} = \frac{1}{8e^{4t}} = K\int_{S_t} dg = K\int_{\delta=e^t}^{\infty} \frac{d\delta}{\delta^5}\int_{\gamma_j=-\frac{\sqrt{2}}{2}}^{\frac{\sqrt{2}}{2}} \frac{d\gamma_1 d\gamma_2}{2}$$

$$= \frac{K}{4e^{2t}}.$$

Hence, $K = \frac{1}{2}$, as claimed.

As usual, if $g = \begin{pmatrix} a_1 & a_2 & a_3 \\ b_1 & b_2 & b_3 \\ c_1 & c_2 & c_3 \end{pmatrix} \varepsilon G$, let $j(g,(z,u)) = (c_1 z+c_2 u+c_3)^{-3}$.

Then if $g = na(g)k, |j(nak,(i,0))|^2 = |j(na,(i,0))j(k,(i,0))|^2 = |j(na,(i,0))|^2 = |j(n,a(i,0))j(a,(i,0))|^2 = |j(a,(i,0))|^2 = \delta^6$ if $a = a(g) = \mathrm{diag}\{\delta,1,\delta^{-1}\}$. Hence,

$\delta=|c_1i+c_3|^{-1}$, $a(g) = \mathrm{diag}\{|c_1i+c_3|^{-1},1,|c_1i+c_3|\}$, and $H(g) = \log a(g) = \mathrm{diag}\{-\log|c_1i+c_3|,0,\log|c_1i+c_3|\}$. Thus, if $\Lambda_z \varepsilon L(A)^*_{\mathbb{C}}$ is chosen so that $\Lambda_z(H_\alpha) = z$ (where $H_\alpha=\mathrm{diag}\{1,0,-1\}\varepsilon L(A)$), then $(\Lambda_z+\rho)H(g) = -(\frac{z}{2}+1)\log|c_1i + c_3|^2 = \log[(|c_1i+c_3|^2)^{-1-\frac{z}{2}}]$ and $\exp(\Lambda_z+\rho)H(g) = (|c_1i+c_3|^2)^{1+\frac{z}{2}}$.

If $a\varepsilon A$, ${}^t\bar{a}=a$; so ${}^t\bar{a}Ha = aHa = H$ and $H^{-1}aH = a^{-1}$, where, as usual

$$H = \begin{pmatrix} & & i \\ & -1 & \\ -i & & \end{pmatrix}.$$

Hence, $H = n_W$ normalizes A but does not centralize it, so represents the non-trivial element of the Weyl group of A. If $n = [\gamma,r]\varepsilon N$, $n_Wn = \begin{pmatrix} 0 & 0 & i \\ 0 & -1 & -i\bar{\gamma} \\ -i & -i\gamma & \frac{|\gamma|^2}{2}-r \end{pmatrix}$, so $\exp(\Lambda_z+\rho)H(n_Wn)= (|1+\frac{|\gamma|^2}{2}-r|^2)^{-1-\frac{z}{2}}$.

Computation of the Function $M(s,\Lambda)$

We use Lemma 3 in Langlands' paper "Eisenstein Series".

Let $\Phi(g)$ be a constant function and form the Eisenstein series

$$E(g,\Phi,\Lambda_z) = \sum_{\Gamma\cap P\backslash\Gamma} \exp(\Lambda_z+\rho)H(\gamma g)\Phi(\gamma g).$$

Then by Langlands' Lemma 3,

$$\int_{\Gamma\cap N\backslash N} E(ng,\Phi,\Lambda_z)dn = \sum_{s\varepsilon W} \exp(s\Lambda_z+\rho)H(g)M(s,\Lambda_z)\Phi(g),$$

where W is the Weyl groups of A and

$$M(s,\Lambda_z) = [\Sigma_{\Gamma\cap P\backslash\Gamma\cap Pn_W N/\Gamma\cap N} \exp(\Lambda_z+\rho)H(p_\gamma)][\int_N \exp(\Lambda_z+\rho)H(n_W n)\,dn].$$

Here, $\gamma = p_\gamma n_W u \varepsilon \Gamma\cap Pn_W N$, where $s\varepsilon W$ is represented by $n_W \varepsilon N_G(A)$.

$M(s,\Lambda_z)$ is analytic for $\mathrm{Re}(\Lambda_z,\alpha)>(\rho,\alpha)$ - that is, for $\mathrm{Re}z > 2$. Thus,

$$\int_N \exp(\Lambda_z+\rho)H(n_W n)\,dn = \frac{1}{2}\int_{\gamma_j, r=-\infty}^{\infty} \frac{d\gamma_1 d\gamma_2 dr}{(|1+\frac{|\gamma|^2}{2}-ir)^{1+\frac{z}{2}}}$$

$$= \frac{\pi}{2}\int_{r=-\infty}^{\infty}\int_{y=0}^{\infty} \frac{2y\,dy\,dr}{(|1+\frac{y^2}{2}-ir|^2)^{1+\frac{z}{2}}} = \pi\int_{r=-\infty}^{\infty}\int_{y=0}^{\infty} \frac{dy\,dr}{(|1+y+ir|^2)^{1+\frac{z}{2}}}$$

$$= \pi\int_{r=-\infty}^{\infty}\int_{y=1}^{\infty} \frac{dy\,dr}{(|y-ir|^2)^{1+\frac{z}{2}}} = \pi\int_{r=-\infty}^{\infty}\int_{y=1}^{\infty} \frac{dy\,dr}{(y^2+r^2)^{1+\frac{z}{2}}}$$

$$= 2\pi\int_{y=1}^{\infty} \frac{dy}{y^{1+z}}\int_{r=0}^{\infty} \frac{dr}{(1+r^2)^{1+\frac{z}{2}}} .$$

If $p = \begin{pmatrix} * & * & * \\ & * & * \\ & & c \end{pmatrix}$ and $u = [\mu,r]$, $pn_W u = p\begin{pmatrix} 0 & 0 & 0 \\ 0 & -1 & -i\bar{\mu} \\ -i & -i\mu & \frac{|\mu|^2}{2}-ir \end{pmatrix}$

$= \begin{pmatrix} * & * & * \\ * & * & * \\ -ic & * & * \end{pmatrix}$. Therefore, if $\gamma = \begin{pmatrix} a_1 & a_2 & a_3 \\ b_1 & b_2 & b_3 \\ c_1 & c_2 & c_3 \end{pmatrix} \varepsilon\Gamma\cap Pn_W N$,

$H(p_\gamma) = \mathrm{diag}\{-\log|c_1|, 0, \log|c_1|\}$; so if

$$F(z) = \Sigma_{\Gamma\cap P\backslash\Gamma\cap Pn_W N/\Gamma\cap N} \exp(\Lambda_z+\rho)H(p_\gamma),$$

we have
$$F(z) = \sum_{Z[i]^*/U} \frac{P(c_1)}{(|c_1|^2)^{1+\frac{z}{2}}} ,$$

where $U = \{i,-i,1,-1\}$ and

$$P(c_1) = \#\left\{\gamma\varepsilon\Gamma\cap P\backslash\Gamma\cap Pn_W N/\Gamma\cap N \,\middle|\, \gamma = \begin{pmatrix} * & * & * \\ * & * & * \\ c_1 & * & * \end{pmatrix}\right\}.$$

LEMMA 2. If c_1,c_2,c_3 are relatively prime Gaussian integers such that $|c_2|^2 = 2\mathrm{Re}\,c_1\bar{c}_3 i$, there exists $\gamma\varepsilon\Gamma$ such that

$$\gamma = \begin{pmatrix} * & * & * \\ * & * & * \\ c_1 & c_2 & c_3 \end{pmatrix}.$$

PROOF. Let $\tilde{e}_3 = \bar{c}_1e_1+\bar{c}_2e_2+\bar{c}_3e_3\varepsilon L$. Then $(\tilde{e}_3,\tilde{e}_3) = 0$, so by Lemma 2 of Chapter IV, there exists $\tilde{e}_2,\tilde{e}_1\varepsilon L$ such that $(\tilde{e}_2,\tilde{e}_2) = -1, (\tilde{e}_3,\tilde{e}_1) = i$, $(\tilde{e}_1,\tilde{e}_1) = 0$, $\tilde{e}_2\perp\tilde{e}_1$, and $\tilde{e}_2\perp\tilde{e}_3$. Let $\delta(e_1) = \tilde{e}_1$, $\delta(e_2) = i^{\varepsilon}\tilde{e}_2$, and $\delta(e_3) = \tilde{e}_3$. Then $\delta\varepsilon\Gamma$ for suitable ε in Z and

$$\delta = \begin{pmatrix} * & * & \bar{c}_1 \\ * & * & \bar{c}_2 \\ * & * & \bar{c}_3 \end{pmatrix}.$$

Thus, if $\gamma = {}^t\bar{\delta}, \gamma\varepsilon\Gamma$ and

$$\gamma = \begin{pmatrix} * & * & * \\ * & * & * \\ c_1 & c_2 & c_3 \end{pmatrix}.$$

COROLLARY. $P(c_1) = \#\{(c_1,c_2,c_3)\varepsilon Z[i]^3 \mid c_1,c_2,c_3$ are relatively prime in $Z[i]$ and $|c_2|^2 = 2\mathrm{Re}\,c_1\bar{c}_3 i\}/\sim$, where $(c'_1,c'_2,c'_3)\sim(c_1,c_2,c_3)$

if and only if $c_1' = c_1, c_2' = c_2+\gamma c_1$, and $c_3' = c_3+c_2 i\bar{\gamma}+c_1(|\frac{\gamma}{2}|^2 i+r)$ with $\gamma\varepsilon(1+i)Z[i]$ and $r\varepsilon Z$.

Let $\hat{P}(c_1) = \#\{(c_1,c_2,c_3)\varepsilon Z[i]^3 \mid |c_2|^2 = 2\mathrm{Re}\, c_1\bar{c}_3 i\}/\sim$. Then $\hat{P}(c_1) = \sum_{d|c_1} P(d)$ and

$$\sum_{Z[i]^*/U} \frac{\hat{P}(c_1)}{(|c_1|^2)^{1+\frac{z}{2}}} = \sum_{c_1\varepsilon Z[i]^*/U} \sum_{d|c_1} \frac{P(d)}{(|d|^2)^{1+\frac{z}{2}}} \times \frac{1}{(|\frac{c_1}{d}|^2)^{1+\frac{z}{2}}} = \sum_{Z[i]^*/U} \frac{P(d)}{(|d|^2)^{1+\frac{z}{2}}} \sum_{Z[i]^*/U} \frac{1}{(|c|^2)^{1+\frac{z}{2}}} = F(z)\zeta_{Q[i]}(1+\tfrac{z}{2}).$$

Also, let $Q(c)$ equal the number of Gaussian integers modulo c whose norm is divisible by $(\mathrm{Re}\,c, \mathrm{Im}\,c)$.

LEMMA 3. $\hat{P}(c_1) = (\mathrm{Re}\,c_1, \mathrm{Im}\,c_1)Q(c_1)$.

PROOF. Let $j = (\mathrm{Re}\,c_1, \mathrm{Im}\,c_1)$ and $\tilde{c}_1 = j^{-1}c_1$. Then $(\mathrm{Re}\,\tilde{c}_1, \mathrm{Im}\,\tilde{c}_1) = 1$; so there exists $d\varepsilon Z[i]$ such that $\mathrm{Re}\,\tilde{c}_1\bar{d}i = 1$. Suppose that $|c_2|^2 = 2\mathrm{Re}\,c_1\bar{c}_3 i = 2j\mathrm{Re}\,\tilde{c}_1\bar{c}_3 i$. Then $2j$ divides $|c_2|^2$; so $c_2 = (1+i)c_2'$ with $|c_2'|^2$ divisible by j. Conversely, if $|c_2'|^2$ is divisible by j, $|(1+i)c_2'|^2 = 2\mathrm{Re}\{c_1\overline{(\frac{|c_2'|^2}{j}d+k\tilde{c}_1)}i\}$; so

$$\{(c_1,(1+i)c_2', \tfrac{|c_2'|^2}{j}d+k\tilde{c}_1) \mid k=0,1,\dots,j-1\}$$

is a full set of $\sim$ inequivalent vectors with fixed c_1 and c_2'. Similarly, we see that

$$\{(c_1,(1+i)c_2', \tfrac{|c_2'|^2}{j}d+k\tilde{c}_1) \mid k=0,1,\dots,j-1; |c_2'|^2\equiv 0 \bmod j\}$$

is a full set of $\sim$ inequivalent vectors with fixed c_1. Hence $\hat{P}(c_1) = jQ(c_1)$, as claimed.

LEMMA 4. Let $c = c'c''$ with $(|c'|^2, |c''|^2) = 1$. Then $\hat{P}(c) = \hat{P}(c')\hat{P}(c'')$.

PROOF. Let $c' = j\tilde{c}', c'' = j''\tilde{c}'', c = j\tilde{c}$ with $j, j', j'' \in Z, \tilde{c}, \tilde{c}', \tilde{c}'' \in Z[i]$ divisible by no rational integer $\neq 1$. Clearly, $j = j'j'', \tilde{c} = \tilde{c}'\tilde{c}''$. Since $(c', c'') = 1$, if $b = b_1c' + b_2c''$, b runs modulo c if and only if b_1 runs modulo c'' and b_2 runs modulo c'. Also, $|b|^2 = j'^2|\tilde{c}'|^2|b_1|^2 + j''^2|\tilde{c}''|^2|b_2|^2 + 2j'j''\mathrm{Re}\, b_1\bar{b}_2\tilde{c}'\tilde{c}''$; so clearly, $j = j'j''$ divides $|b|^2$ if and only if j'' divides $|b_1|^2$ and j' divides $|b_2|^2$.

COROLLARY. If $c = (1+i)^{\varepsilon} 2^{\eta} \prod_{p_j \equiv 3(4)} p_j^{2\eta_j+\varepsilon_j} \prod_{\substack{p_j \equiv 1(4) \\ \mathcal{Y}_j\bar{\mathcal{Y}}_j = p_j}} p_j^{m_j} \mathcal{Y}_j^{k_j}$ with $\varepsilon, \varepsilon_j = 0$ or 1, $\eta, \eta_j, m_j, k_j \geq 0$, then $\hat{P}(c) =$

$$\hat{P}((1+i)^{\varepsilon}2^{\eta}) \prod_{p_j \equiv 3(4)} \hat{P}(p_j^{2\eta_j+\varepsilon_j}) \prod_{\substack{p_j \equiv 1(4) \\ \mathcal{Y}_j\bar{\mathcal{Y}}_j = p_j}} \hat{P}(p_j^{m_j}\mathcal{Y}_j^{k_j}).$$

LEMMA 5. 1) If $\varepsilon=0$ or 1 and $\eta \geq 0$, $Q((1+i)^{\varepsilon}2^{\eta}) = 2^{\varepsilon+\eta}$. 2) If $p \in Z$ is a prime $\equiv 3(4)$, $\varepsilon=0$ or 1, and $\eta \geq 0$, then $Q(p^{2\eta+\varepsilon}) = p^{2\eta}$. 3) If $p \in Z$ is a prime $\equiv 1(4)$ and if $\mathcal{Y}\bar{\mathcal{Y}} = p$ with $\mathcal{Y}$ a prime of $Z[i]$, then $Q(p^k\mathcal{Y}^{\eta}) = p^{\eta}[(k+1)p^k - kp^{k-1}]$ $(k, \eta \geq 0)$.

PROOF. 1) Suppose that $c = 2^{\eta}(1+i)^{\varepsilon}$ with $\varepsilon=0$ or 1 and $\eta \geq 0$ and that b runs modulo c with $|b|^2$ divisible by 2^{η}. Then $b = (1+i)^{\eta}b'$ with b' running modulo $(1+i)^{\varepsilon+\eta}$. Hence $Q(c) = 2^{\varepsilon+\eta}$.

2) If $c = p^{2\eta+\varepsilon}$, then $p^{2\eta+\varepsilon}$ divides $|b|^2$ if and only if $b = p^{\eta+\varepsilon}b'$ with b' running modulo p^{η}.

3) If $b = b'$ modulo p^k, $|b|^2 \equiv |b|^2$ modulo p^k; so p^k divides $|b|^2$ if and only if p^k divides $|b'|^2$. Hence, $Q(p^k\mathscr{Y}^n) = p^nQ(p^k)$. Since $(\mathscr{Y},\bar{\mathscr{Y}}) = 1$, b runs modulo p^k if and only if $b = b_1\mathscr{Y}^k+b_2\bar{\mathscr{Y}}^k$ with b_1 running modulo $\bar{\mathscr{Y}}^k$, b_2 running modulo $\mathscr{Y}^k$. Let $b_1 = \Sigma_0^{k-1}\lambda_j\bar{\mathscr{Y}}^j$ and $b_2 = \Sigma_0^{k-1}\mu_j\mathscr{Y}^j$, where $\bar{\lambda}_j$ and μ_j are elements of a fixed set of coset representatives of the Gaussian integers modulo $\mathscr{Y}$. Then p^k divides $|b|^2$ if and only if

$$b_1\bar{b}_2\mathscr{Y}^{2k}+\bar{b}_1b_2\bar{\mathscr{Y}}^{2k} \equiv 0\ (p^k). \qquad (*)$$

If $\lambda_0\equiv\lambda_1\equiv \dots \equiv\lambda_{k-\nu-1} \equiv 0\ (\bar{\mathscr{Y}})$ and $\lambda_{k-\nu} \not\equiv 0\ (\bar{\mathscr{Y}})$, then (*) holds if and only if

$$\lambda_{k-\nu}\bar{b}_2\mathscr{Y}^{k+\nu}+\bar{\lambda}_{k-\nu}b_2\bar{\mathscr{Y}}^{k+\nu} \equiv 0\ (p^{\nu}) -$$

that is, if and only if $\mathscr{Y}^{\nu}$ divides b_2. Hence, if $\nu=1,\dots,k$, there are $(p-1)p^{\nu-1}p^{k-\nu} = (p-1)p^{k-1}$ possibilities for b, whereas if $\nu=0$, there are p^k possibilities. Thus,

$$Q(p^k) = k(p-1)p^{k-1}+p^k = (k+1)p^k-kp^{k-1}.$$

COROLLARY. $\hat{P}((1+i)^{\varepsilon}2^{\eta})=2^{2\eta+\varepsilon};\hat{P}(p^{2\eta+\varepsilon})=p^{4\eta+\varepsilon}$ if $p\equiv3(4)$; and $\hat{P}(p^k\mathscr{Y}^n) = p^{n+k}[(k+1)p^k-kp^{k-1}]$ if $p = \mathscr{Y}\bar{\mathscr{Y}}\equiv1(4)$.

LEMMA 6. $\displaystyle\Sigma_{Z[i]^*/U} \frac{\hat{P}(c)}{(|c|^2)^{z+1}} = \zeta_{Q[i]}(z)\frac{L(\chi,2z)}{L(\chi,2z+1)}$, where $\chi(n) = (\frac{-4}{n})$.

PROOF. By the Corollary to Lemma 4, the left hand side equals

S_p, where

$$S_p = \begin{cases} \sum_{n=0}^{\infty} \sum_{\varepsilon=0}^{1} \dfrac{\hat{P}((1+i)^{\varepsilon} 2^{\eta})}{(2^{2\eta+\varepsilon})^{1+z}} & \text{if } p=2, \\[2ex] \sum_{\eta=0}^{\infty} \sum_{\varepsilon=0}^{1} \dfrac{\hat{P}(p^{2\eta+\varepsilon})}{(p^{4\eta+2\varepsilon})^{1+z}} & \text{if } p\equiv 3(4), \\[2ex] \sum_{k=0}^{\infty} \dfrac{\hat{P}(p^k)}{(p^{2k})^{1+z}} + \sum_{k=0}^{\infty} \sum_{\eta=1}^{\infty} \dfrac{\hat{P}(\mathcal{Y}^n p^k)+\hat{P}(\mathcal{Y}^n p^k)}{(p^{\eta+2k})^{1+z}} & \text{if } p\equiv 1(4) . \end{cases}$$

Applying Lemma 5, we find that $S_2 = (1-2^{-z})^{-1}$, $S_p = (1-p^{-2z})^{-1}(1+p^{-2z})^{-1}(1+p^{-1-2z})$ if $p\equiv 3(4)$, and $S_p = (1-p^{-1-2z})(1-p^{-2z})^{-1}(1-p^{-z})^{-2}$ for $p\equiv 1(4)$. Hence, $\sum \dfrac{\hat{P}(c)}{(|c|^2)^{z+1}} =$

$$(1-2^{-z})^{-1} \prod_{p\equiv 1(4)} \frac{(1-p^{-1-2z})}{(1-p^{-2z})(1-p^{-z})^2} \prod_{p\equiv 3(4)} \frac{(1+p^{-1-2z})}{(1-p^{-2z})(1+p^{-2z})} =$$

$$\zeta_{Q[i]}(z) \frac{L(\chi,2z)}{L(\chi,2z+1)} .$$

COROLLARY. $M(s,\Lambda_z) = 2\pi \int_{r=0}^{\infty} \dfrac{dy}{y^{1+z}} \int_{r=0}^{\infty} \dfrac{dr}{(1+r^2)^{1+\frac{z}{2}}} \dfrac{\zeta_{Q[i]}(\frac{z}{2})}{\zeta_{Q[i]}(\frac{z}{2}+1)} \times \dfrac{L(\chi,z)}{L(\chi,z+1)}$; in particular, $M(s,\Lambda_z)$ has a simple pole at $z=2$ with residue $24\pi^{-2}$.

PROOF. The first statement follows immediately from the Lemma. The existence of a simple pole is clear from the well-known properties of $\zeta_{Q[i]}(z)$ and $L(\chi,z)$. Finally, using the fact that

$$\operatorname*{res}_{z=1} \zeta_{Q[i]}(z) = L(\chi,1) = \frac{\pi}{4} ,$$

that $\zeta(2) = \frac{\pi^2}{6}$, and that $L(\chi,3) = \frac{\pi^3}{32}$, one finds easily that

$$\operatorname*{res}_{z=2} M(s,\Lambda_z) = 24\pi^{-2},$$

as claimed.

<u>Computation of the Volume</u>

Using the notation of [6], page 236, take V and W to be spaces of constant functions and take $\phi,\Psi \in \mathcal{O}(V,W)$ (that is, the space of continuous functions ϕ on G such that $\phi(n\gamma gk) = \phi(g)$ for $n \in N, k \in K$, and $\gamma \in \Gamma \cap P$, and $\phi|A$ has compact support). Let π be the orthogonal projection of $L^2(\Gamma\backslash G)$ onto the space of constant functions, and let

$$\hat{\phi}(g) = \Sigma_{\Gamma\cap P\backslash\Gamma}\ \phi(\gamma g).$$

Then we have

$$\int_{\Gamma\backslash G} dg \int_{\Gamma\backslash G} \pi(\hat{\phi})\overline{\pi(\hat{\Psi})}\, dg = \int_{\Gamma\backslash G} \pi(\hat{\phi})dg \overline{\int_{\Gamma\backslash G} \pi(\hat{\Psi})dg} = \int_{\Gamma\backslash G} \hat{\phi}(g)dg \overline{\int_{\Gamma\backslash G} \hat{\Psi}(g)dg}.$$

Also,

$$\int_{\Gamma\backslash G} \hat{\phi}(g)dg = \int_{\Gamma\backslash G} \Sigma_{P\cap\Gamma\backslash\Gamma}\ \phi(\gamma g)dg = \int_{\Gamma\cap P\backslash G} \phi(g)dg = \frac{1}{4}\int_{\Gamma\cap N\backslash G} \phi(g)dg$$

$$= \frac{1}{4}\int_{\Gamma\cap N\backslash N\times A\times K} \phi(nak)e^{-2\rho(\log a)}dndadk = \frac{1}{4}\operatorname{vol}(\Gamma\cap N\backslash N)\int_A \phi(a)e^{-2\rho(\log a)}da$$

$$= \frac{1}{4}\ \Phi(\rho)$$

(where $\Phi(\Lambda) = \frac{1}{2\pi}\int_A \phi(a)e^{-(\Lambda+\rho)H(a)}da$).

Hence,

$$\int_{\Gamma\backslash G} dg \int_{\Gamma\backslash G} \pi(\hat{\phi})\overline{\pi(\hat{\Psi})}dg = \frac{1}{16}\,\Phi(\rho)\overline{\Psi(\rho)}.$$

We claim that $\int_{\Gamma\backslash G} \pi(\hat{\phi})\overline{\pi(\hat{\Psi})}dg = \frac{1}{4} \operatorname*{res}_{z=2} M(s,\Lambda_z)\underline{\Phi}(\rho)\overline{\underline{\Psi}}(\rho)$. This will imply that $\mathrm{vol}(\Gamma\backslash G) = 4 \operatorname*{res}_{z=2} M(s,\Lambda_z)^{-1} = \frac{\pi^2}{96}$, and hence that $\mathrm{vol}(F) = \frac{1}{3}(\frac{\pi^2}{64})$.

LEMMA 7 (Langlands). $\int_{\Gamma\backslash G} \hat{\phi}(g)\hat{\psi}(g)dg =$

$$\frac{1}{8\pi}\int_{\mathrm{Re}\Lambda = \Lambda_o} \sum_W M(s,\Lambda)\Phi(\Lambda)\overline{\underline{\Psi}}(-s\Lambda)|d\Lambda| .$$

PROOF. $\int_{\Gamma\backslash G} \hat{\phi}(g)\overline{\hat{\Psi}(g)}dg = \int_{\Gamma\backslash G} \hat{\phi}(g)\sum_{\Gamma\cap P\backslash\Gamma} \overline{\Psi(\gamma g)}dg = \int_{\Gamma\cap P\backslash G} \hat{\phi}(g)\overline{\Psi(g)}dg$

$$= \frac{1}{4}\int_{\Gamma\cap N\backslash N\times A\times K} \hat{\phi}(na)\overline{\Psi(a)}e^{-2\rho(\log a)}dndadk = \frac{1}{4}\int_{\Gamma\cap N\backslash N}\int_A \hat{\phi}(na)\overline{\Psi(a)}e^{-2\rho(\log a)}dadn$$

$$= \frac{1}{8\pi}\int_A\int_{\mathrm{Re}\Lambda = \Lambda_o}\int_{\Gamma\cap N\backslash N} E(na,\Phi,\Lambda)dn|d\Lambda|\overline{\Psi(a)}e^{-2\rho(\log a)}da$$

$$= \frac{1}{8\pi}\int_{\mathrm{Re}\Lambda = \Lambda_o} \sum_W M(s,\Lambda)\Phi(\Lambda)\int_A \exp(s\Lambda-\rho)(\log a)\overline{\Psi(a)}da|d\Lambda|$$

$$= \frac{1}{8\pi}\int_{\mathrm{Re}\Lambda = \Lambda_o} \sum_W M(s,\Lambda)\Phi(\Lambda)\overline{\int_A \exp(s\bar{\Lambda}-\rho)(\log a)\Psi(a)ds}|d\Lambda|$$

$$= \frac{1}{8\pi}\int_{\mathrm{Re}\Lambda = \Lambda_o} \sum_W M(s,\Lambda)\Phi(\Lambda)\overline{\underline{\Psi}(-s\bar{\Lambda})}|d\Lambda| .$$

LEMMA 8. $\int_{\Gamma\backslash G} \pi(\hat{\phi})\overline{\pi(\hat{\Psi})}dg = \frac{1}{4}\operatorname*{res}_{z=2} M(s,\Lambda_z)\underline{\Phi}(\rho)\overline{\underline{\Psi}(\rho)}$.

PROOF. By Langlands [6], page 239, if $f(\Lambda) = (\mu-(\Lambda,\Lambda))^{-1}$ with

$\mu > (\rho,\rho)$ and if $A = \mu-\lambda(f)^{-1}$, then if $\bar{\Psi}(\Lambda) = (\Lambda,\Lambda)\bar{\Phi}(\Lambda)$, $A\hat{\phi} = \hat{\Psi}$.

Therefore, $(A\hat{\phi},1) = (\hat{\Psi},1) = \frac{1}{4}\bar{\Psi}(\rho) = \frac{1}{4}(\rho,\rho)\bar{\Phi}(\rho) = (\rho,\rho)(\hat{\phi},1)$. Hence, as the constant functions are in the space generated by the $\hat{\phi}$, $A1 = (\rho,\rho)1$.

Let $E(x)(-\infty < x < \infty)$ be the spectral resolution of A. Then the constant functions are in the range of $E((\rho,\rho)) - E((\rho,\rho)-0) = E$. We claim that this range consists precisely of the constant functions and get $(E\hat{\phi},\hat{\Psi}) = (\pi\hat{\phi},\pi\hat{\Psi})$.

Choose a and b such that $a > (\rho,\rho) > b$ and $a-b$ is small. Then $\frac{1}{2}\{(E(a)\hat{\phi},\hat{\Psi})+(E(a-0)\hat{\phi},\hat{\Psi})\} - \frac{1}{2}\{E(b)\hat{\phi},\hat{\Psi})+(E(b-0)\hat{\phi},\hat{\Psi})\} =$

$$\frac{1}{2\pi i}\lim_{\delta\to+0}\int_{C(a,b,c,\delta)}(R(\mu,A)\hat{\phi},\hat{\Psi})d\mu \qquad (1)$$

where $C(a,b,c,\delta)$ is the contour shown in Figure 1.

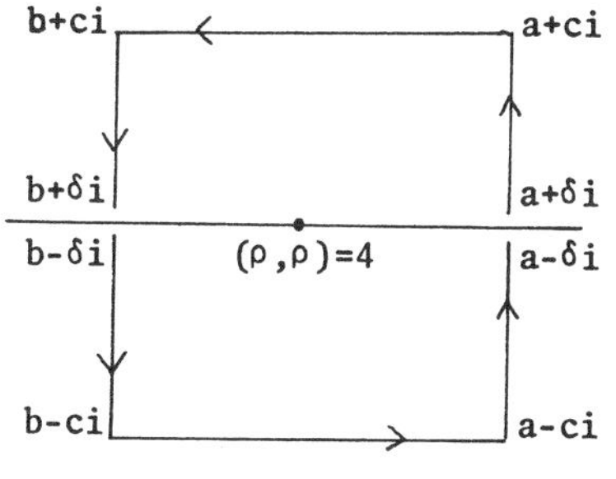

Fig. 1. The Contour $C(a,b,c,\delta)$

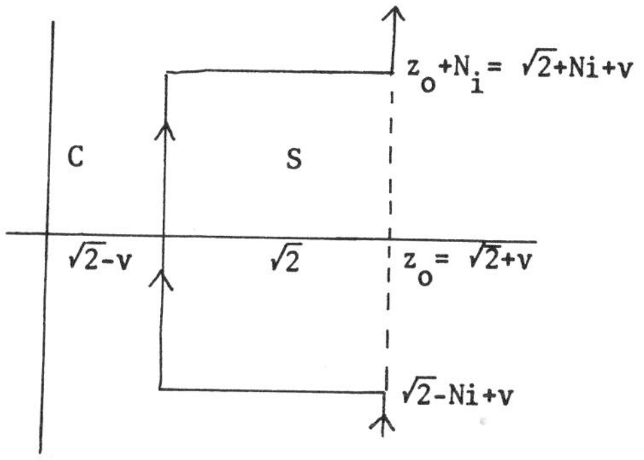

Fig. 2. The Region S

If $\mathrm{Re}\mu > (\Lambda^o,\Lambda^o)$ with $\Lambda^o = \Lambda_{z_o} \in L(A)^*$ and $\Lambda^o(H_\alpha) = z_o > \rho(H_\alpha) = 2$, then

$$(R(\mu,A)\hat{\phi},\hat{\Psi}) = \sum_{s\varepsilon W} \frac{1}{8\pi i} \int_{\mathrm{Re}\Lambda = \Lambda_o} \frac{1}{\mu-(\Lambda,\Lambda)} M(s,\Lambda)\,\Phi(\Lambda)\,\bar{\underline{\Psi}}(-s\bar{\Lambda})\,|d\Lambda|$$

$$= \sum_{s\varepsilon W} \frac{1}{8\pi i} \int_{\mathrm{Re}z=z_o} \frac{1}{\mu-(\Lambda_z,\Lambda_z)} M(s,\Lambda_z)\,\Phi(\Lambda_z)\,\overline{\quad(-s\bar{\Lambda}_z)}\,dz.$$

Since $\Lambda_z = z\Lambda_1$, $(\Lambda_z,\Lambda_z) = z^2(\Lambda_1,\Lambda_1)$; but since $\Lambda_1 = \alpha$, $(\Lambda_1,\Lambda_1) = \alpha(H_\alpha) = 1$ and so $(\Lambda_z,\Lambda_z) = z^2$. Therefore, (1) is equal to

$$\sum_{s\varepsilon W} \lim_{\delta\to+0} \frac{1}{2\pi i} \int_{C(a,b,c,\delta)} d\mu \left\{ \frac{1}{8\pi i} \int_{\mathrm{Re}z=z_o} \frac{\phi_1(z,s)dz}{\mu-z^2} \right\},$$

where $\phi_1(z,s) = M(s,\Lambda_z)\,\Phi(\Lambda_z)\overline{\underline{\Psi}(-s\bar{\Lambda}_z)}$ and $z_o > 2$. Let $\phi_o(s) = \operatorname{res}_{z=2} \phi_1(z,s)$ $(s=1$ or $n_W)$, and let $z_o = v+2$ with $v > 0$, v small. Then $(2-v)^2 < (\rho,\rho) = 4$, so choose $b > (2-v)^2$. Since $\mathrm{Re}(z^2) = (\mathrm{Re}z)^2-(\mathrm{Im}z)^2$, there exists $N > 0$ such that if $\mathrm{Re}z = 2-v$ or $|\mathrm{Re}z-2|\leq v$ and $|\mathrm{Im}z|>N$, then $\mathrm{Re}(z^2) < b-\frac{1}{N}$.

We have $\mathrm{Re}\mu-\mathrm{Re}(z^2)>\mathrm{Re}\mu-b+\frac{1}{N} > \frac{1}{N}$ if z lies in S (see Figure 2) and μ lies in $C(a,b,c,\delta)$; so $\mu-z^2\neq 0$ for μ in a small region containing $C(a,b,c,\delta)$ and $z\varepsilon S$.

Therefore,

$$\frac{1}{2\pi i} \int_{\mathrm{Re}z=z_o} \frac{\phi_1(z,s)}{\mu-z^2}\,dz = \frac{1}{2\pi i} \int_C \frac{\phi_1(z,s)}{\mu-z^2}\,dz + \frac{1}{\mu-4} \operatorname{res}_{z=2} \phi_1(z,s).$$

Hence, (1) equals

$$\sum_{s\in W} \lim_{\delta\to+0} \left(\frac{-1}{16\pi^2}\right) \int_C \left[\int_{C(a,b,c,\delta)} \frac{d\mu}{\mu-z^2}\right] \phi_1(z,s)dz$$

$$+ \sum_{s\in W} \lim_{\delta\to+0} \frac{1}{8\pi i} \operatorname*{res}_{z=2} \phi_1(z,s) \int_{C(a,b,c,\delta)} \frac{d\mu}{\mu-4} .$$

But for z on C, $\lim_{\delta\to+0} \int_{C(a,b,c,\delta)} \frac{d\mu}{\mu-z^2} = 0$ as $\frac{1}{\mu-z^2}$ is holomorphic inside the contour D (Figure 3). Also, $\frac{1}{2\pi i} \lim_{\delta\to+0} \int_{C(a,b,c,\delta)} \frac{d\mu}{\mu-4} = 1$.

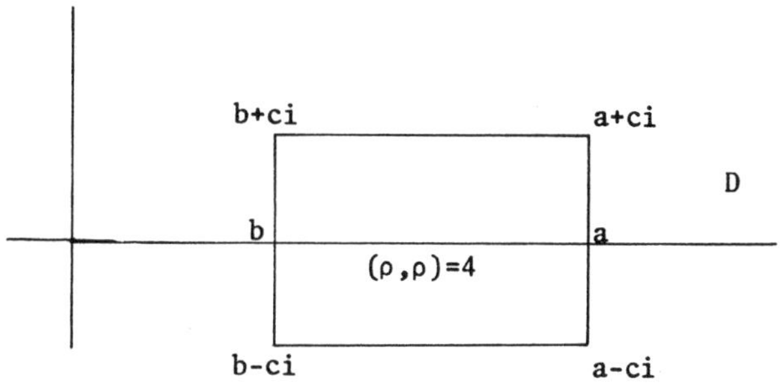

Fig. 3. The Contour D

Therefore, (1) equals $\frac{1}{4}\sum_{s\in W} \operatorname*{res}_{z=2} \phi_1(z,s)$. But if $s = 1$, $\phi_1(z,s)$ is entire, so (1) equals $\frac{1}{4} \operatorname*{res}_{z=2} \phi_1(z,n_W) = \frac{1}{4} \operatorname*{res}_{z=2} M(n_W,\Lambda_z)\underline{\hat{\phi}}(\rho)\overline{\underline{\hat{\Psi}}(\rho)} = (\pi\hat{\phi},\pi\hat{\Psi})$, as claimed.

COROLLARY. $\mathrm{Vol}(F) = \frac{1}{3}(\frac{\pi^2}{64})$.

REMARKS. 1) The proof of Lemma 8 is essentially that of [7], page 146-148, which was carried out only for G a Chevalley group.

2) It is possible to obtain the volume of the fundamental domain using

Borel's Reduction Theory and the results of Chapters IV-VII. For the results obtained imply easily that for $3m$ an integer greater than or equal to 6,

$$\dim A_m(\Gamma) = (3m-1)(3m-2)\frac{\mathrm{vol}(F)}{\pi^2} + \frac{k(m)}{3\times 32},$$

where $k(m)\in Z$, and in particular $k(2) = -10$. Then since $\dim A_m(\Gamma)$ is a non-negative integer, $\mathrm{vol}(F) = \frac{k}{3}(\frac{\pi^2}{64})$ with k a positive integer congruent to one modulo 48. But using Lemma 5 of Chapter II, one sees that the set

$$F_1 = \{(z,u)\in D \mid |\mathrm{Re}\,z| \leq \tfrac{1}{2}, |z|^2 \geq 1, |u|^2 \leq 1, \mathrm{Re}\,u \geq 0, \ \mathrm{Im}\,u \geq 0\}$$

contains a fundamental domain for Γ, and hence $\mathrm{vol}(F) \leq \mathrm{vol}(F_1)$. But by direct computation, one finds that

$$\mathrm{vol}(F_1) \leq \frac{2}{\sqrt{3}-1}\ \frac{\pi^2}{3\times 64}.$$

Hence $k \leq \frac{2}{\sqrt{3}-1}$ and so $k=1$.

Langlands' method, of course, has the advantage of being independent of Selberg's Trace Formula; its use in this Chapter thus provides a check on the validity of the preceding calculations.

CHAPTER IX

CONCLUSION

We have proven the following

THEOREM. If $3m \in Z$ and $m \geq 2$,

$$\dim A_m(\Gamma) = \frac{(3m-1)(3m-2)}{3\times 64} - \frac{1}{12}[3\mathrm{tr}_{Q[i]/Q}(\frac{i^{3m}}{1+i}) + 3(-1)^{3m}+1]$$

$$+ \frac{1}{3\times 128}[10\mathrm{tr}_{Q[i]/Q}\{(3m-1)i^{3m-1} - (3m-2)i^{3m-2}\} + 17(-1)^{3m}(6m-3)]$$

$$+ \frac{17}{128} + \frac{7}{32}\mathrm{tr}_{Q[i]/Q}(\frac{i^{3m}}{1-i}) + \frac{1}{16}\mathrm{tr}_{Q[\zeta]/Q}(\frac{\zeta^{3m}}{1-\zeta^{-1}}) + \frac{1}{36}\mathrm{tr}_{Q[\rho]/Q}(\rho^{3m})$$

$$+ \frac{1}{12}\mathrm{tr}_{Q[i]/Q}(i^{3m})\mathrm{tr}_{Q[\rho]/Q}(\frac{\rho^{3m}}{1-\rho}) - \frac{1}{12}\mathrm{tr}_{Q[i]/Q}(i^{3m+1})\mathrm{tr}_{Q[\rho]/Q}(\frac{\rho^{3m+1}}{1-\rho})$$

$$+ \frac{(-1)^{3m}}{12}\mathrm{tr}_{Q[\rho]/Q}(\frac{\rho^{3m-1}}{1-\rho}) + \frac{1}{9}.$$

By the remark preceding Lemma 7 of Chapter II, this is one less than the dimension of the space of all automorphic forms for Γ of weight m if $3m$ is divisible by 4, and equals the dimension of the space of all forms of weight m otherwise.

We can rewrite our dimension formula as

$$\dim\Lambda_m(\Gamma) = \frac{(3m-1)(3m-2)}{3\times 64} + \frac{3m}{3\times 64}X_m + Y_m + Z_m,$$

where X_m, Y_m, and Z_m are given in Tables 1, 2, and 3.

Using Tables 1, 2, and 3, we find, for instance, that $\dim A_m(\Gamma) = 0$ for $2m=6,7,9,11,13,15,19$, and 23, 1 for $3m=8,10,14,17,21$, and 27, 2 for $3m=12,18$, and 25, 3 for $3m=6$, 22, and 29, 4 for $3m=20$ and 26, 5 for $3m=30$, 6 for $3m=24$, and 7 for $3m=28$.

TABLE 1

VALUES OF X_m

3m mod 4	X_m
0	27
1	-7
2	7
3	-27

TABLE 2

VALUES OF Y_m

3m mod 4	Y_m
0	$\frac{-5}{48}$
1	$\frac{-5}{96}$
2	$\frac{5}{48}$
3	$\frac{5}{96}$

TABLE 3

VALUES OF Z_m

3m mod 24	Z_m	3m mod 24	Z_m
0	$\frac{3}{32}$	12	$\frac{-5}{32}$
1	$\frac{17}{3\times 64}$	13	$\frac{-31}{3\times 64}$
2	$\frac{-17}{96}$	14	$\frac{-41}{96}$
3	$\frac{87}{64}$	15	$\frac{71}{64}$
4	$\frac{-47}{96}$	16	$\frac{-23}{96}$
5	$\frac{11}{64}$	17	$\frac{27}{64}$
6	$\frac{-41}{96}$	18	$\frac{-17}{96}$
7	$\frac{149}{3\times 64}$	19	$\frac{197}{3\times 64}$
8	$\frac{-23}{96}$	20	$\frac{-47}{96}$
9	$\frac{17}{3\times 64}$	21	$\frac{-31}{3\times 64}$
10	$\frac{5}{32}$	22	$\frac{-3}{32}$
11	$\frac{197}{3\times 64}$	23	$\frac{149}{3\times 64}$

REFERENCES

1. Borel, A., Ensembles Fondamentaux pour les Groupes Arithmetiques et Formes Automorphes (Mimeographed Notes).

2. Hasse, H., Zahlentheorie, Akademie-Verlag GmbH, Berlin, 1949.

3. Helgason, S., Differential Geometry and Symmetric Spaces, Academic Press, New York and London, 1962.

4. Harish-Chandra, Automorphic Forms on Semisimple Lie Groups, Springer, Berlin, 1968.

5. Langlands, R. P., The Functional Equations Satisfied by Eisenstein Series (Mimeographed Notes).

6. ____________, Eisenstein Series, in Proc. Sympos. Pure Math., vol. 9, Amer. Math. Soc., Providence, R. I., 1966, pp. 235-252.

7. ____________, The Volume of the Fundamental Domain for some Arithmetical Subgroups of Chevalley Groups, ibid., pp. 143-148.

8. Selberg, A., Automorphic Functions and Integral Operators, Seminars on Analytic Functions, Institute for Advanced Study, Princeton, N. J. and U. S. Air Force, Office of Scientific Research, 1957.

9. Shimizu, H., On Discontinuous Groups Operating on the Product of the Upper Half Planes, Annals of Math., 77 (1963), pp.33-71.

10. Weber, H., Elliptische Functionen und Algebraische Zahlen, Vieweg & Sohn, Braunschweig, 1891.

11. Weiss, E., Algebraic Number Theory, McGraw-Hill, New York, 1963.

The Johns Hopkins University

Memoirs of the American Mathematical Society

Number 159

Chung-Nim Lee
and Arthur G. Wasserman

On the groups $JO(G)$

Published by the

AMERICAN MATHEMATICAL SOCIETY

Providence, Rhode Island

VOLUME 2 · ISSUE 1 · NUMBER 159 (first of 2 numbers) · MAY 1975

ABSTRACT

ON THE GROUPS JO(G)

The purpose of this memoir is to classify real representations of a compact Lie group G under stable J-equivalence. Two representations V,W of G are said to be stably J-equivalent if there exist equivariant maps $S(V \oplus U) \to S(W \oplus U)$, $S(W \oplus U) \to S(V \oplus U)$, both of degree one, for some representation U, where $S(V \oplus U)$ denotes the unit sphere in $V \oplus U$. Denote by JO(G) the quotient group of the representation ring RO(G), modulo stable J-equivalence. It is shown (Theorem 3.20) that V,W are stably J-equivalent if and only if $\dim V^C = \dim W^C$ for every cyclic subgroup C of G such that $C/C \cap G_o$ is a p-group, p a prime where V^C denotes the subspace of V fixed under C, or equivalently if and only if the difference character, $\chi_V - \chi_W$, is constant on each connected component of G which has prime power order G/G_o, and $\sum_{g \in C} \chi_V(g) = \sum_{g \in C} \chi_W(g)$ for every finite cyclic subgroup C of a prime power order. In addition, the above theorem has a localized version at any collection if primes. The sufficiency of these conditions is shown via a Thom-Pontryagin type construction and the necessity follows from simple equivariant transversality arguments. The crucial fact necessary to handle arbitrary compact Lie groups is that for any x in G the centralizer and the conjugacy class of x meet transversally. Some of the interesting consequences of these results are 1) JO(G) is a free abelian group, 2) JO(G) injects into the product $\Pi_p JO(G_p)$ where G_p is the inverse image of a Sylow p-subgroup of G/G_o, 3) if G is connected, then $JO(G) \approx RO(G)$. Moreover, these methods provide another proof of the Atiyah-Tall theorem $JO(G) \approx RO(G)_\Gamma$.

ACKNOWLEDGEMENTS

Conversations with G. Bredon, J. McLaughlin and D. Wigner were very helpful. The authors would also like to thank Denise Cotter for typing the first draft of this paper and Jean Whipple who prepared the final manuscript.

Both authors were supported in part by the National Science Foundation during the preparation of this memoir.

AMS(MOS) 1970 Classification 55E50

KEY WORDS. Compact Lie groups, linear representations, equivariant maps, stable J-equivalence.

ISBN 0-8218-1859-7

ON THE GROUPS JO(G)

by

Chung-Nim Lee and Arthur G. Wasserman

INTRODUCTION

J. F. Adams raised the following question at the 1963 Seattle Conference on Differential Topology [1]: given orthogonal representations V, W of a compact Lie group G, when does there exist an equivariant map $f: S(V) \to S(W)$ of degree k? (S(V) is the unit sphere in V.)

The problem was first considered by Atiyah and Tall in [3]. They define two representations V, W of a finite group G to be J equivalent if there are equivariant maps $f: S(V) \to S(W)$, $h: S(W) \to S(V)$ with both degree f and degree h prime to the order of G; V, W are stably J equivalent if there is a representation U of G such that $V \oplus U$ is J equivalent to $W \oplus U$. Then the groups JO(G) may be defined as the quotient of the real representations RO(G) by the subgroup $T(G) = \{V - W | V \text{ is stably J equivalent to } W\}$. Atiyah and Tall were able to compute the groups JO(G) for G a finite p-group $(p \neq 2)$ in terms of the action of the Adams operation ψ^k on RO(G). Let $RO(G)_\Gamma = RO(G)/WO(G)$ where WO(G) is the subgroup generated by $V - \psi^k V$ where k is prime to the order of G; their result is then $JO(G) \approx RO(G)_\Gamma$ for G a finite p-group $(p \neq 2)$. Subsequently, Snaith extended the result to include the case p = 2 for complex representations [9].

Received by the editors April 24, 1974
Research supported in part by the National Science Foundation

Atiyah and Tall were also able to show that for p-groups, i) stably J equivalent representations are J equivalent, and ii) if there is an equivariant map f: $S(V) \to S(W)$ of degree prime to the order or G then V and W are J equivalent. Hence their work produces a complete answer to the question: given two representation V, W of a finite p-group G, when does there exist an equivariant map f: $S(V) \to S(W)$ with deg f prime to the order of G? (The answer is: if and only if $V \simeq W$ in $RO(G)_{\Gamma}$.)

If k is a collection of primes, we say that representations V,W of a compact Lie group G are J_k equivalent if there are equivariant maps f: $S(V) \to S(W)$, h: $S(W) \to S(V)$ with deg f and deg h units in Z_k, the integers localized away from all primes in k. The notion of J_k equivalence generalizes the notion of J equivalence of Atiyah and Tall (which is the special case k = all primes not dividing the order of G) and applies to arbitrary compact Lie groups. The groups $JO_k(G)$ are defined in the obvious way.

The main object of this paper is to study the group $JO_k(G)$ for G an arbitrary compact Lie group. In Section 1, precise definitions are given and justification is provided for only considering real representations. In Section 2, numerical invariants of J_k equivalence are constructed which distinguish representations V,W which are not stably J_k equivalent. The invariants are basically just the dimensions of the fixed point sets of certain cyclic subgroups $C \subset G$ acting on V,W respectively. The main result of Section 2 states that if V,W are J_k equivalent representations of a compact Lie group G, then for any cyclic subgroup $C \subset G$ with $C/C \cap G_0$ a p-group, $p \notin k$, $\sum_{g \in C} \chi(g) = 0$ where G_0 is the connected component of the identity element in G and $\chi = \chi_V - \chi_W$ is the character of the virtual representation V - W. The proofs in this section use only a simple transversality argument. As an immediate corollary of this result, one has that J_k equivalent representations of a connected group are actually linearly equivalent. Also if G/G_0 is a p-group and $p \notin k$ then the difference character $\chi = \chi_V - \chi_W$ of J equivalent representations V,W is constant on each connected component of G. In Section 3, it is shown that the converse is also true; that is, if $\sum_{g \in C} \chi(g) = 0$ for all cyclic subgroups $C \subset G$ with $C/C \cap G_0$ a p-group, $p \notin k$ then V is stably J_k equivalent to W if $2 \notin k$. An extra condition is needed if $2 \in k$. The proofs in Section 3 just use the construction of Proposition 3.1, and the fact that the equivariant homotopy classes of stable equivariant maps from S(V) to S(W) form a group. It follows quickly that $JO_k(G)$ injects into the product of the $JO_k(G_p)$'s where G_p is a p-Sylow subgroup if G is finite or the inverse image of a p-Sylow subgroup of G/G_0 if G is an arbitrary compact Lie group. Thus using the results of [3], [9], the computation of $JO_k(G)$ is complete for G a finite group. However, combining the results of Section 2 with the techniques of [3], provides a simple proof of the main result: representations V, W of a compact Lie group G are stably J_k equivalent if and only if for every finite cyclic subgroup $C \subset G$ with $C/C \cap G_0$ a p-group, $p \notin k$, $\sum_{g \in C} \chi(g) = 0$. If $2 \in k$, a further condition is necessary.

In Section 4, we shall apply the results of the previous sections to computations, conjectures and counter-examples. Among others, the question of Adams concerning the existence of an equivariant stable map of degree k is reduced to the question for groups G such that G/G_0 is a p-group, p prime (Corollary 4.3). Generalizations of some results of Atiyah and Tall in [3] on existence of equivariant maps are included (Theorems 4.1 and 4.4). Regarding computations, a Künneth type of theorem for the functor $\tilde{JO}_k$ is proved (Proposition 4.7), and a formula for computing the rank of $\tilde{JO}(G)$ for a finite abelian group G is given.

SECTION 1. THE FUNCTOR $JO_k(G)$

Let G be a compact Lie group. A real representation of G of dimension n is a continuous map ϕ: G $\mathbb{R}^n \to \mathbb{R}^n$ satisfying

i) $\phi(e,x) = x$ for all $x \in \mathbb{R}^n$

ii) $\phi(g_1, \phi(g_2,x)) = \phi(g_1g_2,x)$ for all $g_1, g_2 \in G$, $x \in \mathbb{R}^n$

iii) $\overline{\phi}_g = \phi(g,x)$: $\mathbb{R}^n \to \mathbb{R}^n$ is a linear map for each $g \in G$.

Alternatively, we may think of a representation as a continuous homomorphism $g \to \overline{\phi}_g$ from G to GL(n, $\mathbb{R}$). $\mathbb{R}^n$, together with the action of G determined by ϕ, will frequently be denoted by V, W, U, etc., and $\phi(g,v)$ will be shortened to gv. $\mathbb{R}$ will denote the trivial one dimensional representation of G. Two representations V, W are said to be equivalent if there is a linear isomorphism T: V $\to$ W which is equivariant, i.e., T(gv) = gT(v). Since G is compact, every representation has a G invariant inner product. Let S(V) denote the unit sphere of V with respect to some invariant inner product. S(V) is a G-manifold.

Let $RO^+(G)$ denote the set of equivalence classes of representations of G. RO^+ is a semiring with direct sum of representations as addition and tensor product of representations as product. RO(G), the associated ring, is the free abelian group generated by irreducible representations. If V is a real representation of G, the character of V, χ_V: G $\to \mathbb{R}$, is defined by $\chi_V(g)$ = trace of g on V and extends to a monomorphism $RO(G) \to C(G)$ where $C(G)$ denotes the vector space of continuous real valued functions on G satisfying $f(hgh^{-1}) = f(g)$ and $f(g) = f(g^{-1})$ for all $h, g \in G$. We shall frequently identify an element of $RO(G)$ with its character.

If M, N are n dimensional closed, connected, oriented manifolds and $f: M \to N$ is continuous, the degree of f is defined by $f_*([M]) = (\deg f)[N]$ where $[M] \in H_n(M:\mathbb{Z}) \approx \mathbb{Z}$, $[N] \in H_n(N;\mathbb{Z}) = \mathbb{Z}$ are the generators given by the orientations on M, N respectively. Alternatively, if $y \in N$, f can be approximated by a homotopic map h which is differentiable and has y as a regular value; then $h^{-1}(y)$ is a finite set and $\deg f = a_+ - a_-$ where a_+ is the number of points x in $h^{-1}(y)$ for which $dh_x: \tau(M)_x \to \tau(N)_y$ is an orientation preserving isomorphism and a_- is the number of points for which dh_x is an orientation reversing isomorphism where $\tau(M)_x$ and $\tau(N)_y$ denote the tangent spaces.

Modifying the terminology of [3] slightly, we say that representations V, W of a finite group G are <u>weakly J equivalent</u> if there are equivariant maps $f: S(V) \to S(W)$, $h: S(W) \to S(V)$ such that $(\deg f, |G|) = 1$, i.e., deg f is prime to the order of G and $(\deg h, |G|) = 1$. If k is a collection of prime numbers, $k = \{p_1, p_2, \ldots\}$, V, W are said to be J_k <u>equivalent</u> if there are equivariant maps $f: S(V) \to S(W)$, $h: S(W) \to S(V)$ with $\deg f = p_1^{a_1} p_2^{a_2} \ldots p_s^{a_s}$, $\deg h = p_1^{b_1} p_2^{b_2} \ldots p_t^{b_t}$, i.e., deg f deg h are units in the ring $\mathbb{Z}[p_1^{-1}, p_2^{-1} \ldots] = \mathbb{Z}_k$. If $k = \phi$, the empty collection, we require deg f = deg h = 1 and let $J = J_\phi$ and refer to this as J equivalence instead of J_k equivalence. Note that J_k equivalence makes sense for arbitrary compact Lie groups. If G is finite and k is the set of primes not dividing $|G|$, then weak J equivalence and J_k equivalence agree. V, W are said to be <u>stably J_k equivalent</u> if there is a representation U of G such that $V \oplus U$ is J_k equivalent to $W \oplus U$. If V_1 is J_k equivalent to W_1 and V_2 is J_k equivalent to W_2 then $V_1 \oplus V_2$ is J_k equivalent to $W_1 \oplus W_2$ since $S(V_1 \oplus V_2) = S(V_1) * S(V_2)$ (where * denotes the join) and, the equivariant map $f_1 * f_2 \colon S(V_1 \oplus V_2) = S(V_1) * S(V_2) \to S(W_1) * S(W_2) = S(W_1 \oplus W_2)$ satisfies $\deg(f_1 * f_2) = \deg f_1 \deg f_2$ where $f_i \colon S(V_i) \to S(W_i)$ is an equivariant map $(i = 1,2)$, and the product of units is a unit. In particular, if V is J_k equivalent to W then, since U is J_k equivalent to U, $V \oplus U$ is J equivalent to $W \oplus U$, i.e: J_k equivalence implies stable J_k equivalence.

Let $T_k(G) \subset RO(G)$, $T_k(G) = \{V - W \mid V \text{ is stably } J_k \text{ equivalent to } W\}$. Then $T_k(G)$ is a subgroup of RO(G) by the above remarks.

<u>Definition</u> $JO_k(G) = RO(G)/T_k(G)$. The natural projection of RO(G) onto $JO_k(G)$ will be denoted by ν_G. Again, if $k = \phi$ we set $T_k(G) = T(G)$, $JO_k(G) = JO(G)$. $\widetilde{RO}(G)$ will denote the kernel of the dimension homomorphism $\varepsilon : RO(G) \to \mathbb{Z}$, $\varepsilon(v)$ = dimension of V or, alternatively,

$\varepsilon(\chi) = \chi(e)$, e = identity of G. Since $T_k(G) \subset RO(G)$, ε induces a homomorphism ε': $JO_k(G) \to \mathbb{Z}$ and the kernel of ε', $\tilde{JO}_k(G)$, may be identified with $\mathbb{R}\tilde{O}(G)/T_k(G)$. The map: $\tilde{RO}(G) \to \tilde{JO}_k(G)$ will be denoted by $\tilde{\nu}_G$.

<u>Proposition 1.1</u> $RO(G)$, $\tilde{RO}(G)$, $JO_k(G)$, $\tilde{JO}_k(G)$ are contravariant functors from the category of compact Lie groups and continuous homomorphism to the category of abelian groups and ν_G, $\tilde{\nu}_G$ are natural transformations.

<u>Proof</u> If i: $H \to G$ is a continuous homomorphism and $\chi \in RO(G)$ then $i^* \chi = \chi \circ i$. Alternatively, if V is a representation and $\phi : G \times X \to V$ is is the action then an action of H on V, $\Phi : H \times V \to V$ is defined by $H \times V \xrightarrow{ixid} G \times V \xrightarrow{\phi} V$, $\bar{\Phi} = \Phi \circ (ixid)$ and $i^*(V,\phi) = (V,\bar{\phi})$.

If f: $S(V) \to S(W)$ is G equivariant then f is also H equivariant $(S(V) = S(i^*(V)))$ hence $i^*(T_k(G)) \subset T_k(H)$. Thus we have the commutative diagram

$$\begin{array}{ccccccccc}
0 & \longrightarrow & T_k(G) & \hookrightarrow & RO(G) & \xrightarrow{\nu_G} & JO_k(G) & \longrightarrow & 0 \\
 & & \downarrow{\scriptstyle i^*} & & \downarrow{\scriptstyle i^*} & & \downarrow{\scriptstyle i^*} & & \\
0 & \longrightarrow & T_k(H) & \hookrightarrow & RO(H) & \xrightarrow{\nu_H} & JO_k(H) & \longrightarrow & 0
\end{array}$$

There is a similar diagram involving $\tilde{RO}(G)$, $\tilde{\nu}_G$, etc.

Note: If i: $H \subset G$ is an inclusion and V is a representation of G, i^*V will also be denoted by $V|H$.

The above definitions make sense for complex representations, sympletic representations, etc. More generally, suppose we are given subgroups $F_n \subset GL(n, \mathbb{R})$ and maps $F_n \times F_m \to F_{n+m}$ such that

$$\begin{array}{ccc} F_n \times F_m & \longrightarrow & F_{n+m} \\ \cap & & \cap \\ GL(n,\mathbb{R}) \times GL(m,\mathbb{R}) & \hookrightarrow & GL(n+m,\mathbb{R}) \end{array}$$

commutes. Then $RF(G)$, $JF_k(G)$ are defined and there is a commutative diagram

$$\begin{array}{ccc} RF(G) & \xrightarrow{j} & RO(G) \\ \big\downarrow \bar{\nu}_G & & \big\downarrow \nu_G \\ JF_k(G) & \xrightarrow{j} & JO_k(G). \end{array}$$

We are interested in characterizing those $\chi \in RF(G)$ such that $\bar{\nu}_G(\chi) = 0 \in JF_k(G)$. To simplify matters, we shall restrict our attention to real representations. The following proposition shows that that is not a serious restriction.

Proposition 1.2 Suppose that for every real representation U of G there is an F representation $\tilde{U}$ with U a real subrepresentation of $j(\tilde{U})$. Then $j: JF_k(G) \to JO_k(G)$ is a monomorphism. Hence $\bar{\nu}_G(\chi) = 0$ if and only if $\nu_G(j(\chi)) = 0$.

Proof Let $V - W \in RF(G)$ and suppose $j\bar{\nu}_G(V - W) = 0$ in $JO_k(G)$. Then there is a real representation U of G with $U \oplus V$ J_k equivalent to $U \oplus W$. By hypothesis, U is a real subrepresentation of an F representation $\tilde{U}$, $\tilde{U} = U \oplus X$. Hence $X \oplus U \oplus V$ is J_k equivalent to $X \oplus U \oplus W$ and $\bar{\nu}_G(V - W) = 0$.

Examples

1) Let $F_{2n} = F_{2n+1} = GL(n,\mathbb{C})$ and take $\tilde{U}$ to be the complexification of U. A similar argument applies to the symplectic case.

2) Let G be finite, $|G| = N$. If ω is a primitive N^{th} root of unity let $F_{2n} = F_{2n+1} = GL(n, \mathbb{Q}(\omega)) \subset GL(2n, \mathbb{R})$ where $\mathbb{Q}(\omega)$ is the extension field of the rationals containing ω. Then every complex representation U of G is equivalent to an F representation U. Hence we may take $\tilde{U}$ to be an F representation which is equivalent to the complexification of U.

3) Let $F_{2n} = F_{2n+1} = GL(n, \mathbb{Q}(\omega')) \subset GL(n, \mathbb{R})$ where ω' is a primitive r-th root of unity. If G is finite $|G| = N$, the proposition applies by taking $\tilde{U} = tr(U)$ where $tr(U)$ denotes the trace of U as defined in Section 3. Roughly, U is equivalent to a representation in $GL(n, \mathbb{Q}(\omega))$ as in Example 2 and if we regard $\mathbb{Q}(\omega)$ as a subgroup of $GL(\phi(N), \mathbb{Q})$, $\tilde{U}$ is a representation in $GL(n\phi(N), \mathbb{Q}) \subset GL(n\phi(N), \mathbb{Q}(\omega'))$. If G is not finite the proposition does not apply but nevertheless, Corollary 3.12 below can be used to show easily that $JF_k(G)$ injects into $JO_k(G)$.

4) Let $F_n = O(n)$. Since G is compact every representation is equivalent to an orthogonal representation.

5) Let $F_n = SO(n)$ and take $\tilde{U} = U \oplus U$ and use the remark in 4).

SECTION 2. NECESSARY CONDITIONS AND INVARIANTS

In this section we will establish necessary conditions for two representations, V, W of a compact Lie group G to be J_k equivalent. One obvious necessary condition is that dim V = dim W. More generally, for certain subgroups H of G it is shown that $\dim V^H = \dim W^H$ where $V^H = \{v \in V | hv = v \text{ for all } h \in H\}$. If H is abelian and there is a toral subgroup T of G such that $H/H \cap T$ has order p^r for some $p \notin k$ then $\dim V^H = \dim W^H$ if V is J_k equivalent to W. An immediate corollary is that $\nu_G\colon RO(G) \approx JO_k(G)$ if G is connected. More generally, if G has p^r components and $\nu_G(\chi) = 0$ in $JO_k(G)$ where $p \notin k$, then $d\chi \equiv 0$, i.e. χ is constant on the connected components of G and thus χ is induced from a virtual representation of G/G_0 where G_0 denotes the component of the identity element $e \in G$. The main ingredient in the proof that $d\chi \equiv 0$ is Theorem 2.20 which states that the conjugacy class of x in G is transverse to the centralizer of x in G at x.

Let A be an abelian group and let $\alpha\colon RO(G) \to A$ be a homomorphism. If $\alpha(T_k(G)) = 0$ then α is said to be an invariant of J_k equivalence. The invariants we will consider are of the form $\alpha(\chi) = \frac{1}{|H|} \sum_{h \in H} \chi(h)$ where $H \subset G$ is a finite subgroup. Equivalently, if $\chi = V - W \in RO(G)$, $\alpha(\chi) = \dim V^H - \dim W^H$. The proof that α is an invariant (for certain H) is just a simple counting argument but requires some preparation.

If M, P are G-manifolds, $C^k(M; P)$ will denote the C^k maps from M to P with the C^k topology; $C^k_G(M; P) \subset C^k(M; P)$ will denote the G-equivariant maps. If $N \subset P$ is a closed invariant submanifold, then $TR(M; P, N) \subset C^k(M; P)$ will denote those maps transverse regular to N; $TR_G(M; P, N) = TR(M; P, N) \cap C^k_G(M; P)$. D^n will denote the closed unit ball in $\mathbb{R}^n$ with trivial G action.

Proposition 2.1 $TR_G(G \times D^n; P, N)$ is open and dense in $C^\infty_G(G \times D^n; P)$.

Proof $C^\infty_G(G \times D^n; P) \approx C^\infty(D^n, P)$ and $TR_G(G \times D^n: P, N) \approx TR(D^n; P, N)$ where the homomorphisms are given by restricting $f: G \times D^n \to P$ to $f|\{e\} \times D^n$. The proposition follows since $TR(D^n; P, N)$ is open and dense in $C^\infty(D^n; P, N)$ by Thom's theorem [11].

Proposition 2.2 Let M be a compact G manifold and $h: M \to P$ a G-equivariant map. If h is transverse regular to $N \subset P$ in a neighborhood U of the closed invariant subset $C \subset M$ and G acts freely on M - C, then there is an $\tilde{h}: M \to P$ close to h in $C^\infty(M; P)$ with $\tilde{h} \in TR_G(M; P, N)$ and $\tilde{h}$ is equivariantly homotopic to h.

Proof By the slice theorem [4], one can choose a finite number of $G \times D^n_i \subset M$, i = 1, 2, ..., r so that $M - U \subset \bigcup_{i=1}^r G \times D^n_i \subset M - C$. Since $M - \bigcup_{i=1}^r G \times \overset{\circ}{D}{}^n_i$ is compact, there is a small open neighborhood θ

of h in $C_G^\infty(M; P)$ such that all $f \in \theta$ are transverse regular to N in a neighborhood of $M - \bigcup_{i=1}^{r} G \times \overset{\circ}{D}{}_i^n$ and f is equivariantly homotopic to h. The restriction map ζ_i: $C_G^\infty(M; P) \to C_G^\infty(G \times D_i^n; P)$ is open and continuous; hence $\zeta_i^{-1}(TR_G(G \times D_i^n; P, N))$ is open and dense in $C_G^\infty(M, P)$. Thus $\theta \cap \bigcap_{i=1}^{r} \zeta_i^{-1}(TR_G(G \times D_i^n; P, N) = Z$ is nonempty and any $\tilde{h} \in Z$ satisfies $\tilde{h} \in TR_G(M; P, N)$ and $\tilde{h}$ is equivariantly homotopic to h.

Corollary 2.3 Let V, W be representation of $\mathbb{Z}_p$ with dim V = dim W, p prime, and h: $S(V) \to S(W)$ a $\mathbb{Z}_p$ equivariant map. Suppose that $S(W)^{\mathbb{Z}_p} \neq \phi$, and $b \in S(W)^{\mathbb{Z}_p}$. If $h(S(V)^{\mathbb{Z}_p}) = b$ then deg $f \equiv 0 \bmod p$.

Proof Let $* \in S(W)^{\mathbb{Z}_p}$, $* \neq b$. We shall apply Proposition 2.2 with $M = S(V)$, $P = S(W)$, $N = *$, $C = S(V)^{\mathbb{Z}_p}$ to get a map $\tilde{h}$ homotopic to h and transverse regular to $*$; $\tilde{h}^{-1}(*)$ is then a finite invariant subset of $S(V) - S(V)^{\mathbb{Z}_p}$. Hence $\mathbb{Z}_p$ acts freely on $\tilde{h}^{-1}(*)$, and the number of points $\tilde{h}^{-1}(*)$ is divisible by p. Let a^+ be the number of points $x \in \tilde{h}^{-1}(*)$ for which $d\tilde{h}_x$: $\tau(S(V))_x \to \tau(S(W))_*$ is orientation preserving and a^- the number of points $x \in \tilde{h}^{-1}(*)$ for which $d\tilde{h}_x$ is orientation reversing (with respect to fixed orientations on S(V), S(W)). If p = 2, we note that deg h = deg $\tilde{h} = a^+ - a^- = a^+ + a^- - 2a^-$ but $a^+ + a^- \equiv 0 \bmod 2$ as noted above, so we are done. If p > 2, we note that $d\tilde{h}_{gx} = dg_{\tilde{h}(x)}\, d\tilde{h}_x\, dg_{gx}^{-1}$, and since $\mathbb{Z}_p$ preserves orientation on both S(V) and S(W) (p > 2) $d\tilde{h}_{gx}$ preserves orientation if and only if $d\tilde{h}_x$ preserves orientation. Thus a^+ and a^- are both divisible by p; hence deg h = $a^+ - a^- \equiv 0 \bmod p$.

Remark 1 If $S(V)^{\mathbb{Z}_p} = \phi$, the condition on h is vacuous.

Remark 2 Using the equivariant homotopy extension lemma ([12], Lemma 3.2), one need only assume $h|S(V)^{\mathbb{Z}_p}$ is homotopic to the constant map to b in Corollary 2.3.

Corollary 2.4 Let V, W be representation of $\mathbb{Z}_p$, p a prime, with dim V = dim W and let $f_i\colon S(V) \to S(W)$, i = 1, 2, be equivariant maps. Denote by $f^{\mathbb{Z}_p}$ the restriction $f|S(V)^{\mathbb{Z}_p}: S(V)^{\mathbb{Z}_p} \to S(W)^{\mathbb{Z}_p}$. If $f_1^{\mathbb{Z}_p}$ is homotopic to $f_2^{\mathbb{Z}_p}$, then $\deg f_1 \equiv \deg f_2 \bmod p$.

Proof Let $f_i' = f_i * \mathrm{id}\colon S(V) * S(\mathbb{R}^2) \to S(W) * S(\mathbb{R}^2)$. Then $\deg f_i' = \deg f_i$ and $f_1'^{\mathbb{Z}_p}$ is homotopic to $f_2'^{\mathbb{Z}_p}$. Moreover, the equivariant homotopy classes of maps from $S(V) * S(\mathbb{R}^2) = S(V \oplus \mathbb{R}^2)$ to $S(W \oplus \mathbb{R}^2)$ form a group. Letting $h = f_1' - f_2'$ and applying Corollary 2.3 (and the remark following), we have $\deg h = \deg f_1' - \deg f_2' = \deg f_1 - \deg f_2 \equiv 0 \bmod p$.

Remark Two suspensions are required above, the first to provide a fixed base point, the second to get a folding map.

Lemma 2.5 ([10]) Let V_1, W_1 be representations of $\mathbb{Z}_{p^r}$, p prime with $V_1^{\mathbb{Z}_p} = W_1^{\mathbb{Z}_p} = 0$. If $\dim V_1 \leq \dim W_1$, then there is an equivariant map $f\colon S(V_1) \to S(W_1)$. If $\dim V_1 = \dim W_1$, then $\deg f \not\equiv 0 \bmod p$.

Proof Let $n = \dim S(W_1)$. $S(V_1)$, $S(W_1)$ are principal $\mathbb{Z}_{p^r}$ bundles, and $S(W_1)$ is n-universal since $\pi_i(S(W_1)) = 0$ $i < n$. Hence, if $\dim S(V_1) \leq n$, there is a principal bundle map, i.e., an equivariant map $f\colon S(V_1) \to S(W_1)$. Similarly, if $\dim S(W_1) \leq \dim S(V_1)$, there is an equivariant map $f'\colon S(W_1) \to S(V_1)$. Now $f' \circ f$, identity $S(V_1) \to S(V_1)$ satisfy the hypotheses of Corollary 2.4 and thus $\deg (f' \circ f) = \deg f' \deg f \equiv 1 \bmod p$. In particular, $\deg f \not\equiv 0 \bmod p$.

Proposition 2.6 Let V,W be representations of $\mathbb{Z}_p$, p prime with $\dim V = \dim W$. If $f\colon S(V) \to S(W)$ is equivariant and $\deg f \not\equiv 0 \bmod p$ then $\dim V^{\mathbb{Z}_p} = \dim W^{\mathbb{Z}_p}$ and $\deg f^{\mathbb{Z}_p} \not\equiv 0 \bmod p$. Conversely, if $\dim V^{\mathbb{Z}_p} = \dim W^{\mathbb{Z}_p}$ and $\deg f^{\mathbb{Z}_p} \not\equiv 0 \bmod p$, then $\deg f \not\equiv 0 \bmod p$.

Proof Suppose $\deg f \not\equiv 0 \bmod p$. There are three cases to consider.

Case 1 $\dim V^{\mathbb{Z}_p} < \dim W^{\mathbb{Z}_p}$. Let $C_b\colon S(V) \to S(W)$ denote the constant map to $b \in S(W)^{\mathbb{Z}_p}$. Corollary 2.4 applied to C_b, f yield $\deg f \equiv \deg C_b \bmod p$. But $\deg C_b = 0$, so this case is impossible.

Case 2 $\dim V^{\mathbb{Z}_p} > \dim W^{\mathbb{Z}_p}$. Let V_1 (resp. W_1) be the orthogonal complement of $V^{\mathbb{Z}_p}$ (resp. $W^{\mathbb{Z}_p}$) in V (resp. W).

Since $\dim V_1 \le \dim W_1$, Lemma 2.5 applies and there is an equivariant map $h: S(V_1) \to S(W_1)$.

Now $S(V) = S(V^{\mathbb{Z}_p}) * S(V_1)$, $S(W) = S(W^{\mathbb{Z}_p}) * S(W_1)$, and we may apply Corollary 2.4 to f, $f^{\mathbb{Z}_p} * h$ to get $\deg f \equiv \deg (f^{\mathbb{Z}_p} * h) \bmod p$. But $\deg (f^{\mathbb{Z}_p} * h) = 0$ since h is homotopic to a constant map and hence this case is also impossible.

<u>Case 3</u> $\dim V^{\mathbb{Z}_p} = \dim W^{\mathbb{Z}_p}$. As above, we compare f and $f^{\mathbb{Z}_p} * h$ to get $\deg f \equiv \deg (f^{\mathbb{Z}_p} * h) \equiv \deg f^{\mathbb{Z}_p} \deg h \bmod p$. By Lemma 2.5 $\deg h \not\equiv 0 \bmod p$, hence $\deg f \not\equiv 0 \bmod p$ if and only if $\deg f^{\mathbb{Z}_p} \not\equiv 0 \bmod p$.

<u>Theorem 2.7</u> Let V, W be representations of a finite p-group G and $f: S(V) \to S(W)$ an equivariant map with $\deg f \not\equiv 0 \bmod p$. Then $\dim V^G = \dim W^G$ and $\deg f^G \not\equiv 0 \bmod p$.

<u>Proof</u> Let H be a normal subgroup of G of index p. Since f is H-equivariant and $|H|$, the order of H, satisfies $|H| < |G|$ we may assume inductively that $\dim V^H = \dim W^H$ and $\deg f^H \not\equiv 0 \bmod p$. Now, V^H, W^H are representations of $G/H = \mathbb{Z}_p$ and f^H is a G/H equivariant map so Proposition 2.6 applies to yield $\dim (V^H)^{\mathbb{Z}_p} = \dim V^G = \dim W^G$ and $\deg (f^H)^{\mathbb{Z}_p} = \deg f^G \not\equiv 0 \bmod p$.

<u>Corollary 2.8</u> Let V, W be representations of the compact Lie group G and let $f: S(V) \to S(W)$ be an equivariant map of degree $\not\equiv 0 \bmod p$.

If $H \subset G$ is a finite p-group, then $\dim V^H = \dim W^H$.

<u>Proof</u> Theorem 2.7 applies to $V|H$, $W|H$.

<u>Corollary 2.9</u> If $\chi \in \mathbb{R}O(G)$ and $\nu_G(\chi) = 0 \in JO_k(G)$ where $p \nmid k$, then for any finite p-subgroup $H \subset G$ we have $\sum_{g \in H} \chi(g) = 0$.

<u>Remark</u> Corollary 2.8 or 2.9 provides necessary conditions for stable J_k equivalence. In fact, for finite groups if $2 \nmid k$ these conditions are also sufficient as will be shown in §3. For connected groups, however, there are stronger invariants which will now be presented.

<u>Proposition 2.10</u> Let V be a representation of the n-torus T^n with $V^{T^n} \neq 0$. If $h\colon S(V) \to S(V)$ is equivariant and h^{T^n} is homotopic to $C_b^{T^n}$, the constant map : $S(V)^{T^n} \to S(V)^{T^n}$ to $b \in S(V)^{T^n}$ then h is equivariantly homotopic to $C_b\colon S(V) \to S(V)$.

<u>Proof</u> Perform a preliminary homotopy so that $h(S(V)^{T^n}) = b$. Let H be a maximal isotropy group in $S(V) - S(V)^{T^n}$. Then $h^H\colon S(V)^H \to S(V)^H$ is a T^n/H equivariant map and we may apply Proposition 2.2 with $M = S(V)^H$, $N = * \subset S(V)^{T^n}$, $* \neq b$, $C = S(V)^{T^n}$ to get a T^n/H map $\tilde{h}^H$ with $\tilde{h}^H$ transverse regular to $*$. Since $(\tilde{h}^H)^{-1}(*)$ is a finite invariant subset of $S(V)^H - S(V)^{T^n}$ and T^n/H is connected, we conclude that $(\tilde{h}^H)^{-1}(*) = \phi$.

Hence $\tilde{h}^H$ is homotopic to the constant map to b since $S(V)^H - \{*\}$ is equivariantly contractible to b. Applying the equivariant homotopy extension theorem we may now assume $h(S(V)^H) = b$. Next choose a maximal isotropy group in $S(V) - S(V)^H$ and proceed as above. Since there are only a finite number of isotropy groups in S(V) we eventually get $h(S(V)) = b$.

Corollary 2.11 If V is a representation of T^n and $f_i : S(V) \to S(V)$ i = 1, 2, are equivariant maps with $\deg f_1^{T^n} = \deg f_2^{T^n}$, then $\deg f_1 = \deg f_2$. In particular, if $V^{T^n} = 0$, then $\deg f_1 = \deg f_2 = 1$.

Proof As in Corollary 2.4 we consider $f_i' = f_i * \mathrm{id}: S(V \oplus \mathbb{R}^2) \to S(V \oplus \mathbb{R}^2)$. Since the equivariant homotopy classes of maps from $S(V \oplus \mathbb{R}^2)$ to $S(V \oplus \mathbb{R}^2)$ form a group we let $h = f_1' - f_2'$ and note that $\deg h = \deg f_1' - \deg f_2' = \deg f_1 - \deg f_2$. But deg h = 0 by Proposition 2.10.

Corollary 2.12 Let V be a representation of T^n with $\dim V^{T^n} > 1$ and $f: S(V) \to S(V)$ an equivariant map. If $\deg f^{T^n} = r$, then f is equivariantly homotopic to r times the identity map, r(id) and hence $\deg f = \deg f^{T^n}$.

Proof Let h = f - r(id). Then $\deg h^{T^n} = 0$ and h is equivariantly homotopic to a constant map by Proposition 2.10.

Corollary 2.13 Let V be a representation of T^n and $H \subset T^n$. If $f\colon S(V) \to S(V)$ is equivariant then $\deg f = \deg f^H$.

Proof By Corollary 2.12 $\deg f = \deg f^{T^n}$. Also by Corollary 2.12 applied to the T^n/H space V^H and the map f^H, $\deg f^H = \deg(f^H)^{T^n/H} = \deg f^{T^n}$ since T^n/H is also a toral group. Hence, $\deg f = \deg f^H$.

Corollary 2.14 Let V, W be representations of T^n and $f_1\colon S(V) \to S(W)$, $f_2\colon S(W) \to S(V)$ equivariant maps with $\deg f_1 \not\equiv 0 \bmod p$, $\deg f_2 \not\equiv 0 \bmod p$. Then if $H \subset T^n$, $\dim V^H = \dim W^H$ and $\deg f_1^H \not\equiv 0 \bmod p$, $\deg f_2^H \not\equiv 0 \bmod p$.

Proof Consider $f_2 \circ f_1\colon S(V) \to S(V)$. By Corollary 2.13, $\deg (f_2 \circ f_1)^H = \deg(f_2 \circ f_1) = \deg f_2 \deg f_1 \not\equiv 0 \bmod p$. But $\deg(f_2 \circ f_1)^H = \deg f_2^H \deg f_1^H$; hence $\deg f_1^H \not\equiv 0 \bmod p$, $\deg f_2^H \not\equiv 0 \bmod p$, implying trivially that $\dim V^H = \dim W^H$.

Corollary 2.15 If $\chi \in RO(G)$ and $\nu_G(\chi) = 0 \in JO_k(G)$ for any k then $\int_H \chi = 0$ for every $H \subset T^n \subset G$. In particular, if H is finite and $H \subset T^n \subset G$, then $\sum_{h \in H} \chi(h) = 0$.

Proof This is just a restatement of Corollary 2.14 applied to $\chi|T^n$.

Theorem 2.16 Let G be a compact Lie group, $\chi \in RO(G)$ with

$\nu_G(\chi) = 0 \in JO_k(G)$ where $p \notin k$. If $K \subset G$ and there is a torus $T^n \subset G$ such that $K \cap T^n$ is normal in K and of finite index and $K/K \cap T^n$ is a p group, then $\int_K \chi = 0$.

Proof Let $H = K \cap T^n$ and $\chi = \chi_V - \chi_W$. Since $\nu_G(\chi) = 0$ there are maps $f_1: S(V) \to S(W)$, $f_2: S(W) \to S(V)$ with $\deg f_1 \not\equiv 0 \bmod p$, $\deg f_2 \not\equiv 0 \bmod p$. Now V^H, V^W are representations of $N(H)/H$ and f_1^H, f_2^H are $N(H)/H$ equivariant maps where $N(H)$ denotes the normalizer of H in G. Since $K \subset N(H)$, we may, by restriction, regard V^H, W^H as K/H representations f_1^H, f_2^H as K/H maps. Theorem 2.7 applies since K/H is a p-group and hence $\dim (V^H)^{K/H} = \dim V^K = \dim W^K$.

Corollary 2.17 If G is a compact Lie group and $\chi \in RO(G)$ with $\nu_G(\chi) = 0 \in JO_k(G)$ where $p \notin k$ then $\sum_{g \in A} \chi(g) = 0$ for any finite abelian subgroup $A \subset G$ with $A/A \cap T^n$ a p-group; in particular, for any finite cyclic subgroup A with $A/A \cap G_0$ a p-group.

Remark This special case is all we shall require in the sequel. As an example of the power of Corollary 2.17 we have:

Corollary 2.18 If G is a compact, connected Lie group then for any collection k of primes $\nu_G: RO(G) \xrightarrow{\approx} JO_k(G)$ is an isomorphism.

Proof A special case: $G = S^1 = \mathbb{R}/\mathbb{Z}$. Then $RO(S^1)$ is the free abelian group generated by ρ_n, $n = 1,2,\dots$ where $\rho_0: S^1 \to GL(1, \mathbb{R})$

is given by $\rho_0(t) = (1)$ and $\rho_n: S^1 \to GL(2, \mathbb{R})$ is given by

$$\rho_n(t) = \begin{pmatrix} \cos 2\pi nt & \sin 2\pi nt \\ -\sin 2\pi nt & \cos 2\pi nt \end{pmatrix} \quad \text{for } n > 0$$

Let $\chi \in RO(S^1)$ with $\nu_{S^1}(\chi) = 0 \in JO_k(S^1)$. Then $\chi = \sum_{n=0}^{N} a_n \rho_n$, where $a_n \in \mathbb{Z}$, $N \in \mathbb{Z}^+$; we assume $a_N \neq 0$. Let $\mathbb{Z}_q \subset S^1$ and χ_n the character of ρ_n. Then $\sum_{a \in \mathbb{Z}_q} \chi_n(g) = \begin{cases} q \text{ if } n = 0 \\ 0 \text{ if } n > 0 \text{ and } q \nmid n \\ 2q \text{ if } n > 0 \text{ and } q \mid n \end{cases}$. By Corollary 2.17 $\sum_{g \in \mathbb{Z}_q} \chi\,(g) = 0$ for every $\mathbb{Z}_q \subset S^1$. Taking $q > N$ we get $a_0 = 0$; taking $q = N$ we get $a_N = 0$. Therefore $\chi = 0$.

General case Let $\chi \in RO(G)$ with $\nu_G(\chi) = 0 \in JO_k(G)$. If $S^1 \subset G$, $\chi | S^1 = 0$ by the special case above. Since χ is continuous the corollary will be completed by the following well known lemma.

Lemma 2.19 Let G be a compact Lie group. Elements of finite order are dense in G and circle subgroups are dense in G_0 (the connected component of the identity element e in G). Moreover, $B = \{X \in \tau(G_0)_e \mid X$ generates a closed 1-parameter subgroup$\}$ is dense in the tangent space $\tau(G_0)_e$.

Proof Let $g \in G$ and let $\langle g \rangle$ denote the closed subgroup of G generated by g. Then $\langle g \rangle = T^n \times \mathbb{Z}_m$ for some $n \geq 0$, $m \geq 1$. Now $T^n = \mathbb{R}^n/\mathbb{Z}^n$ and we have $\mathbb{Q}/\mathbb{Z}$ dense in $\mathbb{R}/\mathbb{Z}$ ($\mathbb{Q}$ = {rationals}) and hence $\mathbb{Q}^n/\mathbb{Z}^n \times \mathbb{Z}_m$ is dense in $T^n \times \mathbb{Z}_m$; moreover, every $x \in \mathbb{Q}^n/\mathbb{Z}^n \times \mathbb{Z}_m$ is of finite order, hence elements of finite order are dense in G.

To see that circle subgroups are dense, let U be any open set in G_0. Choose $x \in U$ such that x has finite order; then any 1-parameter subgroup containing x is closed; hence circles are dense in G_0.

To show that B is dense in $\tau(G_0)_e$ note first that B is a cone and hence it is sufficient to show $B \cap U$ is dense in U for some open neighborhood U of the origin in $\tau(G)_e$. But for U sufficiently small, the exponential map restricted to U is open and hence the inverse image of a dense set (elements of finite order) is dense; moreover if exp X is an element of finite order, $X \in B$.

Theorem 2.20 Let G be a compact Lie group and let $x \in G$. Let $C(x) = \{g x g^{-1} | g \in G\}$ be the conjugacy class of x in G and let $Z(x) = \{g \in G | gx = xg\}$ be the centralizer of $x \in G$. Then C(x) and Z(x) are closed submanifolds of G containing x and the tangent space $\tau(G)_x$ splits

$$\tau(G)_x = \tau(C(x))_x \oplus \tau(Z(x))_x .$$

Proof C(x) is just the orbit through x of the Lie group G acting by conjugation on the manifold G and hence is a closed submanifold. Z(x) is a closed subgroup of G and hence a closed submanifold. Clearly $x \in C(x) \cap Z(x)$.

Note that Z(x) is the isotropy subgroup of x under the conjugation action; hence $C(x) \approx G/Z(x)$ and $\dim C(x) = \dim G - \dim Z(x)$. Thus, it will be sufficient to show that $\tau(C(x))_x \cap \tau(Z(x))_x = 0$. To that end we first translate everything to the identity element e. For $y \in G$ let $Ly: G \to G$ denote left translation by y, $Ly(g) = yg$. Then

$$dL_{x^{-1}}(\tau(C(x))_x) = \tau(L_{x^{-1}}C(x))_e \text{ and}$$

$$dL_{x^{-1}}(\tau(Z(x))_x) = \tau(L_{x^{-1}} Z(x))_e = \tau(Z(x))_e.$$

Note that $\tau(Z(x))_e = \{u \in \tau(G)_e \mid ad_x u = u\}$ where $ad_y: \tau(G)_e \to \tau(G)_e$ is the differential of the automorphism $g \to ygy^{-1}$. Also $L_{x^{-1}}C(x) = \{x^{-1} gxg^{-1} \mid g \in G\}$ and $\tau(L_{x^{-1}} C(x))_e$ is the set of tangent vectors to curves $\phi(t)$, at $t = 0$, of the form $\phi(t) = x^{-1} g(t)x\, g(-t)$ where $g(t)$ is a 1-parameter subgroup of G. Let $v \in \tau(G)_e$ be the tangent vector at $t = 0$ to the curve $g(t)$; $v = \frac{d}{dt} g(t)|_{t=0}$. Then $\frac{d}{dt}\phi(t)|_{t=0} = ad_{x^{-1}} v - v$. To see this, recall the formula $\exp tX \exp tY = \exp(t(X+Y) + \frac{t^2}{2} [X, Y] + 0(t^3)$ for any vectors $X, Y \in \tau(G)_e$ (see e.g., Lemma 1.8, p. 96 [7]) from which we may conclude $\frac{d}{dt}(\exp tX \exp tY)|_{t=0} = X + Y$. Letting $X = \frac{d}{dt}(x^{-1}g(t)x)|_{t=0} = ad_{x^{-1}}v$, $Y = \frac{d}{dt} g(-t)|_{t=0} = -v$, we have $\frac{d}{dt}(\exp tX \exp tY)|_{t=0} = \frac{d}{dt} (x^{-1}g(t)x\, g(-t)|_{t=0} = \frac{d}{dt}(\phi(t))|_{t=0} = ad_{x^{-1}}v - v$.

Suppose now that $U \in \tau(Z(x))_e \cap \tau(L_{x^{-1}}C(x))_e$. Then $ad_x U = U$ and $U = ad_{x^{-1}} v - v$ for some $v \in \tau(G)_e$; applying ad_x we get $ad_{x^{-1}} v - v =$

$v - ad_x v$ or $2v = ad_x v + ad_{x^{-1}} v$. Since G is compact, $\tau(G)_e$ admits an invariant inner product and ad_x, $ad_{x^{-1}}$ are orthogonal transformations relative to that inner product. Thus $ad_x v = v$ and $U = 0$ as claimed.

Remark D. Wigner has pointed out that Theorem 2.20 is definitely false if G is not compact. For example, let G be the group of 2 x 2 matrices of the form $\begin{bmatrix} a & b \\ 0 & a^{-1} \end{bmatrix}$, $a \in \mathbb{R}$, $b \in \mathbb{R}$, $a > 0$ and $x = \begin{bmatrix} 1 & 1 \\ 0 & 1 \end{bmatrix}$. Then the centralizer $Z(x)$ consists of all matrices of the form $\begin{bmatrix} 1 & c \\ 0 & 1 \end{bmatrix}$, $c \in \mathbb{R}$ while the conjugacy class of x is the subset consisting of $\begin{bmatrix} 1 & c \\ 0 & 1 \end{bmatrix}$ with $c > 0$.

Theorem 2.21 Let G be a compact Lie group such that G/G_0 is a p-group. If $\chi \in RO(G)$ with $\nu_G(\chi) = 0 \in JO_k(G)$ where $p \nmid k$ then $d\chi \equiv 0$, i.e., χ is constant on the connected components of G.

Proof Suppose $d\chi_g \neq 0$. Since $d\chi\colon \tau(G) \to \mathbb{R}$ is continuous and elements of finite order are dense in G, by Lemma 2.19, we may assume the order of g is finite, $g^n = e$. Since χ is constant on conjugacy classes $d\chi_g | \tau(C(g))_g = 0$. By Theorem 2.20, it is sufficient to show $d\chi_g | \tau(Z(g))_g = 0$. Let $X \in \tau(Z(g))_g$ with $d\chi_g(X) \neq 0$. Again, since tangent vectors $Y \in \tau(Z(g))_e$ which generate closed 1-parameter subgroups are dense in $\tau(Z(g))_e$ by Lemma 2.19 and since $d\chi_g$ is continuous, we may assume $dL_{g^{-1}}(X)$ generates a circle subgroup, S^1, of $Z(g)$. Let A be the subgroup generated by S^1 and g. Then A is isomorphic to $S^1 \times \mathbb{Z}_m$ where $m|n$.

Moreover $\chi \in \tau(A)_g$, hence $d\chi|A \neq 0$. Note that m need not be a power of p and hence we have <u>not</u> reduced the general case to the special case of groups of the form $S^1 \times \mathbb{Z}_{p^r}$. We do claim that for any finite $H \subset A$, $\sum_{h \in H} \chi(h) = 0$. We have $H/H \cap G_0 \subset G/G_0$ is a p-group; we need only show $H \cap G_0$ is contained in a toral subgroup of G to be able to apply Corollary 2.17. But $A \cap G_0 = S^1 \times \mathbb{Z}_r$ for some r, $r|m$, and there is a torus in G_0 containing $S^1 \times \mathbb{Z}_r$ (let $x \in S^1$ with $\langle x \rangle = S^1$ and let y generate $\mathbb{Z}_r$; then the closure of a 1-parameter subgroup containing xy is a torus and contains S^1 and y and hence $A \cap G_0$). Thus $H \cap G_0$ is contained in a torus. The following lemma will then complete the proof.

<u>Lemma 2.22</u> Let $\chi \in RO(S^1 \times \mathbb{Z}_m)$ and suppose $\sum_{h \in H} \chi(h) = 0$ for every finite subgroup $H \subset S^1 \times \mathbb{Z}_m)$. Then $d\chi = 0$.

<u>Proof</u> Let $\chi = \chi_V - \chi_W \in RO(S^1 \times \mathbb{Z}_m)$.

<u>Case 1</u> $V = V^{S^1}$, $W = W^{S^1}$. Then V, W may be regarded as representations $\tilde{V}$, $\tilde{W}$ of $S^1 \times \mathbb{Z}_m/S^1 = \mathbb{Z}_m$ with $V = \pi^*\tilde{V}$, $W = \pi^*\tilde{W}$ where $\pi: S^1 \times \mathbb{Z}_m \to \mathbb{Z}_m$ is the projection. Clearly, χ is constant on components in this case.

Case 2 $V^{S^1} = 0 = W^{S^1}$. Since $S^1 \times \mathbb{Z}_m$ is abelian, every irreducible representation V, is one or two dimensional by Schur's lemma; if S^1 operates nontrivially on U, U must be two dimensional and, if K denotes the kernel of the homomorphism $\rho_U\colon S^1 \times \mathbb{Z}_m \to 0(2)$ associated to U, ρ is determined (up to inner automorphism) by K. Moreover, $\frac{1}{|H|} \sum_{h \in H} \chi_U(h)$ $= \begin{cases} 2 \text{ if } H \subset K \\ 0 \text{ otherwise}\end{cases}$. Now let $V - W = \sum_{i=1}^{N} a_i U_i$ where $a_i \in \mathbb{Z}$, $a_i \neq 0$ and U_i is irreducible. Let K_i be the subgroup of $S^1 \times \mathbb{Z}_n$ which determines U_i. Let s be chosen so that K_s is maximal among $K_1, \ldots, K_N$, i.e., $K_s \subset K_i$ implies i = s. Then $\frac{1}{|K_s|} \sum_{h \in K_s} \chi(h) = \frac{1}{|K_s|} \sum_{i=1}^{N} a_i \left(\sum_{h \in K_s} \chi_{U_i}(h)\right) = 2a_s$ by the maximality of K_s; but by assumption $\sum_{h \in K_s} \chi(h) = 0$. Thus $2a_s = 0$ and $\chi = 0$.

General Case Let $V = V^{S^1} \oplus V/V^{S^1}$, $W = W^{S^1} \oplus W/W^{S^1}$. Then $V - W = (V^{S^1} - W^{S^1}) + (V/V^{S^1} - W/W^{S^1})$. By Case 1, it is sufficient to show $V/V^{S^1} - W/W^{S^1} = 0$ which, by Case 2 requires that $\dim (V/V^{S^1})^K = \dim (W/W^{S^1})^K$ for all K such that $(S^1 \times \mathbb{Z}_m)/K \approx S^1$. But $\dim V^K = \dim W^K$ by assumption; hence we need only show $\dim (V^{S^1})^K = \dim (W^{S^1})^K$ for all K such that $(S^1 \times \mathbb{Z}_m)/K \approx S^1$. But $(V^{S^1})^K = V^{S^1 \times \mathbb{Z}_m}$ for any such K since K meets every component of $S^1 \times \mathbb{Z}_m$ and thus any $v \in V$ fixed by K and by S^1 is fixed by $S^1 \times \mathbb{Z}_m$. Let $\tilde{K} \subset S^1 \times \mathbb{Z}_m$ be such that $(S^1 \times \mathbb{Z}_m)/\tilde{K} \approx S^1$ and $\tilde{K}$ is strictly larger than any K_i, i = 1,...N, where $V/V^{S^1} - W/W^{S^1} = \sum_{i=1}^{N} a_i U_i$ as in Case 2 and K_i is the subgroup associated to U_i. Then $\dim (V/V^{S^1})^{\tilde{K}} = \dim (W/W^{S^1})^{\tilde{K}}$ since $\tilde{K} \not\subset K_i$ for any i and hence $0 = \dim V^{\tilde{K}} - \dim W^{\tilde{K}} = \dim (V^{S^1})^{\tilde{K}} - \dim (W^{S^1})^{\tilde{K}} = \dim V^{S^1 \times \mathbb{Z}_m} - \dim W^{S^1 \times \mathbb{Z}_m}$.

<u>Corollary 2.23</u> Let G be a compact Lie group and let G/G_0 be a p-group. If $\chi \in RO(G)$ and $\nu_G(\chi) = 0 \in JO_k(G)$ where $p \nmid k$ then $\chi = \pi^*\tilde{\chi}$ where $\tilde{\chi} \in RO(G/G_0)$, $\pi: G \to G/G_0$. Moreover, for any cyclic subgroup $H \subset G/G_0$, $\sum_{h \in H} \tilde{\chi}(h) = 0$.

<u>Proof</u> By Theorem 2.2, χ is constant on the connected components of G and hence, is of the form $\tilde{\chi} \circ \pi$ for some class function $\tilde{\chi}: G/G_0 \to \mathbb{R}$. Since $\tilde{\chi}(g^{-1} G_0) = \tilde{\chi}(g^{-1}) = \tilde{\chi}(g) = \tilde{\chi}(gG_0)$, $\tilde{\chi}$ can be expressed uniquely as $\tilde{\chi} = \sum_{i=1}^{N} a_i \chi_i$ where $a_i \in \mathbb{R}$ and $\chi_i \in RO(G/G_0)$ is the character of an irreducible representation of G/G_0. Thus $\chi = \tilde{\chi} \circ \pi = \sum_{i=1}^{N} a_i \pi^*\chi_i$. On the other hand, since $\chi \in RO(G)$, χ can be expressed uniquely as $\sum_{j=1}^{M} b_j \chi_j$ where $b_j \in \mathbb{Z}$, $\chi_j \in RO(G)$ is an irreducible character. Since $\pi^*\tilde{\chi}_i$ is irreducible, we have $a_i \in \mathbb{Z}$ and thus $\tilde{\chi} \in RO(G/G_0)$. To prove the second assertion, let $H \subset G/G_0$, H cyclic ($\approx \mathbb{Z}_{p^r}$ say) and generated by gG_0. Since elements of finite order are dense and π is an open map, we may assume that $g^n = e \in G$, where $n = p^s t$, $t \not\equiv 0 \bmod p$, and $s \geq r$. Let $L \subset G$ be the subgroup generated by g^t. Then $L \approx \mathbb{Z}_{p^s}$ and π maps L onto H. Moreover $\frac{1}{|L|} \sum_{g \in L} \chi(g) = \frac{1}{|H|} \sum_{h \in H} \chi(h)$ since χ is constant on connected

components; but $\sum_{g \in L} \chi(g) = 0$ by Corollary 2.17.

There is one more necessary condition to be discussed. If V is a representation of G, let $\omega_1(V) \in \mathrm{Hom}(G, \mathbb{Z}_2)$ be defined by $\omega_1(V)(g) = (\det \circ \zeta)(g)$ where $\zeta: G \to O(n)$ is the homomorphism associated to V and $\det: O(n) \to O(1) = \mathbb{Z}_2$ is the determinant homomorphism. Alternatively, $\omega_1(V)(g) = +1$ if g preserves orientation on V and $\omega_1(V)(g) = -1$ if g reverses orientation on V. Clearly ω_1 extends to a group homomorphism $\omega_1: RO(G) \to \mathrm{Hom}(G, \mathbb{Z}_2)$. Moreover, ω_1 is onto since any $\rho \in \mathrm{Hom}(G, \mathbb{Z}_2)$ may be regarded as a 1-dimensional representation V with $\omega_1(V) = \rho$.

Proposition 2.24 If V, W are representations of the compact Lie group G and if $f: S(V) \to S(W)$ is an equivariant map with $\deg f \neq 0$, then $\omega_1(V) = \omega_1(W)$.

Proof Since $f(gx) = gf(x)$, $\deg (f \circ g) = \deg (g \circ f)$ and thus $\deg f\, \omega_1(V)(g) = \omega_1(W)(g) \deg f$ since the degree of $g: S(V) \to S(V)$ is $\omega_1(V)(g)$. If $\deg f \neq 0$, $\omega_1(V)(g) = \omega_1(W)(g)$, i.e., g preserves orientation on S(V) if and only if g preserves orientation on S(W).

Proposition 2.25 If V, W are representations of a compact Lie group G such that $\dim V^H = \dim W^H$ for every cyclic subgroup H of order 2^r, $r \geq 0$ then $\omega_1(V) = \omega_1(W)$.

Proof Case 1 $G = \mathbb{Z}_2$, $g \in \mathbb{Z}_2$ the generator. We may express V as $V^{\mathbb{Z}_2} \oplus V/V^{\mathbb{Z}_2}$ where $V^{\mathbb{Z}_2}$ is the +1 eigenspace of g and $V/V^{\mathbb{Z}_2}$ is isomorphic to the -1 eigenspace. Then g preserves orientation on V if and only if $\dim V/V^{\mathbb{Z}_2}$ is even. But $\dim V/V^{\mathbb{Z}_2} = \dim V - \dim V^{\mathbb{Z}_2} = \dim V^e - \dim V^{\mathbb{Z}_2}$. By assumption, $\dim V^e = \dim W^e$, $\dim V^{\mathbb{Z}_2} = \dim W^{\mathbb{Z}_2}$ so $\dim W/W^{\mathbb{Z}_2} = \dim V/V^{\mathbb{Z}_2}$ and g preserves orientation on W if and only if g preserves orientation on V.

Case 2 $G = \mathbb{Z}_{2^r}$, $g \in \mathbb{Z}_{2^r}$ a generator. We express V as $V^{\mathbb{Z}_{2^r}} \oplus V^{\mathbb{Z}_{2^{r-1}}}/V^{\mathbb{Z}_{2^r}} \oplus V^{\mathbb{Z}_{2^{r-2}}}/V^{\mathbb{Z}_{2^{r-1}}} \oplus \ldots V^e/V^{\mathbb{Z}_2}$ Note that g preserves orientation on $V^{\mathbb{Z}_{2^{r-i-1}}}/V^{\mathbb{Z}_{2^{r-1}}}$ if $i \geq 1$ since multiplication by $g^{2^{i-1}}$ defines a complex structure on $V^{\mathbb{Z}_{2^{r-i-1}}}/V^{\mathbb{Z}_{2^{r-1}}}$ and g acts complex linearly with respect to that structure. Hence we need only look at $V^{\mathbb{Z}_{2^r}}$, $V^{\mathbb{Z}_{2^{r-1}}}/V^{\mathbb{Z}_{2^r}}$ the +1, -1 eigenspaces of g. As before, by assumption $\dim V^{\mathbb{Z}_{2^{r-1}}} = \dim W^{\mathbb{Z}_{2^{r-1}}}$, $\dim V^{\mathbb{Z}_{2^r}} = \dim W^{\mathbb{Z}_{2^r}}$ and hence $\dim V^{\mathbb{Z}_{2^{r-1}}}/V^{\mathbb{Z}_{2^r}} = \dim W^{\mathbb{Z}_{2^{r-1}}}/W^{\mathbb{Z}_{2^r}}$ and g preserves orientation on V if and only if g preserves orientation on W.

Case 3 G is a compact Lie group, $g \in G$ is of finite order, $g^s = e \in G$. If $s = 2^r \cdot t$ where t is odd, then g^t preserves orientation on V if and only if g preserves orientation on V.

Hence we may assume $t = 1$, and $\langle g \rangle$, the group generated by g is $\mathbb{Z}_{2^r}$, i.e., we have reduced to Case 2.

Case 4 G is a compact Lie group, $g \in G$. Since the determinant is a continuous function we need only recall that elements of finite order are dense in G (Lemma 2.19) and apply Case 3.

Remark The results in this section on p-groups can also be obtained by Smith theory. The equivariant transversality theorem of [12] can also be used to obtain all the results of this section.

SECTION 3. CONSTRUCTIONS AND SUFFICIENT CONDITIONS

In this section we will show that the conditions which are necessary for the existence of an equivariant map $f\colon S(V) \to S(W)$, (of suitable degree) are sufficient for the existence of a stable map, i.e., are sufficient for the existence of an equivariant map $S(V \oplus U) \to S(W \oplus U)$ for some representation U of G. The construction of Proposition 3.1 reduces the problem of constructing a G equivariant map where G is a compact Lie group to the problem of constructing H equivariant maps where H is a finite p-group. The problem for p-groups was solved in [3] (see also [9]) but a simpler proof is given in the spirit of Section 2. Combining these results yields the Main Theorem 3.20.

Proposition 3.1 Let V, W be representations of a compact Lie group G with $\omega_1(V) = \omega_1(W)$. Suppose there is an H equivariant map $f\colon S(V) \to S(W)$ of degree d where H is a subgroup of finite index $[G; H] = r$ say. Then there is a representation U of G and a G equivariant map $f\colon S(V \oplus U) \to S(W \oplus U)$ of degree rd.

Proof Choose U satisfying i) there is an $x \in S(V \oplus U)$ with $G_x = H$, ii) there is a $y \in S(W \oplus U)$ with $G_y = G$, iii) $S(W \oplus U)^H$ is connected; for example, one could take U to be the induced representation of the trivial one dimensional representation of H, if $H \neq G$, since H has finite index in G.

Let $V' = V \oplus U$, $W' = W \oplus U$ and let $f': S(V') \to S(W')$ be given by $f * id: S(V') = S(V) * S(U) \to S(W')$. Then f' is an H equivariant map of degree d and, since $G_{-x} = H$, $f'(-x) \in S(W')^H$ where $-x$ denotes the antipodal point of x. Since $S(W')^H$ is connected, we can perform a preliminary H equivariant homotopy to insure that $f'(-x) = y$ ([12]).

Let D be a slice at x in $S(V')$, that is, D is an H invariant disk about x such that $gD \cap D = \phi$ for all $g \notin H$. There is an H equivariant homeomorphism $\phi: D/\partial D \to S(V')$ satisfying $\phi(x) = x$, $\phi(\partial D) = -x$. If we regard $f' \circ \phi: D/\partial D \to S(W')$ as a map $\tilde{f}: (D, \partial D) \to (S(W'), \{y\})$, there is a canonical extension of $\tilde{f}$ to a G equivariant map $F': (GD, G(\partial D)) \to (S(W').\{y\})$ given by $F'(gv) = gf(v)$ where $g \in G$, $v \in D$. The desired map $F: S(V') \to S(W')$ is the composition of the quotient map $\pi: S(V') \to S(V')/(S(V') - G\overset{o}{D}) = GD/G(\partial D)$ and $F': GD/G(\partial D) \to S(W')$. We must show that the degree of F is rd. By assumptions, there are orientations $[S(V')] \in H_n(S(V'): \mathbb{Z})$, $[S(W')] \in H_n(S(W'); \mathbb{Z})$ such that $f'_*[S(V')] = d[S(W')]$; we must show that $F_*[S(V')] = rd[S(W')]$. Recall that $F = F' \circ \pi$. Choose coset representatives $g_1H, \ldots, g_rH$; then $GD/G(\partial D) \approx \bigvee_{i=1}^{r} D_i/\partial D_i$ where $\vee$ denotes the one point union of the spheres $D_i/\partial D_i$ and $D_i = g_iD$. Choose orientations $[D_i/\partial D_i] \in H_n(D_i/\partial D_i; \mathbb{Z})$ so that $\pi_*[S(V')]$ $= \sum_{i=1}^{r} [D_i/\partial D_i] \in \sum_{i=1}^{r} H_n(D_i/\partial D_i; \mathbb{Z}) = H_n(GD, G(\partial D); \mathbb{Z}) = H_n(S(V'), S(V') - G\overset{o}{D}; \mathbb{Z})$. By definition, $F' \mid (D_i, \partial D_i) = g_1 \circ \tilde{f} \circ g_i^{-1}$ and hence

$F'_*[D_i/\partial D_i] = (g_i)_*\tilde{f}_*(g_i^{-1})_*[D_i/\partial D_i] = (g_i)_*\tilde{f}_*(\omega_1(V')(g_i^{-1}))[D/\partial D] =$

$(\omega_1(V')(g_i^{-1}))(g_i^{-1})_*d[S(W')] = \omega_1(V')(g_i^{-1})\omega_1(W')(g_i)d[S(W')]$. Therefore

$\deg F = d \sum_{i=1}^{r} \omega_1(v')(g_i^{-1})\omega_1(W')(g_i)$. However, $\omega_1(V')(g_i^{-1}) = \omega_1(W')(g_i^{-1})$

by assumption; hence deg F = rd as claimed.

Definition Denote by $[S(V), S(W)]_G$ the G-equivariant homotopy classes of G equivariant maps $f: S(V) \to S(W)$. $\{S(V), S(W)\}_G$ = direct lim $[S(V \oplus U), S(W \oplus U)]_G$ $(U \in RO^+(G))$ will denote the stable equivariant homotopy classes of equivariant maps. Note that the degree homomorphism deg: $[S(V), S(W)]_G \to \mathbb{Z}$ defines a map deg: $\{S(V), S(W)\}_G \to \mathbb{Z}$.

Corollary 3.2 If V, W are representations of a compact Lie group G with $V|G_0 = W|G_0$ and $\omega_1(V) = \omega_1(W)$ then there is an $f \in \{S(V), S(W)\}_G$ with deg f = $|G/G_0|$.

Proof G_0 has finite index in G.

Corollary 3.3 If G is finite, V is stably J equivalent to W if and only if V is stably weakly J equivalent to W.

Proof Recall that V is J equivalent (resp. weakly J equivalent) to W if there are equivariant maps $f: S(V) \to S(W)$, $h: S(W) \to S(V)$ with degree f = degree h = 1 (resp. (deg f, $|G|$) = (deg h, $|G|$) = 1). Stable J equivalence implies stable weak J equivalence since $(1, |G|) = 1$.

Conversely, if V is stably weakly J equivalent to W, then dimension V = dimension W and $\omega_1(V) = \omega_1(W)$ by Proposition 2.24. Hence, by Corollary 3.2 there is a $\phi \in \{S(V), S(W)\}_G$ with deg $\phi = |G|$. Since $\{S(V), S(W)\}_G$ is a group and deg: $\{S(V), S(W)\}_G \to \mathbb{Z}$ is a homomorphism whose image contains the integer $|G|$ and deg f which is prime to $|G|$ the image of deg contains 1.

Let G be a compact Lie group and $\pi: G \to G/G_0$ the projection. If S_p is a p-Sylow subgroup of G/G_0, then $\pi^{-1}(S_p) = G_p$ will be referred to as a "p-Sylow subgroup" of G.

Theorem 3.4 $JO(G) \to \Pi\, JO(G_p)$, $\tilde{JO}(G) \to \Pi\, \tilde{JO}(G_p)$ are monomorphisms where the product is taken over all primes p such that p divides $|G|$.

Proof Let $V - W \in \tilde{RO}(G)$ with $\nu_{G_p}(V - W) = 0 \in \tilde{JO}(G_p)$. We must show that $\nu_G(V - W) = 0 \in \tilde{JO}(G)$. First observe that $\omega_1(V) = \omega_1(W)$; $\omega_1(V - W) \in \mathrm{Hom}(G, \mathbb{Z}_2)$. Since any $s \in \mathrm{Hom}(G/G_0, \mathbb{Z}_2)$ satisfies $s(g) = 1$ if g has odd order $\mathrm{Hom}(G/G_0, \mathbb{Z}_2)$ injects into $\mathrm{Hom}(G_2/G_0, \mathbb{Z}_2)$. But $\omega_1(V - W)\, G_2 = 1$ by Proposition 2.24 since $\nu_{G_2}(V - W) = 0 \in \tilde{JO}(G_2)$; hence $\omega_1(V) = \omega_1(W)$. Suppose $|G/G^0| = p_1^{r_1}\, p_2^{r_2} \cdots p_t^{r_t}$. Since G_{p_1} has index $p_2^{r_2} \cdots p_t^{r_t}$ in G, we may apply Proposition 3.1 to get a map $f_1 \in \{S(V), S(W)\}_G$ with deg $f_1 = p_2^{r_2} \cdots p_t^{r_t}$. Similarly, we get maps f_i: $\{S(V), S(W)\}_G$ with deg $f_i = |G/G_0|/p_i^{r_i}$. Since (deg $f_1, \ldots,$ deg f_t) = 1 and $\{S(V), S(W)\}_G$ is a group there is an $f \in \{S(V), S(W)\}_G$ with deg $f = 1$. Interchanging V and W we have $\nu_G(V - W) = 0 \in JO(G)$.

Corollary 3.5 $\nu_G(\chi) = 0 \in JO(G)$ if and only if $\nu_{G_p}(\chi) = 0 \in JO(G_p)$ for all p.

Theorem 3.6 $\tilde{J}O_k(G)$ injects into $\Pi \tilde{J}O_k(G_p)$.

Proof By precisely the same arguments as in Theorem 3.4 we get maps $f_i \in \{S(V), S(W)\}_G$ with deg $f_i = u_i|G/G_0|/p_i^{r_i}$ where u_i is a unit $\mathbb{Z}_k$ and hence (deg $f_1,\ldots,$deg f_t) = 1 in $\mathbb{Z}_k$. Therefore there exist $a_i \in \mathbb{Z}_k$ such that $\sum a_i$ deg $f_i = 1$, or equivalently integers b_i such that $\sum b_i$ deg $f_i = u$, u a unit in $\mathbb{Z}_k$. Hence the map $\sum b_i[f_i] \in \{S(V), S(W)\}_G$ has degree a unit in $\mathbb{Z}_k$. Thus there is an $f \in \{S(V), S(W)\}_G$ with deg f a unit in $\mathbb{Z}_k$.

We now investigate the groups $\tilde{J}O_k(G_p)$.

Proposition 3.7 If $p \nmid k$ then $\tilde{J}O_k(G_p) = \tilde{J}O(G_p)$.

Proof For any group H there is a natural epimorphism $\tilde{J}O(H) \to \tilde{J}O_k(H)$ so we need only show $V - W = 0 \in \tilde{J}O_k(G_p)$ implies $V - W = 0 \in \tilde{J}O(G_p)$..

Let $h \in \{S(V), S(W)\}_{G_p}$ with deg h a unit in $\mathbb{Z}_k$. Then $\omega_1(V) = \omega_1(W)$ by Proposition 2.24 and $V|G_0 = W|G_0$ by Corollary 2.18; hence there is an $f \in \{S(V), S(W)\}_{G_p}$ with deg $f = |G_p/G_0| = p^r$ by Corollary 3.2.

Since $p \notin k$, (deg h, deg f) = 1 and hence there is an $\ell \in \{S(V), S(W)\}_{G_p}$ with deg $\ell = 1$.

<u>Proposition 3.8</u> If $p \in k$, $p \neq 2$, then $JO_k(G_p)$ injects into $\widetilde{RO}(G_0)$.

<u>Proof</u> We have the diagram

$$\begin{array}{ccc} \widetilde{RO}(G_p) & \longrightarrow & \widetilde{RO}(G_0) \\ {\scriptstyle \tilde{\nu}_{G_p}} \downarrow & & \downarrow {\scriptstyle \tilde{\nu}_{G_0}} \approx \\ \widetilde{JO}_k(G_p) & \longrightarrow & \widetilde{JO}_k(G_0) \end{array}$$

where ν_{G_0} is an isomorphism by Corollary 2.18. If $V - W \in \widetilde{JO}_k(G_p)$ with $V|G_0 = W|G_0$ (and automatically $\omega_1(V) = \omega_1(W)$ since $p \neq 2$) then there is an $f \in \{S(V), S(W)\}_{G_p}$ with deg $f = |G/G_0|$ by Corollary 3.2 and hence $V - W \simeq 0$ in $JO_k(G_p)$ since $|G_p/G_0|$ is a unit in $\mathbb{Z}_k$.

<u>Proposition 3.9</u> If $2 \in k$ then $\widetilde{JO}_k(G_2)$ injects into $\widetilde{RO}(G_0) \times \mathrm{Hom}(G_2, \mathbb{Z}_2)$ via the map $V \to (V|G_0, \omega_1(V))$.

<u>Proof</u> First note that by Proposition 2.24 the map $V \to \omega_1(V)$ from $RO(G)$ to $\mathrm{Hom}(G, \mathbb{Z}_2)$ factors through $JO_k(G)$. If $V - W \in \widetilde{JO}_k(G_2)$ with $\omega_1(V) = \omega_1(W)$ and $V|G_0 = W|G_0$ Corollary 3.2 applies and we have an $f \in \{S(V), S(W)\}_{G_2}$ with deg $f = |G_2/G_0|$, a unit in $\mathbb{Z}_k$.

<u>Corollary 3.10</u> If $\chi \in RO(G_p)$ and $p \in k$ then $\nu(\chi) = 0$ in

$JO_k(G_p)$ if and only if $\chi | G_0 = 0$ and $\omega_1(\chi) = 0$.

<u>Proof</u> Necessity follows from Corollary 2.18 and Proposition 2.24. Sufficiency follows from Propositions 3.8 and 3.9.

We now investigate the groups $JO(G_p)$. The following theorem for finite p-groups provides a complete determination of the kernel of $\nu_{G_p} : RO(G_p) \to JO(G_p)$.

<u>Theorem 3.11</u> If P is a finite p-group and $\chi \in RO(P)$, then $\nu_p(\chi) = 0$ in JO(P) if and only if $\sum_{h \in H} \chi(h) = 0$ for all cyclic subgroups H of P.

The proof will be postponed until the end of this section. We will first show how Theorem 3.11 applies to non-finite groups.

<u>Corollary 3.12</u> If G is a compact Lie group with G/G_0 a p-group, p prime and $\chi \in RO(G)$, then $\chi \in T(G)$ if and only if $\chi = \pi^*\tilde{\chi}$ where $\tilde{\chi} \in T(G/G_0)$, that is, $\pi^*: JO(G/G_0) \to JO(G)$ is injective.

<u>Proof</u> If $\tilde{\chi} \in T(G/G_0)$, then $\pi^*\chi \in T(G)$ since $\pi^*: JO(G/G_0) \to JO(G)$ is a homomorphism and hence $\pi^*(0) = 0$. Conversely, if $\chi \in T(G)$, then $\chi = \pi^*\tilde{\chi}$ and $\sum_{h \in H} \tilde{\chi}(h) = 0$ for all cyclic $H \subset G/G_0$ by Corollary 2.23 and hence by Theorem 3.11 $\tilde{\chi} \in T(G/G_0)$.

<u>Corollary 3.13</u> JO(G) is a free abelian group for any compact

Lie group G.

Proof Since subgroups of free abelian groups are free abelian, it is sufficient to prove the theorem for G/G_0 a p-group by Theorem 3.4. Since $RO(G) \supset RO(G/G_0)$ as a direct summand (the splitting is given by $V \to V^{G_0}$), $RO(G) = RO(G/G_0) \oplus L(G)$ and $JO(G) = RO(G)/T(G) = (RO(G/G_0)/T(G/G_0)) \oplus L(G) = JO(G/G_0) \oplus L(G)$ by Corollary 3.12, it is sufficient to show that $JO(G/G_0)$ is free abelian. But $JO(G/G_0)$ is finitely generated and torsion free since $\nu_{G/G_0}(\chi) = 0$ if and only if $\sum_{h \in H} \chi(h) = 0$ for all cyclic subgroups H of prime power order by Theorem 3.11.

Corollary 3.14 If $2 \notin k$ then $JO_k(G)$ is a free abelian group. If $2 \in k$ then $JO_k(G)$ is the direct sum of a free abelian group and $Hom(G, \mathbb{Z}_2)$.

Proof If $2 \notin k$ then by Theorem 3.6 it is sufficient to consider the case G/G_0 a p-group. If $p \notin k$ then $JO_k(G) = JO(G)$ by Proposition 3.7 and hence is free abelian by Corollary 3.13. If $p \in k$, $p \neq 2$ then $JO_k(G) \subset RO(G_0)$ by Proposition 3.8 and hence is free abelian. If $2 \in k$ we have $\omega_1\colon JO_k(G) \to Hom(G, \mathbb{Z}_2)$ by Proposition 2.24 and we shall construct a splitting $\alpha\colon Hom(G, \mathbb{Z}_2) \to JO(G)$ as follows: let $\zeta_i\colon G \to \mathbb{Z}_2$, $i = 1, \ldots, r$ be a basis for the vector space $Hom(G, \mathbb{Z}_2)$ over the field $\mathbb{Z}_2$. Let V_i be a 1-dimensional representation with $\omega_1(V_i) = \zeta_i$, i.e., regard

ζ_i as a map: $G \to GL(1, \mathbb{R})$. Define $\alpha(\zeta_i) = V_i - \mathbb{R} \in JO_k(G)$ and extend linearly to $\mathrm{Hom}(G, \mathbb{Z}_2)$. To see that α is well defined, we need only show that $2(V_i - \mathbb{R}) = 0 \in JO_k(G)$. Let $H = \text{kernel } (\zeta_i)$. Then $2V_i|H = \mathbb{R}^2$ and $\omega_1(2V_i) = \omega_1(\mathbb{R}^2)$; hence by Proposition 3.1 there is a $f \in \{S(2V_i)\ S(\mathbb{R}^2)\}_G$ with deg $f = 2$. Therefore $2(V_i - \mathbb{R})$ in 0 in $JO_k(G)$ and is well defined. Clearly $\omega_1 \circ \alpha$ = identity. To complete the proof we must show that kernel (ω_1) is free abelian. By Theorem 3.6 we have that kernel (ω_1) injects into $JO_k(G_p)$ and $JO_k(G_p)$ is free abelian if $p \neq 2$ since either $p \notin k$ and Proposition 3.7, Corollary 3.13 apply or $p \in k$ and Proposition 3.8 applies. If $p = 2$ then Proposition 3.9 applied to kernel (ω_1) shows that kernel (ω_1) injects into $\prod_{p \neq 2} JO_k(G_p) \times RO(G_0)$ and hence is free abelian as claimed.

Remark 1 One can determine $\omega_1(\chi)(g)$ as follows: let N = order of g and let $\beta(g) = \frac{1}{N} \sum_{s=1}^{N} (\chi(g^{2s}) - \chi(g^s))$, then $\omega_1(\chi)(g) = (-1)^{\beta(g)}$.

Remark 2 $T_k(G)$ is finitely generated if G/G_0 is a p-group, p prime by Corollary 3.12. However, for arbitrary G, that is no longer the case. For example, if $\tilde{\chi} \in RO(G/G_0)$ vanishes on elements of prime power order and $\chi \in RO(G)$ is arbitrary then $\pi^*(\tilde{\chi}) \otimes \chi \in T(G)$.

We now prove some elementary propositions needed in the proof of Theorem 3.11. For the remainder of this section all groups considered are finite. Denote by $C(G)$ the algebra over $\mathbb{R}$ of all real valued class functions f satisfying $f(g) = f(g^{-1})$, and define an inner product on $C(G)$ by $\langle f, h\rangle = \frac{1}{|G|} \sum_{g \in G} f(g)\, h(g)$. Let $\psi^s\colon C(G) \to C(G)$ be defined by $\psi^s(f)(g) = f(g^s)$, $s \in \mathbb{Z}$.

Lemma 3.15 The ring homomorphisms ψ^s have the following properties:

a) $\psi^1 = $ identity,

b) $\psi^s \circ \psi^{s'} = \psi^{ss'}$,

c) $\psi^s = \psi^{s'}$ if $s \equiv s' \bmod |G|$,

d) if $i\colon H \to G$ is a homomorphism, then $\psi^s i^* f = i^* \psi^s f$,

e) if $(s, |G|) = 1$ and $f \in C(G)$ then $\sum_{g \in G} f(g) = \sum_{g \in G} \psi^s f(g)$,

f) if $(s, |G|) = 1$, then ψ^s is an orthogonal transformation,

g) if $(s, |G|) = 1$ then $\psi^s\,(RO(G)) \subset RO(G)$,

h) if $(s, |G|) = 1$, ψ^s permutes the irreducible representations of G.

Proof a) through d) are trivial. For e) we need only note that $g \leftrightarrow g^s$ is a set isomorphism of G if $(s, |G|) = 1$. f) is a special case of e). For g) it is shown in ([2], §5) that $\psi^s(V) = P_s(\Lambda^1 V, \Lambda^2 V, \ldots, \Lambda^n V)$ where $P_s(\sigma_1, \ldots, \sigma_n) = x_1^s + \ldots + x_n^s$, σ_i the i-th elementary symmetric function of $x_1, \ldots, x_n$, $n > \dim V$, and $\Lambda^i V$ is the i-th exterior power of V. Alternatively, one may regard the complexifications of V as a repre-

in $GL(n, \mathbb{Q}(\omega))$ where ω is a primitive N-th root of unity, $N = |G|$, and ψ^s is an element of the Galois group of the field $\mathbb{Q}(\omega)$ over $\mathbb{Q}$ defined by $\psi^s(\omega) = \omega^s$. Then, if V is associated to the homomorphism $\zeta: G \to GL(n, \mathbb{Q}(\omega))$, $\psi^s V$ is associated to the homomorphism $\overline{\psi}^s \circ \zeta$ where $\overline{\psi}^s: GL(n, \mathbb{Q}(\omega)) \to GL(n, \mathbb{Q}(\omega))$ is induced by ψ^s. For h) let $[\psi^s]$ denote the matrix of ψ^s with respect to the orthogonal basis for C(G) given by the irreducible characters χ_i. By a), b), c), $[\psi^s]^{-1} = [\psi^{s'}]$ where $ss' \equiv 1 \bmod |G|$, and $[\psi^s]^t_{ij} = \frac{1}{|G|}\sum_{g \varepsilon G} \chi_i(g) \chi_j(g^s) =$

$$\frac{1}{|G|}\sum_{g^{s'} \varepsilon G} \chi_i(g^{s'}) \chi_j(g^{s's}) = \frac{1}{|G|}\sum_{g^{s'} \varepsilon G} \chi_i(g^{s'}) \chi_j(g) =$$

$\frac{1}{|G|}\sum_{g \varepsilon G} \chi_i(g^{s'}) \chi_j(g) = [\psi^{s'}]_{ij}$; therefore, $[\psi^s]$ is an orthogonal matrix and also integral by g). Thus $\psi^s(\chi_i) = \pm \chi_j$ for some j. But $\psi^s(\chi_i)(e) = \chi_i(e^s) = \chi_i(e) = \dim \chi_i > 0$ so $\psi^s(\chi_i) = + \chi_j$ and ψ^s is a permutation matrix.

<u>Definitions</u> Let $\Gamma_G = \{s \varepsilon \mathbb{Z} \mid 1 \leq s \leq |G|, (s, |G|) = 1\}$. Γ_G is the multiplicative group of units in the ring $\mathbb{Z}_N$ where $N = |G|$. Let WO(G) be the subgroup of RO(G) generated by the $V - \psi^s V$, $V \varepsilon RO(G)$, $s \varepsilon \Gamma_G$, and let $RO(G)_\Gamma = RO(G)/WO(G)$.

Note that $RO(G)_\Gamma$ is a free abelian group by Lemma 3.15 h.

Let $WO(G)' = \{V - W \varepsilon RO(G) \mid \dim V^H = \dim W^H \text{ for all } H \subseteq G\}$

and $RO(G)_\Gamma{}' = RO(G)/WO(G)'$.

Lemma 3.16 $V - W \in WO(G)$ if and only if $V|C - W|C \in WO(C)$ for all cyclic subgroups C of G, i.e., $RO(G)_\Gamma$ injects into $\prod_{C \subset G} RO(C)_\Gamma$.

Proof We must first show that the restriction map $j_C^*: RO(G) \to RO(C)$ takes $WO(G)$ into $WO(C)$. Let $V - \psi^s V \in WO(G)$. Then $j_C^*(V - \psi^s V) = j_C^* V - j_C^* \psi^s V = j_C^* V - \psi^s j_C^* V$ by Lemma 3.15 d. If $n = |C|$ then $s = na + s'$ where $1 \leq s' < |C|$ and $(s', |C|) = 1$ and $\psi^s W = \psi^{s'} W$ for all $W \in RO(C)$ by Lemma 3.15 c. Hence $j_C^* V - \psi^s j_C^* V = j_C^* V - \psi^{s'} j_C^* V \in WO(C)$. In fact, we have shown that for $H \subseteq G$ $RO(H)_{\Gamma_G} = RO(H)_{\Gamma_H}$ with the obvious notation. Next define the trace $tr_G: RO(G) \to RO(G)^\Gamma$ where $RO(G)^\Gamma = \{\chi \in RO(G) |\ \psi^s \chi = \chi$ for all $s \in \Gamma_G\}$ by $tr_G(\chi) = \sum_{s \in \Gamma_G} \psi^s \chi$. Note that $tr_G\ WO(G) = 0$; hence we may regard tr_G as a map from $RO(G)_\Gamma$ into $RO(G)^\Gamma$. If $\pi: RO(G) \to RO(G)_\Gamma$ is the quotient map, we have $\pi \circ tr_G =$ multiplication by $|\Gamma_G|$ since $\pi(\psi^s \chi) = \pi(\chi)$; therefore tr_G is a monomorphism since $RO(G)_\Gamma$ is free abelian. The restriction monomorphism $RO(G) \to \prod_{C \subset G} RO(C)$ induces a monomorphism $RO(G)^{\Gamma_G} \to (\prod_{C \subset G} RO(C))^{\Gamma_G} = \prod_{C \subset G} RO(G)^{\Gamma_G} = \prod_{C \subset G} RO(C)^{\Gamma_C}$. Note further that for $\chi \in RO(C)_\Gamma$, $C \subset G$, $tr_G \chi | C = (|\Gamma_G|/|\Gamma_C|)\ tr_C(\chi|C)$. Hence, if $\chi \neq 0$ in $RO(G)_\Gamma$, $tr_G \chi \neq 0$ in $RO(G)^\Gamma$ and for some $C_0 \subset G$ $(tr_G \chi)|C_0 \neq 0$ in $RO(C_0)^\Gamma$. But $(tr_G \chi)|C_0 = (|\Gamma_G|/|\Gamma_C|) tr_{C_0} (\chi | C_0)$; hence $\chi | C_0 \neq 0$ in $RO(C_0)_\Gamma$.

Proposition 3.17 For any finite group G, $RO(G)_\Gamma = RO(G)_\Gamma'$.

Proof By Lemma 3.15 e, $\sum_{h \varepsilon H} (\chi(h) - \psi^s\chi(h)) = 0$; hence $WO(G) \subset WO(G)'$, i.e., we have an epimorphism $RO(G)_\Gamma \to RO(G)_\Gamma'$. By considering the diagram

$$\begin{array}{ccccc} 0 & \longrightarrow & RO(G)_\Gamma & \longrightarrow & \prod_{C \subset G} RO(C)_\Gamma \\ & & \downarrow & & \downarrow \\ & & RO(G)_\Gamma' & \longrightarrow & \prod_{C \subset G} RO(C)_\Gamma' \\ & & \downarrow & & \downarrow \\ & & 0 & & 0 \end{array}$$

we see that we need only prove Proposition 3.17 for cyclic groups. Proceeding by induction on the order of C, let $V - W \varepsilon$ kernel of the epimorphism $RO(C)_\Gamma \to RO(C)_\Gamma'$ with dim V as small as possible. If $C' \subset C$, $C' \neq \{e\}$ then $V \approx V^{C'} \oplus V/V^{C'}$, $W \approx W^{C'} \oplus W/W^{C'}$. If $V^{C'} \neq 0$, we are done since $|C/C'| < |C|$ and hence $V^{C'} - W^{C'} = 0 \varepsilon RO(C/C')_\Gamma$ by induction and hence $V - W = V/V^{C'} - W/W^{C'}$ with $\dim V/V^{C'} < \dim V$. Hence, we may assume $V^{C'} = 0$ for all $C' \subset C$, $C' \neq \{e\}$.

Case 1 $C = \mathbb{Z}_2$, $g \varepsilon \mathbb{Z}_2$ the generator. We must have $gv = \pm v$ for all $v \varepsilon V$ but since $V^{\mathbb{Z}_2} = 0$ $gv = -v$. Since $W^{\mathbb{Z}_2} = 0$ and $\dim W^e = \dim V^e$, we have $V \sim W$ in $RO(\mathbb{Z}_2)_\Gamma$.

Case 2 $C = \mathbb{Z}_n$. $n > 2$, $g \varepsilon \mathbb{Z}_n$ a generator. The irreducible representations ρ_j of $\mathbb{Z}_n$ without fixed points outside 0 are all two dimensional and given by

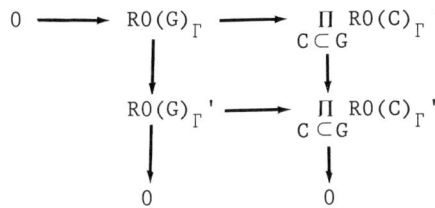

$$\rho_j(g^r) = \begin{pmatrix} \cos\frac{2jr\pi}{n} & \sin\frac{2jr\pi}{n} \\ -\sin\frac{2jr\pi}{n} & \cos\frac{2jr\pi}{n} \end{pmatrix}$$

where $(j, n) = 1$. Given j_1, j_2 with $(j_1, n) = (j_2, n) = 1$, choose $1 \leq s < n$ such that $sj_1 \equiv j_2 \bmod n$. Then $\psi^s\rho_{j_1} = \rho_{j_2}$, i.e., any two irreducible representations of a cyclic group wihtout fixed points outside 0 are Γ-equivalent and hence $V - W \sim 0$ in $RO(C)_\Gamma$.

Proof of Theorem 3.11 If G is a finite p-group then $\pi: RO(G) \to RO(G)_\Gamma$ factors through $JO(G)$ by Corollary 2.8; i.e., $T(G) \subseteq WO(G)' = WO(G)$. We need only show that $WO(G) \subseteq T(G)$ to complete the proof. In fact, for any finite group G, $\nu_G: RO(G) \to JO(G)$ factors through $RO(G)_\Gamma$. Equivalently, we must show that for any faithful irreducible representation V of G and any s with $(s, |G|) = 1$, V is stably J-equivalent to $\psi^s V$. (If V is not faithful, there is a normal subgroup $H \subset G$ with V a faithful irreducible representation of G/H).

Case 1 dim V = 1. Then $G = \mathbb{Z}_2$ and $\Gamma_G = 1$, and $\psi^1 V = V$.

Case 2 dim V = 2, s odd. Let $\zeta: G \to O(2)$ be the homomorphism associated to V, $V = \zeta^* V_2$ where V_2 is the standard representation of $O(2)$ on $\mathbb{R}^2$. Regard V_2 as the complex numbers: if $x \in O(2)$ and $z \in V_2$, then $x(z) = \lambda z$ or $\lambda\bar{z}$ for some $\lambda \in C$, $|\lambda| = 1$. Let $c = \begin{pmatrix} 1 & 0 \\ 0 & -1 \end{pmatrix} \in O(2)$; $c(z) = \bar{z}$. Then every element in $O(2)$ is of the form λ or λc. Define $h_s: O(2)$

$\to O(2)$ by $h_s(\lambda) = \lambda^s$, $h_s(\lambda c) = \lambda^s c$. h_s is easily seen to be a homomorphism. Let $f_s\colon S(V_2) \to S(h_s^* V_2)$ be given by $f_s(z) = z^s$. Then $f_s(\lambda z) = \lambda^s z^s = h_s(\lambda)f(z)$ and $f_s(cz) = \bar{z}^s = h_s(c)f_s(z)$ so f_s is $O(2)$-equivariant and $\deg f_s = s$. Let χ (resp. χ') be the real character of V_2 (resp. $h_s^* V_2$). Then $\chi'(\lambda) = \chi(h_s(\lambda)) = \chi(\lambda^s) = \psi^s\chi(\lambda)$ and $\chi'(\lambda c) = \chi(h_s(\lambda c)) = \chi(\lambda^s c) = 0$ by direct computation. But $\psi^s\chi(\lambda c) = \chi((\lambda c^s) = \chi(\lambda c) = 0$ if s is odd since $(\lambda c)^2 = 1 \in O(2)$. Hence the map $f_s\colon S(V_2) \to S(h_s^* V_2) = S(\psi^s V_2)$ may also be regarded as a G equivariant map $S(\zeta^* V_2) = S(V) \to S(\zeta^* \psi^s V_2) = S(\psi^s V)$. Choosing $s' \in \Gamma_G$ with $ss' \equiv 1 \bmod |G|$, we have a G equivariant map $f_{s'}\colon S(\psi^s V) \to S(\psi^{s'} \psi^s V) = S(V)$; hence V is stably weakly J equivalent to $\psi^s V$ and hence stably J equivalent by Corollary 3.3.

<u>Case 3</u> $\dim V = 2$, s even. Then $|G|$ is odd and $\zeta\colon G \to O(2)$ actually maps into $SO(2)$. Let $h_s'\colon SO(2) \to SO(2)$. Then $h_s'^* V_2 = \psi^s V_2$ and the argument of Case 2 may be repeated for this case.

<u>Case 4</u> $V = i_*^H(W)$ the induced representation of a representation W of the subgroup H, and $\dim W \leq 2$. If $(s, |G|) = 1$, then $\psi^s i_*^H(W) = i_*^H(\psi^s W)$ as is easily checked, and there are H-equivariant maps $f_s\colon S(W) \to S(\psi^s W)$, $f_{s'}\colon S(\psi^s W) \to S(W)$ by Cases 1, 2 or 3. In [3], it is shown that there is a canonical "extension" of f_s, $f_{s'}$ to G-equivariant maps F_s: $S(i_*^H W) \to S(i_*^H \psi^s W) = S(\psi^s i_*^H W)$, $F_{s'}\colon S(\psi^s V) \to S(V)$ with $\deg F_s = s^d$,

$\deg F_{s'} = s'^{d}$ where $d = [G; H]$. Note that $(s^d, |G|) = 1 = (s'^d; |G|)$ and hence by Corollary 3.3 V is stably J equivalent to $\psi^s V$ in this case.

Case 5 V arbitrary. By Brauer's Induction Theorem ([5]) any $V \in RO(G)$ may be expressed as an integral linear combination of induced representations $i_*^H(W_H)$ of representations W_H of subgroups H with $\dim W_H \leq 2$. Hence $V - \psi^s V$ is an integral linear combination of terms of the form $i_*^H(W_H) - \psi^s\, i_*^H(W_H)$ which are in $\ker(\nu_G)$ by Case 4; hence $V - \psi^s V \in \ker(\nu_G)$, i.e., V is stably J equivalent to $\psi^s V$.

Remark For any finite group G we have the diagram

$$\begin{array}{ccc} RO(G) & & \\ \downarrow \pi & \searrow \nu_G & \\ RO(G)_\Gamma & \longrightarrow & JO(G) \end{array}$$

and we have $\pi(\chi) = 0$ if and only if $\sum_{h \in H} \chi(h) = 0$ for all cyclic subgroups H of G; $\nu_G(\chi) = 0 \in JO(G)$ if and only if $\sum_{h \in H} \chi(h) = 0$ for all cyclic subgroups H of G of prime power order.

Remark 2 In the above proof, we proved that $WO(G) = T(G)$ for a p-group G. In other words, we proved the following theorem of Atiyah and Tall (Theorem 2.7, V.[3]) which they proved for a p-group G, p odd prime.

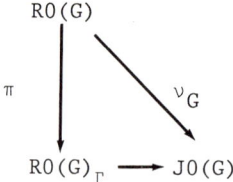

<u>Theorem 3.18</u> (Atiyah and Tall) For a p-group G, p prime, ν_G: $RO(G)_\Gamma \to JO(G)$ is an isomorphism.

The main theorem of Atiyah-Tall in [3] is the complex analog of this theorem which was extended by Snaith [9] to include 2-groups. We shall prove this as a corollary of Theorem 3.18. First some notation. Let $R(G)$, $R(G)_\Gamma$, $J(G)$, $\bar{\nu}_G$ etc. denote the complex analogs of $RO(G)$, $RO(G)_\Gamma$, $JO(G)$, ν_G etc. respectively for any finite group G. In the notation of Section 1, $R(G) = RF(G)$, $J(G) = JF(G)$ where $F_{2n} = F_{2n+1} = GL(n, C)$ (Example 1 following Proposition 1.2). There is the communtative diagram:

$$\begin{array}{ccc} R(G)_\Gamma & \xrightarrow{r'} & RO(G)_\Gamma \\ \downarrow{\scriptstyle \bar{\nu}_G} & & \downarrow{\scriptstyle \nu_G} \\ J(G) & \xrightarrow{r''} & JO(G) \end{array}$$

where r', r'' are induced by r'' $R(G) \to RO(G)$ restricting the ground field to the reals.

<u>Corollary 3.19</u> (Atiyah and Tall [3], Snaith [9]) For a p-group G, p prime, $\bar{\nu}_G$: $R(G)_\Gamma \to J(G)$ is an isomorphism.

<u>Proof</u> By Theorem 3.18, ν_G is an isomorphism. We need only prove that ν_G is injective since it is already onto by definition. By the commutativity of the diagram, it will be sufficient to prove that r' is injective. Denote by c: $RO(G) \to R(G)$ the complexification and by

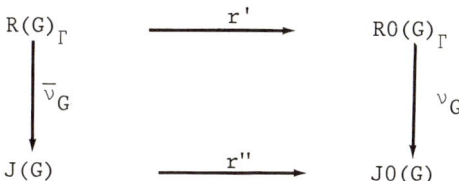

$c'\colon RO(G)_\Gamma \to R(G)_\Gamma$ its induced homomorphism. Then $(c \circ r)(\chi) = \chi + \overline{\chi}$ where $\overline{\chi}$ is the conjugate of χ, and hence $(c' \circ r')\,[\chi] = 2[\chi]$ since χ is stably J equivalent to $\overline{\chi} = \psi^{-1}(\chi)$. Thus if $r'\,[\chi] = 0$ in $RO(G)_\Gamma$ then $(c' \circ r')\,[\chi] = 2[\chi] = 0$ in $R(G)_\Gamma$ and hence $[\chi] = 0$ in $R(G)_\Gamma$ since $R(G)_\Gamma$ is free abelian just as $RO(G)_\Gamma$ is. This proves that r' is injective as desired.

Now we come to the main theorem of the entire paper.

MAIN THEOREM 3.20 Let V, W be representations of a compact Lie group G, and k a set of prime numbers. Then V is stably J_k equivalent to W if and only if the following conditions are satisfied:

1) $\omega_1(V) = \omega_1(W)$

2) $\dim V^H = \dim W^H$ for all finite cyclic subgroups $H \subset G$ such that the quotient group $H/H \cap G_0$ is a p-group, p prime and $p \notin k$.

Proof The necessity follows from Theorem 2.16 and Proposition 2.24.

To prove the sufficiency, we need only consider the case when G/G_0 is a p-group since $JO_k(G)$ injects into $\Pi\, JO_k(G_p)$ by Theorem 3.6. Thus we assume that G/G_0 is a p-group, p prime. Let V, W be representations satisfying 1) and 2). We want to show that V is stably J_k equivalent to W.

If $p \in k$ then Condition 2) implies $V|G_0 = W|G_0$, and hence this fact together with Condition 2) implies that V is stably J_k equivalent to W by Corollary 3.10.

Hence assume that $p \notin k$. In this case, $JO_k(G) \approx JO(G)$ by Proposition 3.7. Condition 2) implies Condition 2') $\dim V^H = \dim W^H$ for all finite abelian subgroups $H \subset G$. If above H is cyclic then 2') certainly holds, since $H/H \cap G_0 \subset G/G_0$ a p-group. Thus 2') holds for all cyclic subgroups of H. But $RO(H)_\Gamma' \to \prod_{C \subset H} RO(C)_\Gamma'$ is injective by Lemma 3.16 and Proposition 3.17 where C denotes a cyclic subgroup of H. This means $\dim V^H = \dim W^H$ as claimed. Notice that we proved in Theorem 2.21 and Corollary 2.23 that Condition 2') implies that $\chi_{V-W} = \pi^*\tilde{\chi}$, where $\tilde{\chi} \in RO(G/G_0)$ and moreover for any cyclic subgroup $H \subset G/G_0$, $\sum_{h \in H} \tilde{\chi}(h) = 0$. By Theorem 3.11, $\tilde{\chi} \in T(G/G_0)$ and hence $\pi^*\tilde{\chi} \in T(G)$ and therefore $\nu(\chi) = 0 \in JO(G)$.

SECTION 4. COMPUTATIONS, CONJECTURES AND COUNTER-EXAMPLES

In this section we return to the original question: given representations V, W of G, when does there exist an equivariant map of degree k? First we show that if G is finite and there is a G equivariant map $f: S(V) \to S(W)$ of degree 1 (resp. of degree a unit in Z_k), then V is stably J equivalent (resp. J_k equivalent) to W, i.e., there is a G equivariant map $h: S(W) \oplus U) \to S(V \oplus U)$ of degree 1 (resp. of degree a unit in Z_k) for some representation U of G. For arbitrary compact Lie groups, an example due to Petrie shows that one must add the condition $V|G_0 = W|G_0$.

Next, the problem of constructing a stable map of degree k is reduced to a problem for groups G such that G/G_0 is a p-group, namely, there exists a stable G map of deg k from S(V) to S(W) if and only if there exist stable G_p maps of deg k from S(V) to S(W) for every prime p (Corollary 4.3). Unfortunately, only partial results have been obtained for p-groups.

Next, it is shown that for groups such that G/G_0 is a p-group, stable J equivalence implies weak J equivalence. It is shown by example that this result is false for arbitrary groups.

Finally, the rank of $\tilde{J}O(G)$ for G an abelian group is shown to be given by the preposterous formula

$$\operatorname{rank}(\tilde{J}O(G)) = \sum_p \left(-\frac{1}{p-1} + \sum_{r=1}^{\infty} p^{-r}|\operatorname{Hom}(G, \mathbb{Z}_{p^r})|\right).$$

Theorem 4.1 Let V, W be representations of a compact Lie group G. Then V is stably J_k equivalent to W if and only if $V|G_0 = W|G_0$ and there exists an $f \in \{S(V), S(W)\}_G$ with deg f a unit in $\mathbb{Z}_k$. In particular, if G is finite and $f \in \{S(V), S(W)\}_G$ with deg f a unit in $\mathbb{Z}_k$, then V is stably J_k equivalent to W.

Proof If $\nu_G(V - W) = 0$ in $JO_k(G)$, then $V|G_0 = W|G_0$ by Corollary 2.18 and there exist stable maps in both directions by definition, so the necessity is clear.

Suppose now that $V|G_0 = W|G_0$ and there is an $f \in \{S(V), S(W)\}_G$ with deg f a unit in $\mathbb{Z}_k$; we must show $\nu_G(V - W) = 0$ in $JO_k(G)$. Since $\omega_1(V) = \omega_1(W)$ by Proposition 2.24, it is sufficient by Theorem 3.20 to show that $\dim V^C = \dim W^C$ for all cyclic subgroups $C \subset G$ with $C/C \cap G_0$ a p-group, $p \nmid k$. Let $C_0 = C \cap G_0$ and $f^{C_0} = f|S(V)^{C_0}$. Now $\dim V^{C_0} = \dim W^{C_0}$ since $V|G_0 = W|G_0$, and since $C_0 \subset T \subset G_0$, $\deg f^{C_0} = \deg f \not\equiv 0 \bmod p$ by Corollary 2.13. Since V^{C_0}, W^{C_0} are representations of the p-group C/C_0, Theorem 2.7 applies and $\dim(V^{C_0})^{C/C_0} = \dim V^C = \dim W^C$.

Remark Theorem 4.1 provides a necessary condition for the existence of a stable equivariant map $f: S(V) \to S(W)$ when G is finite; i.e., there is an f with deg f = 1 if and only if $\nu_G(V - W) = 0 \in JO(G)$.

The hypothesis that $V|G_0 = W|G_0$ cannot be omitted in Theorem 4.1 as the following example shows.

Example (Petrie [8]) Let $G = S^1$ and let $\rho_n : S^1 \to U(1)$ be given by $\rho_n(\lambda) = e^{2\pi i n\lambda}$ and let V_n be the associated representation space. Choose positive integers p, q, a, b so that $-ap + bq = 1$ and define

$f: S(V_p \oplus V_q) \to S(V_1 \oplus V_{pq})$ by

$$f(z_0, z_1) = \frac{(\bar{z}_0^a z_1^b, z_0^q + z_1^p)}{||(\bar{z}_0^a z_1^b, z_0^q + z_1^p)||} .$$

It is easily seen that f is equivariant and deg f = 1 by looking at the regular value (0, 1). However, for any stable equivariant map $h \in \{S(V_1 \oplus V_{pq}), S(V_p \oplus V_q)\}_{S^1}$ we have deg h = 0 by Corollary 2.18. Note that $(V_p \oplus V_q)|H$ is stably J equivalent to $(V_1 \oplus V_{pq})|H$ for every proper subgroup $H \subset S^1$.

Consider now the problem of finding an $f \in \{S(V), S(W)\}_G$ with deg f = k. Since the image of the degree homomorphism deg: $\{S(V), S(W)\}_G \to \mathbb{Z}$ is a principal ideal generated by a unique integer $\delta \in \mathbb{Z}^+ = \{n \in \mathbb{Z} | n \geq 0\}$, there exists an f with deg f = k if and only if $\delta | k$ (δ divides k). Hence, the problem is to determine the integer δ. We may regard $\delta = \delta(V, W) = \delta(V - W)$ as a function $\delta : \widetilde{RO}(G) \to \mathbb{Z}^+$. Our present information on the function δ is contained in the following.

Theorem 4.2 Let G be a compact Lie group and $\chi, \chi' \in RO(G)$. Then

a) $\delta(\chi + \chi') | \delta(\chi) \delta(\chi')$

b) if $i: H \to G$ is a homomorphism, then $\delta(i^*\chi) \mid \delta(\chi)$

c) $\delta(\chi)\ \delta(-\chi) \neq 0$ if and only if $\chi \mid G_0 = 0$, $\omega_1(\chi) = 0$.

If G is finite, then

d) $\delta(\chi) > 0$ if and only if $\omega_1(\chi) = 0$

e) $\delta(\chi) = 1$ if and only if $\nu_G(\chi) = 0 \in JO(G)$

f) $\delta(\chi)$ is a unit in Z_k if and only if $\nu_G(\chi) = 0 \in JO_k(G)$

g) $\delta(\chi) \mid |G|$.

Moreover, for arbitrary G we have

h) $\delta(\chi) \mid \delta(\chi \mid G_0) |G/G_0|$

i) $\delta(\chi) = \text{l.c.m.}\ \{\delta(\chi \mid G_{p_i})\}$ where G_p denotes a "p-Sylow subgroup of G.

<u>Proof</u> a) If $\chi = V - W$, $\chi' = W - U$ and $f \in \{S(V), S(W)\}_G$, $f' \in \{S(W), S(U)\}_G$ then $f' \circ f \in \{S(V), S(U)\}_G$ so deg $(f' \circ f)$ is divisible by $\delta(\chi + \chi')$. Choosing f, f' with deg $f = \delta(\chi)$, deg $f' = \delta(\chi')$ yields the desired result. For b), if $f \in \{S(V), S(W)\}_G$ we may also regard f as being in $\{S(i^*V), S(i^*W)\}_H$ and choosing f with deg $f = \delta(\chi)$ yields the claim. For c), if $\delta(\chi) \neq 0$, then $\omega_1(\chi) = 0$ by Proposition 2.24. If $\delta(-\chi) \neq 0$ also, then Corollary 2.18 says $\chi \mid G_0 = 0$. Conversely, if $\chi \mid G_0 = 0$, $\omega_1(\chi) = 0$, then Corollary 3.2 says $\delta(\chi) \neq 0$, $\delta(-\chi) \neq 0$. d) is a special case of c). e) and f) are restatements of Theorem 4.1 g) and h) follow from Corollary 3.2. For i), $\delta(\chi \mid G_{p_i}) \mid \delta(\chi)$ for each i by b), so l.c.m.

$\{\delta(\chi|G_{p_i})\}|\delta(\chi)$. On the other hand, by Proposition 3.1 there is an $f_i \in \{S(V), S(W)\}_G$ with $\deg f_i = \delta(\chi|G_{p_i})[G; G_{p_i}]$ and hence an $f \in \{S(V),S(W)\}_G$ with $\deg f = \text{g.c.d.} \{\delta(\chi|G_{p_i})[G; G_{p_i}]\}$. Since $\delta(\chi|G_0)\delta(\chi|G_{p_i})$ by b), $\text{g.c.d.} \{\delta(\chi|G_{p_i})[G; G_{p_i}]\} = \delta(\chi|G_0)\ \text{g.c.d.} [\frac{\delta(\chi|G_{p_i})}{\delta(\chi|G_0)}[G: G_{p_i}]\}$. But

$\frac{\delta(\chi|G_{p_i})}{\delta(\chi|G_0)} \Big| \left|\frac{G_{p_i}}{G_0}\right|$ by h) and hence $\text{g.c.d.} \{\frac{\delta(\chi|G_{p_i})}{\delta(\chi|G_0)}[G: G_{p_i}]\} = \Pi \frac{\delta(\chi|G_{p_i})}{\delta(\chi|G_0)}$

Thus $\deg f = \delta(\chi|G_0)\ \Pi \frac{\delta(\chi|G_{p_i})}{\delta(\chi|G_0)} = \text{l.c.m.} \{\delta(\chi|G_{p_i})\}$. Hence $\delta(\chi)|\text{l.c.m.}\{\delta(\chi|G_{p_i})\}$.

Corollary 4.3 If V, W are representations of a compact Lie group G, then there exists a stable G equivariant map: $S(V) \to S(W)$ of degree k if and only if there exists a stable G_p-equivariant map: $S(V) \to S(W)$ of degree k for every prime p dividing the order G/G_0.

Remark Corollary 4.3 reduces the problem of computing δ for an arbitrary finite group to a problem for p-groups. The techniques of Section 2 yield inequalities involving $\sum_{h \in H} \chi(h)$ and therefore necessary conditions for $\delta(\chi) \leq p^r$ say. It is not known whether these conditions are sufficient.

Conjecture If G is a p-group, $\chi \in \widetilde{RO}(G)$, $\omega_1(\chi) = 0$ and $\delta(\chi)|p^r$, $\delta(-\chi)|\ p^r$ then there is a subgroup $H \subset G$ with $[G:H]=p^r$ and $\nu_H(\chi|H) = 0 \in JO(H)$.

We now consider the problem of unstable maps.

Theorem 4.4 (Atiyah and Tall [3]) If G is a p-group, p prime and V, W are stably J equivalent representations, then V is weakly J equivalent to W.

Proof If $\nu_G(V - W) = 0 \in JO(G)$, then $V - W \sim 0$ in $RO(G)_\Gamma$ by Theorem 3.18; hence, we may write $V = \bigoplus_{i=1}^{n} V_i$, $W = \bigoplus_{i=1}^{n} W_i$ where V_i, W_i are irreducible real representations and $W_i = \psi^s(V_i)$ for some $s \in \Gamma_G$. It is sufficient to show that V_i is weakly J equivalent to $\psi^s(V_i)$ since $S(V) = S(V_1)*\ldots*S(V_n)$, $S(W) = S(W_1)*\ldots*S(W_n)$. In the proof of Theorem 3.11 Case 4, we showed that V_i is weakly J equivalent to $\psi^s(V_i)$ if $V_i = i_*^H U$ where $\dim U \leq 2$, hence we need only show that every irreducible real representation of a p-group is induced from a one or two dimensional representation of a subgroup. If $p \neq 2$, then every irreducible real representation of G admits a complex structure, and it is well known (52.1, [5]) that every irreducible complex representation of a p-group is induced from a one (complex) dimensional representation of a subgroup. The following lemma takes care of the case $p = 2$. The proof was provided by J. McLaughlin.

Lemma 4.5 If V is an irreducible real representation of a 2-group G, then V is induced from a one or two dimensional representation of a subgroup.

Proof Let G be a 2-group of the smallest order for which the lemma is false and let V be a representation which is not induced from any subgroup. Then V is a faithful representation since G is minimal. If H is a maximal normal subgroup of G, then $V|H$ is also irreducible, for, if $V|H = V_1 \oplus V_2$ where V_1 is irreducible, we would have $i_*^H V_1 = V$, and by induction, $V_1 = j_*^k W$ where $j: K \subset H$, hence $V = i_*^H V_1 = i_*^H j_*^k W = i_*^k W$.

Case 1 G has a maximal subgroup H which is cyclic. Then dim V ≤ 2 since $V|H$ is irreducible and we may regard V as $i_*^G V$.

Case 2 G has an abelian normal subgroup A which is not cyclic. Let $A_2 = \{g \in A | g^2 = e\}$. Then A_2 is also a normal subgroup of G; in fact, a vector space over $\mathbb{Z}_2$ of dimension > 1. Since A_2 has an odd number of 2-dimensional subspaces, there is an invariant 2-dimensional subspace; i.e., there is a normal subgroup $\mathbb{Z}_2 \times \mathbb{Z}_2 \subset G$. Let H be the centralizer of $\mathbb{Z}_2 \times \mathbb{Z}_2$ in G. Then H has index ≤ 2 in G since Aut $(\mathbb{Z}_2 \times \mathbb{Z}_2)$ (= the symmetric group S_3) has $\mathbb{Z}_2$ as 2-Sylow subgroup. Since the center of H is not cyclic, H does not admit a faithful irreducible representation; hence $V|H$ is not irreducible.

Case 3 G arbitrary 2-group. By Theorem 5.4.10, p. 199 [6] every 2-group falls into Case 1 or 2 above.

Remark It is not true that stably J equivalent representations of a p-group are J equivalent. For example, let $G = \mathbb{Z}_5$ and let $\rho_n: \mathbb{Z}_5 \to U(1) \subset GL(2, R)$ be given by $\rho_n(g^r) = (e^{\frac{2\pi nri}{5}})$ and let V_n be the associated representation space. Then V_1, V_2 are stably J equivalent since $\dim V_1^{\mathbb{Z}_5} = \dim V_2^{\mathbb{Z}_5} = 0$ and $\dim V_1^e = \dim V_2^e = 2$. If $f: S(V_1) \to S(V_2)$ is given by $f(z) = z^2$, then deg f = 2 and f is $\mathbb{Z}_5$ equivariant. On the other hand, if $h: S(V_1) \to S(V_2)$ is any other $\mathbb{Z}_5$ equivariant map, then deg h $\equiv$ deg f mod 5 by Corollary 2.4.

Corollary 4.6 If G is a compact Lie group with G/G_0 a p-group, then V, W are stably J equivalent if and only if V, W are weakly J equivalent, i.e., there are maps $f: S(V) \to S(W)$, $h: S(W) \to S(V)$ with (deg f, p) = (deg h, p) = 1.

Proof By Corollary 3.12, $V = U \oplus \pi^*V'$, $W = U \oplus \pi^*W'$ where V', W' are stably J equivalent representations of G/G_0. By Theorem 4.4, V', W' are weakly J equivalent and hence π^*V', π^*W' are weakly J equivalent and hence V, W are weakly J equivalent.

The following example shows that the hypothesis, G is a p-group (resp. G/G_0 is a p-group) cannot be removed in Theorem 4.4 (resp. Corollary 4.6).

Example Let $G = \mathbb{Z}_{30}$ and $\rho_n: G \to U(1) \subset GL(2, \mathbb{R})$ defined by

$\rho_n(g^r) = (e^{\frac{2\pi nri}{30}})$ and let V_n be the associated representation space. Then $V = V_5 \oplus V_6$ is stably J equivalent to $W = V_3 \oplus V_{10}$ since $\dim V^e = \dim W^e = 4$, $\dim V^{\mathbb{Z}_5} = \dim W^{\mathbb{Z}_5} = 2$, $\dim V^{\mathbb{Z}_3} = \dim W^{\mathbb{Z}_3} = 2$, $\dim V^{\mathbb{Z}_2} = \dim W^{\mathbb{Z}_2} = 2$ and these are th only subgroups of G of prime power order. However, there is no G equivariant map: $S(V) \to S(W)$ since $S(V)^{\mathbb{Z}_6} = \phi$ but $S(W)^{\mathbb{Z}_6} = \phi$ and no G equivariant map: $S(W) \to S(V)$ since $S(W)^{\mathbb{Z}_{10}} \neq \phi$ but $S(V)^{\mathbb{Z}_{10}} = \phi$.

We now turn to the computation of the group $\tilde{J}O(G)$ when G is a finite abelian group.

<u>Proposition 4.7</u> If $G = H \times K$ where H and K are finite groups with $(|H|, |K|)$ a unit in $\mathbb{Z}_k$ then $\tilde{J}O_k(G) = \tilde{J}O_k(H) \times \tilde{J}O_k(K)$.

<u>Proof</u> Consider the diagram

$$\begin{array}{ccc} \tilde{R}O(G) & \xrightarrow{\;i_H^* \times i_K^*\;} & \tilde{R}O(H) \times \tilde{R}O(K) \\ {\scriptstyle \tilde{\nu}_G}\downarrow & & \downarrow{\scriptstyle \tilde{\nu}_H \times \tilde{\nu}_K} \\ \tilde{J}O_k(G) & \xrightarrow{\;i_H^* \times i_K^*\;} & \tilde{J}O_k(H) \times \tilde{J}O_k(K) \end{array}$$

where $i_H\colon H \to G$, $i_K\colon K \to G$ are inclusion maps. Let $\pi_H\colon G \to H$, $\pi_K\colon G \to K$ be the projection maps. If $\chi \in RO(H)$ then $i_H^*(\pi_H^* \chi) = \chi$ and $i_K^*(\pi_H^* \chi) = 0$, hence $i_H^* \times i_K^*\colon \tilde{R}O(G) \to \tilde{R}O(H) \times \tilde{R}O(K)$ is onto. Since $\nu_H \times \nu_K$ is onto, $i_H^* \times i_K^*\colon \tilde{J}O_k(G) \to \tilde{J}O_k(H) \times \tilde{J}O_k(K)$ is also onto. If $C \subset G$ is a cyclic subgroup of order p^r, $p \notin k$, then $C \subset H$ or $C \subset K$. Hence, if $\chi \in \tilde{R}O(G)$ and $i_H^*(\nu_G \chi) = 0$, $i_K^*(\nu_G \chi) = 0$, then $\sum_{g \in C} \chi(g) = 0$ and $\omega_1(i_H^* \chi) = \omega_1(i_K^* \chi) = 0$

by Theorem 3.20. Since Hom $(G: \mathbb{Z}_2) = \text{Hom}(H, \mathbb{Z}_2) \times \text{Hom}(K, \mathbb{Z}_2)$, we have $\omega_1(\chi) = 0$ and hence again by Theorem 3.20, $\tilde{\nu}_G(\chi) = 0 \in \tilde{JO}(G)$. Thus $i_H^* \times i_K^*: \tilde{JO}_k(G) \to \tilde{JO}_k(H) \times \tilde{JO}_k(K)$ is also a monomorphism.

Remark The above theorem is false if G is not finite. Take, for example, $G = S^1 \times \mathbb{Z}_5$, $V_1 \in \tilde{RO}(S^1)$ associated to $\rho_1(t) = (e^{2\pi it}) \in U(1) \subset GL(2, R)$, $W_n \in \tilde{RO}(\mathbb{Z}_5)$ associated to $\rho_n(g^r) = (e^{\frac{2\pi nri}{5}}) \in U(1) \subset GL(2, \mathbb{R})$, $\chi = V_1 \times W_1 - V_1 \times W_2$. Then $\chi | S^1 = 0$ and $\tilde{\nu}_{Z_5}(\chi | \mathbb{Z}_5) = \nu_{Z_5}(W_1 - W_2) = 0 \in \tilde{JO}(\mathbb{Z}_5)$ but $\tilde{\nu}_G(\chi) \neq 0$ since χ is not constant on connected components of G as required by Theorem 2.21.

The requirement that $(|H|, |K|)$ be a unit in $\mathbb{Z}$ is also essential as one can see from the counterexample of $H = \mathbb{Z}_2 = K$ so that rank $\tilde{JO}(\mathbb{Z}_2 \times \mathbb{Z}_2)) = 3$ but rank $(\tilde{JO}(\mathbb{Z}_2) \times \tilde{JO}(\mathbb{Z}_2)) = 2$.

One cannot replace direct product by semi-direct product. If $H = \mathbb{Z}_3 \times \mathbb{Z}_3$, $K = \mathbb{Z}_2 = \{e, g\}$ acting on H by $(a, b)^g = (b, a)$ then rank $(\tilde{JO}(H))$ + rank $(\tilde{JO}(K)) = 4 + 1 = 5$, but rank $\tilde{JO}(H \times_\alpha K) = 4$.

Corollary 4.8 If the finite group $G = \Pi G_p$ (direct product of p-Sylow subgroups), then $\tilde{JO}(G) = \Pi \tilde{JO}(G_p)$. In particular, if G is any finite abelian group, then $\tilde{JO}(G) = \Pi \tilde{JO}(G_p)$.

If G is a finite group, let $\eta(G) = \{H \subset G | H$ is a normal subgroup so that G/H is a cyclic p-group, p a prime$\}$, and $F(\eta(G))$ the free abelian group generated by $\eta(G)$. For $H \in \eta(G)$, define $\beta(H)$ to be the stable J equivalence class of the representation given by the composition $G \to G/H \approx \mathbb{Z}_{p^r} \subset U(1) \subset GL(2, \mathbb{R})$. Notice that the choice of the isomorphism $G/H \approx \mathbb{Z}_{p^r}$ does not effect the equivalence class in $RO(G)_\Gamma$ and hence in $JO(G)$.

Lemma 4.9 For a finite abelian group G, $\beta\colon F(\eta(G)) \to \widetilde{JO}(G)$ is an isomorphism.

Proof Since $\widetilde{JO}(G) = \Pi\, \widetilde{JO}(G_p)$ and $F(\eta(G)) = \Pi\, F(\eta(G_p))$, it is sufficient to consider the case when G is an abelian p-group. If V is a non=trivial irreducible representation of G and $\rho\colon G \to GL(1, \mathbb{R})$ or $GL(2, \mathbb{R})$ is the associated homomorphism, kernel $(\rho) \in \eta(G)$ and $\beta(\text{kernel }(\rho)) = V$ in $\widetilde{RO}(G)_\Gamma = \widetilde{JO}(G)$. Thus β is onto. Suppose $\beta(\sum_{i=1}^{r} a_i H_i) = \nu_G(V - W) = 0 \in \widetilde{JO}(G)$ and let H_{i_o} be the largest group with $a_{i_o} \neq 0$. Then $\dim V^{H_{i_o}} - \dim W^{H_{i_o}} = a_{i_o} \dim\beta(H_{i_o}) \neq 0$ so $\nu_G(V - W) \neq 0 \in \widetilde{JO}(G)$ by Theorem 2.7; hence β is a monomorphism.

Remark The above lemma is clearly false if G is not abelian. If G is any finite simple group which is not cyclic, then $\eta(G) = \phi$. Hence Lemma 4.8 would require that $\widetilde{JO}(G) = 0$ for every such simple group. Even if G is a p-group the above lemma fails. For example, let G be the semi-direct product of $\mathbb{Z}_4$ with $\mathbb{Z}_2$, h a generator of $\mathbb{Z}_4$, g a generator of $\mathbb{Z}_2$ and $h^g = ghg^{-1} = h^3$. Then $\eta(G) = \{H_1, H_2, H_3\}$ where $H_1 = \langle h \rangle$, $H_2 =$

$<g, h^2>$, $H_3 = <gh.\ h^2>$, so rank $F(\eta(G)) = 3$ but rank $\tilde{JO}(G) = 4$.

<u>Note</u> In the above examples, ranks were computed by using the following facts:

a) rank $RO(G)$ = the number of equivalence classes in G where $g \sim g^{-1}$, $g \sim g_1 g g_1^{-1}$

b) rank $RO(G)_\Gamma$ = the number of equivalence classes in G where $g \sim g^s$, $s \in \Gamma_G$, $g \sim g_1 g g_1^{-1}$

c) rank $JO(G)$ = the number of equivalence classes in $G^* = \{g \in G | g^{p^r} = e$ for some prime $p\}$ where $g \sim g^s$, $s \in \Gamma_G$, $g \sim g_1 g g_1^{-1}$.

<u>Proposition 4.10</u> For any finite abelian group G, rank $\tilde{JO}(G) =$

$$\sum_p \left(\sum_{r=1}^{\infty} \frac{|\mathrm{Hom}(G, \mathbb{Z}_{p^r}) - \mathrm{Hom}(G, \mathbb{Z}_{p^{r-1}})|}{(p-1)p^{r-1}} \right) \text{ where } \mathbb{Z}_1 = \{0\}.$$

<u>Proof</u> It is sufficient by Lemma 4.9 to show that the cardinality of $\eta(G) = \sum_{r=1}^{\infty} \frac{|\mathrm{Hom}(G,\mathbb{Z}_{p^r})| - |\mathrm{Hom}(G,\mathbb{Z}_{p^{r-1}})|}{(p-1)p^{r-1}}$ for an abelian p-group.

If $H \in \eta(G)$, then $G/H = \mathbb{Z}_{p^r}$ for some r and hence there is an epimorphism (in fact, $p^{r-1}(p-1)$ epimorphisms): $G \to \mathbb{Z}_{p^r}$ with kernel H. Conversely, any epimorphism $\rho: G \to Z_{p^r}$ has kernel $(\rho) \in \eta(G)$. Since $\mathbb{Z}_{p^{r-1}} \subset \mathbb{Z}_{p^r}$, we may regard $\mathrm{Hom}(G, \mathbb{Z}_{p^{r-1}}) \subset \mathrm{Hom}(G, \mathbb{Z}_{p^r})$ and the number of epimorphisms: $G \to \mathbb{Z}_{p^r})$ is precisely $|\mathrm{Hom}(G, \mathbb{Z}_{p^r})| - |\mathrm{Hom}(G, \mathbb{Z}_{p^{r-1}})|$; hence the proposition.

QA 3 A57 #159

Note If $p^N > |G|$ then $|\mathrm{Hom}(G, \mathbb{Z}_{p^r})| - |\mathrm{Hom}(G, \mathbb{Z}_{p^{r-1}})| = 0$ for $r > N + 1$ and hence the above sum is actually a finite sum.

Corollary 4.11 If G is a finite abelian group then rank $\tilde{JO}(G) =$

$$\sum_p \left(-\frac{1}{p-1} + \sum_{r=1}^{\infty} p^{-r} |\mathrm{Hom}(G, \mathbb{Z}_{p^r})| \right) .$$

Proof Just a rearrangement of the sum in Proposition 4.10.

Note The inner sum is not finite in this formula.

REFERENCES

1. J. F. Adams, Seattle Conference on Differential and Algebraic Topology 1963, Collection of Unsolved Problems.
2. ——————, Lectures on Lie Groups, Benjamin, New York, 1969.
3. M. F. Atiyah and D.O. Tall, Group Representations, λ-rings and the J-homomorphisms, Topology 8 (1969), pp. 253-297.
4. G. Bredon, Introduction to Compact Transformation Groups, Academic Press, New York and London, 1972.
5. C. W. Curtis and I. Reiner, Representation Theory of Finite Groups and Associative Algebras, Wiley (Interscience), New York, 1962.
6. D. Gorenstein, Finite Groups, Harper and Row, New York, Evanston, London, 1968.
7. S. Helgason, Differential Geometry and Symmetric Spaces, Acacemic Press, New York and London, 1962.
8. Ted. Petrie, "Smooth S^1 Actions on Homotopy Complex Projective Spaces and Related Topics," Bull. Amer. Math. Soc. 78 (1972), pp. 105-153.
9. V. Snaith, "J-Equivalences of Group Representations," Proc. Camb. Phil. Soc. 70 (1971), pp. 9-14.
10. N. Steenrod, The Topology of Fibre Bundles, Princeton University Press, Princeton, New Jersey, 1951.
11. R. Thom, Quelques propriétés globales des variétés differentiables, Comm. Math. Helv. 28 (1954), pp. 17-86.
12. A. G. Wasserman, "Equivariant Differential Topology," Topology 8 (1969) (1969), pp. 127-150.

The University of Michigan, Ann Arbor